Theory and Technology of Sheet Rolling

Theory and Technology of Sheet Rolling

Numerical Analysis and Applications

V. L. Mazur
O. V. Nogovitsyn

CISP

CRC Press is an imprint of the
Taylor & Francis Group, an **informa** business

Translated from Russian by V.E. Riecansky

CRC Press
Taylor & Francis Group
6000 Broken Sound Parkway NW, Suite 300
Boca Raton, FL 33487-2742

First issued in paperback 2020

ISBN 13: 978-0-367-57081-1 (pbk)
ISBN 13: 978-0-8153-8706-0 (hbk)

Visit the Taylor & Francis Web site at
http://www.taylorandfrancis.com

and the CRC Press Web site at
http://www.crcpress.com

Contents

Introduction

Technical progress is an integral part of the objective process of the development of society and its productive forces. In metallurgy, technical progress is aimed at solving urgent problems of improving the quality and competitiveness, expanding the range of products, developing new types of products, saving metal, fuel, water, thermal and electric energy, and reducing production waste. Each of these tasks is a complex of complicated problems and unresolved issues. With a view to resolving them, fundamental and applied research is constantly carried out, offering real recommendations in the form of technical and technological solutions, the correct use of which promises additional tangible benefits and the elimination of negative effects. The development of a number of industries is largely dependent on the production of rolled steel from ferrous metals.

The development of technology for obtaining various types of sheet steel, the construction of rolling equipment, systems and means of automation of the production process are subject to a single goal - to ensure the required service properties and production volumes.

The technology of sheet rolling production is a set of knowledge about the methods and means of processing steel from the initial billet to obtaining sheets and strips of specified sizes, surface quality, structure and mechanical properties. The creation of effective technology is always scientific research, which ends with the acquisition of new knowledge. Here there are: the formulation of the problem, the analysis of the initial information, the formulation of the working hypotheses, their theoretical elaboration, organization, planning, performing the experiments, analyzing and summarizing the results, checking the correctness of the accepted hypotheses on the basis of the data obtained, justifying the conclusions, the revealed regularities and forecasts.

The creation of new and improvement of existing technologies at present can not be confined to various semi-empirical approaches based only on production experience, but should be based on a reliable theoretical foundation. It is for this reason that the development of the theory of sheet rolling goes by developing methods for a sufficiently accurate quantitative description of the processes occurring in the metal when it is formed in the gap between the rolls and after deformation. A comprehensive study of the behaviour of the metal in the deformation zone during rolling and the establishment of regularities connecting the parameters of the quality of the strips with the modes of their deformation are possible only with the use of methods of continuum mechanics that allow quantitative estimates of the transformations occurring in a metal. For the numerical analysis of these processes, it is necessary to take into account the basic rheological properties of the metal, the conditions on the surfaces of its contact with the rolls, and many other factors.

At the same time, a significant part of the achievements of the fundamental sciences, especially physical metallurgy, does not yet find proper application in solving new classes of theoretical and applied problems of sheet rolling production. These include the forecasting of the structure and properties of steel, defects in the continuity of sheet metal, stresses, strains and loads in rolling stands.

Thus, due to the multifactorial nature, complex interdependence, multi-stage production processing in metallurgy, the complex development of the thin-sheet rolling technology is impossible without sufficiently rigorous mathematical modelling of the studied processes. The correctness of mathematical models, taking into account the maximum number of influencing factors, including the translation of previously accepted assumptions into the components of mathematical dependences, should be based on the results of the latest research, which are recognized achievements of scientific and technical progress.

An example of solving the problem of improving the technology of sheet metal production is the development of technological modes of its heating, rolling, heat treatment, finishing, ensuring the production of the required quality at minimal cost. Due to mathematical models, it is possible to determine the technology of all operations at the rolling plant in relation to the specified final properties of the products. The inverse problem is very demanded: on the basis of known, actually realized technological parameters of the production process, to predict the final properties of the finished product.

The task of ensuring a high quality of sheet products certainly requires an assessment of the reliability of technological processes of rolling and subsequent processing of metal. Structural features and operating conditions of process equipment can lead to deviations of certain quality indicators of rolled products from specified values. The reliability of technology in the decisive measure determines the possibility of obtaining products with a given level and stability of its quality indicators. Metallurgical enterprises are paying increasing attention to improving the stability of the consumer properties of sheet metal. The reliability of the technological process of rolling production, which ensures the minimum dispersion of the quantitative characteristics of product quality indicators, is also the subject of fundamental scientific research. The construction of mathematical models that take into account the probabilistic nature of production processes in metallurgy opens new possibilities for weakening the influence of instability of the initial parameters and the current disturbances that arise during the technological redistribution, on the quality indicators of the finished product. The mentioned aspects of the analysis of the technological process of sheet metal production are still poorly covered in the technical literature. Known results require further systematization and bringing to practical use. Obviously, such promising methods of researching technological processes as statistical imitation modelling, which allows the most objective analysis and identification of regularities, interrelations between technology parameters and rolled product quality at all stages of its production, should be popularized in the technical literature. The authors hope that this book will greatly contribute to this.

In recent years, mathematical models that adequately describe the stress–strain state of coils of hot-rolled and cold-rolled thin-sheet steel have been developed significantly. The production experience of the advanced metallurgical plants has shown that the modes of cutting the rolled strips into rolls are almost the decisive element in the technology of sheet production. Despite certain successes achieved in the area under consideration, the domestic and foreign experience that is available here is not generalized, not systematized and not brought to a wide audience of specialists in rolling production and in the design of metallurgical equipment. The materials outlined in the book should fill this gap in the literature on metal forming.

And the last. Today for metallurgists, perhaps, there is no more urgent topic than energy saving. This topic is so broad and multifaceted that it requires separate, special publications. Realizing this, the authors of the book nevertheless considered it necessary to draw the attention of scientists, designers and specialists of the rolling shops of metallurgical plants to some fairly simple but very effective measures for the thermal insulation of metallurgical units, for heat and energy savings.

The same applies to the problem of preventing defects in the surface of thin sheet steel. This topic is always relevant for industrial practice. Intensification of technological processes for the production of thin sheets and strips at all stages of the production cycle increases the risk of various flaws and defects that reduce the quality of the finished products. Detailed consideration of the causes, the mechanisms of formation and methods of preventing all known defects in sheet steel due to the enormous amount of materials on this topic was not included in the plan of this book. Nevertheless, understanding the importance of the problem, the authors devoted a special section in the book to the issue of preventing defects 'kinks' ('cross breaks') of thin sheets and strips, one of the most common in industrial practice.

This book is devoted to the development of modern theory of sheet rolling of steels on the basis of modern methods of numerical analysis of the problems of mechanics of continuous media, pressure treatment of metals, physical metallurgy. When working on the book, the authors relied on the well-known works of rolling experts, mechanics experts and metal scientists on a similar and close subject. The book summarizes the scientific and technical results obtained by the authors in carrying out research at the Institute of Ferrous Metallurgy of the National Academy of Sciences of Ukraine and at metallurgical enterprises in Ukraine, Russia, and Kazakhstan. The book also analyzes the materials of publications and the experience of foreign companies that deserve attention. The results of the presented theoretical studies were the scientific foundation of new technical and technological solutions introduced into the production practice.

The authors are very grateful to their colleagues from the Department of Thin Sheet Production of the Institute of Ferrous Metallurgy, with whom they carried out joint research, for many years of creative cooperation and friendship, their tolerance and benevolence. We would especially like to note the decisive role of the first scientific leaders of the Department of Thin Sheet Production A.P. Chekmarev, V.I. Meleshko, A.A. Chernyavsky, A.P. Kachaylov in the formation of a creative atmosphere in the department, creating conditions for professional growth, and comprehensive support of young researchers. The authors of the book will always remember this and appreciate it immensely.

A significant influence on the set of views, ideas, and ideas expressed in the book was provided by scientific seminars held at the Dnepropetrovsk Metallurgical Institute by Academician A.P. Chekmarev and A.P. Grudev, the Head of the Department of Metal Pressure Treatment, as well as cooperation of the authors with leading scientists of the Moscow Institute of Steel and Alloys, the Central Scientific Research Institute of Ferrous Metallurgy, Moscow, the Lipetsk and Donetsk Polytechnic Institutes, the Donetsk Research Institute of Ferrous Metallurgy, the Ural School of Rolling Technology and many other scientific teams and design organizations.

The authors are extremely grateful to the specialists of Zaporozhstal', Magnitogorsk, Kapaganda, Mapiupol', Cherepovetsk and Novolipetsk metallurgical concerns, who contributed their creative work in the production of industrial models and in the introduction of joint development in the production practice. Their help and participation are invaluable.

This book is in fact the second, supplemented and expanded edition of the monograph published in 2010 by V.L. Mazur and A.V. Nogovitsyn 'The theory and technology of thin-sheet rolling (numerical analysis and technical applications)'. The authors believe that supplementing the book with 'fresh' results from the theory and technology of sheet rolling production will strengthen its content and relevance for science and practice.

1

One-dimensional model of the deformation zone

1.1. The deformation zone in the moving coordinate system

The rolling process is investigated as a combination of two processes, carried out by the rolling rolls: the transfer movement of the strip in the gap between the rolls and the process of deformation of metal in the gap between the rolls. The selection of the suitable moving coordinate system (the inertial reference system) makes it possible to separate the movement of the elementary volumes of the rolled metal, caused by deformation, from the transfer movement. This approach makes it possible to carry out the energy and force analysis of the sheet rolling process by a non-traditional procedure.

The rolling process will be investigated in the coordinate system which moves in the rolling direction with the translational velocity, equal to the circumferential velocity of the generating line of the roll V_r (Fig. 1.1). Consequently, the absolute velocity V_a of a material particle of rolled metal will consist of the transfer velocity V_r and the relative velocity V_m caused by the deformation of the strip in the gap between the rolls. In turn, the velocity V_m is a sum of the velocity of the contact point of the rolls surface V_k and the velocity of relative sliding V_{rel} of the deformed metal on the surface of the roll at the given point of its circumference.

The absolutely rigid rolling rolls roll without sliding on the planes β and β′ in the selected moving system of the coordinates. These planes are connected rigidly with the moving coordinate system and pass on the surface of the rolled strip (see Fig. 1.1). The points of contact A and A' of the rolls with the planes β and β′ are the instantaneous centres of rotation of the rolls and, consequently, the instantaneous centres of rotation of their contact surfaces.

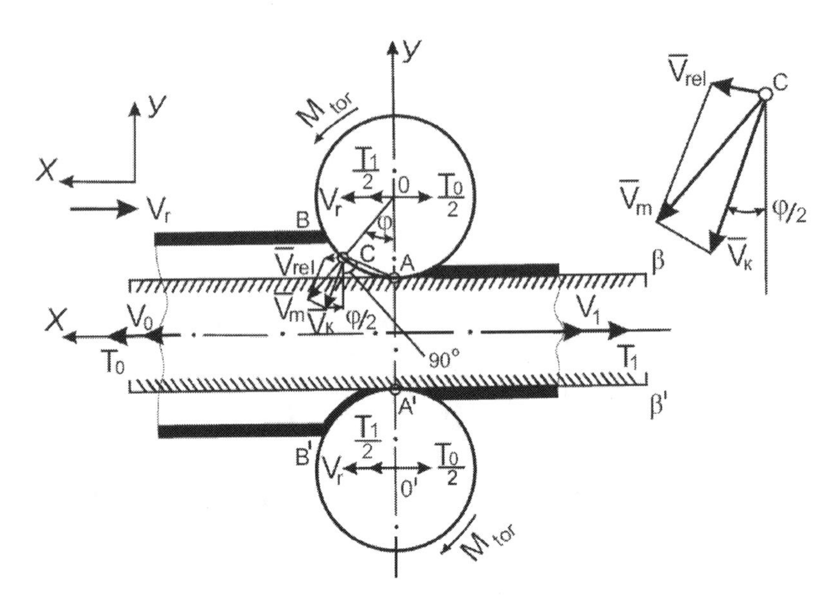

Fig. 1.1. Diagram of the rolling process in the absolutely 'rigid' rolls, represented in the moving coordinate system. Points: **A** – the instantaneous centre of rotation of the rolls; **B** – the point of the beginning of contact of the strip with the roll; **C** the actual coordinate along the contact arc, corresponding to the angle φ; $\overline{\mathbf{V}}_m$ – the relative velocity of the deformed metal at the point C; $\overline{\mathbf{V}}_{rel}$ – the velocity of the relative sliding of the metal; $\overline{\mathbf{V}}_k$ the velocity of the surface of the roll in the relative coordinate system.

Consequently, it may be assumed that at every moment of time the rolling process is identical with upsetting of the metal between the inclined plates, rotating around the points A and A'.

As a result of the elastic deformation of the roll, the centre travels towards the rolling axis by the value Δ_r, Fig. 1.2. The rolling plane β of the non-loaded part of the roll barrel is displaced by the same value. The point of their contact is the instantaneous centre of rotation of the deformed generating line of the roll.

Analysis of the vector of movement of the points of the deformed arc of the roll shows that the loading and deformation of the rolled metal take place in the section X_1, situated in the direction from the section of entry of the metal into the deformation zone to the line of the centres of the rolls (l.c.r). In this section, the velocity of the contact surface of the roll V_k is directed to the rolling axis.

The velocity of movement of the contact surface of the roll V_k behind the lines of the centres of the rolls is directed to one side from the rolling axis. Therefore, in this section (X_0) unloading and elastic restoration of the deformed roll take place, and it may also

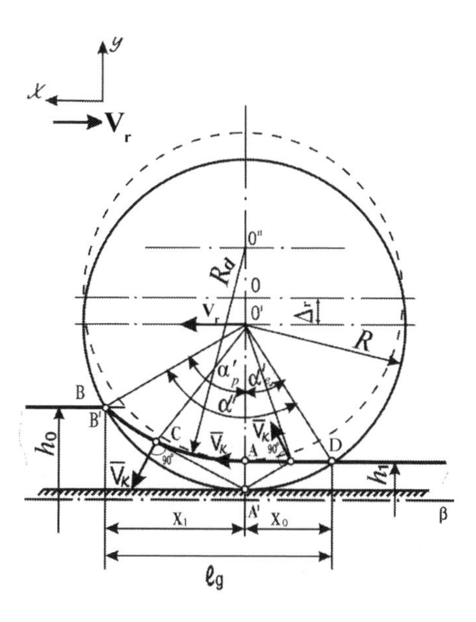

Fig. 1.2. Diagram of the rolling process represented in the moving coordinate system and taking the deformation of the roll into account: R – the radius of the roll; R_d – the radius of the deformed contact arc of the roll; Δ_r – displacement of the centre of the roll as a result of elastic deformation of its generating line; O – the centre of the generating line of the roll; O' – the centre of the generating line of the roll, displaced as a result of elastic deformation; O'' – the centre of the circumference of the deformed contact arc; B, A – the points of contact of the surface of the rigid roll with the deformed strip on the line of the centres of the rolling rolls (A) and at entry into the contact zone (B); B', A', D – the points of the contact of the surface of the deformed roll with the deformed strip on the line of the centres of the rolling rolls (A'), at entry into the contact zone (B'), D – the points of contact of the surface of the deformed roll with the deformed strip on the line of the centres of the rolling rolls (A'), at entry into the contact zone (B'), D – at exit from the contact zone; α', α'_p, α'_e – the central angle of the deformed contact arc (α'), in the plastic deformation zone (α'_p) and in the zone of elastic restoration of the generating line of the roll (α'_e); V_r – the velocity of the moving coordinate system, equal to the circumferential velocity of the rolls; C – the actual point of contact of the deformed arc of the roll with the rolled strips; V_k – the velocity of the contact point C in relation to the instantaneous centre of rotation of the roll A'. The dotted line indicates the circumference of the roll prior to flattening of the arc contact. The dashed line indicates the conventional surface on which the roll 'rolls'.

be assumed that the rolled metal changes from the plastic to elastic state.

This hypothesis has been confirmed by the data obtained by the authors of [1] in experiments consisting of the investigation of contact stresses in the process of rolling 08kp and St3 steels. It was shown that in rolling the 08kp steel with a reduction of 26.9% there are three deformation zones: the elastic zone at entry into the deformation zone with a length of 0.64 mm (or 7.1% of the total length of the contact arc), a plastic zone 6.6 mm long (72.7%), and an elastic zone at exit from the deformation zone 1.83 mm long (20.2%). The geometrical length of the contact arc without elastic deformation of the roll $\left(\sqrt{R\cdot\Delta h}\right)$ would be 5.95 mm which corresponds to the length of the plastic zone determined in these experiments. For the case of rolling St3 steel with a small reduction (9.2%) the lengths of the sections of the deformation zone were respectively 0.34 mm (4.7%); 5.79 mm (80.5%); 1.07 mm (14.8%), and $\sqrt{R\cdot\Delta h}$ = 5.8 mm.

These results indicate that the elastic zone of the deformation region (X_0) starts immediately behind the line of the centres of the rolls, and the length of the plastic deformation zone is practically equal to the calculated value $l = \sqrt{R\cdot\Delta h}$, where R is the radius of the roll.

These conclusions are very important for the calculation of contact stresses in the deformation zone, the rolling force and the torque on the roll.

1.2. Differential equation of rolling

The relatively simple mechanism of metal flow in sheet rolling and the small thickness of rolled metal in comparison with the diameter of the working rolls and width of the sheet make it possible to proposed a number of simplifications which greatly simplify the determination of the stress state of metal in the deformation zone. The main assumption is the assumption of the constancy of stresses and strains along the height and width of the sheet in any vertical section of the deformation zone and ignorance of the inertia force.

Taking this into account, the differential equation of the equilibrium of an infinitely thin flat element in the vertical section of the deformation zone (Fig. 3) has the form:

$$\sigma\cdot h-(\sigma+d\sigma)\cdot(h+dh)+2p_r\frac{dx}{\cos\varphi}\sin\varphi-2\tau_V\frac{dx}{\cos\varphi}\cos\varphi=0, \quad (1.1)$$

where σ are longitudinal stresses, p_r and τ_v are the contact normal and tangential stresses, h is the height of the cross section; φ is the central angle.

Neglecting the infinitely small values of the second order, equation (1.1) can be written in the form:

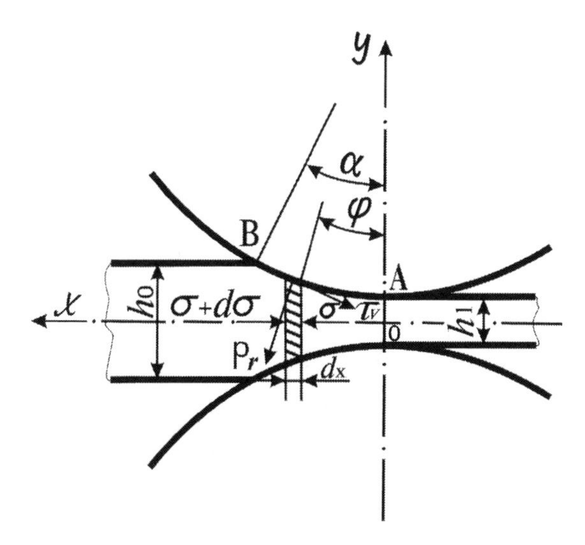

Fig. 1.3. The scheme of the one-dimensional deformation zone.

$$d\sigma - (p_r - \sigma) + \frac{2tg\varphi}{h}dx + \frac{2\tau_V}{h}dx = 0. \qquad (1.2)$$

Equation (1.2) includes three unknown quantities: p_r, σ and τ_v. To determine them, equation (1.2) should be supplemented by two independent relationships. One of them is represented by the well-known description of the contact tangential stresses:

$$\tau_v = f \cdot p_r, \qquad (1.3)$$

where f is the friction coefficient.

The differential equilibrium equation which holds for both the plastic (X_1) and elastic zone (X_0) of the deformation zone, is integrated using the following equation connecting the stresses:

$$p_r - \sigma = T, \qquad (1.4)$$

where T is the stress intensity.

In the plastic deformation zone X_1 according to the plasticity condition one obtains:

$$T = 2K, \qquad (1.5)$$

where K is the plastic shear resistance.

The stress intensity in the elastic zone X_0 changes from $2K$ to 0. The stress intensity distribution in the elastic zone is similar to elliptical, as assumed in solving the elastic contact problem:

$$T = 2K\sqrt{1-\left(\frac{X}{X_0}\right)^2}, \tag{1.6}$$

where $0 < X < X_0$.

As a result of the simultaneous application of the conditions (1.5) and (1.6) in the rolling equation makes the proposed approach to studying the deformation zone to differ greatly from the well-known methods in which the equation (1.2) is solved for the plastic zone which also includes the section behind the line of the roll centres.

In a number of works [2, 3] the contact stresses in the elastic regions of the deformation zone are finally determined by the geometrical method after calculating the contact stresses in the plastic region of the deformation zone. The length of the elastic regions is determined as the result of elastic recovery of the strip. The contact stress curve typical of this approach is shown in Fig. 1.4.

The rolling equation (1.2) will be transformed further. After substituting the relationships (1.3) and (1.4) into (1.2), and taking into account that $dx = R\ \cos\varphi d\varphi$, equation (1.2) acquires the following form:

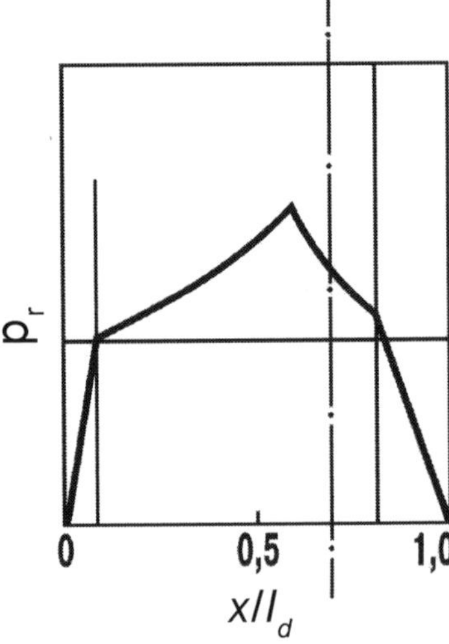

Fig. 1.4. The curve of contact normal stresses p_r, calculated by the traditional method. Notation: x – the actual coordinate; l_d – the length of the deformation zone; dot-and-dash line – the line connecting the centres of rotation of the rolls.

$$\frac{d\sigma}{d\varphi}=\left[T\pm\frac{f\cdot(\sigma+T)}{\text{tg}\ \varphi}\right]\frac{2R\sin\varphi}{h}, \tag{1.7}$$

where $h = h_1 + 2R(1-\cos\sigma)$ for $0 < \sigma < \alpha_p$); $h = h_1$ for $0 < \varphi < \alpha_e$), where α_p, α_e are the central angles of the contact arc for the loading and unloading regions in the deformation zone, see Fig. 1.2.

The solution of equation (1.7) in the longitudinal stresses σ makes it possible to solve the problem of defining the boundary conditions in the entry and exit sections of the deformation zone because the value of σ in these sections is equal to the rear σ_{rear} and front σ_{front} tension of the strip, respectively.

1.3. Numerical solution method

The differential equation (1.7) is the linear equation of the first order with variable coefficients. In the general form it can be written as

$$\frac{d\sigma}{d\varphi}=f(\varphi,\sigma) \tag{1.8}$$

The Euler and Runge–Kutta methods are the widely used methods of numerical solution of the ordinary differential equation. The Euler method is simpler but the error of the results in this method is larger. However, the accuracy of the method for the investigated problem is satisfactory.

The procedure for application of the method may be described as follows. Selecting the sufficiently small step $\Delta\alpha$, a series of $N + 1$ equally spaced points is constructed on the contact arc AB:

$$\varphi_i = i\cdot\Delta\alpha,\ i=0,1,\ldots,N,$$

where $\Delta\alpha = \alpha/N$.

The required integral curve $\sigma = \sigma(\varphi)$ is replaced by a broken curves with the tips $M_1(\varphi_i, \sigma_i)$ whose terms M_i, M_{i+1} are straight lines with the angular coefficients:

$$\frac{\sigma_{i+1}-\sigma_i}{\Delta\alpha}=f(\varphi_i,\sigma_i). \tag{1.9}$$

From (1.9) we obtain the recurrent formula for calculating σ_{i+1}:

$$\sigma_{i+1}=\sigma_i+\Delta\alpha\cdot f(\varphi_i,\sigma_i),\ i=0,1,\ldots,n-1. \tag{1.10}$$

The distribution of stress p_r along the length of the contact arc is determined by the method of successive calculations using the relationships which have the following form for the creep zone:

$$\begin{aligned}
&\sigma_i = \sigma_{i+1} + \left[-2K_{i+1} + \frac{f(\sigma_{i+1} + 2K_{i+1})}{\operatorname{tg}\varphi_{i+1}} \right] \frac{2R\sin\varphi_{i+1}}{h_1 + 2R(1-\cos\varphi_{i+1})} \Delta\alpha, \\
&i = n, n-1, \ldots, 0 \\
&\sigma_N = -\sigma_{\text{rear}}, \\
&p_i = 2K_i - \sigma_i, \\
&\tau_i = f \cdot p_{i,}
\end{aligned} \tag{1.11}$$

and for the creep zone:

$$\begin{aligned}
&\sigma_{i+1} = \sigma_i + \left[2k_i + \frac{f(\sigma_i + 2k_i)}{\operatorname{tg}\varphi_i} \right] \frac{2R\sin\varphi_i}{h_1 + 2R(1-\cos\varphi_i)} \Delta\alpha, \\
&i = 0, 1, \ldots, n \\
&\sigma_0 = -\sigma_{\text{front}}, \\
&p_{i+1} = 2k_{i+1} - \sigma_{i+1}, \\
&\tau_{i+1} = f \cdot p_{i+1,}
\end{aligned} \tag{1.12}$$

Changing i from N to 0, we determine all values of p_i using the relationships (1.11). Subsequently, starting at $i = 0$ we determine the values of $p_i + 1$ from the relationships (1.12) to intersection with the curve, calculated previously for the creep zone.

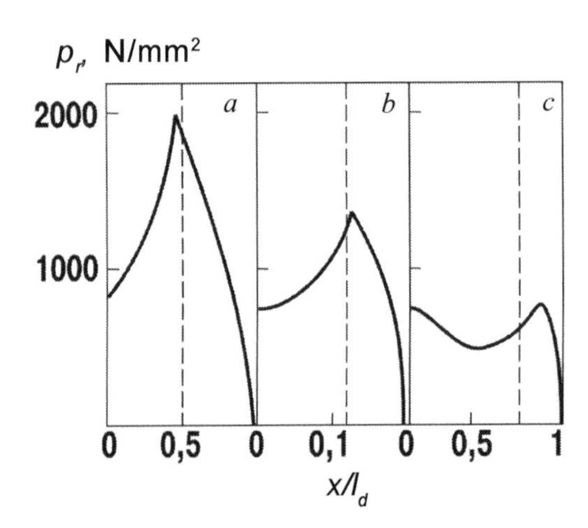

Fig. 1.5. Curves of contact normal stresses calculated by the proposed method at different reductions (R = 300 mm; h_1 = 0.25 mm; f = 0.02): a – ε = 5%; b – ε = 30%; c – ε = 55%. The dashed line indicates the lines, connecting the centres of the rolls.

As an example, Fig. 1.5 shows the curves of normal contact stresses calculated by the proposed method, the calculations were carried out for three cases of rolling a strip of finite thickness h_1 = 0.25 mm in the sixth stand of the 1400 continuous cold-rolling mill with the relative reductions of 5, 30 and 55%. The characteristic feature of these curves is the convex form of the curve at exit from the deformation zone, and the concave form at entry. On the whole, the form of the calculated curves is very similar to the available experimental data [1, 4] rather than to the curves obtained by the conventional methods (see Fig. 1.4).

1.4. Rolling force

The rolling force is determined by solving a system of equations:

$$\begin{cases} P = B\dfrac{1}{n-1}\sum_{i=1}^{N} p_i \cos\varphi, & (1.13) \\ l = \sqrt{R\Delta h + X_0^2} + X_0, & (1.14) \\ X_0 = 16\dfrac{1-\nu^2}{\pi E}R\dfrac{P}{Bl}, & (1.15) \end{cases}$$

where B is the strip width.

The system of equations (1.13)–(1.15) is solved by the iteration method. The convergence of the iteration procedure is sufficiently high, 3–5 iteration cycles. Under certain conditions in rolling (higher values of R, f, K; low value of h) the iteration process starts to diverge. This indicates that in these conditions the process of production of the strip of the required thickness cannot be realised, and, consequently, the system of equations (1.13)–(1.15) does not have any solution.

The accuracy of calculating by the proposed method and the required calculation time are determined by the integration step $\Delta\alpha$ or by the number of divisions of the deformation zone, N. It is well-known [5] that the Euler method in the first step generates an error whose order is equal to $0(\Delta\alpha^2)$ and in subsequent calculation steps $0(\Delta\alpha)$. The following procedure was used for selecting the number of divisions of the deformation zone. Equation (1.12) was solved by the numerical method taking into account all the assumptions made in the analytical solutions. Comparison of the results yielded the error of the numerical methods. At the number of divisions N = 50 the

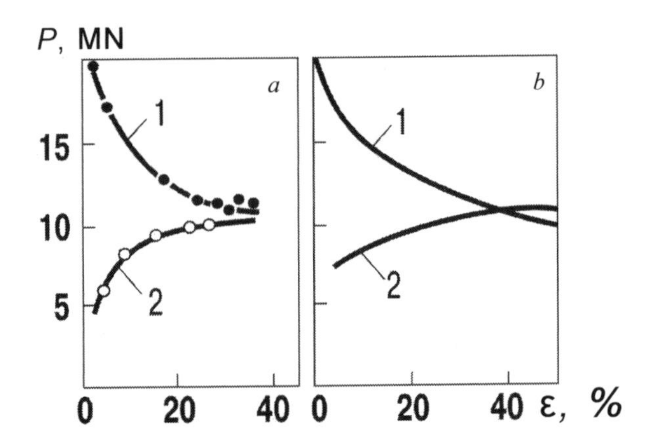

Fig. 1.6. Experimental (*a*) and calculated (*b*) dependences of rolling force *P* on the relative reduction ε in rolling strips in the sixth stand of the 1400 mill: 1 – h_1 = 0.25 mm; 2 – h_1 = 0.50 mm.

error of calculating the total pressure was 0.4%, and the maximum error in the determination of the specific pressures, which form in the neutral section, did not exceed 0.78%.

Thus, the Euler method ensures sufficiently high accuracy of solving the rolling equation for the appropriate selection of the integration step.

The results obtained by the described mathematical model of the rolling force are of considerable importance for the theory and practice of the cold rolling of thin strips. The practical experience with the production of thin sheet steel creates problems which cannot be predicted with sufficient accuracy and explained by theory. These problems include, for example, the anomalous effects observed in cold rolling of thin strips. For example, when mastering the technology of rolling tin plate in the 1400 six-stand mill of the Karaganda Metallurgical Works it was established that the stable rolling process in the final stand of the mill is not possible at relative reductions smaller than 30...40%.

The calculation of the dependence of the rolling force on the relative reduction in the last (sixth) stand of the mill in rolling of strips with a final thickness of 0.25 mm shows that as a result of reducing the thickness of the strip arriving in the rolls with a decrease of reduction the rolling force may increase quite steeply (Fig. 1.6 *b*).

This principal fact, which was previously not known in the theory of thin sheet rolling, was confirmed by the experiments carried out by the authors of the book in the 1400 industrial six-stand mill (Fig.

1.6 *a*). In the experiments, the reduction in the stand 6 was gradually decreased from 38 to 2% as a result of the redistribution of the reductions between the stands of the mill. The rolling force, the thickness of the strip prior to and behind the sixth stand and the strip tension were recorded in an automatic recording device. The complexity of the experiment was represented by the fact that in the case of rolling a strip with a final thickness of 0.25 mm at small reductions the rolling process was unstable as a result of the large increase of the rolling force. Identical experiments for strips with a final thickness of the strip of 0.5 mm were not associated with any difficulties because the rolling force decreased with a decrease of the reduction and the process was stable.

There are several hypotheses for explaining the effect of the increase of the rolling force with a decrease of reduction as a result of reducing the thickness of the strip at entry into the gap between the rolls of the final stand of the mill: decrease of the extent of cold working of the steel strip; decrease of the deformation resistance of the metal as a result of increasing its temperature in the deformation zone; increase of the thickness of the lubricant layer between the roll and the strip. It is assumed that the 'anomalous' dependence is determined by the change of the stress state of the metal in the deformation zone. The form of the curves, presented in Fig. 1.5, shows that the increase of the reduction at certain rolling conditions (high values of R and K, small values of f and h) results in a small gradient of the increase of the contact normal stresses in the creep zone (in a certain section of the gradient is negative).

1.5. The torque in the roll

There are theoretical methods for calculating the torque in the roll in the rolling of strips with and without tension based on the summation of the moments from the contact tangential stresses in the deformation zone in relation to the centre of the roll:

$$M_{\text{tor}} = BR\int_0^{\alpha} d\varphi. \tag{1.16}$$

The mathematical model of the contact tangential stresses (1.3) does not yield sufficiently reliable results. This is due to the fact that the areas of the stress curves in the backward creep and forward creep zones are comparable and this results in large areas in the calculations if the neutral section is not determined accurately.

The energy analysis of the rolling process in the moving coordinate system shows [6, 7] that the torque in the roll is equal to the moment of the forces acting on the roll in relation to the instantaneous centre of rotation. For the absolutely rigid rolls the equation for the torque has the following form:

$$M_{\text{tor}} = 2BR^2 \int_0^{\alpha} \left(p_r \cos\frac{\varphi}{2} + \tau_V \sin\frac{\varphi}{2} \right) \sin\frac{\varphi}{2} d\varphi + \frac{T_0 - T_1}{2} R, \quad (1.17)$$

where T_0, T_1 is the total rear and front tension of the strip, respectively.

For the case of cold rolling we can use the equivalence of the torque in relation to the roll axis in relation to the torque of the forces and its instantaneous centre of rotation.

In the case of elastic deformation the centre of the roll O approaches the rolling axis by the value Δ_r and occupies a new position O' (Fig. 1.2). The value Δ_r can be determined from the equation:

$$\Delta_r = 2\frac{P}{B}\cdot\frac{1-\nu^2}{\pi E}. \quad (1.18)$$

The vector of the velocity of any point of the deformed contact arc is inclined under the angle Θ in relation to the axis OY:

$$\Theta = \frac{\pi}{2} - \operatorname{arctg}\frac{R_g \sin\varphi' - X}{R_g\left(1-\cos\varphi'\right) + \Delta_b} \quad (1.19)$$

here $R_g = l^2/\Delta h$ is the radius of the deformed arc of contact of the roll with the rolled strip;

$$\varphi' = \varphi \cdot R/R_g.$$

The sum of the projections of the contact forces $q(\varphi)$ on the vector V_k at the arbitrary point C of the contact arc is determined from the equation:

$$q(\varphi) = p_r(\varphi)\cos\left(\Theta - \varphi'\right) + \tau_v(\varphi)\sin\left(\Theta - \varphi'\right). \quad (1.20)$$

The arm of the momentum r_{AC} (the length of the segment AC) of the force $q(\varphi)$ in relation to the instantaneous centre of rotation A' is determined from the expression:

$$r_{AC}(\varphi) = \left[\Delta_b + R_g\left(1 - \cos\varphi'\right)\right] / \cos\left(\frac{\pi}{2} - \Theta\right). \tag{1.21}$$

The torque on the roll is equal to the sum of the moments of the forces $q(\varphi)$ with the arm $r_{AC}(\varphi)$ and the reactions of the framework of the rolling stand to the effect of the front T_1 and rear T_0 tensions of the strip in relation to the instantaneous centre of rotation of the circumference of the roll A:

$$M_{\text{tor}} = BR\int_0^{\alpha} q(\varphi)\cdot r_{AC}(\varphi)d\varphi + \frac{T_0 - T_1}{2}R. \tag{1.22}$$

According to the equations (1.20)–(1.22) the magnitude of the torque momentum is controlled (it is 2–4 orders of magnitude greater than τ_v) by the normal contact stresses p_r. This circumstance makes the equation (1.22) more reliable than the equation (1.16).

The average deviations of the results of calculating the torque from the experimental data, obtained by the authors of the book in the 2030 industrial five-stand rolling mill and by the authors of [8] in a laboratory mill did, not exceed 10%.

The above methods of calculating the rolling force and the torque can be used to estimate these parameters in the investigation of complicated rolling cases, for example, in rolling welded joints with characteristic differences in the thickness and the properties of the sections of contacting strips and also the welded joint (see chapter 7).

2

Mathematical model of wide-strip hot rolling of steel

2.1. Modeling the structure of the steel during hot rolling

Composite components of the mathematical model of the process of hot rolling of wide-strip steel are: the model of deformation; the methods for the calculation of the process parameters (rolling force, torque on the rolls, rolling power); the method of determining the strip temperature; the method of controlling deformation of metals, etc. In the last two or three decades, the researchers have paid considerable practical and scientific attention in the calculation of the size of the austenite grain during heating of the metal before rolling, cold working of the grains during deformation and its further relaxation due to softening of the metal in pauses between deformation cycles.

2.1.1. Model of austenizing steel in heating

The grain size of austenite formed during heating for rolling a metal depends on the chemical composition of the steel, temperature and isothermal holding time. The growth of austenite grains in the steel is a thermally activated complex process which consists of a series of simpler processes depending on factors that have different physical nature.

The crystal growth of austenite is defined by the tendency of the system to minimize energy and is expressed in the minimum surface area and the minimum grain boundary energy. Changing the heating temperature and the phase composition of the steel, the presence of impurities and their subsequent dissolution change the grain growth mechanism. The decisive factor, especially after dissolution of carbonitrides, is the grain boundary migration with movement to the centre of curvature. The speed of movement of the boundary G during grain growth can be determined by the well-known formula:

$$G = \frac{dD}{d\tau} = k_1 / D, \tag{2.1}$$

where D – the current diameter of the austenite grain; τ – time; k_1 – a coefficient.

After integrating equation (2.1), we obtain:

$$D^2 - D_0^2 = \frac{k_1}{2}\tau, \tag{2.2}$$

where D_0 – the austenite grain diameter in time $\tau = 0$.

Given this the square of the austenite grain diameter is directly proportional to the time τ, and its rate of growth is given by:

$$k_1 = k_0 \cdot \exp\left(-\frac{Q_a}{RT}\right), \tag{2.3}$$

where k_0 – constant, Q_a – activation energy of grain growth. R – the Boltzmann constant, T – temperature, K.

The number of particles of the second phase and the amount of the elements dissolved in the austenite influence the migration of grain boundaries and thus the activation energy for grain growth Q_a. Assuming that the activation energy is proportional to the concentration of the second phase and the dissolved elements, it is defined by the following expression:

$$\begin{gathered} Q_a = a + b\cdot \mathrm{Al_{un}} + c\cdot \mathrm{V_{un}} + d\cdot \mathrm{Ti_{un}} + f\cdot \mathrm{Nb_{un}} + \\ + m(d\cdot \mathrm{Al_d} + c\cdot \mathrm{V_d} + D\cdot \mathrm{Ti_d} + f\cdot \mathrm{Nb_d}) \end{gathered} \tag{2.4}$$

where $\mathrm{Al_{un}}$, $\mathrm{V_{un}}$, $\mathrm{Ti_{un}}$, $\mathrm{Nb_{un}}$ – the mass fractions of the elements that make up the undissolved carbides and nitrides; $\mathrm{Al_d}$, $\mathrm{V_d}$, $\mathrm{Ti_d}$, $\mathrm{Nb_d}$ – the mass fractions of elements dissolved in austenite; a, b, c, d, e, f – coefficients.

The mass content of the elements included in the undissolved carbides and nitrides was calculated using the following relations:

$$\begin{gathered} \mathrm{Al_{un}} = \mathrm{Al_s} - \mathrm{Al_d}(\mathrm{N}),\ \mathrm{V_{un}} = \mathrm{V_s} - \mathrm{V_d}(\mathrm{N}), \\ \mathrm{Ti_{un}} = \mathrm{Ti_s} - \mathrm{Ti_d}(\mathrm{N}),\ \mathrm{Ti_{un}} = \mathrm{Ti_s} - \mathrm{Ti_d}(\mathrm{N}), \end{gathered} \tag{2.5}$$

where $\mathrm{Al_s}$, $\mathrm{V_s}$, $\mathrm{Ti_s}$, $\mathrm{Nb_s}$ – content in the steel of the microalloying elements.

Coefficients a, b, c, d, e, f and m in equation (2.4) were calculated from the experimental data on the size of austenite grains, given in [9], where low-carbon steels with high a manganese content (1.24–1.60%) and microalloyed with aluminum, vanadium and niobium, (Table 2.1) were investigated.

Additionally, the calculations were carried out using the experimental data on the effect titanium on the austenite grain size shown in [10]. In the study by the authors of this book, the first stage included the determination of the weight fraction of the elements dissolved and undissolved in austenite in dependence on the heating temperature of steel. The influence of the heating temperature on the amount of undissolved steel carbide- and nitride-forming elements (*a*) and the activation energy of grain growth (*b*) shown in Fig. 2.1.

According to the calculation results the vanadium nitride is completely dissolved in austenite already at a temperature of 1050°C; dissolving of the aluminum nitride is completed at 1150°C; niobium carbides in the undissolved condition may be found at a temperature above 1200°C (Fig. 2.1 *a*).

In the second stage, the experimental values of the austenite grain size for various steels and heating temperature calculated using the equations (2.2) and (2.3) were used to calculate the activation energy of grain growth (Q_a). Thereafter, the least squares method was employed to determine the coefficients a, b, c, d, f and m in the equation (2.4). It was determined that a = 64500; b = 5000; c = 15000; d = 75000; f = 175000 kcal/mol · °C/%.

Coefficient m is 0.1. This indicates that the undissolved carbides, and nitrides of aluminum, vanadium, titanium and niobium influence the activation energy of grain growth 10 times more effectively than dissolved. Figure 2.1 *b* shows the effect of heating temperature on the activation energy of grain growth of the steels listed in Table 2.1

Table 2.1 Chemical composition of the investigated steels

Steel No.	The content of chemical elements,%								
	C	Si	Mn	P	S	Al	Nb	V	N
1	0.06	0.05	1.58	0.012	0.012	0.030	–	–	0.0056
2	0.06	0.05	1.60	0.011	0.013	0.028	0.17	–	0.0061
3	0.06	0.05	1.26	0.012	0.015	0.032	0.10	–	0.0052
4	0.06	0.05	1.24	0.013	0.012	0.032	–	0.12	0.0055
5	0.06	0.05	1.26	0.012	0.012	0.029	0.05	0.06	0.0048

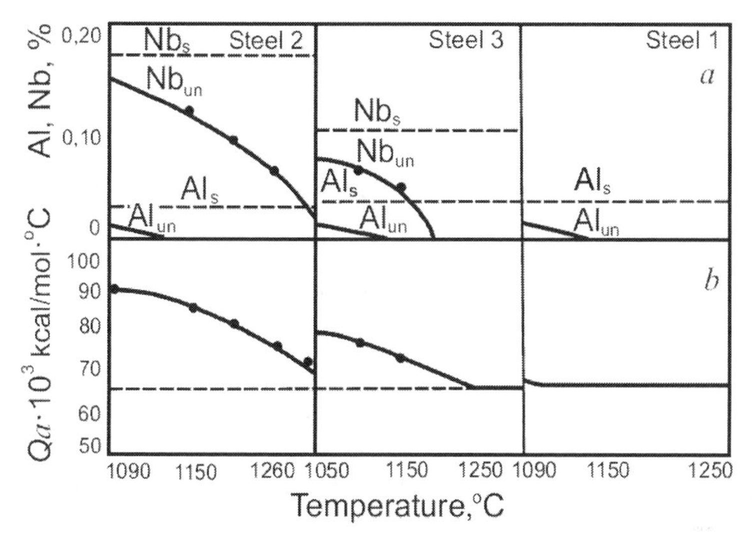

Fig. 2.1. Influence of the heating temperature of steel by the amount of dissolved carbide and nitride-forming elements (a) and the magnitude of activation of the grain growth (b)

and Fig. 2.2 shows the calculated and experimental dependences of the austenite grain size on the heating temperature.

Increasing temperature leads to a decrease in the restraining effect of the carbonitride particles on the growth of austenite grains, expressed in reducing the activation energy of grain growth up to the temperature of complete dissolution of the phases. Further increase in temperature does not change the activation energy and the austenite grain increases rapidly in proportion to the increase in the heating temperature. The results comparing the calculated values with the experimental data showed a real ability to predict the austenite grain size and the austenite chemical composition using mathematical modelling.

2.1.2. Mathematical model of the austenitic structure upon deformation under isothermal conditions

Deformation is accompanied by the hardening of austenite grains and its subsequent relaxation (recrystallization) during the time between passes. These processes are described mathematically using model of deformation and recrystallization of austenite developed by C.M. Sellars, J.A. Whiteman [11] S. Liska, J. Wozniak [12] and others.

The model of dynamic recrystallization includes the calculation of the Zener–Hollomon parameter (temperature-compensated strain rate):

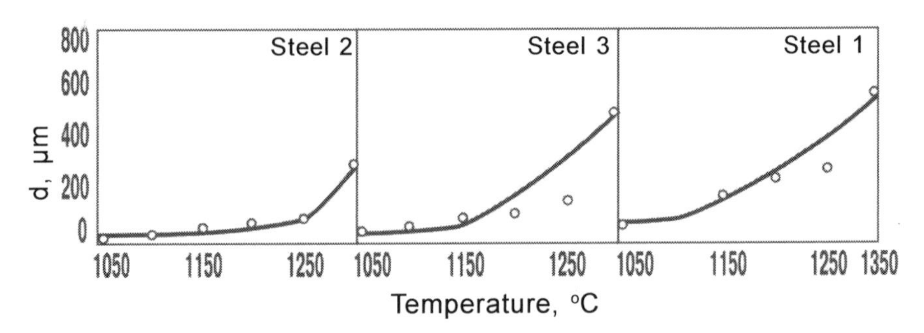

Fig. 2.2. Dependence of the austenite grain diameter d on the heating temperature of steels of various chemical compositions. Solid lines – calculation by the described procedure; points – experimental data [9],

$$Z = \dot{\varepsilon} \exp(Q_u / RT); \tag{2.6}$$

The degree of deformation at which the dynamic recrystallization starts:

$$\varepsilon_k = \left\{ a_1 D_0 \cdot \left[\dot{\varepsilon} \exp \frac{Q_d}{RT} \right] \right\} a_2; \tag{2.7}$$

the share of the dynamically recrystallized structure:

$$X_d = 1 - \exp\left(f_1 \left[\frac{\tau}{\tau_{d0.95}} \right] \right). \tag{2.8}$$

where $\tau_{d0.95} = C_1 D_0^{C_2} Z^{C_3}$;

the grain size after dynamic recrystallization:

$$D_d = d_1 Z^{-d_2}. \tag{2.10}$$

The model of static crystallization includes calculation of the following parameters of the structure:

The time to 50% of static crystallization:

$$\tau_{0.5} = g_1 \varepsilon^{g_2} D_0^{g_3} [\exp(Q_{st} / RT)] Z^{g_4}; \tag{2.11}$$

the share of the statically recrystallized structure:

$$X_{st} = 1 - \exp[-\beta(\tau / \tau_{0.5})^{n_{st}}]; \tag{2.12}$$

the size of statically recrystallized grains:

$$D_{st} = S_1 \varepsilon^{S_2} D_0^{S_3} Z^{S_4}; \tag{2.13}$$

the grain size in secondary recrystallization:

$$D_w^{n_w} - D_{st}^{n_w} = A\tau_{pk} \exp(-Q_w / RT). \tag{2.14}$$

In the above formulas the following notations are used::

a_1, a_2, a_3, a_4, A – coefficients and exponents;
c_1, c_2, c_3, c_4 – coefficients and exponents;
d_1, d_2 – coefficients and exponents;
D – austenite grain diameter;
D_0 – the original austenite grain diameter;
D_d – the diameter of dynamically recrystallized austenite grains;
D_{st} – the diameter of statically recrystallized austenite grains;
D_w – the austenite grain size after recrystallization;
f_1, f_2 – coefficients and exponents;
$g_1 - g_4$ – coefficients and exponents;
n_{st}, n_w – exponents;
Q_u – activation energy of hot deformation;
Q_{st} – activation energy of static recrystallization;
Q_w – activation energy of recrystallization;
R – the gas constant;
$S_1 - S_4$ – coefficients and exponents;
τ – duration of static recrystallization;
τ_{pk} – duration of recrystallization;
$\tau_{d0.95}$ – the time required for dynamic recrystallization of 95% volume of deformed austenite;
$\tau_{0.5}$ – the time required for static recrystallization of 50% of deformed austenite;
T – temperature;
$\dot{\varepsilon}$ – strain rate;
ε – the degree of deformation.

The values of the constants appearing in a generalized model of the deformed structure of austenite are summarized in Table 2.2 for carbon [11], low-alloy [12] and alloyed (austenitic) [13] steels.

Notes to Table 2.2:
1) $K = 0.152\ \dot{\varepsilon}^{0.15}$
2) $f(o) = 9/7\ (0.1/C) + 17/5\ Cr + 73\ Mo + 53.7V$
3) Instead $Z \rightarrow \dot{Z} = 14.9 \ln \dfrac{R}{8.5 \times 10^{-9}}$.

Analysis and synthesis of these techniques allowed us to build a generalized model of the formation of the austenite structure of carbon and low-alloy steels during deformation described below.

The share of the recrystallized austenite grains after deformation is calculated by the Avrami equation:

Table 2.2 The values of constants recrystallization in formulas (2.6)–(2.14).

Process	Formula	Parameter	Carbon steel [11]	Low-alloyed steel [12]	Alloyed steel [13]
Deformation	(2.6)	Q_u, J/mol R	312000 8.2	312000 8.2	414000 8.2
Dynamic recrystallization	(2.7)	Q_d, J/mol a_1 a_2 a_3 a_4	– – – – –	312000 3.42×10 0.5 0.15 1.0	351240 4.55×10^{-18} 0.5 1.0 K[1)]
	(2.8)	f_1 f_2	– –	–2.996 2.0	–2.966 1.25
	(2.9)	c_1 c_2	– –	1.06×10^{-5} 0.0	2.802×10^{-16} 0.8
	(2.10)	d_1 d_2	– –	1.80×10^{3} 0.15	$6.813\text{x}10^{6}\varepsilon^{1/6}$ 0.111
Static recrystallization	(2.11)	Q_{st}, J/mol g_1 g_2 g_3 g_4	480000 3.54×10^{-21} –4 2 –0.375	300000+f(o)[2)] 2.5×10^{-19} –4 2 0	351240 1.45×10^{-11} –2 2 –0.375
	(2.12)	β n_{st}	2.996 2	2.996 2	2.996 0.5–2.0
	(2.13)	S_1 S_2 S_3 S_4	25 –1 0.5 –0.64[3)]	0.5 –1 0.67 0	2.24 –0.5 1 –0.006
Secondary recrystallization	(2.14)	Q_w, J/mol n_w A	400000 10 3.84×10^{32}	420000 6 8×10^{24}	420000 10 5.86×10^{31}

$$X = 1 - \exp[-\beta(\tau/\tau_r)^n], \quad (2.15)$$

where $\beta = 0.69$, if the time of recrystallization is replaced by the time to 50% recrystallization ($\tau_r = \tau_{0.50}$), or $\beta = 2.996$ at $\tau_r = \tau_{0.95}$; n is an exponent which depends on the deformation parameters, the original structure of the metal and the steel chemical composition. In a number of studies it is accepted that $n = 2$.

The duration of primary recrystallization $\tau_{0.50}$ of deformed austenite for ordinary and low-alloy steels is calculated by:

$$\tau_{0.5} = 2.5\cdot10^{-19}\cdot d_0^2\varepsilon^{-4}\exp\left(\frac{Q_a}{RT}\right), \quad (2.16)$$

where d_0 is the average austenite grain diameter before deformation, μm; ε is the degree of deformation in the pass; Q_a is the effective activation energy of primary recrystallization.

The effective activation energy, as previously mentioned, depends on the chemical composition of the steel. In [12], this dependence is represented by the formula:

$$Q'_a = 300 + 9.7\ln(0.1/\mathrm{C}) + 17.5\mathrm{Cr} + 57.3\mathrm{V}, \tag{2.17}$$

where C, Cr, V is the content of the chemical elements in steel, wt.%.

To account for the influence of carbonitride-forming elements Ti and Nb in our work, this formula is supplemented to the following:

$$Q_a = Q'_a + 350\mathrm{Ti} + 600\mathrm{Nb}. \tag{2.18}$$

The coefficients at Ti and Nb are calculated on the basis of the experimental curves $\tau_{0.5} = \tau(\mathrm{Ti, Nb})$, obtained in [14] by reverse recalculations using the formula (2.16). In the same study it is noted that the degree of softening $R_\sigma = 2\%$ corresponds to the beginning of static recrystallization, i.e. given that $X = R_\alpha^2$, recrystallization can be considered as beginning at $X = 5\%$. There it is also determined that the effect of the chemical elements on the retardation of static recrystallization increases in the sequence: solid solution of vanadium, titanium, niobium.

The average austenite grain diameter after primary recrystallization (d_r) was calculated according to the formula derived from a geometric model shown in [15]:

$$\frac{d_0}{d_r} = A\varepsilon^{2/3}, \tag{2.19}$$

where d_0 is the average diameter of the austenite grain prior to current strain ε. The development of this model allowed us to obtain the formula:

$$d_r = 3.28 d_0^{0.32} \varepsilon^{-0.66}. \tag{2.20}$$

Calculations of the average austenite grain diameter (d) in secondary recrystallization by the well-known formula [11]:

$$d^n = d_r^n + A_c \tau \exp\left(-\frac{Q_a}{RT}\right), \tag{2.21}$$

where τ is the time after the completion of primary recrystallization, have shown that this dependence does not reflect the influence of the recrystallized grain diameter (d_r) on the growth of austenite grains during recrystallization, as d_r enters it as a summand. Therefore, in the following text, preference is given to a formula in which d_r is a factor:

$$d = d_r\left(1 + A_{c1}\ln\frac{\tau}{\tau_r}\right), \tag{2.22}$$

where A_{c1} is a factor.

The advantage of the dependence (2.22) is also that it includes the time course of primary recrystallization (τ_r), thus reflecting the inextricable connection between the two stages of recrystallization – primary and secondary.

2.1.3. Features of modelling the parameters of the structure of austenite under isothermal multiple deformation

In the pauses between two adjacent reduction passes of the strip depending on the length of the delay, the deformation parameters, temperature and the chemical composition of the steel the primary recrystallization of austenite can not be completed. This causes some difficulty in predicting the austenitic steel structure. Solution of the problem is associated with the correct definition of values d_0 and ε, which should be substituted into the formulas (2.16) and (2.20) to calculate the austenite structure after each deformation.

The change in the average grain diameter of austenite during primary recrystallization was studied in [15] and is shown in Fig. 2.3.

This dependence is described by the following formula:

$$d = d_0 - (d_0 - d_r)X^a, \tag{2.23}$$

where $a = 0.123$, X is the share of recrystallized grain size calculated according to the formula (2.15).

For consideration in the next deformation cycle the preceding strain hardening the actual degree of deformation (ε_i) is replaced with an equivalent degree of deformation $\left(\varepsilon_i^*\right)$:

$$\varepsilon_i^* = \varepsilon_i + \Delta\varepsilon_{i-1}, \tag{2.24}$$

where $\Delta\varepsilon_{i-1}$ is the degree of deformation equivalent to the hardening of austenite remaining before the reduction in the i-th pass.

The value $\Delta\varepsilon_{i-1}$ was calculated using the following formula:

$$\Delta\varepsilon_{i-1} = \varepsilon_{i-1}(1 - X_{i-1}). \tag{2.25}$$

where ε_{i-1} is the degree of deformation at the i-1st pass; X_{i-1} is the degree of recrystallization after the i-1st pass.

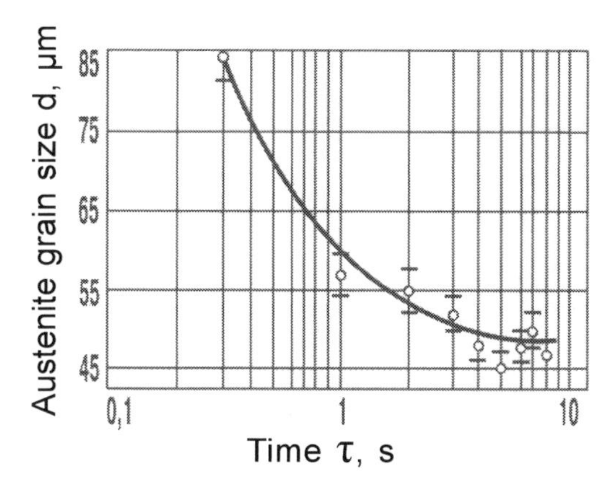

Fig. 2.3. Change of the austenite grain size in the primary recrystallization of low-carbon steel [15].

2.1.4. Formation of the austenite structure in isothermal conditions

The metal temperature in hot rolling is not constant. Throughout the process it generally monotonically decreases. We consider a broken cooling curve which consists of a set of isothermal holding periods at temperatures T_1, T_2,... in the time intervals t_1, t_2,... . Since the duration of primary recrystallization at temperatures T_1, T_2,... is equal to, respectively, the values τ_{r1}, τ_{r2},..., it is assumed that at each temperature T_i part of the deformed austenite recrystallizes proportional to the relation t_i/τ_{ri} recrystallizes.

By the time $\tau_i = \Sigma t_i$ the share of the recrystallized grains is determined by the relation:

$$X_i = 1 - \exp\left[-\beta\left(\sum_i \frac{t_i}{\tau_{r_i}}\right)\right], \tag{2.26}$$

and grain growth during secondary recrystallization at this point of time can be determined by the formula:

$$d_i = d_r\left[1 + A_c \mathrm{h}\left(\sum_i \frac{t_i}{\tau_{r_i}}\right)\right]. \tag{2.27}$$

Comparative analysis of the available experimental data [16] concerning 3sp steel, [17] relating to low-alloy steel with vanadium and niobium, and the results of calculation of the austenitic structure indicate good agreement between the experimental and calculated values of the average diameter of the austenite grains. This suggests that the proposed mathematical model of structure formation of austenite during repeated deformation in isothermal conditions is acceptable and sufficiently reliable.

2.1.5. Calculation of isothermal decomposition diagrams of austenite

Construction of the diagrams of isothermal transformation of austenite makes it possible to predicts the structure formation under continuous cooling of rolled products. The continuous cooling process consists of a set of isothermal holding periods at temperatures T_1, T_2, T_3,... for periods τ_1, τ_2, τ_3. It is assumed that isothermal transformation incubation periods at temperatures T_1, T_2, T_3,... are equal to t_1, t_2, t_3,..., respectively and that part of the incubation period equal to t_i / τ_i is used at each temperature of stepped cooling. Transformation starts when the sum of these relationships becomes equal to $\left(\sum t_i / \tau_{r_i} = 1\right)$.

Analyzing numerous works devoted to assessing the legality of the additivity principle to calculate the start time of transformation, the authors concluded that it is fulfilled for the polymorphic and pearlite transformation. The additivity principle was subsequently extended to the entire interval of the phase transformation.

To represent the development of transformation during continuous cooling on the isothermal diagram it was assumed that the degree of transformation of austenite to ferrite, pearlite or bainite obeys the Avrami equation:

$$X(t) = 1 - \exp\left[-\beta_X \left(\frac{t}{t_x}\right)^K\right], \tag{2.28}$$

where $X(t)$ is the amount of the resulting phase; t is the current time, t_x is the duration of conversion of volume X.

The constants β_x and K are calculated for each phase of the specific chemical composition of the steel under the condition that the start of the transformations (X = 1%) corresponds to the time t_s, and the end of the transformation corresponds to t_f. Then from equation (2.28) we have:

$$K = \frac{6.127}{\ln(t_s / t_f)}. \tag{2.29}$$

The key parameters in this model are the values of the start time (t_s) and the end (t_f) of the transformation at a predetermined temperature. We use the results of [18], in which the isothermal diagram consists of five curves in the coordinates log (t) – T, the start of ferrite formation (F_s), the start of pearlite formation (P_s), the end of pearlite transformation (P_f), the start (B_s) and the end (B_f) of bainite transformation, and line start (M_s) and finish (M_f) of the martensitic transformation (Fig. 2.4).

For the mathematical description of each curve we use the following relationships [18]:

$$\frac{S-S_O}{S_N-S_O}=\frac{1}{\exp(-1/2)}\cdot\frac{(U-U_O)^{1/2}}{(U_N-U_O)^{1/2}}\cdot\exp\left[-\frac{1}{2}\cdot\frac{U-U_O}{U_N-U_O}\right]^2, \quad (2.30)$$

here $U=\dfrac{1000}{T}; U_O=\dfrac{1000}{T_O}; U_N=\dfrac{1000}{T_N}$; $S=\ln(t)$; $S_O=\ln(t_O)$; $S_N=\ln(t_N)$;

S_O, T_O, S_N, T_N are the coordinates of the reference points.

The values S_O, T_O, S_N, T_N are functions of the chemical composition and austenite grain size of the steel and can be described by the empirical dependence [18]:

$$Y_1=\sum_{k=1}^{8}\left[B_{l.k}+B_{l,k+8}(0.8-C_2)\right]C_k, \quad (2.31)$$

where $B_{l,k}$ – empirical coefficients; C_k – parameters reflecting the content of chemical elements in steel and austenite structure:

C_k	C_1	C_2	C_3	C_4	C_5	C_6	C_7	C_8
Values	1	%C	%Mn	%Si	%Ni	%Cr	%Mo	$2^{N\gamma/2}$

Note: Nγ – the austenite grain size.

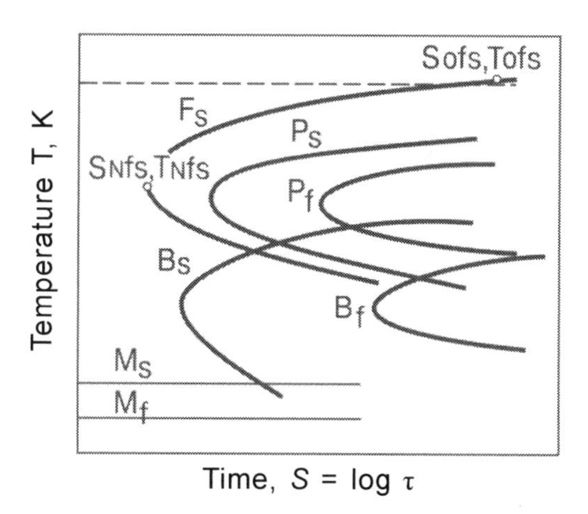

Fig. 2.4. Chart diagram of isothermal decomposition of austenite [18].

The empirical coefficients $B_{l,k}$ were calculated by the multiple regression analysis of reference points S_O, T_O, S_N, T_N, from more than 100 charts of the isothermal decomposition of austenite in the handbook [19]. The multiple correlation coefficients of the obtained dependences were quite high (0.65–0.80), which allows to calculate the diagram of the isothermal decomposition of austenite steels in the range of chemical compositions:

Element	C	Mn	Si	Cr	Ni	Mo	Nγ
Content in steel,%	0.10–0.60	0.20–2.00	0.20–2.50	<2.00	<3.00	<0.83	2–8

The experimental and calculated isothermal diagrams for 20KhG and 20KhGS2 steels are shown in Fig. 2.5. Comparison of the curves indicates good correspondence between the results of calculations and the experimental data that allows to build the diagram of isothermal decomposition of austenite for steels in the broad range of chemical composition, and also using the principle of additivity to predict the phase transformations of austenite under continuous cooling of strips.

2.1.6. Method of calculation of thermokinetic diagrams of of austenite breakdown

The phase composition of the steel is determined in volume fractions V_i (I = 1–5), where V_1 – volume fraction of austenite, V_2 – ferrite, V_3 – perlite, V_4 – bainite, V_5 – martensite. At each step of the transformation the following equation must be satisfied:

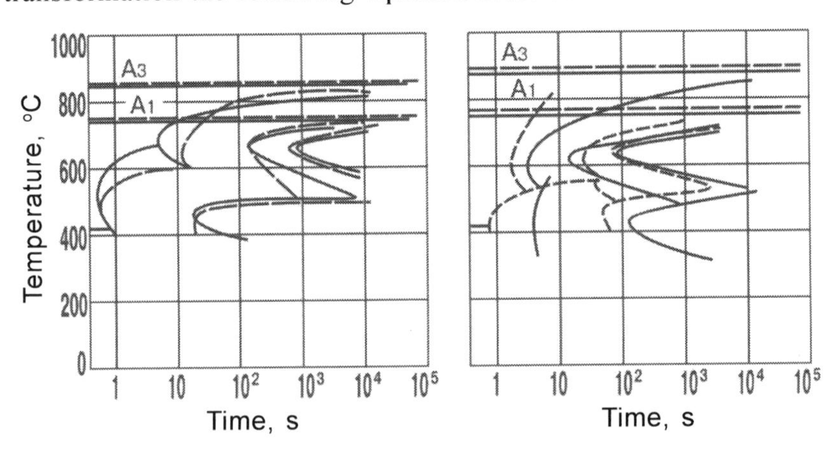

Fig. 2.5. Calculated (solid lines) and experimental (dashed) diagrams of isothermal decomposition of austenite: *a* – 20KhG steel [19], *b* – 20KhGS2 steel (according to M.F. Evsyukova data),

$$\sum_{i=1}^{5} V_i = 1.$$

The basis of the calculation method is the Avrami equation. The time to the beginning and end of austenite transformation to the i-th phase (t_{si}; t_{fi}) is calculated by the formula (2.30), and the order of K_1 is determined by formula (2.29). Values of the volume fraction of the i-th structural component at time $t + \Delta t$ is calculated from the relation:

$$V_i(t+\Delta t) = X_i(t+\Delta t) + [V_1(t) + V_i(t)]V_{mi},$$

where $X_i(t + \Delta t)$ is the fictitious volume fraction of the i-th phase in the 'austenite + i-th phase' mixture, defined by the equation:

$$X_i(t+\Delta t) = 1 - \exp\left[0.01\left(\frac{t_1 + \Delta t}{t_s}\right)^{K_1}\right],$$

where $t_1 = \left[\dfrac{-t_s \ln(1 - X_1)}{0.01}\right]^{1/K_1}$.

The martensitic transformation occurs at temperatures (M_s, M_f) at a high rate speed and is subject to other relationship, different from equation (2.28). At a temperature $M_f < T < M_s$ the amount of martensite formed from the austenite untransformed at this time is proportional to the difference (M_s–T). Based on the recommendations of [18] the amount of martensite is calculated by the relation:

$$V_s = \left(1 - \sum_{i=2}^{5} V_i\right)\left[1 - \exp\left(-0.011(M_s - T)\right)\right].$$

The volume of residual austenite is calculated at the next time interval:

$$V_1(t+\Delta t) = 1 - \sum_{i=2}^{5} V_i(t).$$

If $V_1(t + \Delta t) = 0$, then the austenite breakdown is considered complete.

The carbon concentration in pearlite for hypoeutectoid steels is higher than in the steel as a whole [20]. The carbon content in the austenite remaining after polymorphic and pearlitic transformations was calculated by the formula:

$$C_{\text{aust}} = C_{\text{steel}} / (1 - V_2) - 0.82V_3.$$

At the start of bainitic and martensitic transformations the carbon concentration of the untransformed decayed austenite is in general different from the carbon concentration in the steel. Probably, this fact was the reason why a number of researchers noted the fact that the calculation of the bainite transformation using the isothermal diagrams showed a higher rate of the transformation in comparison with experiment. Experience in our research gives indicates that the bainitic transformation takes place with the austenite enriched with carbon. Analysis of the thermokinetic diagrams shows that with decreasing cooling the martensitic transformation rate also starts at low temperatures, which is a consequence of carbon enrichment of the untransformed austenite.

The algorithm described here was used to construct the diagrams of austenite breakdowm of the St3sp and 09G2s steels represented in the coordinate systems: lg(t)–T, lg (V_{cool})–X (Fig. 2.6).

The method of calculating the average ferrite grain size (d_0) was selected on the basis of the following consideration. The ferrite grain size depends on the austenite grain size, the degree of work

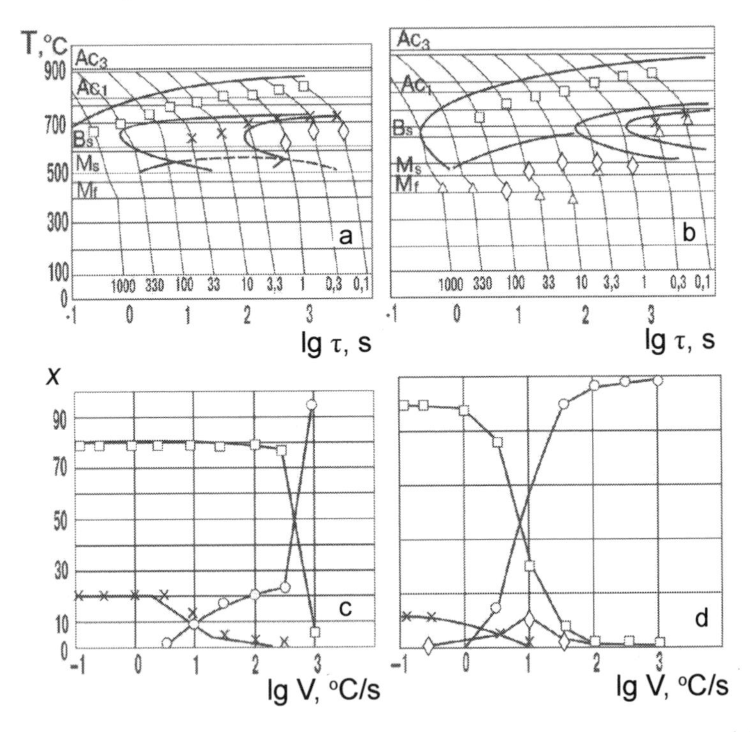

Fig. 2.6. Calculated thermokinetic diagrams of breakdown of St3sp (*a*, *c*) and 09G2S (*b*, *d*) steels. Figure notes: □ – ferrite, x – perlite; ◊ – bainite; ○ – martensite; Δ – the end of austenite breakdown.

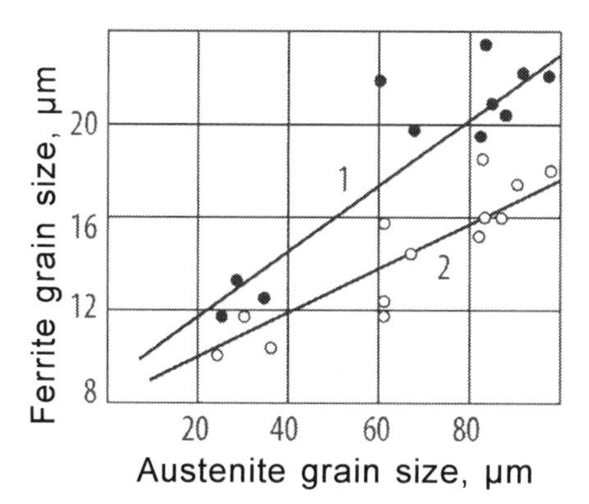

Fig. 2.7. Experimental (points) [22] and calculated (solid lines, using equation (2.32)) values of the ferrite grain size of 08pc steel: 1 – cooling rate 5 deg/s, 2 – cooling rate 16 deg/s.

hardening of austenite and the temperature of the polymorphic transformation. Experience shows that these provisions correspond satisfactorily to the formula derived in [21]:

$$d = \left[5.51 \cdot 10^{10} d_{ef}^{1.75} \exp\left(-\frac{21430}{T_{0.05}} \right) V_f \right]^{1/3}, \tag{2.32}$$

$$\text{where } d_{ef} = d/(1+2.4 \cdot 10^{-16} \cdot \rho^{1.154}) \tag{2.33}$$

is the effective austenite grain size; ρ is the density of dislocations, m^{-2}, $T_{0.05}$ is the temperature at which 5% of ferrite is formed; V_f is the volume fraction of ferrite in the steel.

Figure 2.7 shows the results of calculation of the ferrite grain size by formula (2.32) compared with experimental data [22].

The method of calculating the spacing of pearlite plates was assuming that the value of S_0 decreases with decrease of the transformation temperature. For example, the formula derived in [23] describes this feature quite efficiently:

$$S_O = \left[18 \sum_i \frac{V_p(t_i)}{A_C - T_i} \right] / V_p, \tag{2.34}$$

where $V_p(t_i)$ is the proportion of pearlite formed at temperature T_i.

Figure 2.8 compares the results of the calculation of the distance between the pearlite plates using formula (2.34) with the data given in [20].

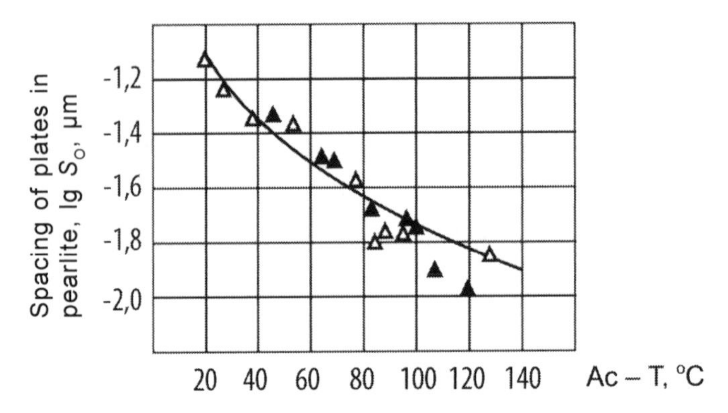

Fig. 2.8. Experimental [20] and calculated (formula (2.34), the solid line) values of pearlite plate spacing: black dots ▲ – steel containing 0.78% C, light points Δ – 0.80% C.

Upon slow cooling of rolled strips, for example, in a roll the recrystallization of ferrite and spheroidization of plate pearlite take place. Changes in the morphology of pearlite lead to softening of the steel due reducing the share of plate pearlite in the structure:

$$\Pi' = V_p(1-\gamma P), \tag{2.35}$$

where Π' is the conditional content in the structure of hardening (plate) perlite; γ is a coefficient, P is the fraction of spheroidized pearlite.

The percentage of spheroidized pearlite was calculated using the Avrami-type equation:

$$P = 1 - \left[\exp\left(-\frac{\tau}{\tau_p}\right)^k\right], \tag{2.36}$$

where k is a constant; τ is time; τ_p is the time to full spheroidization of pearlite.

A.P. Chernyshov in his thesis found that for low-carbon and low-alloy steels the time to full pearlite spheroidization depends on temperature and pearlite plate spacing during the pearlite transformation:

$$\tau_p = \exp\left(\frac{15062}{T} - \frac{0.228}{S_O} - 12.728\right), \tag{2.37}$$

where T – temperature, K, S_O – plate spacing, μm.

The ferrite grains inherit the recrystallized state of austenite and can also be subjected to deformation when the latter occurs at temperatures below Ar_3. In describing the processes of ferrite recrystallization researchers proceeded from the fact that they are similar to the primary and secondary recrystallization of austenite. It is recognized that the recrystallization of ferrite grains is more intense at the same temperature due to the fact that the activation energy of recrystallization of ferrite Q_a is higher than austenite recrystallization activation energy Q_r. We assumed that $Q_a/Q_r = 1.1$.

2.1.7. Evaluation of the model of mechanical properties

Carbon and low-alloy steels in the hot-rolled condition are polyphase mixtures which may include in various ratios the primary phases (F), perlite (P), bainite (B) and martensite (M), the secondary phases in the form of carbides, nitrides of microalloying metals (Al, V, Ti, Nb), and non-metallic inclusions (oxides, sulphides). If the steel is considered as a natural composite of the above components, its mechanical properties (tensile strength σ_B, yield stress σ_T) can be represented as the sum of the strength properties of each component multiplied by its volume fraction in the steel:

$$\sigma = \sigma_F V_F + \sigma_P V_P + \sigma_B V_B + \sigma_M V_M + \sum \sigma_{Di} V_{Di}, \tag{2.38}$$

where σ_F, σ_P, σ_B, σ_M and V_F, V_P, V_B, V_M are the strength and volume fraction of ferrite, pearlite, bainite, martensite; σ_{Di}, V_{Di} are strength and volume fraction of the i-th of the secondary phase.

The properties of each structural component in formula (2.38) are determined by the chemical composition, morphology and dispersion of the structure and its strain hardening. The strength of the ferritic component σ_F is represented by the sum of three terms:

$$\sigma_F = \sigma_F^s + \sigma_F^g + \sigma_F^{st}, \tag{2.39}$$

where σ_F^s, σ_F^g and σ_F^{st} are the solid solution, grain boundary and strain hardening of the ferrite.

The hardening of ferrite by solid solution alloying was calculated by the additive expression:

$$\sigma_F^s = \sum K_{iF} \cdot C_i, \tag{2.40}$$

where K_{iF} are empirical coefficients; C_i is the volume content of the i-th chemical element, %.

When calculating using equation (2.40) the value C_i should represented by the concentration of the alloying element dissolved in ferrite, which in a general case differs from the content of this element in the steel. For ferritic–pearlitic steels it was assumed in this work that Mn, Si, Ni, P are mostly dissolved in ferrite and the chemical elements present in the steel C, N, Cr, Mo, V, Ti, Al are included in the composition of the carbide and nitride phases.

The dependence of the strength characteristics of ferrite on the size (diameter) of the ferrite grains was calculated by the Hall–Petch type equation:

$$\sigma_F^g = K_d \cdot d^{-1/2}, \tag{2.41}$$

where K_d – factor; d – average diameter of ferrite grains, calculated by formula (2.32).

In controlled and hot rolling, when not only the austenite grains, but also the ferrite grains formed during austenite breakdown are deformed, the hardening of ferrite due to its deformation was calculated by the chosen formula:

$$\sigma_F^{st} = (a_F^{st} + c_F^{st} d^{-1/2}) \cdot \Delta\varepsilon^{c_F}, \tag{2.42}$$

where a_F^{st}, b_F^{st}, c_F^{st} are the empirical coefficients; $\Delta\varepsilon$ is the residual deformation of recrystallized ferrite.

The contribution of the pearlite phase to the strength of low-carbon steels was taken into account by the following formula:

$$\sigma_P = K_P \cdot P, \tag{2.43}$$

where K_p – empirical coefficient, P – the volume fraction of pearlite in steel.

In addition to the volume fraction of pearlite in the steel it was also necessary to consider its morphology. The conversion of the plate pearlite to globular occurs, for example, in slow cooling of the hot-rolled strip coil, which reduces the strength properties of the steel. The relationship of globularization of pearlite with the strength properties is described by formula (2.35).

The effect of the bainitic component of the structure of the steel was evaluated using a linear dependence of the strength in the range of the temperature of the start of bainitic transformation B_s to the temperature of the start of martensite formation M_s:

$$\begin{aligned}\sigma_B &= (\sigma_F V_F + \sigma_P V_P)/(V_F + V_P) + (\Sigma((B_s - T_i)/(B_s - M_s)\cdot \\ &(\sigma_M - (\sigma_F V_F + \sigma_P V_P))/(V_F + V_P)\cdot \Delta V_{Bi})/V_M,\end{aligned} \tag{2.44}$$

where T_i is the temperature of bainitic transformation in the i-th time; ΔV_{Bi} is the bainite volume increment for the i-th time interval.

The main factor determining the mechanical properties of martensite is the carbon content in martensite.

In this work, the formula derived in [18] was used as the basis:

$$\sigma_M = a_M + c_M \cdot \sqrt{C}, \tag{2.45}$$

where a_M, c_M are empirical coefficients, C is the carbon content in martensite.

Carbonitride hardening σ_D depends on the volume fraction and dispersion of second phases. However, the authors were unable to find sufficient information to predict the particle size and the distances between them, as well as functional dependences of their impact on the physical and mechanical properties of steel. Therefore, at this stage of the studies an empirical approach was used to calculate carbonitride hardening:

$$\sigma_D = \sum K_{Di} \cdot Me_i, \tag{2.46}$$

where K_{Di} and Me_i are empirical coefficient and the volume fraction of the i-th microalloying additive.

The plastic and toughness properties of the steel are strongly influenced by the presence of non-metallic inclusions – oxides and sulphides in the hot-rolled products, primarily their volume fraction and particle size. The prediction of the values of these parameters is very difficult and should already be done at the stage of steel melting and casting. Therefore, they are less correlated with the components of the chemical composition and structure of the steel than the strength properties.

Calculations of the relative elongation of steel in the tensile tests were based upon the formula proposed in [24]:

$$\delta = ((a_\delta + c_\delta C) / \sigma_c)^{N_\delta}, \tag{2.47}$$

where a_δ, c_δ, N_δ are empirical constants, C is the carbon content in steel, σ_c is the calculated tensile strength of steel.

Room temperature impact toughness was calculated by the formula derived by G.K. Shcherbak in [24]:

$$KCU_{+20} = \sigma_c \cdot \delta / K_{KCU}. \tag{2.48}$$

The toughness of the steel at temperatures below room temperature T_{test} was calculated using the following formula:

$$KCU_{test} = KCU_{+20} - ((KCU_{+20} - KCUt_{br})/(20 - t_{br})) \times T_{test} \tag{2.49}$$

where t_{br} is the brittle fracture temperature; $KCUt_{br}$ is the impact toughness at the brittle fracture temperature.

The brittle fracture temperature was calculated using the well-known formulas that link it with the chemical composition and structure of steel:

$$t_{xp} = a_{xp0} + \sum a_{xpi} C_i + K_{xp} \cdot d^{-1/2}. \quad (2.50)$$

Thus, for selected models of physical and mechanical properties are determining the chemical composition of the steel, the volume of the primary and secondary phases, dispersion and morphology of their structure.

The suitability of the described model of the mechanical properties of rolled steel was evaluated in predicting the strength properties of the steel sheet after hot rolling mill on a continuous wide-strip mill. The structure of the model includes the calculation of the temperature and deformation parameters of the process in each strip rolling mill stand, the calculation of the temperature of the strip on the run-out table and in a roll; calculation of the structural state of austenite (degree of recrystallization, grain size, strain hardening, the chemical composition of austenite), the calculation of the volume fractions of structural steel components in phase transformations of deformed austenite, calculation of the ferrite grain size and pearlite plate spacing, calculation of the extent of spheroidization of perlite (Fig. 2.9).

Test calculations were performed for the cases of rolling strips of St3sp, 09G2SF, 09G2FB steels at different rolling modes: normal rolling (T_{finish} = 880–890°C), controlled rolling (T_{finish} = 730–780°C) and warm rolling (T_{finish} = 530–700°C). The conditions of computational experiments are consistent with the experimental data given in [25]. The chemical composition of the investigated steels in shown in Table 2.3. Comparison of calculated and experimental values of the strength properties of steel is shown in Fig. 2.10.

The calculation results show that the proposed mathematical model satisfactorily describes the properties of steels of different chemical composition, including alloyed with carbonitride hardening and subjected to various modes of thermomechanical processing. Additionally, the mathematical model of the mechanical properties was tested on actual data of mechanical tests of 820 batches of St3sp, 09G2S, 17GS and 10KhSND steel sheets. A specially designed software was used for the calculations of ultimate and yield strength. Correlation analysis showed that the model

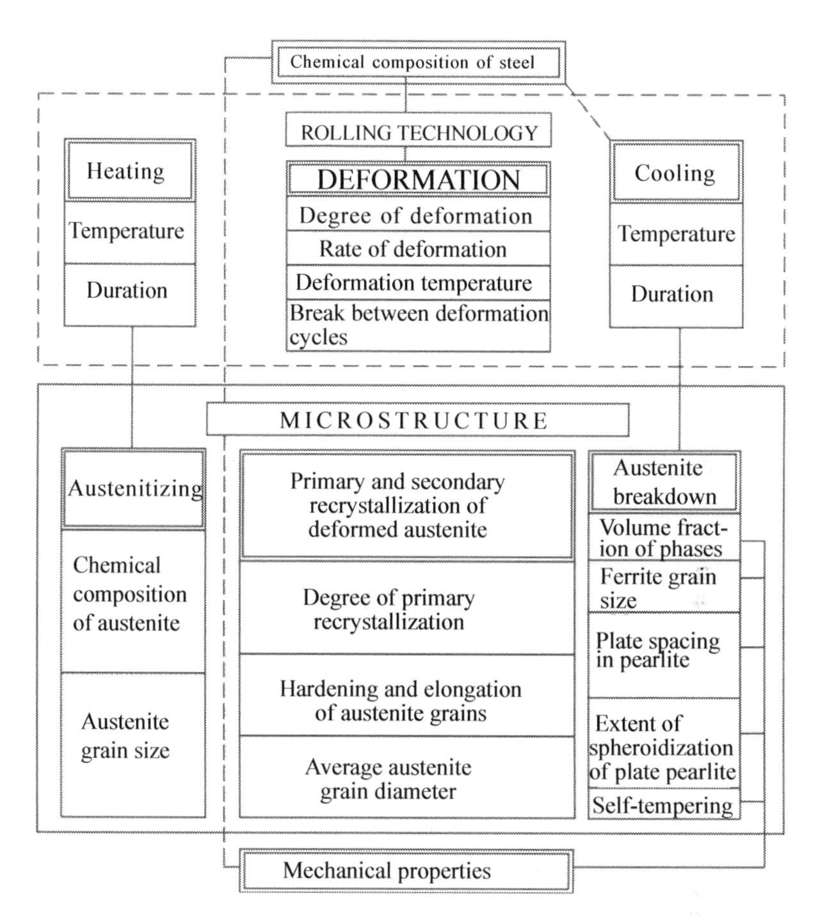

Fig. 2.9. A block diagram of a mathematical model predicting the microstructure and mechanical properties of the steel during hot rolling.

describes with sufficient accuracy the strength characteristics of the properties of a whole class of ferritic–pearlitic steels: the correlation coefficient between the measured and calculated values of the yield stress was 0.86 in the range 240–540 N/mm^2, for tensile strength its value was 0.89 in the range 390–690 N/mm^2. The standard deviation calculated from the experimental data (error of the model) for the yield strength was 23.4 N/mm^2 and for tensile strength 19.5 N/mm^2.

2.2. Deformation resistance

Deformation resistance σ is the stress required for transferring the deformed solid to the plastic flow state. The deformation resistance

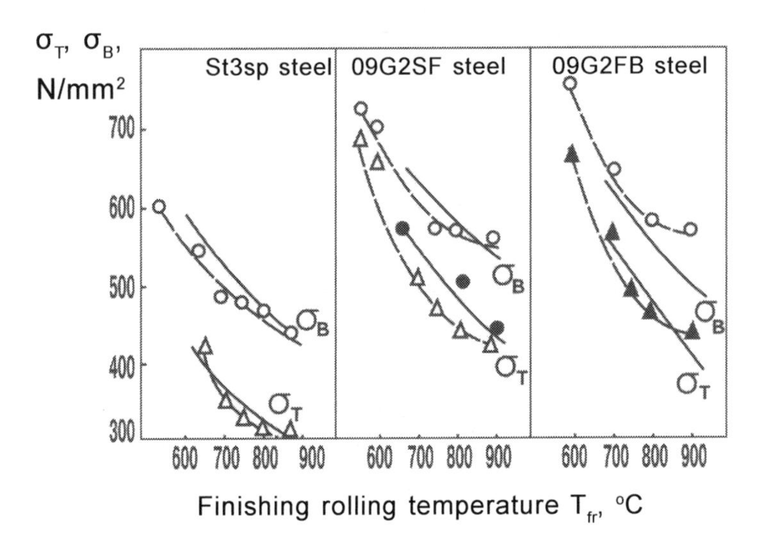

Fig. 2.10. Comparison of the results of calculation of the strength properties of St3sp, 09G2SF and 09G2FB steels (solid lines) with experimental data (points and dotted lines) [25].

Table 2.3. Chemical composition of steels

Steel grade	Mass fraction of elements,%							
	C	Si	Mn	S	P	Al	Cr	Ni
09G2FB	0.09	0.28	0.61	0.004	0.020	0.056	0.06	0.04
09G2SF	0.11	0.,58	1.44	0.009	0.017	0.077	0.03	0.03
St3sp	0.14	0.23	0.46	0.028	0.020	0.005	0.06	0.04

Table 2.3 (continued)

Steel grade	Mass fraction of elements,%					
	Cu	Ti	V	N	Nb	Ca
09G2FB	0.04	0.013	0.080	0.009	0.031	0.003
09G2SF	0.07	0.013	0.081	0.010	–	–
St3sp	0.13	–	–	–	–	–

of a metal at high temperatures depends on strain ε, strain rate $\dot{\varepsilon}$ and deformation temperature T and can be described by the following equation:

$$\sigma = A\varepsilon^a\dot{\varepsilon}^b \exp(-cT). \tag{2.51}$$

The dependence of type (2.51) was proposed on the basis of the experimental investigations of the deformation resistance using special testing machines (plastometers).

The hardening of hot metal during deformation $\sigma = \sigma(\varepsilon)$ is investigated at a rate $\dot{\varepsilon}$ = const and constant test temperature T = const.

This is ensured by the special profile of the cam which activates the bar for upsetting the specimen, and by using special containers heated together with the specimen.

A method of calculating the profile of the cam of the plastometer ensuring the constant strain rate in upsetting the specimen is described below.

The formulation of the problem for testing the metal with the strain rate $\dot{\varepsilon}$ is generally described by the relationship

$$\dot{\varepsilon} = \frac{d\varepsilon}{dt} = \frac{dh}{h}\frac{1}{dt}, \qquad (2.52)$$

where h – height of the sample.

Using the equation (2.52) it is possible to determine the function of the time dependence of the height of the upset specimen $h(t)$ using the given law of the variation of the strain rate $\dot{\varepsilon}(t)$. If it is assumed that the speed of rotation of the flywheel of the plastometer remains constant at the moment of deformation, then the law $h(t)$ can be used to calculate the required profile of the cam.

For a flywheel rotating at a constant velocity, the following equation applies:

$$d\Theta = \omega dt, \qquad (2.53)$$

where Θ is the angle of rotation of the flywheel, rad; ω is the angular velocity (ω = const), rad/s.

According to (2.52) we have:

$$\dot{\varepsilon}(\Theta) = \frac{dh}{h(\Theta)} \cdot \frac{\omega}{d\Theta}. \qquad (2.54)$$

After integrating equation (2.54):

$$-\int_{h_0}^{h} \frac{dh}{h} = \frac{1}{\omega}\int_{0}^{\Theta} \dot{\varepsilon}(\Theta)d\Theta.$$

After integration, we obtain the law of the change of the height of the sample as a function of the rotation angle of the flywheel:

$$h(\Theta) = h_0 \exp\left(-\frac{1}{\omega}\int_0^{\Theta}\dot{\varepsilon}(\Theta)d\Theta\right). \tag{2.55}$$

It follows from equation (2.55) that the extension of the cam $\Delta l(\Theta)$ ensuring the required degree of deformation can be determined as the difference between the initial h_0 and the required $h(\Theta)$ height of the sample

$$\Delta l(\Theta) = h_0 - h(\Theta). \tag{2.56}$$

For modelling the constant strain rate ($\dot{\varepsilon}$ = const) from (2.55) we obtain:

$$h(\Theta) = h_0 l^{\frac{\dot{\varepsilon}}{\omega}\Theta}.$$

Denoting $\dot{\varepsilon}/\omega = \lambda$, from the previous equation it follows:

$$h(\Theta) = h_0 l^{-\lambda\Theta}. \tag{2.57}$$

The values of λ are determined from (2.57) for $h(\Theta) = h_1$:

$$\lambda = \frac{1}{\Theta_k}\ln\frac{h_0}{h_1}, \tag{2.58}$$

where h_1 is the final height of the specimen; Θ_k is the central cam angle.

Now, for the given strain rate it is possible to determine the required angular velocity of the flywheel ω:

$$\omega = \frac{\dot{\varepsilon}}{\lambda}, \tag{2.59}$$

and the profile (height over the generating line of the flywheel) should correspond to the dependence:

$$\Delta l = h_0\left[1 - \left(\frac{h_0}{h_1}\right)^{\Theta/\Theta_k}\right]. \tag{2.60}$$

Two processes take place simultaneously during hot plastic deformation: hardening and softening. The first stage is the change of the stress with increasing strain (up to ε = 0.2–0.4) and it corresponds to the cold working phase which is described by the equation (2.51). To determine the coefficients A, a, b, c, included in equation (2.51), it is necessary to formulate efficiently the experimental examination of the relationship of the deformation resistance σ with the variables ε, $\dot{\varepsilon}$ and T. These investigations are multifactorial experiments in which one of the variables is varied and all other variables remain constant.

In order to apply the method of mathematical experiment planning, the dependence (2.51) is converted to the linear form by logarithmic transformation:

$$Y = b_0 + b_1 x_1 + b_2 x_2 + b_3 x_3, \tag{2.61}$$

where are $x_1 = \ln \varepsilon$; $x_2 = \ln \dot{\varepsilon}$; $x_3 = T$; $b_1 = a$; $b_2 = b$; $c_3 = c$.

Having the linear dependence (2.61), it is quite easy to use the well-developed method of experimental planning which makes it possible to obtain the reliable values of the coefficients A, a, b, c at the minimum number of experiments.

The rolling process in both reversing and continuous mills is carried out by multiple deformation of the strip with breaks between reductions. The metal is softened in these periods, and the degree of softening of the metal can be total or partial, depending on the duration of the break, strip temperature and strip deformation conditions. The softening between the deformations is investigated in plastometers of special design which are suitable for multiple deformation of the same sample with different breaks between reductions [26, 27].

The typical dependences of the deformation resistance of low-carbon steel (0.17% C) on the strain in two reductions of the same sample with a 1.2 s break between the reductions is shown in Fig. 2.11.

By measuring the value $\Delta\sigma = \sigma_1 - \sigma$ it is possible to obtain the dependence of the extent of softening $\Delta\sigma$ on the duration of the break Δt at the given deformation parameters and temperature. The extent of softening R_σ is described by the relationship

$$R_\sigma = \frac{\Delta\sigma}{\sigma_1 - \sigma_0} = \frac{\sigma_1 - \sigma}{\sigma_1 - \sigma_0}, \tag{2.62}$$

where σ_0, σ_1 is the deformation resistance at the beginning and end of the loading curve; σ is the deformation resistance after the break between the reductions.

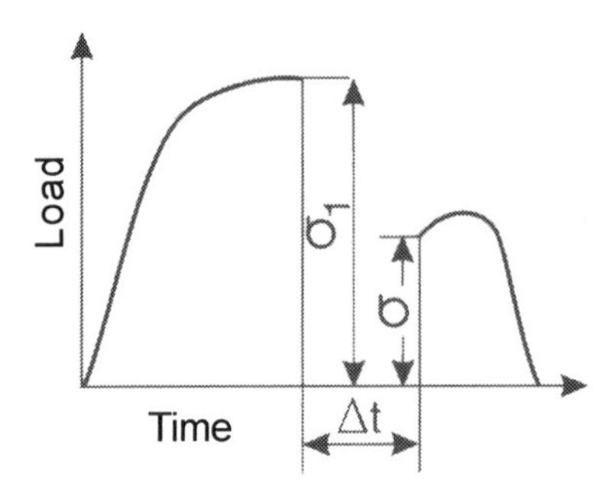

Fig. 2.11. Diagram for the determination of the extent of softening of steel by the double loading method.

The process of hardening and softening of the steel in hot deformation is associated with the cold working and recrystallisation of austenite. It is assumed that the magnitude of softening is proportional to the volume (*X*) of recrystallised austenite.

$$R_\sigma = \sqrt{X}. \tag{2.63}$$

The adequacy of the model (2.63) was evaluated according to the date in [28], in which the plastometric method was used to study the softening of the metal between deformation cycles using the dependence on (2.62).

The experimental data were the results of torsion tests on bars of the St3 and 08Cr18Ni10Ti steels. Tests were performed by double loading at temperatures ranging from 800 to 1200°C, the degree of deformation during the loading cycle $\varphi = 2\pi$ corresponded to –0.34, the strain rate $\dot{\varepsilon} = 0.283$ 1/s. The selected degree and strain rate provided an adequate level of strain hardening at the start of the break between the reductions τ_2, equal to 5 s.

The results of the calculations of R_σ and X and also their experimental values are listed in Table 2.4. Comparison of the quantitative data presented in Table 2.4 and in Fig. 2.12 indicates sufficient reliability of the method of calculating the degree of recrystallization of deformed austenite.

Among the investigated steels, the highest susceptibility to softening was shown by the St3 steel. At a break of 5 s and a temperature of 900°C the St3 steel completely softens (Fig. 2.12, curve 1). The 08Cr18Ni10Ti steel is characterized by low tendency to temper softening, particularly in the temperature range 800–900°C and at 1000°C austenite recrystallizes only by 50% (curve 2 in Fig. 2.12).

Comparison of these quantitative data indicates sufficient reliability of the method of calculating the degree of recrystallization of deformed austenite by formula (2.15).

Table 2.4. Comparison of calculated values of the degree of recrystallization of austenite X with the experimental data of softening steel [28] ($\varepsilon = 0.34$, $\dot{\varepsilon} = 0.283$ 1/s, $\tau_2 = 5$ s)

t,°C	St3 steel				08Cr18Ni10Ti steel			
	Experiment		Calculations		Experiment		Calculation	
	R_σ	X	R_σ	X	R_σ	X	R_σ	X
800	0.85	0.72	0.88	0.77	0.10	0.01	0.17	0.03
900	1.00	1.00	1.00	1.00	0.25	0.06	0.26	0.07
1000	1.00	1.00	1.00	1.00	0.50	0.25	0.60	0.36
1100	1.00	1.00	1.00	1.00	0.70	0.49	0.83	0.69
1200	1.00	1.00	1.00	1.00	0.95	0.90	1.00	1.00

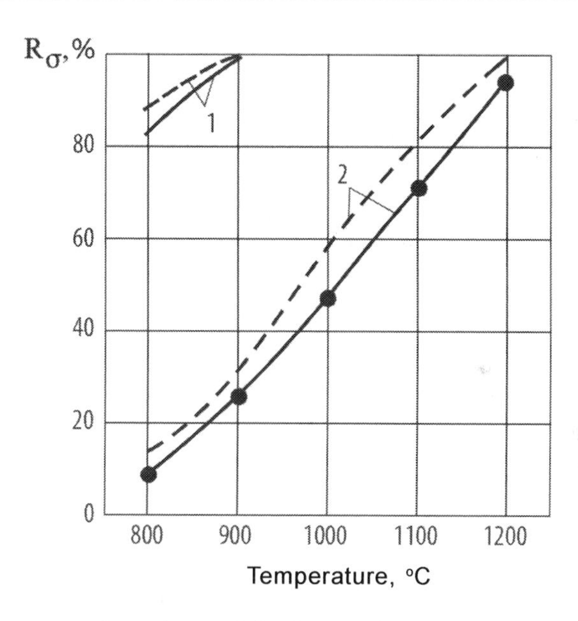

Fig. 2.12. Temperature dependence of the degree of softening R_σ of steel: 1 – carbon steel St3; 08-12Cr18Ni9TI type stainless steel; solid curve – experimental data [28], dotted curve – calculation results.

Thus, the problems of predicting the deformation resistance of metal and steel microstructure are closely linked and cannot be separated. Recent work on the mathematical modelling of deformation resistance during hot rolling of steels also noted the need for using the parameters describing the recrystallization of the deformed structure [29, 30].

2.3. Strip temperature

The prediction of the strip temperature between the passes in reversing or continuous rolling and also during cooling after rolling was carried out using the numerical method of solving the heat conductivity problem for an unlimited plate:

$$\begin{gathered}\frac{\partial T(X,\tau)}{\partial \tau}=\frac{\lambda}{\rho}\frac{\partial^2 T(X,\tau)}{\partial X^2}+\Phi_{\gamma\to\alpha},\\ \tau>0;\ -h/2<X<h/2;\\ T(X,0)=f(X),\end{gathered} \tag{2.64}$$

$$\lambda\frac{\partial T(h/2,\tau)}{\partial X}=\alpha_{u}[T_c-T(h/2,\tau)], \tag{2.65}$$

$$\lambda\frac{\partial T(-h/2,\tau)}{\partial X}=\alpha_{lo}[T_c-T(-h/2,\tau)], \tag{2.66}$$

where are $\Phi_{\gamma\to\alpha}$ is the source member, characterising the release of heat during the phase transformation of the steel; h is the strip thickness; α_u and α_{lo} are the coefficients of heat transfer on the upper and lower surface of the strip; T_c is the temperature of the cooling medium.

When solving the temperature problem, the thermophysical parameters, such as specific heat c and the heat transfer coefficient α control the accuracy of temperature prediction. In most cases, the values of these parameters are selected from the empirical dependences.

To calculate the heat transfer coefficient in cooling of the strip in air, it is recommended to use the following equation

$$\alpha_V=a_V\left(\frac{T^{2.05}}{11400}+1.5V^{0.8}\right),\frac{Bm}{Mt^2K}, \tag{2.67}$$

where a_V is the correction coefficient; V is the speed of movement of the strip, m/s.

The first term in this equation reflects the removal of heat from the strip as a result of radiation, the second term – as a result of convection.

The following dependence was used to calculate the heat transfer coefficient in cooling the strip with water:

$$\alpha_w = (\alpha + a_w 28Q_w), \frac{Bm}{Mt^2K}, \tag{2.68}$$

where a_w is the correction coefficient which takes into account the method of cooling (from top, at the bottom) and the type of cooling system (for example for the upper spraying equipment a_w = 1.0; for the lower a_w = 0.5–0.7); Q_w is the specific consumption of water.

Equation (2.68) includes both the parameters of cooling the strip in air and water. It is thus possible to produce monotonic changes in the boundary conditions with a decrease of the consumption of cooling water to minimum, i.e. when $Q_w \to 0$.

The specific heat of the carbon steel was calculated using the equation derived in processing the data in [31]:

$$C = 4.19\left[117 + 5.8\frac{T}{100} - 0.21\left(\frac{T}{100}\right)^2\right], \frac{\text{J}}{\text{kg}\times\text{K}}. \tag{2.69}$$

$$\Phi_{\gamma\to\alpha}(T,\tau) = \frac{H_{\gamma\to\alpha}}{C_p}\frac{\partial(V_\text{F} + V_\text{P} + V_\text{B})}{\partial t}, \tag{2.70}$$

where V_F, V_P, V_B are the volume fractions of ferrite, pearlite and bainite in austenite breakdown.

The rate of absorption and release of heat in the phase transformations is inversely proportional to the temperature difference (Ac_3–Ac_1). These points are distributed most closely in high-carbon steels (C = 0.6–0.8%). For example, for a steel with a carbon content of 0.65% Ac_3–Ac_1 = 20°C, and for low carbon steel (0.08% C) Ac_3–Ac_1 = 160°C.

The temperature phenomena in metal during heating and cooling were investigated on high-carbon steel sheets (0.65% C, 0.25% S, 1.06% Mn). Specimens were heated in a muffle electric furnace and then cooled at different rates. The metal temperature was measured with a chromel–aluminium thermocouple secured to the specimen. Evaluation of the inertia of the thermocouple showed that it is capable of recording cooling rates to 80°C/s.

The variation of the temperature of the specimen in the electric furnace, heated to a temperature of 900°C, is shown in Fig. 2.13. The heating curves shows clearly a horizontal section corresponding to the transformation of α-iron to austenite at a temperature of 710°C.

Figure 2.14 shows the cooling curve of these samples. The samples were cooled by three methods: in still air, with an air jet and with a jet of a water–air mixture. The samples cooled in still

The following must with taken into account in the determination of the specific heat of hot-rolled carbon steel. The currently available experimental data indicate that the specific heat of the carbon steels in a specific temperature range is higher and these values greatly differ from the values in the vicinity of the boundaries of the temperature range. The physical meaning of this phenomenon may be described as follows. The specific heat is a physical quantity characterising the properties of the metal to increase the temperature when it receives thermal energy. The dimension of the specific heat (J/g · K) indicates that to increase the temperature of 1 kg of metal by 1 K it is necessary to supply a heat of 1 J. The method of measuring specific heat is also based on this. However, this method is not suitable for the metals undergoing phase transformations.

In heating to temperature Ac_1 the carbon steel is in fact α-iron which at temperatures $T > Ac_1$ stars to transform to γ-iron (austenite). The phase transformations is completed at temperature Ac_3. The $\gamma \rightarrow \alpha$ transformation process is endothermic i.e., heat is absorbed during this process. This circumstance is also the reason for increasing the specific heat in the temperature range ($Ac_3 - Ac_1$).

The cooling of carbon steel which is in the austenitic state ($T > Ac_3$) to the temperature Ar_3 of the start of the $\gamma \rightarrow \alpha$ transformation takes place without any thermal anomalies. At $T < Ar_3$ ($Ar_3 < Ac_3$) the transformation of austenite to α-iron starts to take place. This process is accompanied by the release of heat. The enthalpy of transformation of austenite to ferrite, pearlite and bainite according to the literature sources is equal to 76.6 MJ/kg.

The release of heat during the phase transformation of supercooled austenite was taken into account in the temperature problem (2.64) by the source term $\Phi_{\gamma\rightarrow\alpha}$. This term is determined by the relationship:

$$\Phi_{\gamma\rightarrow\alpha}(T,\tau) = \frac{H_{\gamma\rightarrow\alpha}}{C_P} \frac{\partial(V_F + V_P + V_B)}{\partial t}, \qquad (2.70)$$

where V_F, V_P, V_B are the volume fractions of ferrite, pearlite and bainite in austenite breakdown.

The rate of absorption and release of heat at phase transformations is indirectly proportional to the temperature difference ($Ac_3 - Ac_1$). The smallest spacing between these points is found in high-carbon steels (C = 0.6–0.8%. For example, in a steel with 0.65% C $Ac_3 - Ac_1 = 20$°C, and for low carbon steel (0.08% C) $Ac_3 - Ac_1 =$ 160°C.

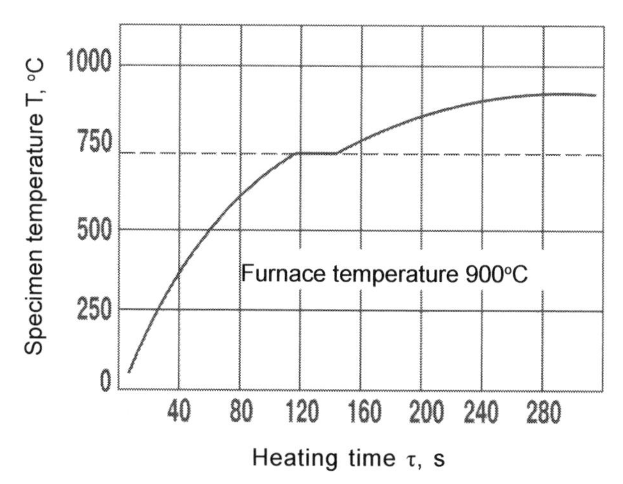

Fig. 2.13. Heating curve of a specimen of 65G sheet steel.

The temperature phenomena in the metal in heating and cooling were investigated using low-carbon sheet steel (0.65% C, 0.25% S, 1.00% Mn). Specimens were heated in a muffle electric furnace and then cooled at different rates. The temperature of the metal was measured with a Chromel–Alumel thermocouple bonded to the specimen. Evaluation of the inertia of the thermocouple showed that it is capable of recording the cooling rate to 80°C/s.

The variation of the temperature of the specimen in the electric furnace heated to a temperature of 900°C is shown in Fig. 2.13. The heating curve shows clearly the horizontal section, corresponding to the transformation of α-iron to austenite at a temperature of 710°C.

Figure 2.14 shows the cooling curve of these specimens. The specimens were cooled by three methods: in still air, with a compressed air jet and with a jet of a water–air mixture. The characteristic feature of the specimens cooled in still air and with the air jet was the increase of temperature (by 100–120°C) after cooling to 600–650°C. This phenomenon was caused by the generation of heat during the transformation of austenite to pearlite. The accelerated cooling of the specimens with the air-jet mixture was characterised by martensitic transformation in which, as is well-known, there is no release of additional heat.

Thus, the results show that as in the determination of the deformation resistance, the calculation of the temperature of the rolled strip is closely linked with the calculation of phase transformations in steel.

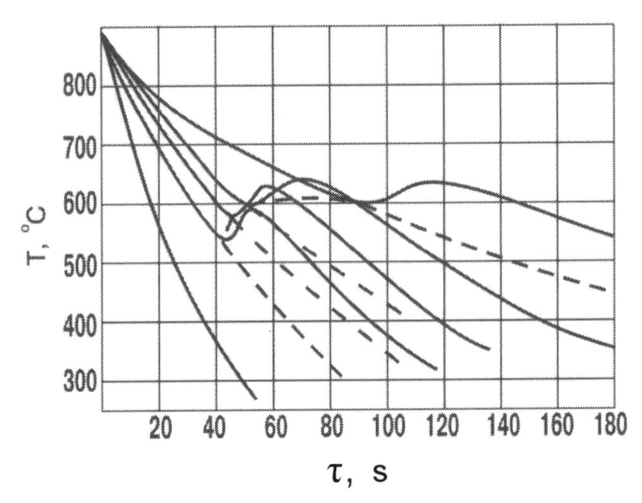

Fig. 2.14. Cooling curves of specimens of 65G sheet steel: solid lines – the actual cooling curves; dashed curves – the conventional cooling curves without considering the release of heat during the pearlitic transformation of austenite.

It should be noted that in heating and cooling the low-carbon steels there were no changes in the form of the heating and cooling curves because the rate of absorption and release of heat during the phase transformation is very low in this case so that there is no thermal effect in the phase transformation.

3

Control of the formation of the microstructure and mechanical properties of rolled steel

The current trend in the market of sheet steels is the expansion of their range while tightening the quality requirements, including the structure and mechanical properties. These requirements can be fully satisfied by the organization and control the formation of the structure and properties of rolled steel in the production line of the mill. Responding quickly to market demands, thus substantially reducing the time and cost for development of new types of metal, allows computer-aided design of the steel chemical composition and processes of rolling and cooling of strips (sheets). Simulation of the hot rolling process allows to solve these problems. Using the computer program it is possible, on the one hand, to determined more accurately the power process parameters (load on the rolling mill), and on the other hand – the temperature–strain state of rolled metal, and finally, to predict the final structure and mechanical properties of rolled products. Therefore, since the end of the twentieth century extensive investigations started all over the world on the development of mathematical models.

In the mid-90's the American Iron and Steel Institute (AISI), together with the U.S. Department of Energy (DOE) and 14 North American steelmakers financed the development of a model of the changes in the microstructure and mechanical properties. In 2001 the INTEG Corporation undertook the task of improving the model and supported by AISI, DOE and five North American steelmakers created a program HSMM (Hot Strip Mill Model, version 6.0). The program allows the user to simulate the processing of steel from its exit from the heating furnace to exit from the mill, tracking and modelling the point at the front end, in the middle and at the end of rolled strip plate along its length for the major brands of carbon and alloy steels with the addition of V, Nb, Ti. The final mechanical properties: tensile strength, yield strength and relative elongation are calculated. The steelmakers, who supported the project, use the program in a variety of practical applications [32].

Voest-Alpine Industrieanlagenbau and Voest-Alpine Stahl Linz developed a computer system for controlling the quality of hot-rolled strip, called VAI-Q Strip. This system was installed in 1997 on the seven-stand hot strip mill of the Voest-Alpine Stahl in Linz. The prediction section of this system allows one to quickly make an accurate assessment of the mechanical properties of hot-rolled strips immediately after coiling on the underground coiler. Later the VAI-Q Strip system was extended to be able to provide the fully automatic operational control of the strip quality parameters. The main model used in the system is a model that describes the rolling process and the corresponding changes of the microstructure. A considerable amount of work has been done to create a model of the phase transformation and for the calculation of the ferrite grain size, modeling the kinetics of the formation of precipitation of particles in the formation of carbides and nitrides of titanium, niobium and vanadium. In the first stage system was used to reduce the number of tensile tests. This was followed by work to solve the optimization problem of the chemical composition of the structural steel, in particular, reduction of the manganese content steel, which has resulted in a considerable saving of alloying elements. Finally, the system was applied to control the coiling temperature to compensate for possible deviations of the mechanical properties from the given level [33].

The Novolipetsk Metallurgical Concern (NLMC) and the Lipetsk State Technical University (LSTU) developed and used in the wide-strip hot rolling mill an automated system for predicting the structure and mechanical properties of hot rolled steel (SPSMPHS) used for obtaining information in the real time mode for the calculated values of the parameters of the structure and mechanical properties [34, 35]. The system is used for the audit of the existing technologies and subsequent correction of the rolling and cooling conditions of strips to guarantee producing the require mechanical properties. It is planned to widen the range of problems which can be solved, in particular, the transfer of the statistical control of the mechanical properties of rolled strip plate to the control by SPSMPHS;

modernisation of the mathematical facilities of the control system for accelerate cooling of strips.

There are examples of the use of mathematical models of the formation of the structure and mechanical properties of thick plates of low-alloy steels at high temperature (T = 950–1050°C) and controlled (T = 830–850°C) rolling and subsequent cooling (accelerated or in air) [36]. Using the model that includes the calculation of the austenite grain size in deformation and the ferrite grain size when cooling, the temperature–deformation conditions were determined for rolling the sheets of 17G1S-U and 14G2 steels on a 3600 reversing mill of Azovstal.

The Institute of Pressure Working of the Freiburg University of Mining and Technology developed a comprehensive model of the rolling of profiled steel and wire rods [37]. The rolling model allows to determine all the process parameters whereas the plastic deformation in the roll gap is determine the methods of finite element or finite differences. The model of the structure determines the average grain size and recrystallization parameters of fifteen steel grades, the cooling model – the phase transformation kinetics. The relationship between the structure and properties have become established on the basis the experimental data obtained by the Institute and are presented in the form of equations. The mathematical modelling results are given in the form of tables and graphs.

The SMS company developed a system for controlling the rolling and cooled CRCT (Controlled Rolling and Cooling Technology). The system is used in the rolling of high-strength and high-alloy steels [38].

In [39] there is an overview of the literature data on the mathematical modeling of the thermomechanical parameters, phase transformations and recrystallization of the steel in hot rolling and cooling of sheets and profiles. These models allow one to predict the final structure and mechanical properties of the finished products. Some results of the application of the models in the existing mills are shown.

Despite significant advances in the development of mathematical modelling of the structure and properties of hot-rolled steel, the examined examples are limited to either the estimate the adequacy of models and the ability to predict the mechanical properties of steel by the known parameters of rolling and cooling, or, in the best case, the adjustment of these modes in order to obtain rolled products with the predetermined properties. In our opinion, the range of applications of the considered mathematical models can be significantly expanded. Therefore, our task in the development of mathematical models was to establish the laws that could be used to develop the theory and technology of hot rolling.

3.1. Modelling of microstructure and mechanical properties in the rolling mill

Hot rolling in a mill is a complicated multiple deformation process during which the temperature of rolled metal, the degree and speed of deformation of rolled steel in individual passes (stands) and the break time between deformation cycles change in a wide range. Figure 3.1 shows the scheme of formation of the microstructure of the metal in the wide-strip hot rolling mill stand (WSHRMS).

The roughing mill group of the wide-strip hot rolling mill usually consists of individual stands with the distance between the stands greater than the length of rolled strip plate. In the stands used at present the last three roughing stands are combined in a continuous group.

The finishing mill group consists of 6–7 four-roll stands in which rolling takes place at the same time. The currently available strip hot rolling stands are characterised by high rolling speed (up to 21 m/s) and a large weight of rolled slabs (up to 36 t). The duration of the break between reductions in two adjacent stands of the finishing

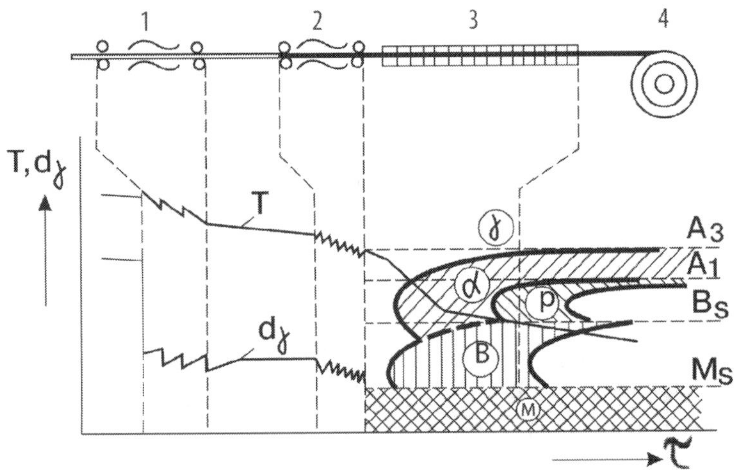

Fig. 3.1. Scheme of formation of the microstructure of steel in a hot strip mill. Figure notes: 1 — roughing stands; 2 — finishing train; 3 — outlet roller conveyor with the installation for water cooling of strips; 4 — coiler.

group changes from several second to several tenths of a second in the final periods, reaching the minimum values of 0.3 s in the high-speed stand.

The end part of the mill includes the roller table and from 2 to 6 coilers, combined in 1–2 groups with 2–3 coilers in each group. The distance from the runoff table to the coilers is 93–300 m. The runoff table is fitted with cooling devices for accelerated cooling of the strips.

The principal flow diagram of the mathematical model of predicting the microstructure and mechanical properties of the strips is shown in Fig. 3.2. The initial parameters of the model include the chemical composition of a specific melt grade and technological parameters for running control: the temperature to which the slab is heated, the deformation conditions in the individual stands of the mill, rolling temperature in different sections of the mill, the finish rolling temperature, the temperature of metal in front of and behind accelerating cooling equipment, the working conditions of the accelerated cooling system (consumption, pressure and temperature of cooling water, the number of cooling sections), etc.

The calculations of the conditions of rolling 2.5–12 mm strips of 3sp steel in the wide-strip hot rolling mill were carried out using the initial data corresponding to actual experimental values. Examples of the calculation of the rolling parameters and austenite structure for the strips with a thickness of 2.5 and 12.00 mm are presented in Table 3.1.

Analysis of the results of calculation of the formation of he austenite structure, including the data in Table 3.1, shows that the primary recrystallisation of austenite is not completed in the final stands of the finishing group of the mill. In the case of thin strips this is associated with the relatively low temperature of the strip in the final stands of the mill. Until now it has been assumed that the cold working of austenite remains unchanged only in rolling of thin strips.

With increase of the thickness of the rolled strips the final diameter of the austenite grains increases. At the start of the phase transformation the austenite grain size of the strips 12.0 mm thick is 1.5 times greater (d = 38.5 μm) than in the 2.5 mm thick strips (d = 25.6 μm)

After rolling, the hot rolled strip is cooled in two stages: fast – on the runoff table and slowly – in a roll. This cooling affects the structure transformation processes which in the steel strips occur

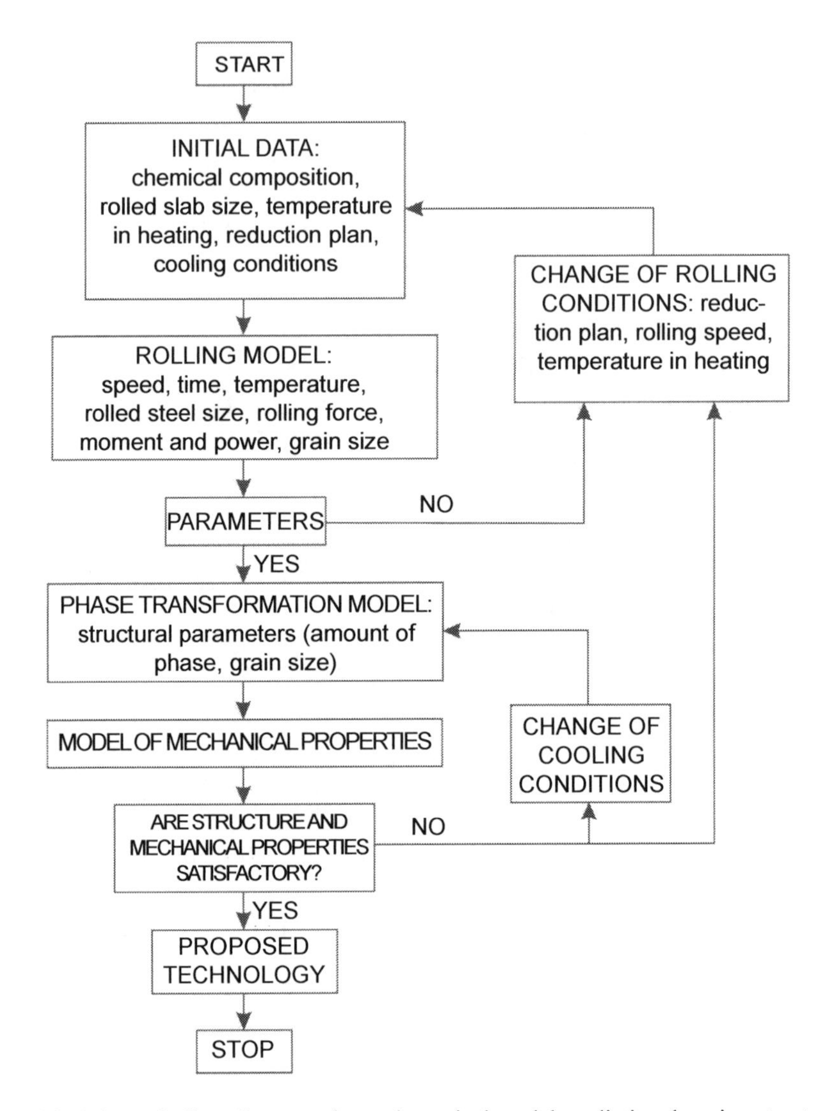

Fig. 3.2. Schematic flow diagram of a mathematical model predicting the microstructure and mechanical properties of hot rolled strips.

mainly in the coil. Thus, the calculations showed that the phase transformation in the steel 3sp strips starts at a temperature of 750–760°C (1% austenite breakdown). Passage through the runoff roller table (roller table length 250 m) results in the formation of the entire volume of ferrite and austenite. The calculation isothermal diagram of the 3sp steel with the cooling curves of the strip on the runoff table and in the coils is shown in Fig 3.3.

Table 3.1 Temperature–rate deformation modes and indicators of recrystallisation of austenite in rolling steel 3sp strips with a thickness of 2.5 mm (numerator) and 12.0 mm (denominator) in the 2000 finishing group mill

Index	No. of stand							
	0	1	2	3	4	5	6	7
Strip thickness, mm	34.0 44.1	18.2 28.4	10.8 21.5	6.9 18.0	4.7 16.0	3.6 13.5	2.8 12.8	2.5 12.0
Rolling speed, m/s	0.6 1.2	1.2 1.9	1.9 2.4	3.0 2.9	4.4 3.2	5.8 3.8	7.4 4.1	8.3 4.3
Strip temperature, °C	1014 990	987 973	966 959	946 946	917 933	899 922	852 911	838 900
Average austenite grain diameter, μm	–	49.0 36.3	39.3 32.1	34.5 29.0	30.9 38.9	26.9 38.2	26.3 31.4	25.6 38.5
Degree of recrystallization	–	1.00 1.00	1.00 1.00	1.00 1.00	1.00 1.00	1.00 1.00	0.85 0.30	055 0.72

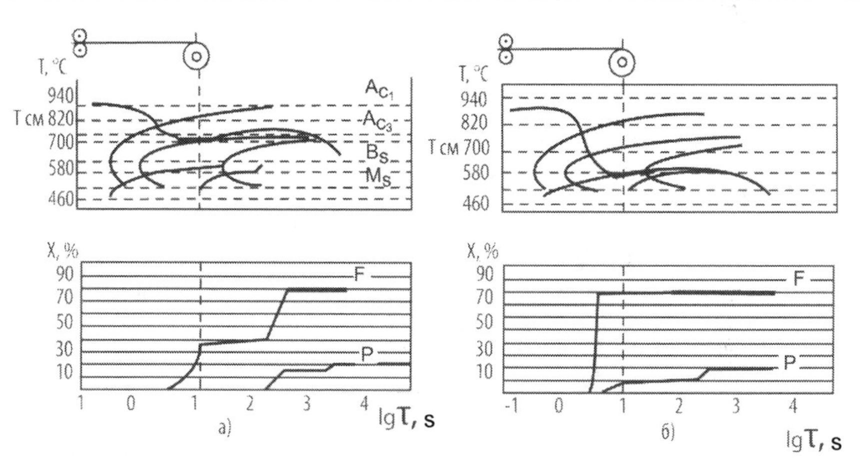

Fig. 3.3. Austenite breakdown and formation of a ferritic–pearlitic structures at high (a) and low (b) coiling temperature of steel St3sp strips.

Slow cooling steel in a coil leads to its softening compared with sheets of the same thickness, rolled in a reversing mill. The difference in the cooling rates in the outer and inner turns of the strip in a coil leads to an uneven distribution of the structure and mechanical properties along the length of the strip (Figure 3.4). Thus, the tensile strength of the outer end of the strip, not adjacent to the coiled strip, exceeds the tensile strength of the strip in the middle of the coil by 65 N/mm². This difference for the first strip turn adjacent to the coil amounts to 48 N/mm². The difference in the

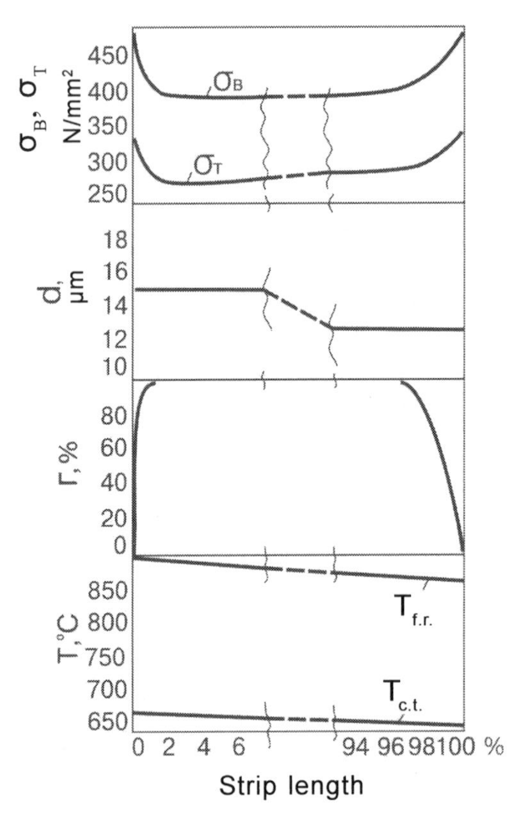

Fig. 3.4. Change in the mechanical properties (σ_B, σ_T), ferrite grain diameter (*d*), spheroidized pearlite fraction (Γ) along the length of the hot-rolled steel 3sp strip with a thickness of 4 mm. Rolling conditions: rolling speed 9.4 m/s (no acceleration); values of $T_{f.r.}$ (finishing rolling temperature) and $T_{c.t.}$ (coiling temperature) are shown in the lower part of the figure.

properties of the outer and inner turns of the carbon steel strip is due, according to the calculations, to varying degrees of lamellar pearlite spheroidization. We have shown that there is a range of coiling temperatures in which the difference in the degree of spheroidization of pearlite has a maximum. So for carbon steels (C = 0.14–0.22 %), this maximum is in the range 600–700°C. For low-alloy steels, for example 09G2S, the maximum is in the coiling temperature range 500–600°C.

As already mentioned, the mathematical model for the prediction of the microstructure and mechanical properties of rolled strip plate can be used for solving the following problems: selection of the chemical composition of steel and thermomechanical treatment conditions, resulting in the optimum properties of rolled strip;

design of mills and cooling systems; control of the temperature–deformations conditions of rolling and accelerated cooling to stabilise the properties of rolled strip plate within the rolled steel unit range (sheet, strip, etc), melts, and also a single production type.

The developed model was used to examine the effect of changes (within the grade composition) of the contents of the main chemical elements (C, Mn and Si) in widely used steels 3sp, 09G2S and 17GS on their strength properties (σ_T and σ_B) [40].

The carbon, manganese and silicon contents of the studied steels was varied in the ranged permitted by the relevant standards. The temperature–rate and deformation hot rolling modes and also the cooling conditions of the strip were varied in the range possible for the second generation wide-strip mills using slabs with a thickness of 250 mm.

The results of calculations for middle sections along the length of the strips showed that the change of the chemical composition within the grade range has a strong effect on the tensile properties of hot-rolled steel. Increase of the content one of the hardening elements (C, Mn and Si) from the minimum to maximum value leads to a marked increase in the strength properties. For the low-alloy steels, the most powerful influence on the strength properties is exerted by the changing manganese content. Thus, the increase of the Mn content from the minimum to the maximum permissible value at constant carbon and silicon increases the strength characteristics by 60–70 N/mm^2. The strength characteristics of the St3sp steel are significantly influence by the changes of the carbon and silicon content. Since in practice different metal heats are characterized by the simultaneous change of the content of several elements in the chemical composition, it is customary to express the composition of the steel by the widely used indicator C_C – carbon equivalent. Changing the content of the hardening elements within the range of the grade composition can cause the variation of the yield stress from 67 N/mm^2 for the 09G2S steel up to 117 N/mm^2 for the 3sp steel.

A series of computational experiments was carried out to assess the technological feasibility of controlling the properties of strips in the wide-strip hot rolling mill. The effect of the temperature of heating the slabs, the finishing rolling and coiling temperatures and the cooling rate on the runoff table as the basic technological parameters determining the structure and properties of the steel was studied.

Calculations showed that the change in the heating temperature of the slabs of the examined steels significantly affects the amount of austenite grain before rolling. Thus, with increasing heating temperature from 1100 to 1300°C, the austenite grain size in mild steel increases from 100 to 400 μm. At the same time, calculations showed that the austenite grain size after roughing rolling and the strength properties of the examined steels are almost independent of the temperature to which the slab is heated.

Increasing the finishing rolling temperature ($T_{f.r.}$) in the range 830–930°C at constant coiling temperature ($T_{c.t.}$) slightly reduces the strength properties (Fig. 3.5 *a*). However, it should be noted that the rolling end temperature increase is provided by increasing the rolling speed, and the constant coiling temperature is provided by increasing the cooling rate (V_{cool}) on the runoff table of the mill thus reducing the temperature of beginning and end of the $\gamma \rightarrow \alpha$ transformation and improving the tensile properties. Increasing the cooling rate to some extent compensates for the negative impact of $T_{f.r.}$ on the strength characteristics of the steel associated with the degree of recrystallization of the steel with increasing deformation temperature.

Figure 3.6 shows the results of computing experiments for rolling of 10 mm strips of St3sp and 09G2S steels in the conditions of the

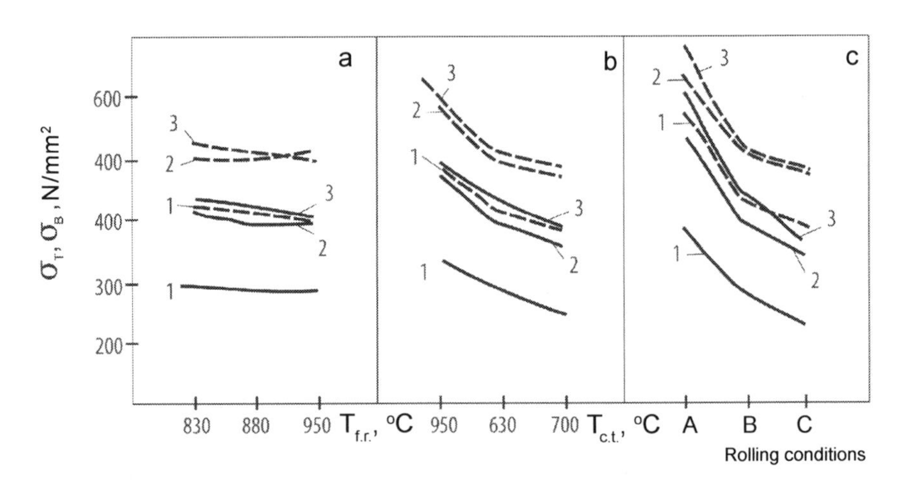

Fig. 3.5. Effect of changes of the rolling temperature conditions on the strength properties of 8 mm strips of 3sp (curves 1), 17GS (curves 2) and 09G2S steel (curves 3). Rolling conditions: **a** – $T_{c.t.}$ = 650°C; **b** – $T_{f.r.}$ = 880°C; **c** – mode A – $T_{f.r.}$ = 830°C; $T_{c.t.}$ = 530°C; V_{roll} = 4.5 m/s; mode B – $T_{f.r.}$ = 880°C; $T_{c.t.}$ = 650°C; V_{roll} = 6.0 m/s; mode C – $T_{f.r.}$ = 930°C; $T_{c.t.}$ = 750°C; V_{roll} = 8.6 m/s. Dashed line — σ_B (tensile strength); solid lines – the yield stress σ_T

wide-strip hot rolling mill. According to the calculation results, the decrease of the coiling temperature of the strips from 750 to 550°C increases the strength properties: yield stress (σ_T) and tensile strength (σ_B) increase by 50–100 N/mm^2, with the temperature $T_{c.t.}$ having a stronger effect on the properties of the 3sp steel (by approximately by a factor of 2).

The strengthening of the steel with the decrease of the coiling temperature takes place as a result of both refining of the ferrite grains to 7–12 μm, and as a result of decreasing the degree of spheroidization of plate-shaped pearlite. In the 3sp steel these two hardening mechanisms have the identical effect on the strength properties. However, as a result of the low pearlite content, the 09G2S steel is only slightly affected by the spheroidization of

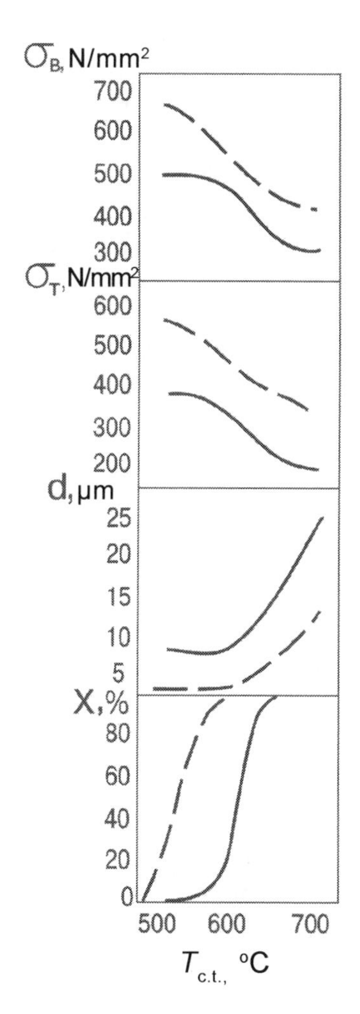

Fig. 3.6. Influence of coiling temperature $T_{c.t.}$ on the mechanical properties (σ_T, σ_B), the diameter d of ferrite grains and pearlite spheroidization degree X in the middle part along the length of the hot rolled strip: solid lines – 3sp steel; dashed – steel 09G2S.

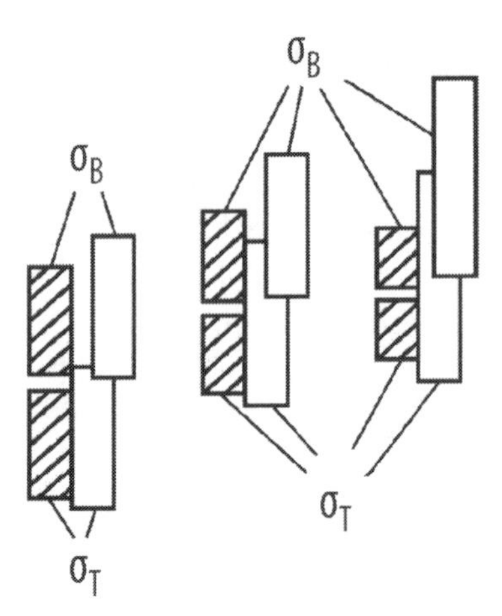

Fig. 3.7. Comparison of the ranges of changes of the strength properties (σ_T, σ_B) of hot-rolled 8 mm thick strips of steels of different grades (3sp, 17GS, 09G2S) caused by fluctuations of the chemical composition of the steels (cross hatched columns) at $T_{c.t.}$ = 650°C; $T_{f.r.}$ = 880°C and the changes of the rolling parameters $T_{c.t.}$ and $T_{f.r.}$ (open columns).

pearlite. Therefore, at $T_{c.t.}$ below 600°C at which the diameter of the ferrite grains is almost completely constant, the degree of hardening of the steel rapidly decreases. The decrease of $T_{c.t.}$ to 600–650°C increases the impact toughness of the steel. With a further decrease of the coiling temperature the impact toughness remains almost constant, and often shows a tendency to decrease.

The greatest effect of ferrite grain refinement and hardening of steel is achieved with a combination of temperature $T_{f.r.}$ close to Ar_3 and accelerated cooling (Fig. 3.5 c).

Mathematical modelling has shown that by changing the technological modes of rolling strips in the wide-strip hot rolling mill it is possible to compensate the influence of chemical composition fluctuations permissible by the standard for the investigated steels on the strength characteristics of rolled sheets (Fig. 3.7).

3.2. Automatic control system of the properties of rolled steel in wide-strip hot rolling mill

The currently available systems for controlling the hot rolling process in the continuous mills can be used to apply a wide range of technologies as a result of the redistribution of deformation in the individual stands of the mill, the change of rolling speed, regulation of the cooling rate of the strips, differentiation of cooling along the length of the strips and the runoff roller table. The main function of these systems is to ensure that the required temperatures of the end of rolling and coiling of the strips into a coil are reached. However, the aim of control is to ensure the guaranteed level of the mechanical properties of rolled steel but this cannot be achieved to the complete extent. This is a result of the changes of the chemical composition of the steel of a specific grade from melt to melt.

In the actual conditions of metallurgical processes there are cases of instability of the melting technology and chemical composition of the steel of the same grade. This results in large variations of the mechanical properties of hot-rolled steel. Weakening of the requirements on the chemical composition of the steel in the new standards increases the probability of variations of the mechanical properties in transition from one melt to another.

Concept of the system

The effect of changes in the chemical composition of the steel on the mechanical properties of rolled material can be eliminated to a certain extent by correcting the rolling technology and cooling conditions of the strips (Fig. 3.7). The control of the rolling technology in order to ensure the required level of the properties of rolled material is an important requirement for the systems of controlling the hot rolling processes. As indicated by the publications [32–39], these systems are used in the hot rolling stands of a number of companies and are usually the subject of know-how. The application of these systems requires the availability of the appropriate mathematical models of prediction and the algorithms of controlling the properties of rolled materials.

The programme software, proposed by the authors for the automated systems of controlling the properties of rolled steel, is based on the mathematical model of predicting the structure and mechanical properties of carbon and low-alloy steels, described in chapter 2. The include the following programs: calculation of the temperature – deformation and energy – force rolling parameter; calculation of the austenite grain size taking into account deformation of austenite and recrystallisation; calculation of dislocation density; calculation of the volume of the structural components in the phase transformations of hot-deformed austenite; calculation of the dispersion of ferrite and pearlite; calculation of the strength and ductility properties of rolled materials.

Usually, the mechanical properties of the rolled steel σ_T and σ_B are restricted at the lower value. Taking into account the non-monotonic nature of the dependences of the properties on the finish rolling and coiling temperatures, it is quite easy to determine the temperatures which ensure that the resultant properties satisfy the requirements:

$$\sigma_T\left(T_{\text{f.r.}},T_{\text{c.t}}\right)>[\sigma_T];\ \sigma_B\left(T_{\text{f.r.}},T_{\text{c.t}}\right)>[\sigma_B], \tag{3.1}$$

where $[\sigma_T]$, $[\sigma_B]$ are the boundary values (minimum allowable) specified by the customer.

The control process is realised as follows. After melting the steel, determination of the chemical composition of the steel, casting, the resultant slabs are sent to the storage of the sheet rolling shop where the values of the mass fractions of the chemical composition of the given batch of the blanks are added to the computer system. The values of the parameters of the basic technology of rolling the strips of the given grade and applications, including the finish rolling temperature and the coiling temperature of the strips, were previously inputted to the appropriate database. The values of the mechanical properties of the rolled sheets, required by the customer, are also included in the computer system.

Prior to rolling a metal batch with the given requirements on the chemical composition, a program is activated which on the basis of the mathematical relationships calculates the values of σ_T and σ_B and verifies the condition (3.1).

If the condition (3.1) is satisfied, the rolling of the strips is carried out by the basic technology. If the condition (3.1) is not satisfied, the programme carries out the computing experiments in which the finishing rolling temperature and the coiling temperature of the strips are reduced and the new values of σ_T, σ_B are calculated, and the condition (3.1) is verified.

If the condition (3.1) is satisfied, the strips are rolled by the basic technology. If the condition (3.1) is not satisfied, the programme carries out the computing experiments in which the finishing rolling temperature and the coiling temperature of the strips are reduced and the new values of σ_T, σ_B are calculated, and the condition (3.1) is verified.

If the condition (3.1) is fulfilled, then to roll the given batch of the strips, the system issues in the 'adviser' mode the values of the parameters of the corrected technology. However, if the condition

(3.1) is not satisfied, it is necessary to continue the correction of the rolling and coiling temperature. Correction should be carried out until the condition (3.1) is satisfied. If the finishing rolling and coiling temperatures are lower than the permissible values, determined previously for the specific production conditions, the calculation experiment is ended by recommendations for re-defining the given melt for the rolling of strips for another order with lower requirements on the mechanical properties.

The Mekhsvoistva automated control system (ACS) is a program and technical complex for several personal computers, connected into the computing network. The composition of the ACS includes the automated working sites (AWS): 1) the counter of the number of slabs; 2) the manager (producer) of the production office; 3) the engineer (laboratory technician) of the mechanical testing laboratory; 4) the research engineer; 5) the station for controlling the 2000 mill, and also the server computer, in which the above AWSs are technically, information- and program-combined.

To control the reliability of the supplied information, a guide of the maximum content of the chemical elements for the individual steel grades has been introduced. To ensure the functioning of the system, the guides of the basic rolling technologies and requirements of the standards and the technical conditions (TC) for the mechanical properties of the rolled sheets have been introduced into the system.

For the efficient operation of the Mekhsvoistva automated control system (ACS), it is important to have the Mechanical Test Laboratory AWS which is essential for introducing into the system reports about the measurement values of the parameters of the mechanical properties of the rolled strips.

Testing the programme for calculating the mechanical properties

The ACS database for the sheet rolling shop of the metallurgical company provided the files of information for the chemical composition, the parameters of rolling technology and the results of mechanical tests of four steel grades (3sp, 09G2S, 17GS), with the volume of 820 reports (batches). Using the specially developed program, calculations were carried out of the mechanical properties (yield strength, tensile strength, relative elongation, impact toughness). The calculated values of the mechanical properties were compared with the results of the mechanical tests by statistical and correlation–regression analysis. To evaluate the matching of the correlation relationship between the calculated and measured values and the error of the model, the parameters of the distribution of the measured parameters of the properties and the difference (X_c–X_m) of the calculated X_c and measured X_m values were compared.

The resultant data show that the tested mathematical model of the mechanical properties has the following characteristics:

- the model describes with a sufficiently high accuracy the strength properties of the entire group of the low pearlite steels (the correlation coefficient 0.86–0.89);
- the accuracy of the model decreases in the calculation of the ductility and toughness properties of the steel, which is fully regular, because these properties depend strongly on the cleanness of steel and form in the stages prior to the rolling stage; nevertheless, the calculated values of relative elongation correlate with the measured values with the correlation coefficient of 0.59 for the combined sample; the calculated value of impact toughness KCU-40 correlated with the measured values with the correlation coefficient $r = 0.33$.

The concept of the system developed at the sheet rolling shop (SRS) proved to be efficient and is used in production.

3.3. Special features of production of hot-rolled strip plate for continuous cold rolling mills

Rolled strip plate for cold rolling mild steel strips

The mechanical properties of rolled strip plate and the magnitude of its hardening, the capacity of the steel to deform in cold deformation without fracture, and the ductility properties of the cold-rolled sheets depend on the structure and the properties of the metal, formed in hot rolling. The practical experience with sheet rolling production shows that the hot-rolled metal, used as the rolled strip plate, should have a ferrite–cementite or ferrite–pearlite structure with the grain size number 6–8, and the size of the globules of the structurally free cementite should not exceed size number 3.

As shown previously, the structure of the rolled steel forms basically during the process of deformation in the final stands of the hot rolling mill and during cooling of the strips on the runoff table. In the determination of the optimum temperatures of the finish of rolling and coiling and of the high-speed cooling conditions of the strips, it is necessary to take into account the fact that different wide-strip hot rolling mills are characterised by different lengths of the runoff roller tables, different lengths, power and the position of the cooling (spraying) systems. In the mills with a long runoff-table (more than 200 m), the formation of the favourable structure is completed at a relatively large distance from the coiler, and the coiling temperature of the strips is well below the Ar_3 point and equals 500–550°C. In the rolling mills in which the runoff roller table is relatively short (approximately 100 m) the optimum coiling temperature is 670–680°C. In both cases, the microstructure of the rolled steel is approximately the same: the ferrite grain size is approximately the size number 7–8, and the grain size number of structurally free cementite is 1–2. Consequently, the optimum coiling temperature of the strips in different wide-strip hot rolling mills should be determined out using a differentiated approach taking into account the length of the runoff table and also the length, position and the power of the spraying system positioned on it.

Cold-rolled strips with a thickness of 0.5 mm and less are the most widely used product range. The production of these strips is the existing mills is sometimes difficult due to poor strain capability of the metal in the cold rolling mill. The rolling capacity of the strip, i.e., the production of the cold-rolled sheet of the required final thickness and also the stability of the rolling processes are determined, in addition to the technological possibilities of equipment and cold rolling conditions, by the mechanical properties of rolled strip plate. Therefore, the production of cold-rolled thin strips, the stability of technology, the quality and yield of products depend on using the rolled strip plate with the highest ductility properties.

The cold-rolled strips for mass applications with a thickness of 0.5–0.55 mm and less are produced using the hot-rolled coiled steel with the thickness of 2.2–2.5 mm made of 08 low-carbon steel. The mechanical properties of rolled strip plate vary in a wide range. Table 3.2 presents the results of random samples of the mechanical properties and structure of the steel after hot rolling in a wide-strip hot rolling mill.

To ensure the high strain capacity properties and produce the required structure and properties of the thin strips, it is necessary to have a microstructure ensuring low strength and high ductility properties by selecting the optimum hot rolling conditions.

The effect of finishing rolling temperature $T_{f.r.}$ and coiling temperature $T_{c.t.}$ in the 1700 wide-strip hot rolling mill with a 100 m runoff table on the microstructure and mechanical properties of the 08 steel is shown in Table 3.3.

Table 3.2. Mechanical properties and microstructure of 08kp steel with a thickness of 2.5 mm after hot rolling

Batch No.	σ_T, N/mm^2	σ_B, N/mm^2	δ_4, %	Hardness, HRB	Ferrite grain size
1	310	380	32.6	61.4	8.75
2	255	340	23.0	53.0	7.5
3	360	340	39.0	73.0	9.5
4	307	388	31.0	60.1	9.0

Table 3.3. Effect of temperature $T_{f.r}$ and $T_{c.t.}$ on the structure and properties of rolled 08 steel

Experiment No.	$T_{f.r.}$, °C	$T_{c.t.}$, °C	σ_B, N/mm^2	σ_T, N/	δ_4, %	Hardness, HRB, units	Ferrite grain size	Cementite grain size
1	825	600	380	280	37	62	9.0–9.5	1 c
2	845	600	390	290	39	68	9.0–9.5	1 c
3	825	670	350	260	35	58	7.0–8.0	4 a
4	845	670	355	260	35	58	8.0	3.5 a

It can be seen that the change of the finishing rolling temperature in the selected range has no significant effect on the properties and microstructure of rolled steel. In the investigated case, the deformation ends in the finishing rolling temperature range adjacent to the point Ar_3 (Fig. 3.8) where the curve of the variation of the ferrite grain size has a distinctive minimum and a small gradient.

In contrast to the finishing rolling temperature, the coiling temperature in this range has a strong effect on the mechanical properties and microstructure of rolled steel (Table 3.3). The best results for the cold rolling of 2.5 mm rolled strip plate in the investigated wide-strip hot rolling mill were obtained at the coiling temperatures of 660–680°C. To ensure these coiling temperatures, spraying of the strip was either interrupted or one of the final sections of the spraying system was activated. The increase of the coiling temperature to 660–680°C decreased the yield strength by 30–50 N/mm^2, the hardness by 5 –10 HRB units, and the average force in subsequent cold rolling by approximately 15–25%. At this temperature $T_{c.t.}$ the ferrite grain size increased to the grain size number 7–8. The negative effects of the high coiling temperature (660–680°C) include the presence in the structure of structurally free cementite (SFC) with the grain size number of 3–4.

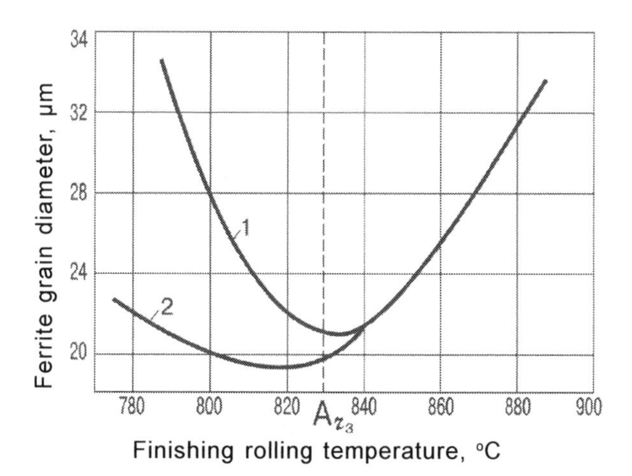

Fig. 3.8. Dependence of the average ferrite grain diameter on the finishing rolling temperature of 2.8 mm thick strips of steel 08kp ($T_{c.t.}$ = 590–610°C) [41]: 1 – internal, 2 – outer coils turns.

Strip plate for the production of cold-rolled strip for deep drawing

The main condition for the efficient stamping of sheet metal is the formation of a uniform structure which forms during hot rolling [42, 43]. The most important parameter is the ferrite grain size (grain size number 7–8) and the size of small inclusions of the structurally free cementite (size number 1–2).

The rolling of thin strips (2.0–2.2 mm) in many wide-strip hot rolling mills ends in most cases in the temperature range 820–840°C at a relatively high rolling speed with considerable (taking into account the reduction in the previous stands) hardening of the strip. These circumstances lead to changes of the austenite grains. The rapid cooling of the strips in the runoff roller table of the mill to $T_{c.t.}$ = 580–620°C results in the formation of a fine-grained structure (grain size number 9–11) and the hardened metal. In addition to this, the non-metallic inclusions are stretched in the unfavourable direction along the rolling direction. This metal is difficult to deform in subsequent cold rolling, and after annealing and skin rolling has low drawing properties.

Increase of the temperature of rolled strip plate increases the finishing rolling temperature, decreases the rolling speed, increases the austenite grain size and also the tendency to form a structure with different grain sizes. The observed processes dictate the need to slow down the development of selective recrystallization which leads

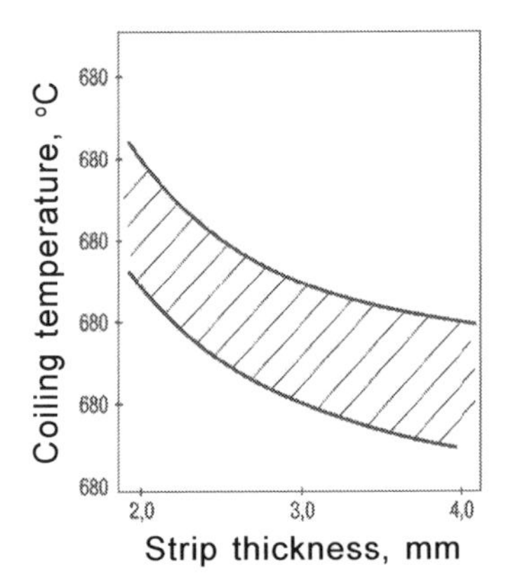

Fig. 3.9. Coiling temperature of strips of different thicknesses in 1700 wide-strip hot rolling mill with a runoff table ~100 m long, in the production of cold-rolled sheet intended for deep drawing.

to the formation of large irregular grains. This can be achieved by rapid cooling on the strips on the runoff table, by selecting rational cooling modes. Calculations and subsequent experiments showed that in the 1700 wide-strip hot rolling mill with a runoff table ~100 m long the strips 2.0–2.2 mm in thickness must be coiled at $T_{c.t.}$ = 630–660°C and the 4.0 mm thick strips at $T_{c.t.}$ = 590–620°C. The coiling temperature of rolled strips of various thicknesses should be determined using the dependence shown in Fig. 3.9

High-carbon steel strip plate

The cold rolling of strips of high-carbon steels in the currently available multi-stand and reversing mills is associated with considerable difficulties. This is caused not only by the high deformation resistance of the steel as a result of a higher carbon content but, most importantly, by the low ductility of these steels so that cold rolling is accompanied by the formation of tears at the edges of the strips, leading to fracture is [44]. The formation of the cracks and fracture of the strip can also take place after cold rolling: the low-ductility high-carbon steel may fracture in uncoiling the coils and multiple bending of the strips in the mechanisms of the pickling system

The ductility of carbon steel depends on the dispersion of the pearlitic structure. For example, in the cold rolling of strips of high-carbon steel after preliminary annealing with the structure of a coarse-globular or coarse-plate pearlite cracks form at the edges of the strip after reaching the total reduction of 20–40% [44, 45]. In rolling the same steel with a sorbitic structure the total reduction of 80% can be achieved without preliminary annealing of the rolled strip plate [45].

The structure of the high-carbon steel, its strength (σ_T, σ_B) and ductility (δ) properties also depend strongly on the hot rolling conditions of the strips: the finishing rolling temperature $T_{f.r.}$, cooling rate V_{cool} and the coiling temperature of the strip $T_{c.t.}$. The rolling and cooling conditions of the strips should ensure in particular the formation of a specific austenitic structure of the steel prior to the phase transformation and the optimum phase transformation temperature.

The high-carbon steels are characterised by very low values of the critical cooling rate (for 65G steel it is V_{cr} = 10°C/s). This results in the likely occurrence of the martensitic transformation on the runoff table and formation of a brittle quenched structure. This circumstance requires the determination of the rolling conditions of the high-carbon steels which would ensure that the pearlitic transformation takes place in the coiled strip, with the maximum rate and at the minimum duration of the incubation period.

Production experiments did not show any apparent impact of the coiling temperature (cooling rate) on the ductility properties of hot-rolled strips of high-carbon steels. It was found that the ductility properties of high-carbon steel are influenced mainly by the finishing rolling temperature. Application of low finishing rolling temperatures forms a more dispersed pearlitic structure of the metal. Its mechanical properties increase correspondingly. Accelerated cooling of the strips leads to the formation of a non-uniform structure and properties of the strip since the non-uniform distribution of cooling water over the area of the strip leads to the formation of regions of coarse-plate pearlite, intermediate structures and martensite which in further processing of the hot-rolled sheets may cause fracture of the strips.

Production experiments established the extreme dependence of the ductility properties (δ_5) of 65G rolled steel on the finishing rolling temperature (Fig. 3.10). The temperature of the end of deformation and coiling temperature of the strips were varied by changing the rolling speed. The highest ductility properties in hot rolling were obtained in the conditions $T_{f.r.}$ = 800°C and $T_{c.t.}$ = 570°C. A decrease or increase of $T_{f.r.}$ and $T_{c.t.}$ impaired the ductility properties of 65G steel.

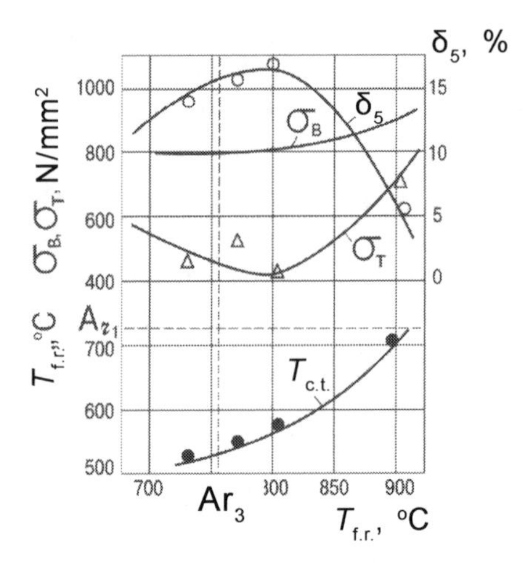

Fig. 3.10. Effect of the finishing rolling temperature of strips in the industrial rolling mill conditions on the properties of 65G steel sheets.

4

Stability and reliability of the hot rolling process

4.1. Increasing the reliability of the analysis of the sheet rolling process

In the industrial mills conditions the physical constants describing the properties of the rolled material and the conditions of external friction in the deformation zone, as well as the temperature, energy-power and kinematic parameters of the process are the statistical distributions with specific characteristics: average values, standard deviation, etc. To select rational technological solutions, it is often more important to use the information about the dispersion parameters rather than their average values. Such information is crucial, for example when analyzing the stability of the rolling process of steel sheets. Therefore, the actual task of the rolling theory is to identify and implement opportunities for a wider use of probabilistic approaches in solving technical and technological tasks in sheet rolling production.

As noted above, the magnitude of deformation resistance of steel σ is calculated using the relationship (2.51). To determine the coefficients of this dependence it is necessary to carry plastometric tests. The available data in this area are summarized in the monographs [26, 46] applied to steels with specific chemical compositions. However, the extension of these results to the entire steel grade, as is usually done in practice, can lead to errors, since even within the requirements of the standards the chemical composition and properties of the same grade can vary significantly due to the acceptable spread of the content of elements. In addition, in steels of similar grades the ranges of the possible content of elements often overlap. Therefore, speaking about the deformation resistance of the steel of some grade one should indicate at least the average value of the deformation resistance and the average standard deviation of the distribution of this quantity.

Attempts to express the deformation resistance of steels in the form of regression equations reflecting the dependence of σ on the content of each element in steel often do not provide reliable results due to the complex and multifaceted effect of the composition on the properties. It seems more appropriate to construct the dependences of σ on the integrated indicators of chemical composition (e.g., carbon and electronic chemical equivalents), generalizing the linear effects and interaction of chemical components in the steel composition. Promising in this regard is the use of the electronic equivalent of the chemical composition of steels and alloys Z^y, proposed by E.V. Prikhod'ko.

The histograms of the distribution of the content of the main elements in steel 60, constructed according to the analysis of compositions of 699 heats, are shown in Fig. 4.1 [47]. The histograms of the distribution of values of Z^y, calculated from the actual chemical analyzes of heats of steel 60 (Fig. 4.1), are shown in Fig. 4.2 *a*. Constructed by modelling the steel composition by the Monte Carlo method using the statistical characteristics of the distributions of the content of the individual elements (Fig. 4.1), the histograms of the distribution of Z^y are shown in Fig. 4.2 *b*. Verification showed that the distribution of actual values of Z^y (Fig. 4.2 *a*) and the resulting simulation (Fig. 4.2 *b*) coincide with the reliability of at least 95%. These results suggest a fundamentally important conclusion that, knowing the statistical characteristics of the distributions of the contents of individual elements in the steel, one can get by modelling close to the true distribution of the electronic chemical equivalent of the composition of the steel Z^y. With this distribution and the dependence of σ on Z^y, we can get the distribution of σ for a given temperature–rate and deformation conditions.

Figure 4.2 shows the distribution of the resistance to deformation of steel 60 at T = 800°C, u = 10 s^{-1} and ε = 0.4, obtained by the simulation using the Monte Carlo method and the dependence:

$$\sigma = -22052 + 72229Z^y - 58359\left(Z^y\right)^2. \tag{4.1}$$

Figure 4.3 shows histograms of the distribution of the values of Z^y

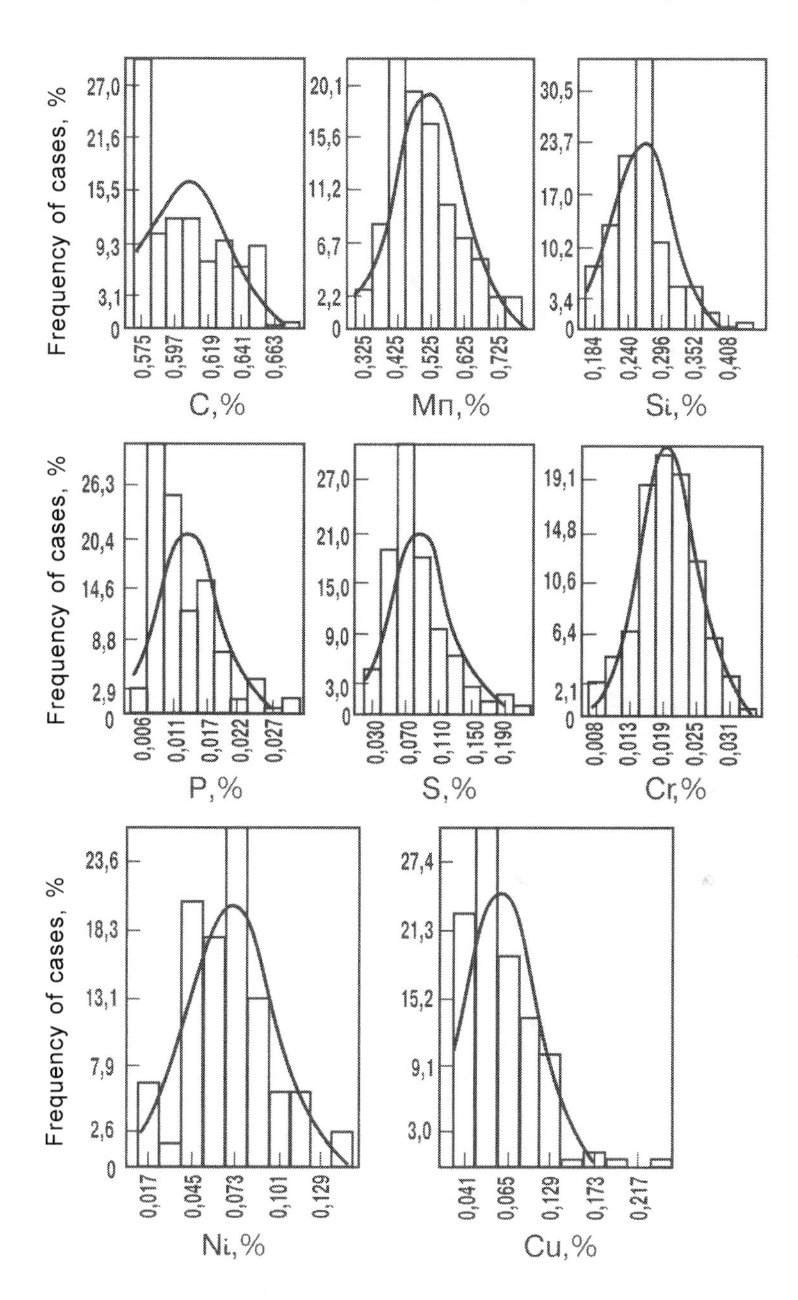

Fig. 4.1. Histograms of the distribution of the content of chemical elements in the composition of steel 60 (sample size 699 heats).

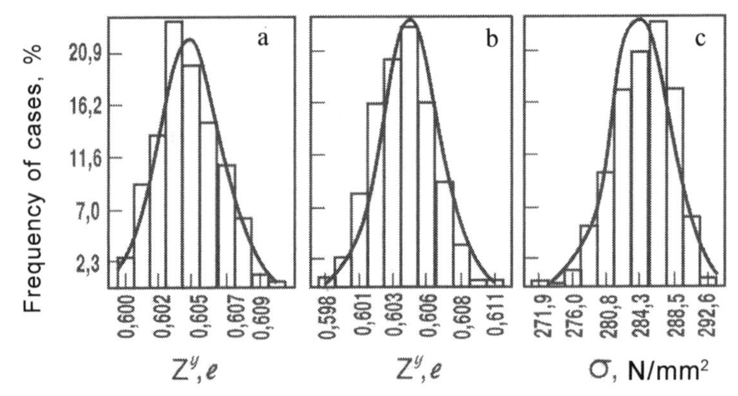

Fig. 4.2. Histograms of the distributions of the chemical electronic equivalent Z^y of the composition and the resistance to deformation σ of steel 60. Figure notes: a – distribution of the actual values of Z^y; b – distribution of Z^y obtained by simulation using the Monte Carlo method; c – the distribution of σ for T = 800°C, u = 10 s^{-1}, ε = 0.4.

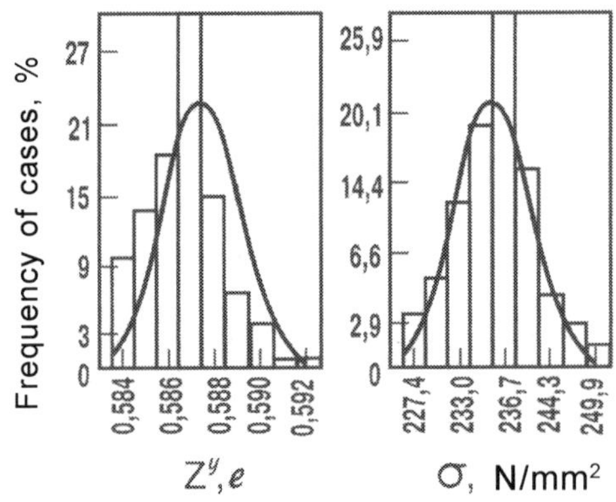

Fig. 4.3. Histograms of the distributions of chemical electronic equivalent Z^y and resistance to deformation σ at T = 800°C, u = 10 s^{-1}, ε = 0.4 for steel 3sp.

and σ for steel 3sp. Analysis of these data shows that the range of varying the value of σ for these modes of deformation is 25 N/mm^2 for steel 3sp and 21 N/mm^2 for steel 60. Immediately, we note that the mechanical properties (yield strength and tensile strength) of the finished rolled products change in a wider range. This suggests that, along with the variability of the chemical composition, the spread of the properties of the rolled products is also significantly affected by the instability of the temperature–speed and deformation modes of the rolling process and by the cooling conditions of the deformed metal.

4.2. Stability assessment of the quality of sheet rolled products

The main indicators of the quality of hot rolled strips are the accuracy of the dimensions and the level of the mechanical properties of the metal. The existing standards stipulate the tolerances on the thickness and width of strips and deviations from the flatness. Reducing the field of the size scatter of the strips saves a significant amount of metal. The metal intended for further cold rolling is subject to very high requirements compared with the standard values. The norms stipulated, for example, in ISO 19903 do not always ensure the stability of the process of cold rolling and high quality of the finished products. The additional requirements relate to the cross-sectional profile of the hot-rolled strips.

Experience has shown [48–51] that the most rational form of the cross-section of hot rolled strips is lenticular. The convexity of the profile should be 0.05–0.12 mm, depending on the width, thickness and destination of the cold rolled strips. So, the strip plate for tin-plate 700–900 mm wide should have the convexity no more than 0.05 mm, for automobile body sheets – 0.08 mm. The maximum convexity of strip plate with a thickness up to 3 mm thick and up to 1250 mm wide is 0.08 mm, and for the thickness exceeding 3 mm and width from 1250 to 1500 mm – 0.12 mm. In summary, it can be assumed that the rational value of the convexity must be between two boundary values (Fig. 4.4), given by the expression:

$$\delta h_{\max} = 0.04 + 0.08 \cdot 10^{-7} B^2 h; \quad \delta h_{\min} = 0.01 + 0.08 \cdot 10^{-7} B^2 h, \quad (4.2)$$

where B is the width of strip plate, mm; h is its thickness, mm.

An important indicator is the wedge shape of the strips. It should not exceed 1% of the thickness of the strips (per 1 m width). The profile of the strips must be symmetrical. Local thickening should not exceed 0.3 of the convexity of the cross-sectional profile.

The longitudinal thickness difference of the strip plate have a strong effect on the stability of the cold rolling process and the rolling conditions of the strips with the welded joints in continuous mills. The recommended allowable longitudinal thickness difference is not more than 0.1 mm.

The results of the study the accuracy of geometrical dimensions of hot rolled strips are given in [51].

Longitudinal thickness differences of hot-rolled strips usually equal 0.1–0.3 mm, transverse differences 0.01–0.15 mm. These values

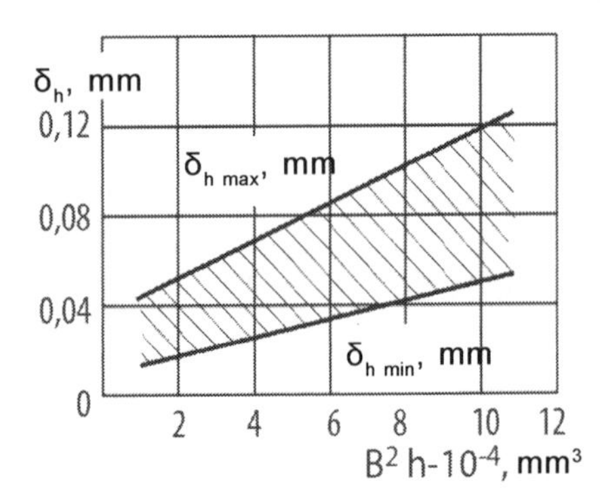

Fig. 4.4. Rational convexity value δh of strip plate for cold rolling depending on the width B and thickness h of hot rolled strips.

basically satisfy the requirements for hot rolled strips, however, they are not always acceptable to for strip plate used in cold rolling shops. The dimensional accuracy of the hot-rolled strips is determined mainly by the value their longitudinal thickness difference due to thickening of the ends, and thickness variations within the rolled batch.

A similar pattern appears when considering differences in the width of the strips. Deviations from the desired value of the width are observed typically on the end portions of the rolled strips. The width of the strips (excluding end sections) often varies within ±5 mm in 90–93% of their length.

An important indicator of quality is the flatness of the strips. High values of warped strips correspond as a rule to wider strips. The flatness of the cold rolled strips depends upon the shape of the cross section of strip plate. The stock with a symmetrical double convex 'lens-shaped' profile of the cross-sectional with a convexity to 0.12 mm gives the highest resistance to the distortion of the flatness of the cold rolled strips.

As noted above, the mechanical properties of steel products in wide-strip mills are unstable within a batch or a melt. The main reason for the instability of the mechanical properties is the variability of the chemical composition. The second most important factor of instability mechanical properties of bands is rolling end temperature variation. Figure 4.5 shows the dependence of the mechanical properties of thin (1.8 mm) hot-rolled strips of steel 10kp of the finishing rolling temperature $T_{f.r.}$. It can be seen that changes $T_{f.r}$ even within 10°C can substantially affect the structure and properties of the hot rolled metal. Figure 4.6 *a* shows the average

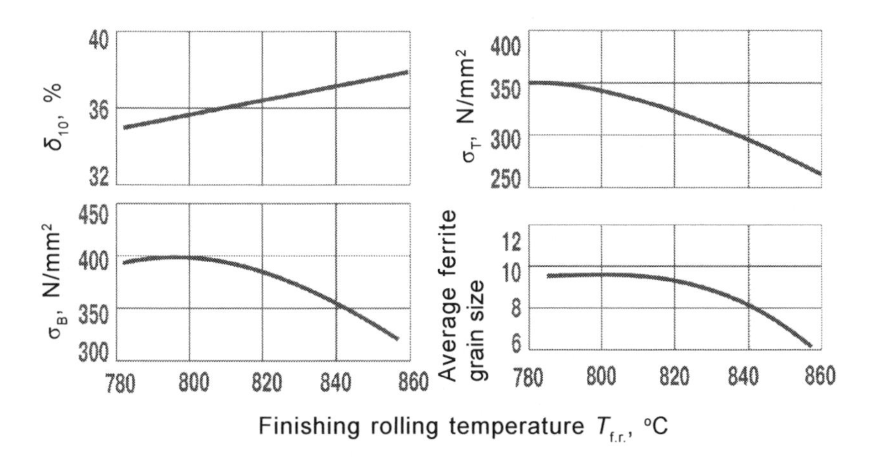

Fig. 4.5. Dependence of the mechanical properties of 1.8 mm thick strips of steel 10kp on the finishing rolling temperature. Coiling temperature is 600–620°C.

standard deviations of the finishing rolling temperature of the strips. The data were obtained under conditions when the automatic rolling temperature control system was switched off. The range of finishing rolling temperature fluctuations in rolling the strips of the same size was ±(20–40)°C for the entire range of experimental data.

Variation of the finishing rolling temperature along the length of the strip and between the strips in a batch are determined mainly by temperature fluctuations of the rolled products on the intermediate roller table, and between the strips of different batches by the differences in the rolling speed. For example, Figs. 4.6 *b* and *c* show the mean values and standard deviations of the temperature of rolled products T_p and initial speed V_{init}.

4.3. Effect of heating conditions of slabs on the stability rolling technology for sheets and strips

The main objective of heating is to provide the desired temperature for rolling slabs with a minimal spread within a batch of slabs (melt). However, structural features of continuous furnaces in some cases make it difficult to solve this problem.

In slab-heating furnaces the metal is heated on three sides: the top and on narrow ends. Wide ends of the consecutively positioned slabs contact each other. The bottom surface of the slab is in contact with either the monolithic hearth or with glide tubes that covered with skid pushers in many furnaces. The instability of heating is

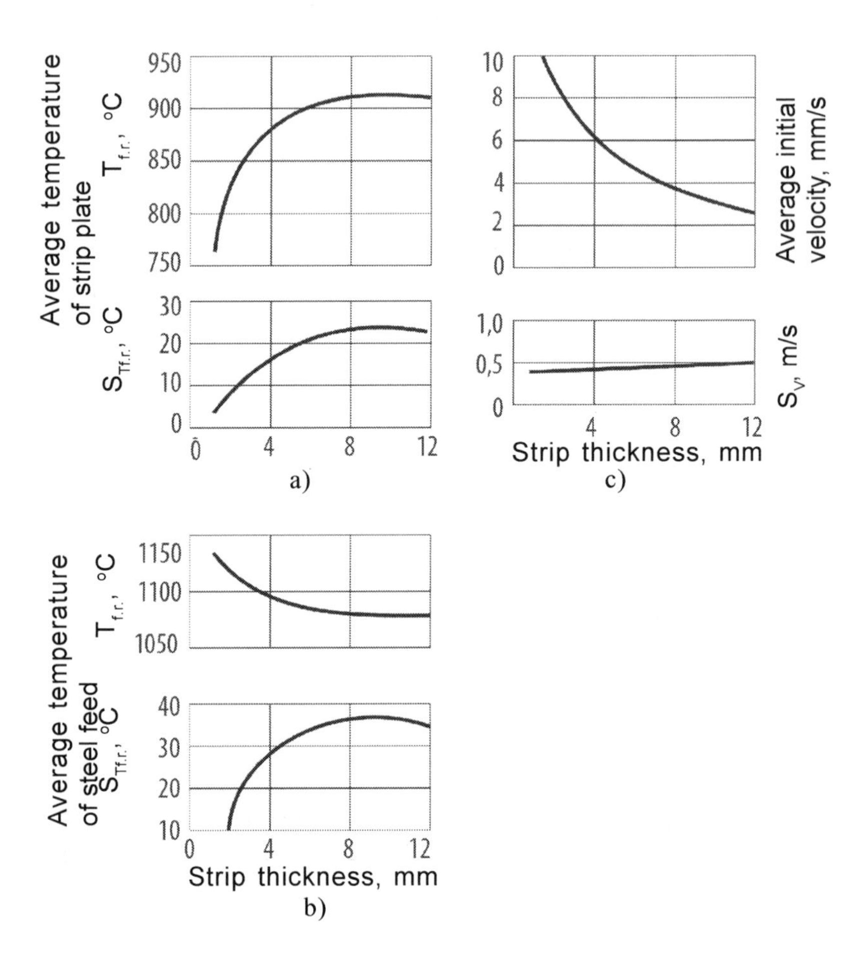

Fig. 4.6. Mean values and standard deviations for the final rolling temperature SHSGP 2000 $T_{f.r.}$ bands of different thicknesses (a); workpiece temperature T_{pl} for roughing train when rolling strips of different thicknesses (b); initial velocity V_{in} bands of different thicknesses (c).

manifested primarily by the presence of skid marks on the slab, which destabilize the rolling process in the wide-strip hot rolling lines. The difference of temperatures of the metal in the area of the marks and outside the marks can reach 50–200°C. Furthermore, there is a temperature gradient in the cross section of the slabs. It was established experimentally that the temperature distribution in the thickness of the slabs T_{sl} is characterized by a coefficient of variation $V_{T_{sl}}$ equal to 3 –12% at the maximum value T_{sl} on the upper surface of the slab.

The above-mentioned factors act in the case of a single-row or multi-row symmetric method of slab arrangement. In the 'chess' arrangement of the slabs the nature of the distribution of T_{sl} along their length is different because of preferential heating conditions of the free end portions of the slabs. The temperature gradient along the heated slabs within the same heat may be 5% or more, while in the case of single-row arrangement the temperature variation does not exceed 1–2%. Experience shows that the use of furnaces with integrated upper and lower heating substantially (1.5–2 times) equalizes the temperature in the longitudinal and transverse sections of the slabs.

Among the factors determining the stability of slab heating are qualitative indicators of fuel used, which is either natural gas or a mixture of blast furnace coke gases. The heat of combustion of these fuel species is substantially different: for natural gas is 5 times higher than for the blast furnace coke gas. Changing the characteristics of, for example, natural gas may lead to a change of its heat of combustion 1.5–2 times. Variations of the content of the components of the blast furnace coke mixture impair heating stability.

Analyzing the results of measurements in the ShSGP 1700 wide-strip hot rolling mill of temperature T_{sl} of more than 500 slabs immediately after their leaving the heating furnaces leads to the following conclusions [48]. For each furnace T_{sl} is a random variable, and the law of its distribution corresponds to normal. The statistical characteristics of T_{sl} for the ShSGP 1700 furnaces are presented in Table 4.1 where $\bar{X}_{T_{sl}}$ is the average temperature of the slab, °C; $S_{T_{sl}}$ is the standard deviation, °C.

Checking multiple sets of T_{sl} values for te slabs of hot and cold charges allowed to establish that these sets belong to the same group. This explains the rather large value of $S_{T_{sl}}$. The instability of the heating mode of the slabs was primarily caused by the poor state of furnace equipment and the fuel quality variability – the blast furnace coke mixture.

Uniform heating for of the metal for rolling a good basis for the stability of the rolling process in the ShSGP wide-strip hot rolling mill. To solve this problem in its entirety during hot rolling it is also necessary to provide a stable duration of transport of metal in the mill line from the heating furnaces to coilers.

Duration τ of the transportation of the strip plate in the ShSGP 1700 wide-strip hot rolling mill was measured in three characteristic areas: first – from each of the four heating furnaces to to the vertical

Table 4.1. Statistical characteristics of the slab temperature output from the furnace SHSGP 1700

Furnace number	$\bar{X}_{Tsl}$, °C	S_{Tsl} °C
1	1260	6
2	1259	9
3	1256	6
4	1265	10

descaling unit of the roughing mill train; the second – in the intervals between the stands of the roughing mill train, the third – on the intermediate roller table. Sample sizes at each site consisted of 136–180 slabs. The sample sets verified by the χ^2 showed a normal distribution law for the duration of the movement of slabs and strip plate in all areas.

Analysis of the duration of transport of slabs of different sections (210×910, 195×1060, 180×1530 mm) and strip plate with thicknesses 40, 35, 37 mm at each site of the ShSGP 1700 wide-strip hot rolling mill revealed the equality of their empirical distribution centres and the equality of variances. This makes it possible to generalize the results of analysis of a limited number of sizes of slabs and strip plate for the whole range of the mill.

The statistical characteristics of the distributions of the duration of transport of the slabs and strip plate are shown in Tables 4.2 and 4.3. Here and later in this chapter: $\bar{X}$ – average value, S – standard deviation, V – the coefficient of variation.

The decrease of the coefficients of variation of the transport duration of the slabs from the furnaces Nos. 2–4 was caused by the change of the average values of $\bar{X}_\tau$ during the time of passage of the slabs of the distance between the adjacent furnaces, which is ~11 s.

The significant two-fold greater scatter of the coefficient of variation for the duration of the transport of the strip plate on the roller conveyors before the horizontal descaling unit through the 1st and 3rd roughing stands is explained by the maintenance of the required amount of waiting time between the strip plate in the finishing group stands in violation of the tempo of exit of the slabs from the furnaces or the rolling speed.

The revealed patterns allow with the reliability of 95% to indicate possible ranges of the duration of transporting the slabs and strip plate($\bar{X}_\tau + S_\tau$) at all the sites studied: the first section:

Table 4.2 Statistical characteristics of duration τ of transport of slabs from reheating furnaces from the vertical descaler in the roughing group of the ShSGP 1700 mill

Furnace No.	$\bar{X}_\tau$, s	S_τ, s	V_τ
1	13.7	1.16	0.085
2	24.7	0.99	0.040
3	35.8	1.08	0.030
4	47.1	1.13	0.024

Table 4.3 Statistical characteristics of duration τ transportation peals for roughing train to intermediate roller of SHSGP 1700 (*h/des – horizontal descaler)

Index	Transportation to stand						At the intermediate roller table
	h/des*	1st	2nd	3rd	4th	5th	
$\bar{X}_\tau$, s	15.7	16.6	22.2	19.9	25.2	34.6	46.8
S_τ,c	0.97	0.82	0.53	1.01	0.77	1.05	7.03
V_τ	0.062	0.049	0.028	0.05	0.031	0.030	0.043

from the furnace number 1 – 13.2÷16 s; from furnace number 2 – 23.7÷25.7 s; from furnace 3 – 34.6÷37 s; from furnace 4 – 45.8÷48.4 s; in the second portion – 130÷140 s; in the third section – 45÷51 s.

For the second (interstand area of the roughing group) and third (intermediate roller table) sites taking into account the additivity of the variance of the normal distribution of several variables, the total accumulated dispersion can reach ~10 s^2, the maximum scatter duration transporting the metal through the three considered areas is ~20 seconds.

4.4. Reliability of hot strip rolling technology

4.4.1. Evaluation of the reliability of hot rolling technology

There are two areas of research and analysis of the reliability of different processes [52]. According to the first one the desired result is obtained by methods of mathematical statistics. In this area attention is paid to the properties of the material characteristics of the technological system, the modes and conditions of its functioning. This direction is most time-consuming. The objectivity of the conclusions is considered proportional to the number of considered factors and depends on the validity of the accepted hypotheses.

Therefore, despite the sufficient accuracy of the results obtained, this approach is used relatively rarely.

According to the second direction the process reliability can be determined without studying the influence of various factors – by direct measuring the properties achieved, quality and comparing them with the desired (permissible) values. Due to the relative ease of implementation, this direction is used quite widely, for example in analyzing and predicting the level of the geometrical characteristics or mechanical properties characteristics of rolled products.

Consideration of both methods allows to ensure the objectivity of the obtained results and suggest the following sequence of stages of research:

1. Assessing the impact of various factors on the reliability of the process;
2. Determination of reliability indices of each element of the system;
3. Development and implementation of organizational measures to improve the process.

The first stage is realized usually by using standard methods of mathematical statistics. The second stage use one of the methods of the theory of reliability of the process, including preliminary theoretical calculation of parameters and reliability of the system under study, modelling 'failures', experimental study. The main of them – a method of modelling failures – is based on a comparison by each quality indicator of a pair of functions T_p and C_n. Here T_p are the requirements imposed on the process; C_n is the ability of the process to implement these requirements.

For a complex dynamic system, such as the ShSGP wide-strip hot rolling mill, it is quite difficult to formulate the concept of 'failure'. Based on the recommendations in [52], a failure is the boundary value of the controlled parameter. The lower boundary is a complete failure (loss of product quality), the upper limit ensures failure-free operation of the process (ensuring 100% the required quality of products), and any intermediate value is associated with the possibility of the loss of quality.

Having said that, in order to study the reliability of the process of hot rolling of sheets, a method was developed to determine the relationship of the statistical characteristics of random factors with energy–power and temperature parameters of the rolling process, and also with indicators of the quality of hot-rolled strips.

We emphasize once again that the research of the process, exposed to random factors, should be based on the procedure for determining the probability characteristics of its output parameters. In practice, guided by the assumption of a normal distribution of these parameters, we are usually limited to finding the moments of these distributions, mathematical expectation, variance, skewness, etc. In most cases, this is not sufficient for the understanding of the observed phenomena, the formulation and implementation of proposals to improve the reliability of the production process due to deliberate effects on the technological regimes.

That is why the method of analysis of reliability of the rolling process of strip on the wide-strip hot rolling mills was proposed and is based on the definition of the relationship of the statistical characteristics of operating random factors and output parameters (energy–power, temperature parameters of the rolling process, indicators of the quality of finished steel products). The accuracy of the adopted techniques was investigated in the condition of the ShSGP 1700 wide-strip hot rolling mill. The temperature of the strips was measured behind the stands No. 2 and 5 of the roughing mill. The thickness of the finished strip is measured with an error ±(0.036–0.1) mm. The current of the main engines of the stands of the roughing and finishing groups was also recorded. The investigated thickness range was 2.2–10 mm, the strip width 1050–1520 mm, rolling speed 2.0–18.5 m/s, the finishing rolling temperature 760–940°C.

The results of comparing the experimental and calculated values are presented Fig. 4.7. The rms deviation of the calculated values from the experimental values: for finishing rolling temperature did not exceed 5°C; for rolling power 0.02 MW; transverse unevenness (thickness differences) was within the error of the instrument.

Additionally, to verify the accuracy of the algorithm the values calculated by the model of the initial speed and acceleration for the final stand of the finishing mills of the ShSGP 1700 and 2000 trains, providing the experimentally determined values of the finishing rolling temperature of the strips $T_{\text{f.r.}}$ were compared. The error in the initial speed was in the range of 0.2 to 0.5 m/s, that is 5%. The error in the determination of acceleration was not greater than 0.02 m/s^2. Thus, we can assume that the agreement between the calculated and experimental data was satisfactory.

In the actual rolling process the energy–power parameters can change not only as a result of targeting (reduction schedule change,

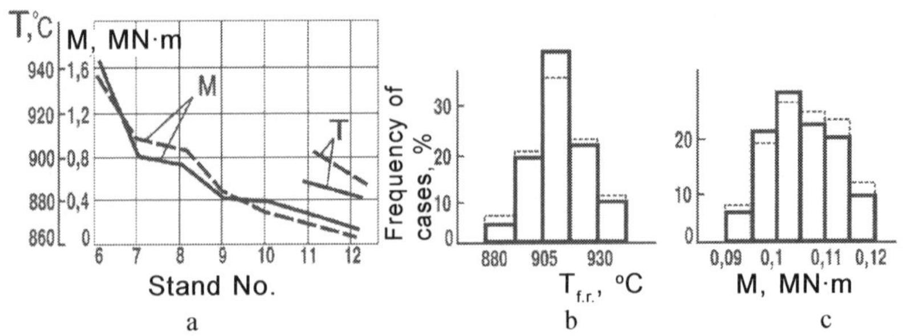

Fig. 4.7. Comparison of experimental (solid lines) and calculated (dashed lines) values of rolling moment *M* and the metal temperature *T*: a – determined values; b, c – stochastic parameter values in the 12th stand.

correction of heating temperature or rolling speed), but also as a result of the unpredictable impact caused by the stochastic nature of the change technological factors. In particular, in metal deformation force *P*, moment *M*, power *N* and temperature *T* can change due to the deviation of the values of the initial technological factors from baseline values at which the temperature–deformation rolling modes were set up. In turn, the inconstancy of the rolling force causes changes of elastic deformation of the stand and, hence, changes in the thickness and speed of the strip at the exit from the roll gap. Torque variation on the rolls leads to additional stress in the drive line of the stand, and the change of the rate of exit from the roll gap band in the stands of the continuous finishing mill leads to changes in the strip tension.

These deviations of the parameters of the rolling process in the system lead to difficult to predict changes to the quality of the finished strips: longitudinal and lateral variations in thickness, flatness and mechanical properties. Therefore, the results of the study of the statistical characteristics of the initial technological factors are important, as they provide the most succinct information on the nature of changes of input perturbations in the ShSGP wide-strip hot rolling mills. The sets of values of the initial technological factors can be obtained for both the specific sections of the length of the strip (beginning, middle, end), and for different strips within the same heat as well as for the strips of the same size, rolled at different periods of operation of the mill.

Characteristic sites of action of the factors enabled their differentiation on the basis of the 'site of action' along the lines of the stand as follows:

1) factors acting at the 'entry' into the stand: slab thickness H_{sl}, width of the slab B_{sl}, length of the slab L_{sl}, the slab heating temperature T_{sl} (these factors are considered in section 4.4.2);

2) factors acting in the line of the stand:

• in the section of the roughing group: thickness of strip plate in the stands H_i, duration τ_i of transportation of strip plate in the sites of the roughing group, thickness of strip plate H_{plate}, strip plate temperature T_{plate};

• at the site of the intermediate roller table: transport velocity of strip plate V_p;

• at the site of the finishing group: strip thickness in the stands H_i, rolling speed in the last stand V_n, the working roll radii R_i, grind profile of the working rolls ΔR_{grind}.

For a comprehensive characterisation of the stability of hot strip rolling we must consider the values of both statistical characteristics of the distributions parameters of this process: the expectation $\bar{X}$ and standard deviation S.

The strip plate temperature $\bar{X}_{Tplate}$ is the main technological factor affecting the stability of the rolling process in the investigated section of the stand. With other conditions being equal, increasing values of $\bar{X}_{Tplate}$ from 1070 to 1150°C increase the stability of the rolling force P and S_p is reduced. This in turn leads to the stabilization of the thickness of the rolled strip and improves flatness. Similarly, with increasing $\bar{X}_{Tplate}$ the values of torque M and rolling power N are stabilised. At the same time, the change of $\bar{X}_{Tplate}$ in the specified range slightly destabilizes the finishing rolling temperature: $S_{Tf.r.}$ increased by 0.5–0.7°C. The influence of $\bar{X}_{Tplate}$ increases from the first to the last stand of the finishing mill group, i.e. the proportion of $\bar{X}_{Tplate}$ in the final value of RMS deviations increases: for rolling force about 1.6 times; for the rolling moment 2.3 times; for the rolling power 1.5 times. In the last stand, the share of influence of $\bar{X}_{Tplate}$ on S_P, S_M, S_N, $S_{Tf.r.}$ ranges from 15 to 30% of the final value.

The effect of pair interaction of the factors file $\bar{X}_{Tplate}$ and S_{Tplate} applies to all stands of the finishing mill. Their influence on the value S_i power parameters exceeds 30% for the first stand and decreases to 10–12% for the latter. Increasing $\bar{X}_{Tplate}$ from 1050 to 1150°C leads to a slight decrease in the influence of S_{Tplate} on $S_{Tf.r.}$. Thus, at S_{Tplate}, equal to 15°C, $S_{Tf.r.}$ with increasing magnitude of $\bar{X}_{Tplate}$ in the specified range is reduced by less than 0.3°C.

These results suggest that $\bar{X}_{Tplate}$ is the most important factor in the control action in the site of the finishing stands of the ShSGP mill.

The stabilizing influence of the factor of 'constant rolling' $V{\cdot}H$ on the energy–power parameters of the process in different stands of the finishing mill is up to 15% of their mean square deviations. This factor is stronger the higher the absolute value. This leads to the conclusion about the possibility of effective control of the stability of the rolling process by correcting the deformation–speed mode and, primarily, by changing the rolling speed.

The scatter in the values of the filling rolling speed S_{V_f} in the last stand of the finishing mill significantly destabilizes all energy and power parameters of the rolling process only in the last two stands. The strength of the influence of this factor is within 2–5% of the value of S_i of the power parameters of the rolling process.

The destabilizing influence of the spread of values of the radii of the working rolls of the finishing stands S_{Ri} appears only on the values S_i of rolling forces and moments. Indicator S_{Ri} of the mean standard deviation refers to a group of factors whose influence increases with increasing numbers of the stands in the finishing mill group. The fraction of the influence of S_{Ri} in the final value of the standard deviation of rolling force and torque is 2–7% in the first and 10–12% in the last stand of the finishing group. Pairwise interactions S_{Ri} with other technological factors are insignificant.

The most significant factor that has a destabilizing effect on the value of the transverse thickness difference is the temperature T_p of strip plate. Changing this temperature by 50–85°C increases the value of $S_{\delta h}$ 1.5–1.6 times.

Factor S_{Ri} also has a destabilizing effect on $S_{\delta h}$, although its share does not exceed 10% of the final value of $S_{\delta h}$. The effect of S_{Ri} increases 1.5 times from the roughing group N–4 to the stand N, where N is the number of stands in the ShSGP wide-strip hot rolling mill. The decrease of the values of S_{Ri} to 0.010–0.015 mm (corresponds to the difference of the temperatures of the rolls supplied for grinding, of 30–45°C) allows to eliminate destabilizing impact of S_{Ri} on $S_{\delta h}$.

The finishing rolling temperature $T_{\text{f.r.}}$ is affected mostly by the temperature and thickness of strip plate, the transport speed of strip plate from the roughing to finishing train, the rolling speed in the last stand of the finishing group. Statistical analysis of temperature, thickness and transport speed of strip plate on the intermediate roller table are shown in Table 4.4. The frequency distribution of these parameters is close to normal.

Table 4.4 Results of statistical analysis of process parameters of hot strip rolling in ShSGP 2000 (n – sample size)

Options	n	$\bar{X}$	S	Excess	Asymmetry
T_{plate}, °C	87	1123.2	12.8	2.0	–0.04
H_{plate}, mm	193	35.10	0.25	2.8	0.65
V_{plate}, m/s	105	3.47	0.35	2.9	–0.4

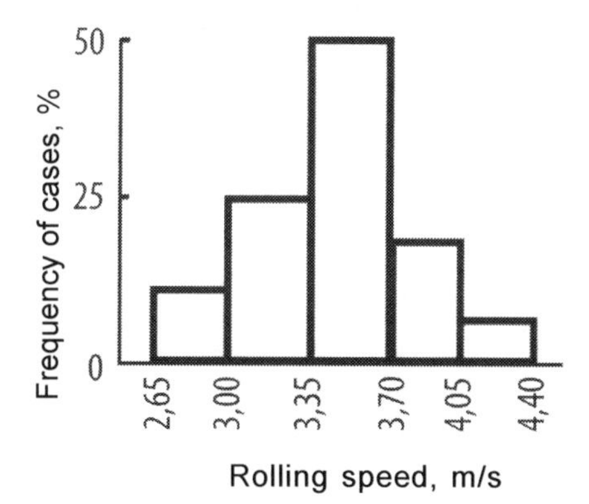

Fig. 4.8. Histogram of the frequency of distribution of the average transport speed of strip plate for intermediate roller table of the ShSGP 2000 mill.

As an example, Fig. 4.8 shows a histogram of the distribution of the values of the average speed of strip plate on the intermediate roller table. Maximum thickness variations of the strips in the stands of the finishing group were 0.18–0.65 mm in the first stands and 0.11–0.32 mm in the final ones.

A comparison of the calculated distribution of the finishing rolling temperature strips 2 mm thick with the experimental values (Fig. 4.9) indicates that the calculations take into account quite sufficiently the influence of the random nature of the process parameters. The actual average finishing rolling temperature is 849°C with the standard deviation of 4.5°C, calculated temperature 854°C at S = 13.3°C.

Studies have shown that the dependence of the standard deviation of the finishing rolling temperature $S_{T_{f.r.}}$ from the standard deviation of the initial thickness $S_{H_{plate}}$, temperature $S_{T_{plate}}$ of the strip plate, the thickness of the strip in the stands S_{h_i} is linear, and for the standard deviation of the average transport speed on the roller table $S_{V_{plate}}$ it is close to linear.

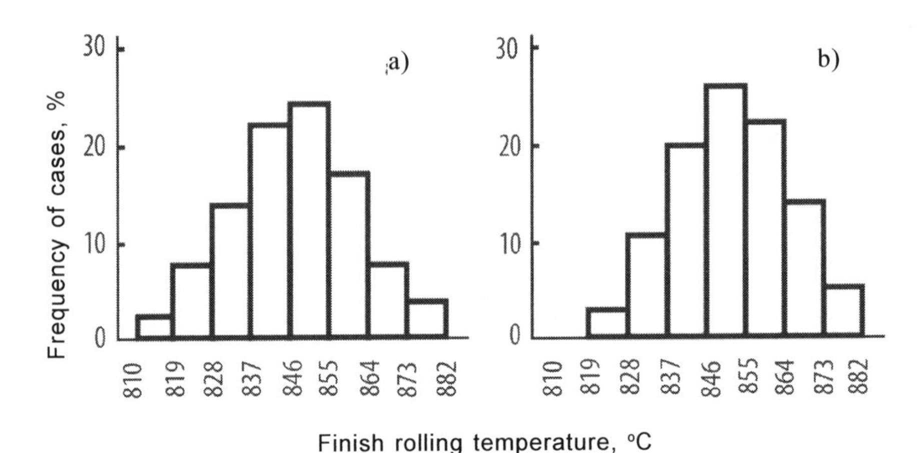

Fig. 4.9. Experimental (a) and calculated (b) histograms of the final rolling temperature strips 2 mm thick.

Variations in the thickness of strips in the stands of the finishing mills have no influence on the finishing rolling temperature at the final strip thickness greater than 4 mm.

Strip plate temperature fluctuations affect the finishing rolling temperature more substantially. At the standard deviation of the temperature of strip plate of 15°C the standard deviations the finishing rolling temperature of the strips with a thickness 2 and 8 mm were 4.8 and 8°C. That is, with increasing thickness of the rolled strips the fluctuations of the temperature of strip plate on the finishing rolling temperature increase.

The standard deviation of the transport speed of 0.5 m/s leads to the standard deviation of the finishing rolling temperature of 3°C for the 2 mm strips and 4.8 °C for the 8 mm strips. The increase of the temperature of strip plate from 1100 to 1150°C increases the effect of S_{Vplate} on $S_{T_{f.r.}}$ (at S_{Vp} = 0.5 m/s $S_{T_{f.r.}}$ increases by 0.5–0.7°C). Increase in the average transport speed from 3.5 to 5 m/s significantly affects the impact of S_{Vplate} on $S_{T_{f.r.}}$. The standard deviation of the finishing rolling temperature decreases by 1.5–2.4°C.

Changes in temperature, thickness and transport speed of the strip plate, and also of the thickness of strips in the stands of the finishing mill with a seven-stand finishing group cause changes in the finishing rolling temperature within ±(20–25)°C. With increasing thickness of the rolled strips the influence of standard deviations of the transport speed and the initial temperature of the strip plate on the standard deviation of the finishing rolling temperature increases, and the effect of variations in the thickness of strips in the stands of the finishing mill decreases. Reducing the scatter of the thickness of the strip plate does not cause any significant reduction in the range of finishing rolling temperatures. Thus, the standard deviation of the thickness of 0.25 mm does not prevent the finishing rolling temperature. To ensure the mean square deviation of $S_{Tf.r.}$ of not more than 5°C, it is necessary to ensure that the mean square deviation of the average transport speed of strip plate to the intermediate roller table does not exceed 5% of the mean values and the standard deviation of the initial temperature of strip plate S_{Tplate} does not exceed 5–6°C.

The heredity of the transmission of disturbances in the mill line defines the requirements for narrowing the ranges of permissible fluctuations of heating temperature, slab thickness, the duration of transportation of strip plate in the mill train. The transfer of disturbances through the rolled strip from the roughing to finishing stands of the group regulates the thickness of slabs, reduction modes and strip plate thickness depending on the destination of the finished hot-rolled strip. In the stands of the finishing group the final nature and values of the quality of completed strips are formed. Therefore, the conditions of operation of the previous section of the stand determined the deformation and speed rolling conditions to stabilize the introduced disturbances.

More detailed results of studies in this area are presented in our paper [48].

4.4.2. Influence of design features of wide-strip mills on the reliability of the rolling process and the quality sheet steel

The composition, location of equipment and technology for wide-strip hot rolling mills operating now at various metallurgical plants are different. The above method of differentiation factors on the basis of the 'site of action' along the lines of the stand suggested that the strength of their influence on the reliability of the whole process is interconnected with design characteristics of each of the sections of the stand.

The length of the main sections of the ShSGP – roughing train, intermediate roller table, finishing train and run-off roller tables are in the following ranges: roughing train – 81–228 m; intermediate roller conveyors – 42–132 m; finishing group – 36–48 m; run-off roller table – 99–206 m. From eight ShSGP wide-strip hot rolling mills the deformation of strip plate in a continuous three-stand roughing subgroup is recommended. Three ShSGPs in the roughing train used reversible rolling, rolling in four mills in the roughing group is carried out by the sequential scheme. The number of stands in the roughing group, taking descalers into account is 3–7, in the finishing group 6–7.

Initial assessment of the impact of design features on the reliability ShSGP on the rolling process reliability was carried out for areas that have the greatest differences – namely roughing stands. Based on materials of [53], comparison was made for the ShSGP 1700 mill, with the largest length of the roughing train.

The following provisions were considered in the investigations. The equations derived by regression analysis were used only in cases where the values of variable factors did not go beyond the variation range. The data on the stochastic process are most informative when the size of the variation range is expanded as much as possible. In this case, the dependence are adequate to the process in the entire range of factors. The boundaries of the ranges of varying the factors in computing experiments obtained on the basis of production data [48] are given in Table 4.5.

In the table: L_{sl}, H_{sl}, T_{sl} – the length, thickness and temperature of the slab; H_p – the thickness of strip plate; H_i – the actual metal thickness, rolled in the i-th stand; τ_i – the transportation duration of the strip plate in the i-th gap between the stands.

In the investigations the conditions of independence of the variable factors as well as of their statistical characteristics were adhered to. For the mathematical expectation $\bar{X}_i$ and the standard deviation S_i this is ensured by the properties of the normal distribution law of the random variable.

Application of the planned experiment reduced the total number of calculations and simplified mathematical description of the results.

Table 4.5 The values of the factors used in computing the experiment for estimating process parameters roughing stands

Factor levels	$\bar{X}_{Lsl}$, m	S_{Lsl}, m	$\bar{X}_T$, °C	S_T, °C	$\bar{X}_{Hsl}$, mm	S_{Hsl}, mm	$\bar{X}_{Tplate}$, mm	S_{Hi}, mm	$S\tau_i$, s
Lower	7.01	0	1200	0	170	0	30	0	0
Upper	9.5	0.3	1270	12	200	0.95	40	0.60	1.10
Zero	8.30	0.15	1235	6	185	0.475	35	0.30	0.55
Variation interval	1.0	0.15	35	6	15	0.475	5	0.30	0.55

Given the number of variable factors, the nature of the connections between them, according to [53], a fractional factorial experiment FFE of the type 2^{19-13} was carried out to obtain the regression equations of the form:

$$y = b_0 + b_1X_1 + b_2X_2 + ... + b_nX_n,$$

which we adopted to describe the relationship of the statistical characteristics of the considered factors.

To provide more control during hot strip rolling it is necessary to obtain data on the effect of the statistical characteristics of technological factors on the parameters of the technology within the group of the stands and assess their relevance to the sections of the stand. Therefore, numerical experiments were implemented so as to obtain the desired relationships for each stand of the group.

Verification by the Fisher criterion showed the adequacy of the derived linear equations with respect to the actual conditions for all the considered process parameters of hot rolling in the roughing train of the ShSGP wide-strip hot rolling mill. In determining the mathematical expectations of the unknown functions of several random variables the arguments were only the mathematical expectations of the variable factors of the rolling process, and for the values of the standard deviation of the unknown functions also their S_i values. This is consistent with the properties of the statistical characteristics of a random function.

Ensuring the stability and reliability of any process is oriented to restricting the scatter of the technological factors. Therefore, in our case, special attention in probabilistic analysis was paid to the study of the relationship of S_i of the random factors and the parameters of the rolling process. For values of $\bar{X}_i$ we determined the location of the centre of distribution in the range of allowable parameters of the process for the corresponding values of the factors. The degree of influence of the factors on the values of the standard deviations was assessed by the magnitude and sign of the corresponding coefficient in the regression equation. The concept of 'the stabilizing factor' used in the subsequent stages means that an increase in the absolute value of its statistical characteristics (expectation) decreases the S_i value of the corresponding rolling parameter. The 'destabilizing' factor acts in the opposite direction.

Temperature of heating slabs $\bar{X}_{T\text{sl}}$ is a major stabilizing factor for all power parameters of the rolling process in the roughing stands. The effect of $\bar{X}_{T\text{sl}}$ on the stability of rolling parameters in different

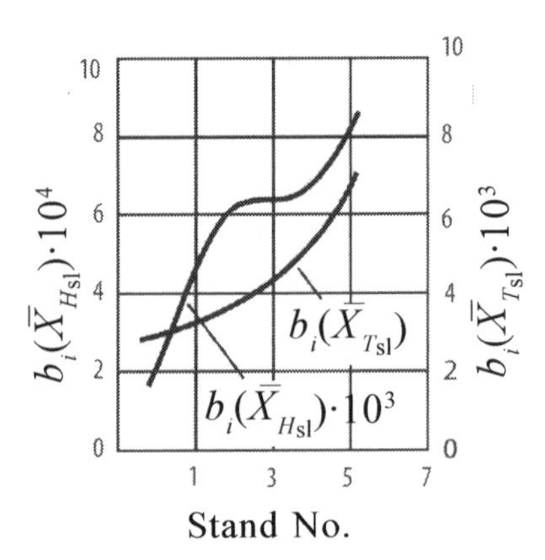

Fig. 4.10. Influence of $\bar{X}_{Tsl}$ and $\bar{X}_{Hsl}$ on the corresponding coefficients b_i in different stands of the roughing mill group.

stands of the group differs. The general view of this dependence is shown in Fig. 4.10. In the investigated conditions an increase of the number of the stand N increases the value b_i at $\bar{X}_{Tsl}$. For the whole range of values of the technological factors the influence of $\bar{X}_{Tsl}$ is maximum and equals up to 40% of the total value of S_i of the parameters of the rolling process.

The stabilizing effect of $\bar{X}_{Tsl}$ on the values of the standard deviations of force, rolling torque and power is virtually independent of the thickness of the slabs, changing in the actual range (170–200 mm), and of the thickness of the strip plate in the range from 30 to 40 mm. The values of the products $b_i \cdot \bar{X}_{Tsl}$ are equal to the absolute 'contribution' of the factor to the final value of S_i of the power parameters of the process for real rolling conditions (Table 4.6).

Deviations of the temperature of heating the slabs S_{Tsl} are the main destabilizing factor. Its share of influence in the final value of S_{Pi}, S_{Mi}, S_{Ni}, S_{Ti} reaches 40%. It is necessary note that for the whole range of values $\bar{X}_{Hsl}$ and $\bar{X}_{Hp}$ the significance of the influence of S_{Tsl} does not change, and this effect applies to all stands of the roughing mill. Deformation of the strip plate in the roughing stands stabilizes their temperature. The values of S_{Ti} decrease with the reduction of strip plate.

Influence of the thickness of slabs on the parameters of the rolling process is ambiguous and depends on the specific conditions of rolling. $\bar{X}_{Hsl}$ is a destabilizing factor for force, torque and power in rolling, and it is a stabilizing factor for the temperature of the strip plate. The $\bar{X}_{Hsl}$ influence amounts up to 10% of the total value of S_i of

Table 4.6. Values of the product $b_i \cdot \bar{X}_{T\text{sl}}$ in the stands 1 and 5 of the roughing mills of ShSGP 1700 in the equations for calculating the average standard deviation of force P, moment M, power N and rolling temperature T*

Stand No.	S_p, MN	S_m, MH·m	S_N, MW	S_T, °C
	v_p, %	v_m, %	v_N, %	v_T, %
First	0.37–0.38	0.049–0.051	0.098–0.102	0.62–0.67
	3.9–4.2	3.7–3.9	3.7–4.0	1
Fifth	0.85–0.88	0.135–0.140	0.91–0.94	9.10–9.40
	4.8–5.9	7.1–8.3	7.3–8.2	1

* The numerator – the value of the standard deviation, the denominator – the coefficient of variation.

the parameters of the rolling process. The form of the dependence in Fig. 4.10 is close enough to the distribution of the reduction mode in the stands of the roughing group of the ShSGP.

Increasing the thickness of the slab in the range from 170 to 210 mm when producing strip plate of the same thickness increases by 14–20% the instability of the force, moment and rolling power in the stands of the whole group. This is because an increase in total (in the group) and unit (in each stand) degree of deformation results in stronger adverse effects of factors that indirectly affect the change in the power parameters, namely the relative thickness difference, the degree of wear of working rolls, the value of 'accumulation' of the deformation in vertical rolls, etc.

Under the same conditions, an increase $\bar{X}_{H\text{sl}}$ increases on average by 18% the temperature stability of deformed metal. This effect is due to the change the ratio of the heat balance terms in the deformation zone and primarily with the changes in the conditions of heating strip plate by deformation work. The heat losses by the metal arising during its transportation and deformation in previous stands of the group are compensated by the additional temperature increase with increasing unit reduction in each subsequent stand.

Slab thickness deviations $S_{H\text{sl}}$ and adjustment accuracy S_{Hi} of the roughing stands destabilize all power parameters of the rolling process. However, the degree of their influence: variations of slab thickness significantly influence the spread of values in one stand (horizontal scale breaker). Disturbances caused by large values of S_{Hi} are transferred along the group to subsequent one or two stands. Moreover, if we consider the stands Nos. 1, 2, the effect of S_{Hi} is transmitted from only one previous stand, and if the stands Nos. 3–5 are considered, then from two stands. With decreasing H_i the same values of S_{Hi} cause a large scatter of the power parameters of rolling. For the rolling temperature the influence of this factors considered insignificant.

Increasing the **duration of transport of the strip plates** between the roughing stands $S\tau_i$ enhances the destabilizing effect on all power parameters of the rolling process and as regards the strength of the effect is equivalent to the effect of S_{Hi}. For example, in the conditions of the ShSGP 1700 wide-strip hot rolling mill the greatest destabilizing effect on the rolling temperature of the stand No. 5 is exerted by the values of $S\tau_i$ in the sections between the stands 1 (between the horizontal descaler and the stand No. 1), 3 (between stands Nos. 2–3), 4 (between the stands Nos. 3–4), 5 (between stands Nos. 4–5). This effect is increased from the 1st to the 5th interval by more than 10 times. A change in the share of the influence $S\tau_i$ in the sections of the roughing group on S_{T5} of deformation temperature in the stand No. 5 is shown in Fig. 4.11.

The destabilizing factors are deviations of the length of the slab S_{Lsl} and the thickness of the strip plate S_{Hplate} but their effect on the energy-power parameters is shown only in the stand No. 5, and S_{Lsl} significantly affects only the rolling of rear ends of the strip plate.

The effect of $\bar{X}_{Hplate}$, $\bar{X}_{Hsl}$ and $\bar{X}_{Tsl}$ on the mathematical expectations of the force, torque, power and temperature in the rolling stands of the roughing group (excluding the descaler) increases in transition from the stand No. 1 to the stand No. 5 approximately 1.1–1.3 times.

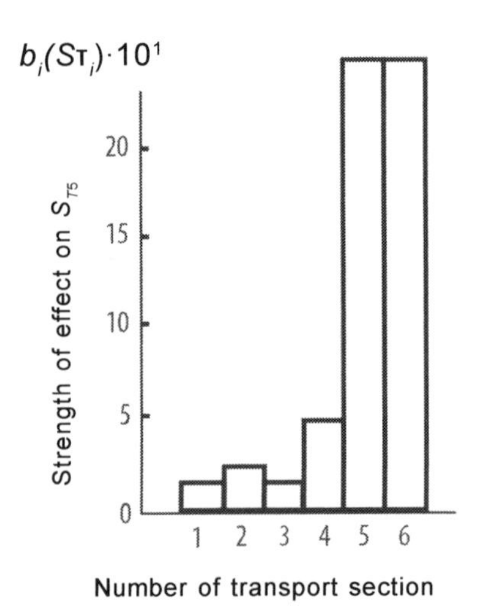

Fig. 4.11. Impact of $S\tau_i$ on of sections roughing mill stands on S_{T5}.

$\bar{X}_{Lsl}$ influences the mathematical expectation values of the rolling parameters of only the rear section of the strip plate only rear ends, $\bar{X}_{Hplate}$ significantly affects only the stand No. 5. The value of the standard deviation of the distributions of the power parameters along the length of the strip plate 1.35 times when rolling the rear ends of the strip plate.

The magnitude of S_i of the final distributions for the power parameters is from 5 to 12% of the value of their mathematical expectation. The maximum share of the influence (up to 40%) on the value of $\bar{S}_i$ is exerted by the statistical characteristics of slab temperature ($\bar{X}_{Tsl}$ and S_{Tsl}). It should be noted that with the increasing $\bar{X}_{Tsl}$ the influence of S_{Tsl} on S_i of the power parameters decreases.

The results allow to conclude that the nature of the change of the standard deviation values of the power parameters of the stands of the roughing group is similar to changes of the relative reduction. So, the value of ε_i of the horizontal descaler–stand No. 2 section increases by 10–16%, in the stand No. 2–stand No. 4 section it remains virtually on one level, but in the stand No. 4–stand No. 5 section it increases by 4–9%. This explains the sharp 'jumps' of magnitude of S_i of the power parameters of rolling in the first and third sections of the roughing group.

The proposed method of numerical experiments using the Monte Carlo method was used to analyze other layouts of the ShSGP wide-strip hot rolling mills. Possible schemes of arrangement and location of the roughing stands of the ShSGP are shown in Fig. 4.12. The simulation resulted in the following conclusions about the impact of structured (layout) of the wide-strip mills for the strip rolling process stability.

Rolling strip using reversing passes in the roughing mill stands

The known rolling schemes include the possibility of three–seven passes in the reversing stand (stands). The availability during the rolling process of the 'roll bite' and 'release' phases of the metal reduces the reliability of the process in this scheme of equipment layout. The quantitative patterns of action of both stabilizing destabilizing factors change. The effect of the stabilizing factor $\bar{X}_{Tsl}$ in three reversing passes is the same as in the case considered by rolling in the above scheme. If five or more reversing passes are made the effect of $\bar{X}_{Tsl}$ increases by 5–15% equals up to 45 % of

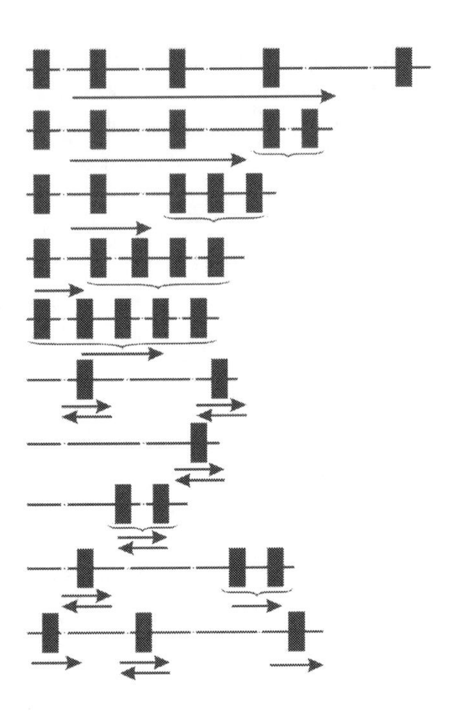

Fig. 4.12. Layout of stands of the roughing mill. Arrows indicate the direction of rolling. Curly brackets denote the stands connected in a continuous subgroup

the total value of S_i of the parameters of the rolling process in the reversing stand. In the case of the rolling pattern with the reversible five passages with subsequent deformation of strip plate in one to three stands, the value of b_i in these stands rapidly increases, and the 'contribution' of the factor to the value of S_i of the parameters in the last rolling mill group can reach 50–53%.

In the rolling pattern with reversing passes in the roughing train the destabilizing effect of deviations of the strip plate transportation time $S\tau_i$ increases and this agrees with our received data, reflecting the relationship of $S\tau_i$ with the thickness variation of the strip plate. It was established that the influence of $S\tau_i$ on S_i when using only reversing passes is similar to that shown in Fig. 4.11 and almost independent of the number of passes.

In the rolling pattern with three reversing passes and subsequent deformation of metal in one or three additional consecutive stands there occurs an additional destabilizing effect in the transport section of strip plate from the reversing to the next stand. The share of influence of $S\tau_i$ on this site is 2–5 times higher than in the previous transportation sections. Increasing the number of passes in the reversing stand aggravates the destabilizing effect of $S_{\tau i}$ in the given roughing stand. The S_i value of all energy–power parameters is almost constant along the length of strip plate and increases by 5–7% for the transition from rolling the front to rear ends. It was found that the effect of $S\tau_i$ may be obscured due to redistribution of the relative reduction in both the passes in the reversing stand and the whole group of the roughing stands.

The effect of other process factors is the same as discussed above for rolling in the ShSGP 1700 with inline stands (without reversing pass).

Rolling of strip plate using a continuous subgroup of stands

The layout of equipment with continuous stand subgroups (see Fig. 4.11) has certain advantages compared to all other layouts of the roughing stands in ensuring the stability of the rolling process. Thus, the action of $S\tau_i$, formed in the first metal transport sections, is already weakened in one or two stands of the continuous subgroup, thereby reducing the total value of S_i of the rolling process parameters in the final stand of the group at ~30 %. The stabilizing action of $\bar{X}_{Tsl}$ becomes stronger, reaching 35–40% of the final value of S_i in the stands of the continuous subgroup.

At the same time, in the discussed arrangement of the stands of the roughing group of the ShSGP the destabilizing effect of S_{Tsl} and S_{Ni} increases compared with other schemes. Unlike the rolling schemes of the mills with a consecutive arrangement of stands (without reversing of strip plate) and with reversing passes the influence of S_{Hi} is passed along the entire group of the stands with an abrupt increase of 1.2–1.4 times in the stands of the continuous subgroups. However, overall the total values of S_i of the rolling process parameters in the roughing stands of the continuous subgroup are lower than the previously discussed arrangement of the mills.

At the intermediate roller table of the stand there forms a narrowly directed destabilizing effect – the value of the statistical characteristics of the temperature of strip plate changes (X_{Tp} and S_{Tp}), which affects the stability of the strip rolling regime in the finishing stands. The major destabilizing factor is the temperature variation of strip plate S_{Tplate} entering the intermediate roller table. Its influence ranges from 30 to 50% of the final value of S_i for all power parameters of rolling stands of the finishing ShSGP group with the greatest influence on the stability of the process at the first

Table 4.7. Values of the product $b_i S_{Tplate}$ for the stands of the finishing group of ShSGP for average standard deviations of force P, moment M and rolling power N

Stands	S_p, MN	S_m, MN·m	S_N, MW
First	1.15–2.3	$6\cdot10^{-2}$–$1.2\cdot10^{-1}$	$2.5\cdot10^{-1}$–$5.2\cdot10^{-1}$
Last	0.5–1.0	$4.5\cdot10^{-3}$–$9\cdot10^{-3}$	$1\cdot10^{-1}$–$5\cdot10^{-1}$

stand of the finishing mill. The role of this factor from the first to the last stand of the finishing mill group weakens 2–2.7 times indicating the stabilizing effect of the finishing stands on the temperature of rolled strips.

It is noteworthy that in the finishing group of the mill the degree of influence of S_{Tplate} on S_i of force, torque and rolling power is almost an order of magnitude greater than a similar effect of the factor S_{Tsl} for the roughing stands. At the same time, as regards the influence of the rolling temperature on S_i, it is less than S_{Tsl}. The values of $b_i S_{Tplate}$ are equal to the values of the 'contribution' of the factor to the final value of S_i of the process parameters for the actual rolling conditions of the ShSGP 1700 are shown in Table 4.7.

For real values of S_{Tplate} its independent effect on the S_i of rolling temperature is insignificant.

The spread of values of the duration of transportation of strip plate on intermediate roller table $S\tau_p$ has a destabilizing effect on the force, rolling torque and rolling power in all rolling stands of the finishing mill. The share its influence can reach 20% of the final value of S_i of the power parameters of the rolling process in the first stand and decrease to 1–5% in the last stand. The noted character of the influence of $S\tau_p$ indicates a weakening of this factor. Both factors considered above are characterized by an increase of the destabilizing effect with increasing length of the intermediate roller table: L_p change from 42 to 120 m increases S_{Tplate} and $S\tau_p$ by 60–70%.

The studies also point to the possibility of stabilizing the parameters of the rolling process in the finishing group through the use of an intermediate rewinder (IRW) or screening devices. The change in S_{Tp} along the length of the strip plate is almost completely levelled (reduced by 80–85%), and the impact of $S\tau_i$ also becomes negligible (<10% of the final value of S_i of the power parameters of the rolling process).

The heredity of the transmission of perturbations in the mill line defines the requirements for narrowing the ranges of permissible fluctuations of heating temperature and slab thickness and the duration of transportation of strip plate in the mill line. Transmission of perturbations through the rolled strip from the finishing to roughing stands mill group must be consider when choosing the thickness of the slab, reduction schedules and strip plate thickness depending on the destination of finished hot-rolled strips. Their quality forms in the stands of the finishing group. Therefore, the selected deformation and speed rolling conditions should ensure the fullest possible levelling of the introduced perturbation.

Most preferable from the viewpoint of ensuring the rolling process stability is the layout with deformation of strip plate in the roughing group of minimum length. Reversing passes impair the process stability in rolling in the ShSGP, but in conjunction with the continuous subgroup lead to a variant equivalent to the method of successive rolling in the roughing stands. From the standpoint of stabilizing the temperature–rate and deformation parameters of the process of hot rolling strips it is desirable that the intermediate roller table of the ShSGP has the minimum allowable length and is equipped with shielding devices. It is also effective to use an intermediate rewinder.

5

Asymmetric strip rolling

5.1. Features and possibilities of asymmetric rolling

The rolling process in real industrial conditions often runs with elements of geometric, speed, temperature and other asymmetry even when it is nominally considered symmetric. This assertion is based primarily on the results of the above probability estimates of the distributions of input and output parameters of the rolling process, conditionally called symmetric. Furthermore, in addition to the frequent deviations arising accidentally in the rolling parameters from the set values a certain asymmetry of this process is deliberately created in order to achieve the desired effect on its energy–force conditions or the quality parameters of rolled metal.

Therefore, the problems of the theory and technology of the process of asymmetric rolling have been the subject of many studies by renowned scientists specializing in the field of metal forming. It is not possible to refer to all work in this direction, we will mention just the most prominent names and significant publications. In their own research the authors of this book relied primarily on the results of theoretical and experimental studies of A.A. Korolev, V.N. Vydrin, V.P. Polukhin, M.Ya. Brovman, V.A. Nikolaev, V.G. Sinitsyn V.S. Gorelik, V.N. Skorokhodov, A.I. Grishkov, their colleagues, professionals of metallurgical plants in Russia, Ukraine and Kazakhstan. In this case the subjects of research are the least understood or controversial issues of the theory and technology of asymmetric rolling.

According to the authors of this book the most complete classification of different types of the asymmetry of the rolling process was proposed by M.Ya. Brovman [54]. Comprehensive results of own experiments and numerous experimental data contained in the technical literature were most fully compiled and systematized by V.A. Nikolaev [55] and V.G. Sinitsyn [56]. Of the types mentioned by M.Ya. Brovman in this book, both theoretically and experimentally investigations were carried out into the rolling process with geometric, high-speed, friction (for different coefficients of friction at the contact surfaces in the deformation zone from the upper and lower rollers) asymmetry, and also the asymmetry due to different mechanical properties of the rolled metal, in particular, because of different temperatures in the strip thickness. The main results of the work performed, proposals for the use of asymmetry effects to improve the rolling process and improve the quality of steel, as well as areas for further development of the theory of asymmetric rolling on the basis of the current capabilities of the method of slip lines, numerical analysis and computer technology are as follows.

Asymmetry of the rolling process in the current conditions is considered not so much as an inevitable consequence of imperfections of the process which should be removed, but as a means of influencing its parameters, as well as the quality of rolled metal [57–61].

So, the research by M.Ya. Brovman showed that in rolling in the mill 3000 of thick carbon steel sheets with a width of 1800–3100 mm due to the mismatch of the speeds of the rolls the rolling force equal to 27–40 MN can be reduced by 1.5–2.0 MN. Thus the thickness of the rolled sheet can be controlled by 0.20–0.25 mm only by the degree of asymmetry of the rolling process.

The results of studies by V.S. Gorelik together with experts of the Donetsk Polytechnic Institute and the Ilyich Iron and Steel Works at Mariupol' and the Azovstal Iron and Steel Works , showed the ability to control the geometric parameters of the strips (bending, flatness and thickness differences) in roughing and finishing stands of wide-strip hot rolling mills (WSHRM) using the asymmetry effects of the rolling process. Experiments were carried out in the stands with the individual and overall (through a gear cage) drive of the work rolls on WSHRM 2000, 1700 (roll asymmetry) and thick plate mills 3000 and 3600 (speed asymmetry) when rolling carbon and low-alloy steel strip plates with a thicknesses of 50–160 mm (roughing stands) and thicknesses of 2–16 mm (in the finishing mill). It was found that bending of the strip plates when exiting the rolls is associated with the uneven heating of the slab through their thickness, tilting of the strip plate at entry into the rolls and the unequal stiffness of the transmissions and drives of the rolls. The sign of the curvature of the strip plate exiting the rolls is defined by the parameters of the deformation zone: in the 'high' deformation zone the strip plate bends on the driven roll, at 'low' – the drive roll. In the non-stationary rolling stage it is possible to purposefully control bending of the strip plate by acting on the relationship roll speeds and loads according to a specific law. In wide-strip hot rolling mills asymmetric rolling with a 1–3% mismatch of the peripheral speeds of the upper and lower work rolls is used to improve the accuracy of the geometrical dimensions of rolled strips. In the stationary stage of strip rolling thicknesses of 2–16 mm in implementing asymmetric modes the longitudinal thickness difference decreased by 20–65%, the transverse difference by 15–30%, and the flatness improved by 50–60%.

These figures obviously refer to the particular case of specific experiments and do not have a general nature. However, the potential possibilities of the asymmetry of the rolling process are illustrated convincingly.

Descriptions of the process of asymmetric rolling using approximate theories for various reasons are imperfect. In [90], for example, it is shown not possible to predict experimentally in this way the dependence of the sign of the curvature of strips on the degree of deformation of strips rolled in the asymmetric conditions, the presence of several extrema on the corresponding curve, etc.

A lot of attention has been paid to the issues of the influence of the mismatch of the angular velocity of the rolls and the differences in their diameters on the output parameters of the rolling process. Theoretical studies [57–60] established and laboratory and industrial experiments [61] confirmed the non-monotonic character of the changes in some output parameters of the asymmetric rolling process even at the monotonic variation of its input parameters. For example, at the angular velocity mismatch or different diameters of the rolls the exiting end of the strip with increasing degree of deformation bends first on the roll with the lower linear speed, and then on the roll with a greater linear speed, and further, this process is repeated with decreasing amplitude. The same 'oscillation' character has a change of curvature of the exiting end of the strip at a fixed compression depending on changes of the mismatch of the roll angular velocity or the difference in the roll diameters.

Wide-strip hot rolling mills often roll strips unevenly heated in thickness. This is due to different conditions in heating furnaces in heating the upper and lower surfaces of slabs, with different cooling conditions of the sides of the strip plate during their movement motion on the roller table. When rolling the metal the strip coming out of the rolls is curved, whereby the front end of it can jam in the delivery guide or between the rolls of a roller table, creating an emergency situation. In this connection, it is interesting to determine the radius of curvature of the strip exiting from the rolls in dependence on the temperature difference of the upper and lower surfaces of the deformable metal, as well as the choice of the strain–rate conditions of asymmetric rolling, which even at a significantly uneven temperature field provide exit of the straight strip from the rolls. As an example, we consider the solution of this problem in the approximate formulation using the results based on the characteristic analysis of the asymmetric rolling process. We keep the notations and solutions given in [59].

The scheme of deformation of the strip, the slip line field and a hodograph are shown in Fig. 5.1. Areas adjacent to the rolls rotate as a rigid body with angular velocities ω_u and ω_l. Here u and l – indices of the upper and lower rolls respectively. After rolling, the strip also rotates as a rigid body with angular velocity Ω, so the output slip lines are part of the circles. Each of them is multiplied by the scale factors and appears twice on the hodograph with a constant difference of the radii of curvature. If the strip unevenly heated in thickness

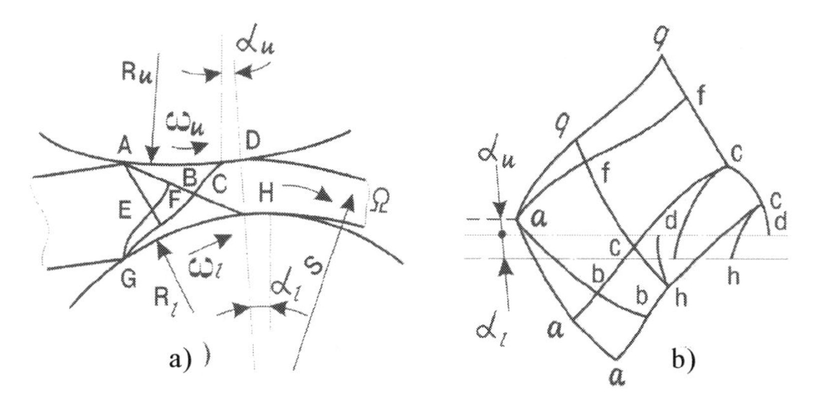

Fig. 5.1. Field of slip lines (a) and hodograph (b) in asymmetric hot rolling. Notations are given in the text,

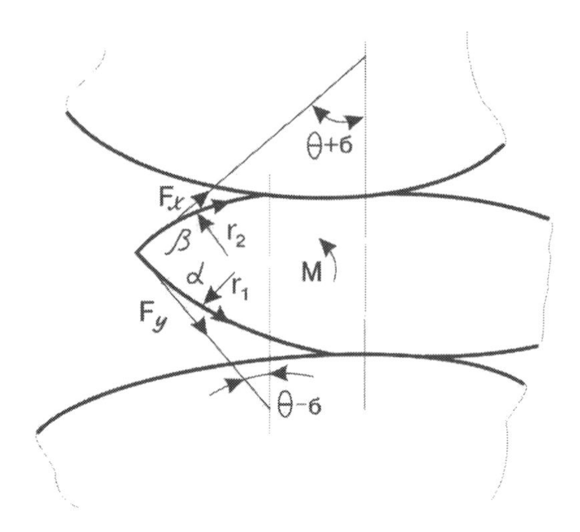

Fig. 5.2. Output slip lines and the forces acting on the exiting end of the strip. The notations are given in the text.

does not feel the impact of forces and moments at the exit, or they are small, the radii of curvature of the circular arcs of the slip lines (Fig. 5.2) satisfy the condition:

$$r_{\beta} / r_{\alpha} = k_{u} / k_{l} , \tag{5.1}$$

where r_{α}, r_{β} are the radii of curvature of the output α and β slip lines; k_u, k_l are the limiting shear stresses of the strip material on the upper and lower surfaces respectively. If we denote $k_u / k_l = \mu_t$, then

$$r_{\alpha} = r ; r_{\beta} = \mu_{t}^{r} . \tag{5.2}$$

When μ_t = 1.0–1.4 the expression connecting strip thickness h and the parameters r and μ_t takes the form:

$$h = r\left(\mu_{t} + 1\right)\left(1 - \sqrt{2}/2\right). \tag{5.3}$$

In the case where the strip is heated uniformly, $\mu_t = 1$, $h = r\left(2 - \sqrt{2}\right)$.

Consider the effect of the relationship between the angular velocities ω_u and ω_l, the roll radii R_u and R_l and the temperatures of the layers of the strip t_u and t_l on the radius and the direction of bending of the strip at the exit of the rolls. We assume for definiteness that $R_u\omega_u > R_l\omega_1$ (it will be shown later that this condition only simplifies the reasoning, and in the final formulas is immaterial). The strip can exit the rolls by one of the three options (Fig. 5.3):

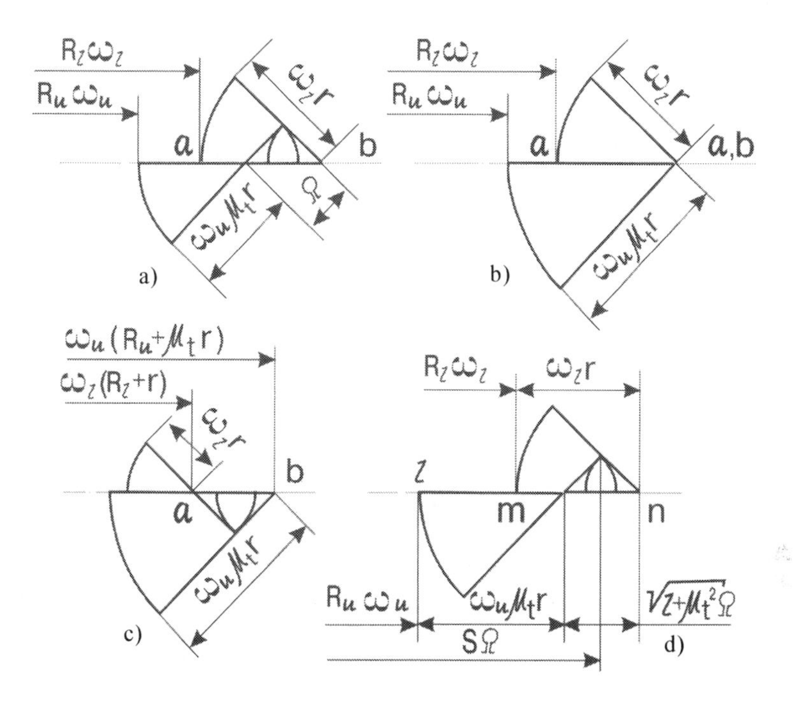

Fig. 5.3. Exit options of the strip from the rolls. Legend – in text.

I – *ob* > *oa* (Fig. 5.3 *a*), i.e. $\omega_l(R_l + r) > \omega_u (R_u + \mu_t r)$, the rolled strip is bent toward the bottom roll;

II – *ob* = *oa* (Fig. 5.3 *b*), i.e. $\omega_u(R_u + r) = \omega_l(R_l + \mu_t r)$, the strip remains straight with a significant difference in the parameters;

III – *ob* < *oa* (Fig. 5.3 *c*), i.e. $\omega_l(R_l + r) < \omega_u(R_u + \mu_t r)$, the strip is bent towards the upper roller.

If *S* is the average radius of curvature of the strip after rolling, then $S\Omega$ is the linear velocity of the material points on the contact line of circles with radii *r* and μ_t. For the output of the hodograph (Fig. 5.3 *d*) the following relations will hold:

$$R_l\omega_l - R_u\omega_u = \left[\mu_t\omega_u + \sqrt{1+\mu_t^2\Omega}\right] r - \omega_l r;$$

$$R_l\omega_l - R_u\omega_u = \mu_t\omega_u r\left(\omega_l + \sqrt{1+\mu_t^2\Omega}\right) r;$$

$$\text{whence } \Omega = \frac{\omega_l\left(R_l + r\right) - \omega_u\left(R_u + \mu_t r\right)}{\sqrt{1+\mu_t^2}}, \qquad (5.4)$$

and the negative value of Ω corresponds to bending towards the upper roller. Further for the options I and III

$$S\Omega = R_l\omega_l + \omega_l r - \Omega r / \sqrt{1+\mu_t^2}$$

$$S\Omega = R_u\omega_u + \mu_t\omega_u r + \Omega r / \sqrt{1+\mu_t^2}$$

where

$$2S\Omega = \omega_l\left(R_l + r\right) + \omega_u\left(R_u + \mu_t r\right) \qquad (5.5)$$

Substituting into the expression (5.5) the value of Ω from the dependence (5.4), we obtain:

$$S = \frac{r\sqrt{1+\mu_t^2}\left[\omega_l\left(R_l + r\right) + \omega_u\left(R_u + \mu_t r\right)\right]}{2\left[\omega_l\left(R_l + r\right) - \omega_u\left(R_u + \mu_t r\right)\right]}; \qquad (5.6)$$

where $r = h\left(2+\sqrt{2}\right)/\left(\mu_t + 1\right)$. (5.7)

In the case where the strip temperature is the same along the height $\mu_t = 1$, the formulae (5.6) and (5.7) take the following form:

$$S = \frac{r\left[\omega_l\left(R_l + r\right) + \omega_u\left(R_u + r\right)\right]}{\sqrt{2}\left[\omega_l\left(R_l + r\right) - \omega_u\left(R_u + r\right)\right]}; \qquad (5.8)$$

$$r = h\left(1 + 1/\sqrt{2}\right). \qquad (5.9)$$

In the formulae (5.6) and (5.8), the positive value of S corresponds to the curvature of the strip to the lower roll, negative – to the top roll, and the earlier restriction of the greater peripheral speed points of the lower roll is now immaterial. Finally, it should be borne in mind that the vanishing of the denominator to zero in (5.6) and (5.8) means that the strip remains straight at exit from the rolls.

To assess the range of applications of the formulas the authors of [59] used the experimental data characterizing the kinematic parameters of the process of asymmetric hot rolling and, in particular, bending of the end of the strip coming out of the rolls.

Comparison of calculated and experimental values of the curvature of the rolled strip showed that the obtained approximate solution makes it simple to determine the direction and curvature of the strip end coming out of the rolls, taking into account differences in the diameters of the rolls, and misalignment of the angular velocities and uneven heating the strip height. However, the calculations using these formulas do not reflect all the features of the curvature of the strip: bending direction is determined by the sign of the denominator in (5.8) and is independent of the amount of reduction, although in reality this dependence exists. The calculation results gives values give close to real curvature values only for large reductions. Throughout the reduction range the calculation results are in satisfactory agreement with the experimental data only at small (1–4%) differences in the diameters and mismatched roll angular velocities. Only small differences in the parameters of the rolling process the direction of bending of the strip remains constant throughout the entire reduction range.

At a significant difference in the roll diameters and (or) mismatch of the angular velocities the nature of strip bending (from the experimental data) becomes oscillatory: at a small reduction the strip is bent to one of the rolls, when the reduction is increased – to another. Unfortunately, the resulting formula does not reflect this feature. The curvature calculated with the help of these equations has a constant sign and corresponds to the experimental values when at reductions of 20–40%.

Despite these shortcomings, these formulas are of some interest for practice. They can be used for small mismatches of the angular velocity of the rolls and differences of their diameters. Furthermore, for large values of these ratios, the results of the calculation (in absolute value) well characterize the amplitude of the deviation of the strip to the upper or lower roll.

General results can be obtained using approaches based on the fundamentals of the theory of plasticity which taking into account to the maximum extent all the factors affecting the process of asymmetric rolling and do not contain any *a priori* assumptions about the nature of its course. Moreover, for reliable determination of the energy–power and kinematic asymmetric rolling process parameters the methods of the theory of plasticity should use high-speed computing. This approach to the mathematical description of the process of asymmetric rolling by methods of slip lines and non-linear programming was developed[1] and presented in our papers [57–61]. Before describing all the mathematical calculations of this solution we denote the basic postulates and approaches to the solution.

As will be shown below, the use in this problem of the flow theory leads to the necessity of solving a system of two (for the plane strain problem) differential equations of hyperbolic type, which is carried out by the slip lines. Through the application of the matrix operators the differential equations are reduced to a system of non-linear algebraic equations describing the field of slip

[1]Jointly with E.V. Binkevich and A.K. Golubchenko.

lines in the deformation zone. Solution is obtained by the nonlinear programming methods, thereby eliminating the time-consuming process of successive approximations, normal for this kind of problems. The designed method[2] also allows without additional complications to solve the inverse problem of finding the values or boundaries of the change of the initial values, providing the required output parameters (forces, moments on the rolls, the curvature of the rolled strip, metal properties, etc.). In the end it can be used to control the rolling parameters by adjusting the degree of asymmetry of the process.

As an illustration of the possibilities of the solution obtained in our paper [57] using the theory of the slip lines attention will be given to the effect of the angle of entry of slabs into the rolls on the parameters of the asymmetric rolling process, and in particular, bending of the front end of the rolled strip exiting the deformation zone. A specific numerical solution for the case of rolling when the slab is sent to the rolls under an angle to one of them showed that the small (up to 5°) angles of inclination of the slab and without rear tension the strip at the exit from the deformation zone is bent to the same roller. Rear tension reduces the amplitude of bending the strip leaving the rolls without changing its sign. A detailed analysis of this case of asymmetric rolling is given below.

In [60] there are the results of the numerical solution by the examined method of the problem of changing the curvature of the rolled strip in dependence on the variation of the radii of the upper and lower rolls and also in dependence upon the relative reduction ratio at the mismatch of the angular velocities of the upper and lower rollers with the same radii. As will be seen, the results of calculation correspond to the available experimental data [61] for the change of the sign of the curvature depending on the ratio of the radii of the rolls.

Formulation of the problem, the algorithm and the results of calculation of energy–force and kinematic asymmetric rolling process parameters by the slip lines method will be examine the solutions, analysis, calculations and generalizations described in detail in [61].. At the same time we preserve the notations adopted in [61].

[2]Developed by E.V. Binkevich.

5.2. Calculation of process parameters of the asymmetric rolling by the method of slip lines

5.2.1. Matrix-operator version of the method of slip lines

Given the complexity of the process of asymmetric rolling, its analytical description requires the use of modern methods of the theory of plasticity with the widespread use of numerical algorithms and computational tools. As noted above, in calculations of the symmetric process sufficiently reliable results can be obtained using the method of slip lines. The application of the matrix-operator variant of this method makes it possible to use it efficiently in the calculation of asymmetric processes.

The grids of slip lines for rolling sheets and strips are usually very complex. Their construction requires solving indirect type problems when initially the form of neither slip lines not their image on the plane of velocities (hodograph) is known. Beginning with the work of Alexander and until recently the grids slip lines were built step by step, starting from some initial slip line, using a method developed by Hill. If the boundary conditions are such that the initial shape of the slip line can not be established, then the trial and error method should be used to determine it, and then, knowing it, the entire grid of slip lines should be constructed. This procedure is quite laborious even solving symmetric rolling problems, and the presence of asymmetry makes it practically impossible to use this method in the classic version. Ewing proposed a method for constructing grids of slip lines based on the decomposition of the radii of curvature into double power series. Further development of Ewing's ideas and the introduction of the system of matrix operators allowed to create a method in which the principles of superposition of grids developed by Hill is implemented using simple matrix operations [61][3].

Below are the basic equations of the planar plastic flow and the ratio of the matrix–operator method used to solve problems of asymmetric rolling.

5.2.2. Basic equations of planar plastic flow

Deformation is considered planar (planar plastic flow) when the material particles move only in planes perpendicular to a certain fixed

[3]In the book [61] used as the basis in the explanation of the material in this section, L.V. Binkevich used sections of his dissertation.

direction, e.g., the z-axis of the Cartesian coordinate system x, y, z, and displacements do not change from plane to plane. The state of the body is determined by the displacement speeds

$$v_z = 0;\ v_x = v_x(x, y);\ v_y = v_y(x, y); \tag{5.10}$$

and strain rates

$$\begin{gathered} \dot{\varepsilon}_x = \frac{\partial v_x}{\partial x};\ \dot{\varepsilon}_y = \frac{\partial v_y}{\partial y};\ 2\dot{\varepsilon}_{xy} = \frac{\partial v_x}{\partial y} + \frac{\partial v_y}{\partial x}, \\ \dot{\varepsilon}_z = \dot{\varepsilon}_{yz} = \dot{\varepsilon}_{zx} = 0, \end{gathered} \tag{5.11}$$

where v_x, v_y, v_z are the components of the velocity vector along the x, y, z axis; $\dot{\varepsilon}_y, \dot{\varepsilon}_x, \dot{\varepsilon}_{xy}\ \dot{\varepsilon}_z, \dot{\varepsilon}_{zx}, \dot{\varepsilon}_{yz}$ are the components of the strain rate tensor.

Since the strain rates ε_z, ε_{zx}, ε_{yz} are equal to zero, then for an isotropic material obviously the tangential stresses τ_{xy}, τ_{yz}, are also equal to zero so that the stress state determined by the components σ_x, σ_y, σ_{xy}, σ_z, and stress σ_z is one of the principal stresses. The remaining principal stresses lying in the plane xy, are defined as usual by the characteristic equation and are:

$$\sigma_{1.2} = \frac{\sigma_x + \sigma_y}{2} \pm \frac{1}{2}\sqrt{(\sigma_x - \sigma_y)^2 + 4\tau_{xy}^2} \tag{5.12}$$

According to the flow rule for the rigid-perfectly plastic body

$$\dot{\varepsilon}_{ij} = \lambda \frac{\partial f}{\partial x_{ij}}; \tag{5.13}$$

where $\dot{\varepsilon}_{ij}$ is the strain rate tensor; λ is the undetermined factor; f is the load surface.

For arbitrary σ independent of hydrostatic stress, the flow condition because of the equality $\varepsilon_z = 0$ gives

$$\frac{\partial f}{\partial \sigma} = 0,\ \frac{\partial f}{\partial \sigma_z} = 0; \tag{5.14}$$

which at the principal axes, together with the yield condition, gives three equations

$$\frac{\partial f}{\partial \sigma_1} + \frac{\partial f}{\partial \sigma_2} = 0;\quad \frac{\partial f}{\partial \sigma_3} = 0;\quad f(\sigma_1, \sigma_2, \sigma_3) = 0. \tag{5.15}$$

Joint satisfaction of these equations is possible only if

$$d\sigma_1 = d\sigma_2. \tag{5.16}$$

After integration, taking into account the expressions (5.12), this condition takes the form

$$\left(\sigma_x-\sigma_y\right)^2+4\tau_{xy}^2=4k^2, \tag{5.17}$$

where $k = \tau_0$ is the value of the shear stress when tested in pure shear, corresponding to the transition of the material from the rigid to plastic state.

Thus, all possible yielding conditions for plane rigid–plastic deformation are reduced to (5.17), which is based on the associated law (5.13) and leads to the relations

$$\dot{\varepsilon}_x=2\lambda\left(\sigma_x-\sigma_y\right),\ \dot{\varepsilon}_y=2\lambda\left(\sigma_y-\sigma_x\right),\ \dot{\varepsilon}_{xy}=4\lambda\tau_{xy}, \tag{5.18}$$

and in the calculation of derivatives the term $4\tau_{xy}^2$ is represented as $2\tau_{xy}^2+2\tau_{yx}^2$.

Eliminating the undetermined multiplier λ from (5.18), we obtain

$$\frac{\sigma_y-\sigma_x}{\tau_{xy}}=\frac{\dot{\varepsilon}_x-\dot{\varepsilon}_y}{\dot{\varepsilon}_{xy}};\ \dot{\varepsilon}_x+\dot{\varepsilon}_y=0. \tag{5.19}$$

The second expression in (5.19) is the condition of incompressibility, and the first expression means matching of the areas with maximum shear stress and maximum shear rate. In fact, the components of bivalent tensor in the Cartesian coordinates (x, y), rotated relative to the original (x, y), are defined by

$$\sigma_{p'q'}=l_{p'i}l_{q'j}\cdot\sigma_{ij};\varepsilon_{p'q'}=l_{p'i}l_{q'j}\cdot\varepsilon_{ij}, \tag{5.20}$$

where $l_{n'm}$ are the cosines between m'-th axis of the new system and the n-th axis of the old system.

In the planar case, when the new system of coordinates (t, n) is rotated relative to the old (x, y) by an angle Ψ, we have

$$\cos(t,x)=\cos(n,y)=\cos\Psi,$$
$$\cos(t,y)=\cos(n,x)=\sin\Psi.$$

The stress components in the new system will be:

$$\sigma_t=\frac{\sigma_x+\sigma_y}{2}+\frac{\sigma_x-\sigma_y}{2}\cos 2\psi+\tau_{xy}\sin 2\psi;$$
$$\sigma_n=\frac{\sigma_x+\sigma_y}{2}-\frac{\sigma_x-\sigma_y}{2}\cos 2\psi-\tau_{xy}\sin 2\psi;$$
$$\sigma_{tn}=\tau_n=\frac{\sigma_x-\sigma_y}{2}\sin 2\psi+\tau_{xy}\cos 2\psi. \tag{5.21}$$

The direction of the axis t with the maximum tangential stress is determined from the maximum of the last expression $\partial\tau_n / \partial\psi = 0$, which gives

$$2\text{tg}2\theta = \frac{\sigma_y - \sigma_x}{\tau_{xy}}, \tag{5.22}$$

where θ is the angle Ψ at $\tau_n = \tau_{\max}$.

The same equation with the replacement of $\sigma_x, \sigma_y, \tau_{xy}$ by $\varepsilon_x, \varepsilon_y, \varepsilon_{xy}$, will also hold for the angle with the maximum shear rate. Equating the right sides of these two expressions for the angle θ, we obtain the first formula of (5.19).

At the axes t, n, rotated with respect to x, y through angle $\Psi = 0$ in view of (5.21) (5.22) (5.17) (5.18):

$$\tau_n = \sigma_n = \frac{1}{2}\left(\sigma_x + \sigma_y\right) = \sigma; \quad \tau_{tn} = \pm k,$$
$$\dot{\varepsilon}_t = \dot{\varepsilon}_n = \dot{\varepsilon}_z = 0. \tag{5.23}$$

In this case $\dot{\varepsilon}_z = 0$ for the initial assumptions, and the normal strain rates $\dot{\varepsilon}_t$ and $\dot{\varepsilon}_n$, which are expressed through $\dot{\varepsilon}_x, \dot{\varepsilon}_y, \dot{\varepsilon}_{xy}$ as well as σ_n, σ_t, σ_{nt}, through σ_x, σ_y, τ_{xy}, are equal to zero due to the incompressibility condition.

From equations (5.23) it follows that the stress state at $\Psi = 0$ is reduced to the imposition of the plane hydrostatic pressure σ on the plane shear stress with intensity k, while the instantaneous deformation of the element is reduced to pure shear or sliding in the direction of the maximum shear stress.

The equilibrium conditions in the absence of mass forces give two equations in the xy axes:

$$\frac{\partial\sigma_x}{\partial x} + \frac{\partial\tau_{xy}}{\partial y} = 0; \quad \frac{\partial\sigma_y}{\partial y} + \frac{\partial\tau_{xy}}{\partial x} = 0, \tag{5.24}$$

so that for the eight unknowns $\sigma_x, \sigma_y, \tau_{xy}, \dot{\varepsilon}_{xy}, \dot{\varepsilon}_y, \dot{\varepsilon}_y, v_x, v_y$ there are eight equations (5.11), (5.17), (5.19), (5.24).

All these equations are valid in the zone of plastic deformation, and the equations (5.17), (5.24) close the problem in stresses. So that at the appropriate boundary conditions, it can be statically determinate. In the rigid zone the strain rates must vanish, and as regards the stress state, in addition to the equilibrium equations (5.24), we can only say that

$$\left(\sigma_x - \sigma_y\right)^2 + 4\tau_{xy}^2 \le k^2 \tag{5.25}$$

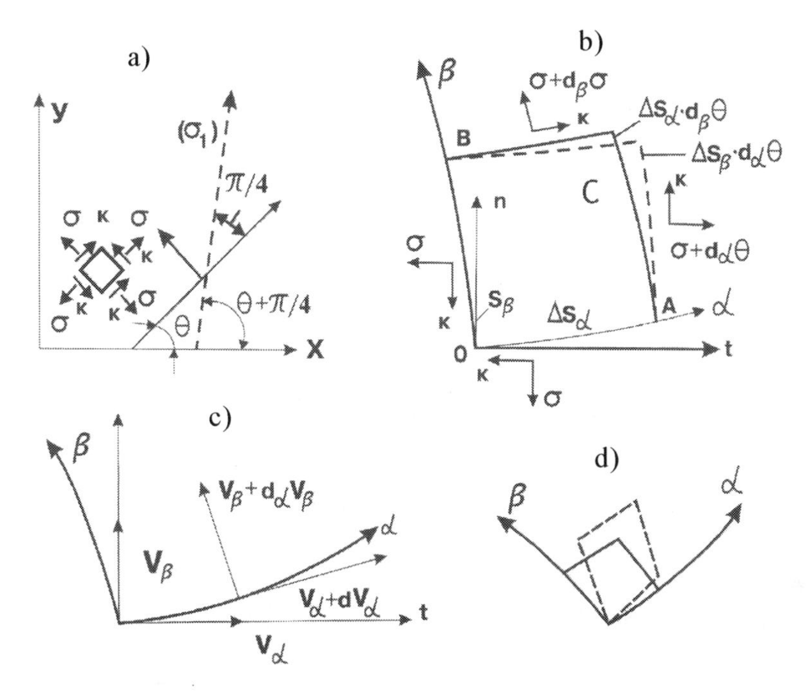

Fig. 5.4. Basic relations for the field of slip lines: **a** – relative positioning of slip lines and main directions; Hencky theorem; **b** – equilibrium of the element defined by the slip lines; **c** – constancy of the extensions of the slip lines; Geiringer ratio; **d** – the nature of the deformation of the unit cell of the field of slip lines.

and the problem of voltage becomes unlocked.

5.2.3. Relations along the slip lines

We introduce the maximum shear stress trajectories, i.e. lines whose tangents at each point coincide with the direction as the maximum shear stress. As noted, these lines are simultaneously the slip lines (lines of maximum shear rate) and (due to the fact that for them $\psi = 0$ (Figs. 5.4 *a*)) satisfy the equation

$$\frac{dy}{dx} = tg\,\theta. \tag{5.26}$$

According to the law of pairing of the shear stresses slip lines will also have lines orthogonal to them

$$\frac{dy}{dx} = -ctg\,\theta. \tag{5.27}$$

The family of slip lines (5.26) with be given the name α-lines, (5.27) the β-lines and we consider the equilibrium conditions of the OASV element in curvilinear coordinates α, β (Fig. 5.4 *b*). By virtue of a special kind of the stress state in the grid of slip lines in the projection onto the tangent to the line α (at 0), we obtain

$$-\sigma\Delta S_\beta - k\Delta S_\alpha + (\sigma + d_\alpha\sigma)(\Delta\sigma_\beta + d_\alpha\Delta S_\beta) + k(\Delta S_\alpha + d_\beta\Delta S_\alpha) -$$
$$k(\Delta S_\beta + d_\alpha\Delta S_\beta)d_\alpha\theta - (\sigma + d_\beta\sigma)(\Delta S_\alpha + d_\beta\Delta S_\alpha)d_\beta = 0,$$

where d_α marks the increment of the value along the line α. Discarding the values of higher order (such as $d_\alpha\sigma(\Delta S_\beta)$ or $d_\beta\sigma(\Delta S_\alpha)d_\beta\theta$ and taking into account (Fig. 5.4 *b*) that $d_\alpha\Delta S_\beta = \Delta S_\alpha d_\beta\theta$, $d_\beta\Delta S_\alpha = -\Delta S_\beta d_\alpha\theta$, we get $(d_\alpha\sigma - 2kd_\alpha\theta)\Delta S_\beta = 0$

or
$$d_\alpha(\sigma - 2k\theta) = 0. \tag{5.28}$$

This implies that along each α-line the combination σ-2*k*θ should remain constant and may change from one line to another. A similar situation holds for the combination σ-2*k*θ along the β line. Thus, we have the Hencky relations

$$\sigma - 2k\theta = \xi;\ \sigma + 2k\theta = \eta, \tag{5.29}$$

where ξ is constant along the line α, η – along the line β.

Relations (5.29) include the parameter θ, defining the geometry of the slip grid, in connection with which the latter has a number of specific properties (5.17).

According to formula (5.23) the normal components $\dot{\varepsilon}_n$ and $\dot{\varepsilon}_t$ of the strain rate tensor in the local Cartesian coordinate system (*t*, *n*), which coincides at the given point with the tangent to slip lines α and β, vanish. If the grid of the slip lines was not curved, it would follow that the elements of the α- and β-lines do not get any longer. In reality, this means that only the projections of the elements of the slip lines of the line α on the axis *t* and line β on the axis *n* do not grow longer (Fig. 5.4 *c*). Writing these conditions for an element ΔS_α, we obtain

$$v_\alpha = v_\alpha + d_\alpha v_\alpha - (v_\beta + d_\alpha v_\beta)d_\alpha\theta.$$

After discarding the terms of higher order

$$d_\alpha v_\alpha - d_\alpha v_\beta = 0.$$

Similarly, for the element ΔS_β

$$d_\beta v_\beta - d_\beta v_\alpha = 0.$$

Thus, for the displacement velocities in the grid of slip lines in we have the Geiringer relationships

$$dv_\alpha - v_\beta d\theta = 0 \text{ along the } \alpha\text{-line};$$
$$dv_\beta - v_\alpha d\theta = 0 \text{ along the } \beta\text{-line}. \tag{5.30}$$

In contrast to the Hencky relations, the Geiringer equations are not integrable in a general case. In the particular case where one set of lines, e.g. α, consists of straight lines, along them v_α = const = 0. That is, straight slip lines are cannot be extended. Obviously, $C = c(\theta)$ so that everywhere in the grid the velocity components have the form

$$v_\beta = \varphi(\theta) + \psi(S_\alpha); \quad v_\alpha = \varphi(\theta),$$

where S_α is the coordinate along the α-lines. Unknown functions φ and ψ must be found from the boundary conditions.

From a practical point of view it is interesting to note that the square element in the system (αβ) is subject to a positive shift ($\tau = k > 0$). So it tends to stretch out into a rhomb in the direction of the bisectrix of the coordinate angle (Fig. 5.4 *d*). This allows us to properly orient the grid of slip lines in concrete problems.

5.2.4. Formulation of boundary-value problems

From formulas (5.17), (5.22) and (5.23) it follows that σ_x, σ_y, τ_{xy} in the *xy* axes are related to the parameters σ and τ in the grid of slip lines by the dependences

$$\sigma_x = \sigma - k\sin 2\theta,$$
$$\sigma_y = \sigma + k\sin 2\theta,$$
$$\tau_{xy} = k\cos 2\theta. \tag{5.31}$$

Using them, we can get the values of normal and tangential stresses on the area arbitrarily inclined to the axes *X, Y*. If the tangent to this area forms the angle ψ with the *X* axis, then by virtue of (5.21)

$$\sigma_n = \sigma + k\sin 2(\theta - \psi),$$
$$\tau_{tn} = \tau_n = k\cos 2(\theta - \psi). \tag{5.32}$$

Thus, if along a given line *L* we define the normal σ_n and tangential τ_n stress components, the parameters σ and θ along *L* can be considered known

$$\theta = \psi \pm \arccos\frac{\tau_n}{k} + m\pi,$$
$$\sigma = \sigma_n - k\sin 2(\theta - \psi). \tag{5.33}$$

According to the Hencky equations if the line L on the given section σ_n, τ_n does not coincide with the slip line (Cauchy problem), the solution for the parameters σ, τ is uniquely defined in the triangle defined by the line segment L and the slip lines α and β. Indeed, each of the two slip lines of this triangle extends to the line L, and therefore the corresponding constants ξ and η and, consequently, in view of (5.29) the values of σ, θ are determined on them. If the line segment L coincides with the slip line, then the solution in its neighborhood can not be defined. The lines for which the solution of the Cauchy problem is not possible bear the name of the characteristics of this system of equations.

The slip lines are characteristics of the system of Hencky equations (5.29) or, equivalently, the non-linear system of equations (5.24), (5.25). They also are the characteristics of the system of Geiringer equations (5.30) for the displacement velocities. Indeed, when defining the functions υ_α, υ_β, for example, along the lines α, increase of these functions in the direction β remains uncertain, since for these two values there is only one equation of the system (5.30). Since two slip lines pass through each point on the plane of independent variables, all these systems of equations are hyperbolic.

To find a solution in the region adjacent to the characteristic, it is necessary to impose additional conditions, in particular, specify the desired functions also on another characteristic (Riemann problem) or on the characteristic and non-characteristic line (the mixed problem).

In technological problems the boundary conditions are more complex. The boundary of the deformable region consists of several (in a general case – the surfaces) at which different boundary conditions are set. Usually there are three types of boundaries: region of the content of the tool and the body, the boundary of the rigid and deformable areas, and the boundary of the deformable body free from the stresses (5.11).

In connection with this, the solution of problems is, firstly, reduced to the solution of any major problem, and breaks down into a chain of solutions of such problems. And secondly, it is associated with the need to ‘design’ the field of characteristics allowing satisfaction of the boundary conditions, changing along the length of the contour of the deformable body. Let us consider the latter circumstance. We determine the radius of curvature of the slip lines by the equalities (Fig. 5.4 *c*):

$$R=\frac{\partial S_\alpha}{\partial\theta}\quad S=\frac{\partial S_\beta}{\partial\theta}, \tag{5.34}$$

where ∂S_α, ∂S_β are the elements of the lines α, β. These values are positive if the corresponding centres of curvature are in positive directions of the lines α, β.

Considering the curvilinear quadrangle OACD as elemental, we obtain

$$dS=dS_\alpha\,, dR=-\,dS_\beta\,, \tag{5.35}$$

which in view of (5.34) gives

$$dS+Rd\theta=0(\text{on}\,\alpha);\; dR-Sd\theta=0(\text{on}\,\beta). \tag{5.36}$$

According to (5.35), moving along the line α in the direction of the concavity of the line β, we will arrive at the point where $S = 0$, i.e., the return point. Because the characteristics can not be continued beyond the envelope (the locus of return points), the radii of curvature do not change their sign in the continuity region.

Due to breaks in the boundary conditions, there are situations when the radius of curvature of a family of lines, such as α, shows a discontinuity in going through some characteristics of another family, such as β. On such a line the derivative of the average pressure in the direction α is discontinuous, and hence there are also discontinuities in the derivatives of the normal stresses. Indeed, the Hencky equation (5.29) can be rewritten as

$$\frac{\partial\sigma}{\partial S_\alpha}=\frac{2k}{R};\quad \frac{\partial\sigma}{\partial S_\beta}=\frac{2k}{S}.$$

As you can see, the jump $R(S)$ causes a jump $\frac{\partial\sigma}{\partial S_\alpha}\left(\frac{\partial\sigma}{\partial S_\beta}\right)$. According to relations (5.31) it follows that jumps $\frac{\partial\sigma}{dS_\alpha},\frac{\partial\sigma}{\partial S_\beta}$ cause jumps of the derivatives σ_h, σ_u. Thus, the discontinuity of the curvature of the characteristic of some family is accompanied by discontinuities of derivatives of stress (weak stress break) in the direction of this line.

The jump in the radius of curvature R on the line β (respectively S on the line α) remains constant as we move along this line. Indeed, let R_1 and R_2 are the values of R before and after the line break. Along β (5.36) have $dR_l = Sd\theta$, $dR_2 = Sd\theta$. Subtracting these equations term by term and taking into account the fact that S is continuous, we obtain $d(R_2 - R_1) = 0$ along β. Or $R_2 - R_1 = \text{const}$ along β.

Thus, if there was a jump of curvature it does not disappear, and the grid of the characteristics is usually in the form of a quilt. At the boundaries of patches, i.e. at the characteristics, there is a discontinuity of the curvature of the other family of the characteristics. This leads to the concept of the design of the field characteristics or mutual location of areas where the curvatures of the characteristics are continuous.

Due to the ambiguity of the solutions based on the flow theory, the application of the method of slip lines involves mandatory use of the successive approximation procedure and consists of the following steps: setting up (with the accuracy up to some geometrical parameters) the form (structure) of the field of slip lines, the solution of boundary-value problems for individual cells of this field (i.e. finding all kinematic and static parameters in each of them), checking (the final stage) fulfillment of the conditions ensuring the physical feasibility of solutions, repetition (if necessary) of the entire cycle starting with the change of the initial geometric parameters that characterize the field.

For all the diversity of approaches to the solution of boundary value problems (geometric method, numerical integration, using the Riemann method), this procedure remains unchanged and is the weakest link when solving by the method of slip lines more or less complex problems of plastic flow and, in particular, in the construction of algorithms, oriented to the control of the process parameters associated with the plastic deformation of metals.

These shortcomings are not found in the method based on the use of matrix operators to describe the field of slip lines, allowing it to reduce the problem of finding it to the solution to the non-linear matrix equation. This method can also be easily algorithmised.

5.2.5. The matrix–operator method of constructing slip line fields

As we have seen, in plane deformation the equations for the stress distribution and velocity distributions are hyperbolic. The characteristics determine the trajectories of the velocity of the maximum angular strains and the orthogonal grid of the slip lines (Fig. 5.5 *a*). Any point P within the grid can be determined using either

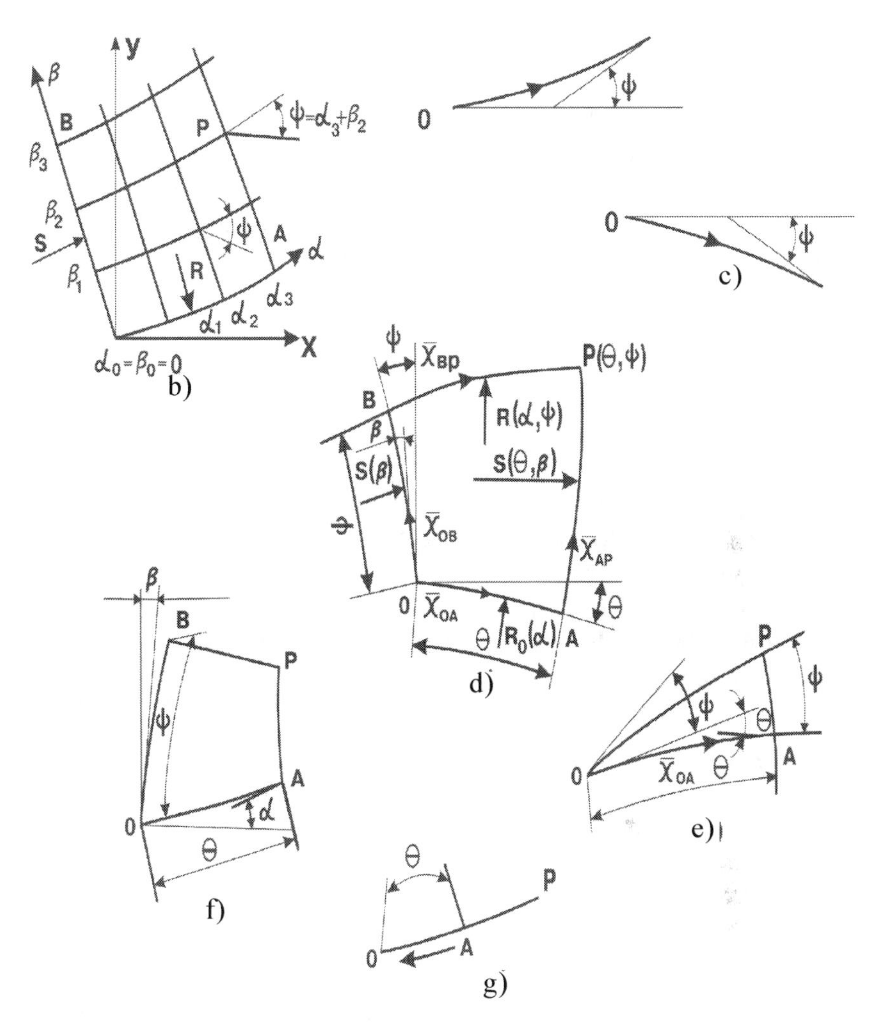

Fig. 5.5. Elements of matrix representations of fields of slip lines: **a** – grid of slip lines; **b** – **c** – directions of rotation of the tangent corresponding to the positive (b) and negative (c) values of the radius of curvature; **d** – the quadrilateral cell with convex basic characteristics; **e** – the same basic with the concave basic characteristics; **f** – centered fan; **g** – shift of the point along the slip line.

the Cartesian (xy) or characteristic ($\alpha\beta$) coordinates. In the second case the coordinates of point P are represented by the angle ϕ (in the anticlockwise direction) between the directions of the respective slip lines at points 0 and P. In Fig. 5.5 a the point P has the coordinates α_3, β_2.

When constructing the matrix operators the slip lines are described using their radii of curvature and, moreover, the so-called moving coordinates. In so doing the following definitions are made. The starting point 0 (Fig. 5.5 *b*, *c*) is indicated on the slip line and the direction on the curve is always considered to be positive from the starting point and the slope is always considered positive and independent of the direction of rotation of the tangent. The radius of curvature of the slip line ρ is given by

$$\frac{1}{\rho}=\pm\left|\frac{d\phi}{dS}\right|,$$

where ϕ is the slope of the tangent to the local tangent at the starting point, S is the arc length.

In this case, the sign of the radius of curvature is positive if ϕ increases along the slip line counterclockwise (Fig. 5.5 *b*), and negative if ϕ increases as we move along the slip line in the clockwise direction (Fig. 5.5 *c*).

With this rule of the signs the second Hencky theorem for a pair of divergent slip lines (Fig. 5.5 *d*) has the form

$$\frac{dS}{d\alpha}=-R\frac{dR}{d\beta}=-S. \tag{5.37}$$

On line OA $\phi=\alpha$, α increases counterclockwise so that R is negative; on line OB $\phi=\beta$, angle β increases counterclockwise, so that S is positive. According to (5.37) the radii R and S increase during movement along the β- and α-lines respectively, which corresponds to the geometric picture: the absolute value of $R(\alpha, \psi)$, $S(\theta, \beta)$ for divergent slip lines is larger than for respectively $R(\alpha)$, $S(\beta)$.

For a pair of converging slip lines (Fig. 5.5 *e*) R is positive, S is negative, the radii decrease with increase of the distance from the baselines, and the second Hencky theorem has the form

$$\frac{\partial S}{\partial\alpha}=R;\quad \frac{\partial R}{\partial\beta}=S. \tag{5.38}$$

Thus, the equation for the curvature radii can be represented in general form

$$\frac{\partial R}{\partial\beta}=\lambda_1 S;\quad \frac{\partial S}{\partial\alpha}=\lambda_2 R, \tag{5.39}$$

where $\lambda_1=\lambda_2=\pm 1$ depending on the relative orientation of the curves α, β.

The equations (5.39) are hyperbolic and the solution of the initial characteristic problems for them can be represented as the double power series

$$R(\alpha,\beta)=\sum_{m,n=0}^{\infty}\left[\alpha_n\alpha_{m+n}(\lambda_1\lambda_2\beta)_m+\lambda_1 b_n(\lambda_1\lambda_2\alpha)_m\beta_{m+n+1}\right],$$

$$S(\alpha,\beta)=\sum_{m,n=0}^{\infty}\left[\lambda_2\alpha_n\,\alpha_{m+n+1}(\lambda_1\lambda_2\beta)_m+b_n(\lambda_1\lambda_2\alpha)_m\beta_{n+n}\right], \tag{5.40}$$

where the expressions containing Greek letters with subscripts denote the powers of the appropriate quantities of the form $\varphi_m=\frac{\varphi_m}{m!}$; a_n, b_n – coefficients of the power series.

For the case shown in Fig. 5.5 *d*, $\lambda_1=-1$, $\lambda_2=-1$ the point *P* has the coordinates $\alpha=\theta$, $\beta=\psi$, and the expressions (5.40) for the radii of curvature of the slip lines, passing through the point *P*, take the form

$$\begin{aligned}R(\theta,\psi)&=\sum_{m,n=0}^{\infty}\left[\alpha_n\frac{\theta^{m+n}}{(m+n)!}\frac{\varphi^m}{m!}-b^n\frac{\theta^m}{m!}\frac{\varphi^{m+n+1}}{(m+n+1)!}\right],\\ S(\theta,\psi)&=\sum_{m,n=0}^{\infty}\left[-\alpha_n\frac{\theta^{m+n+1}}{(m+n+1)!}\frac{\varphi^m}{m!}+b_n\frac{\theta^m}{m!}\frac{\varphi^{m+n}}{(m+n)!}\right].\end{aligned} \tag{5.41}$$

The expressions (5.40) on the basis (emanating from the accepted starting point) slip lines takes the form

$$R(\alpha,0)=\sum_{m,n=0}^{\infty}a_n\frac{\alpha^n}{n!};\quad S_0=\sum_{n=0}^{\infty}b_n\frac{\beta^n}{n!}. \tag{5.42}$$

If, by analogy with (5.42), the curvature radii of arbitrary α and β lines extending through point *P* are represented as

$$R(\alpha,\psi)=\sum_{n=0}^{\infty}r_n(\psi)\frac{\alpha_n}{n!};\quad \underset{0}{S}(\theta,\beta)=\sum_{n=0}^{\infty}s_n(\theta)\frac{\beta_n}{n!}, \tag{5.43}$$

then from (5.41) for the coefficients r_n, s_n we obtain

$$\begin{aligned}r_n(\psi)&=\sum_{m=0}^{\infty}a_{n-m}\frac{\psi^m}{m!}-\sum_{m=n+1}^{\infty}b_{m-n-1}\frac{\psi^m}{m!},\\ s_n(\theta)&=\sum_{m=0}^{\infty}b_{n-m}\frac{\theta^m}{m!}-\sum_{m=n+1}^{\infty}a_{m-n-1}\frac{\theta^m}{m!}.\end{aligned} \tag{5.44}$$

The advantage of these dependences is that they can be represented in the matrix form.

We turn first to the centred fan (Fig. 5.5 *e*). The initial values to describe it are: the starting point 0, the slip line OA and the inner direction OA. The radius of curvature of the slip line OA is a series $R(\alpha)=\sum_{n=0}^{\infty} a_n \alpha_n$, or in matrix form

$$R(\alpha)=[\alpha_0\alpha_1...\alpha_n]\begin{bmatrix}\alpha_0\\ \alpha_1\\ \cdot\\ \cdot\\ \cdot\\ \alpha_n\end{bmatrix}=A\bar{X}_{\mathrm{OA}}, \qquad (5.45)$$

where $A = [\alpha_0\ \alpha_1...\alpha_n]$ is the row matrix of the powers of the coordinates, $\left[\frac{\alpha_n}{n!}\right]$, $\bar{X}_{\mathrm{OA}}=[\alpha_0\alpha_1...\alpha_n]^{\mathrm{T}}$ is the vector of coefficients of the power series for the slip line OA, the order of the letters in the index indicates the inner direction (from the initial point O to A). The radii of curvature of the two remaining boundary slip lines OP and AP of the centred fans are given by (5.43). So the coefficient vectors for them will be

$$\bar{X}_{\mathrm{OP}}=\begin{bmatrix}r_0\\ r_1\\ \cdot\\ \cdot\\ \cdot\\ r_n\end{bmatrix};\quad \bar{X}_{\mathrm{AP}}=\begin{bmatrix}s_0\\ s_1\\ \cdot\\ \cdot\\ \cdot\\ s_n\end{bmatrix}. \qquad (5.46)$$

The coefficients $r_n(\psi)$ and $s_n(\theta)$, necessary to determine $R(\alpha\ \psi)$ and $S(\theta\ \beta)$, can be decomposed and represented as a matrix. We show it for the coefficients $r_k(\psi)$.

We have

$$\bar{X}_{\mathrm{OP}}=\begin{bmatrix}s_0\\ s_1\\ \cdot\\ \cdot\\ \cdot\\ s_n\end{bmatrix}=\begin{bmatrix}a_0\,\psi_0\\ a_1\,\psi_1+a_0\,\psi_0\\ \cdots\\ \cdots\\ \cdots\\ \sum_{m=0}^{\infty}a_{n-m}\,\psi_m\end{bmatrix}. \qquad (5.47)$$

This expression can be rewritten as

$$\overline{X}_{OP} = \begin{bmatrix} \psi_0 & 0 & 0 \\ \psi_1 & \psi_0 & 0 \\ . & . & . \\ \psi_n & \psi_{n-1} & \psi_{n-2} \end{bmatrix} \begin{bmatrix} a_0 \\ a_1 \\ . \\ a_n \end{bmatrix} = \mathrm{P}_\psi^* \, \overline{X}_{OA} \tag{5.48}$$

Proceeding similarly with the coefficients $S_n(\theta)$, we obtain

$$\overline{X}_{AP} = \begin{bmatrix} \theta_1 & \theta_2 & . & \theta_n \\ \theta_2 & \theta_3 & . & \theta_{n+1} \\ . & . & . & . \\ \theta_n & \theta_{n+1} & . & \theta_{n+n} \end{bmatrix} \begin{bmatrix} a_0 \\ a_1 \\ . \\ a_n \end{bmatrix} = Q_\theta^* \, \overline{X}_{OA}\,. \tag{5.49}$$

after finding $\overline{X}_{OA}$ and $\overline{X}_{OB}$ we can find the radius of curvature at any point of these lines.

The approach used when considering the centred fan can be directly applied to a rectangular grid (Fig. 5.5 *d*). Taking $\overline{X}_{AP}$ and $\overline{X}_{OP}$ as the baselines, for the radii of curvature $R(\alpha\ \psi)$ $S(\theta\ \beta)$ of the lines AP and BP, passing through the point P, we obtain

$$\begin{aligned} \overline{X}_{AP} &= P_\theta^* \overline{X}_{OB} + Q_\theta^* \overline{X}_{OA}\,, \\ \overline{X}_{BP} &= P_\psi^* \overline{X}_{OA} + Q_\psi^* \overline{X}_{OB}\,. \end{aligned} \tag{5.50}$$

Construction of slip lines on the concave side of the base slip lines in the centred fan can be performed using the equations (5.49) and (5.50). For Fig. 5.5 *f*, taking OP as the base line, we obtain

$$\begin{aligned} \overline{X}_{OA} &= \left[P_\psi^*\right]^{-1} \overline{X}_{OP}\,, \\ \overline{X}_{BP} &= Q_\theta^0 \overline{X}_{OA} = Q_\theta^* \left[P_\psi^*\right]^{-1} \overline{X}_{OP}\,. \end{aligned} \tag{5.51}$$

For the quadrilateral constructed on the concave base lines (Fig. 5.5 *d*), we obtain

$$\begin{aligned} \overline{X}_{AP} &= P_\theta^* \overline{X}_{OB} - Q_\theta^* \overline{X}_{OA}\,, \\ \overline{X}_{BP} &= P_\psi^* \overline{X}_{OA} - Q_\psi^* \overline{X}_{OB}\,. \end{aligned} \tag{5.52}$$

To construct the fields of slip lines in the whole region of deformation we require operators which allow to change the internal direction of the characteristics, move the starting point, set the boundary conditions and calculate the coordinates of points and forces on the slip lines. They are constructed using the properties of the slip lines and the representations (5.40). Table 5.1 presents data on the effect of matrix operators, obtained by different authors. Links to relevant publications are given in [61]. Table 5.2 lists detailed expressions for the main operators and auxiliary matrices.

Table 5.1. Action of matrix operators

No.	Operation	Converted element	Operator
1	Start offset (Fig. 5.5 *g*)	OA → OP	S_0
2	Changing the direction of reference (Fig. 5.5 *g*)	OA → AO	R_0
3	Construction of the field of slip lines on the convex side of the centred fan (Fig. 5.5 *e*)	OA → OP	P_ψ^*
		OP → OA	$P_\psi^{*-1} = P_{-\psi}^*$
		OA → AP	Q_ψ^*
		OA → PO	$P_{0\psi} = R_0 P_\psi^*$
		AO → PA	$Q_{0\psi} = P_\psi Q_0^*$
4	Construction of the field of slip lines in the quadrangle on the convex sides (Fig. 5.5 *d*)	→ BP	$P_\psi^* \overline{X}_{OA} + Q_\psi^* \overline{X}_{OB}$
		→ PB	$P_{0\psi} \overline{X}_{OA} + Q_{0\psi} \overline{X}_{OB}$
		→ AP	$P_0^* \overline{X}_{OB} + Q_0^* \overline{X}_{OA}$
		→ PA	$P_{0\psi} \overline{X}_{OB} + Q_{0\psi} \overline{X}_{OA}$
5	Construction of the field of slip lines in the quadrangle on the concave sides (Fig. 5.5, *e*)	→ BP	$P_\psi^* \overline{X}_{OA} - Q_\psi^* \overline{X}_{OB}$
		→ PB	$P_{0\psi} \overline{X}_{OA} - Q_{0\psi} \overline{X}_{OB}$
		→ AP	$P_0^* \overline{X}_{OB} - Q_0^* \overline{X}_{OA}$
		→ PA	$P_{0\psi} \overline{X}_{OB} - Q_{0\psi} \overline{X}_{OA}$

6	Construction of the field near a smooth boundary (absence of friction on the OP or OB) (Fig. 5.5, *g*; 5.6 *a*)	OA → PA	$T_{00}^{-1} = -P_{00} - Q_{00}$
		PA → OA	$T_0 = -P_{00} - Q_{00}$
		OA → BA	T_0^{-1}
7	Construction of the field near a flat surface with friction OP (Fig. 5.6 *b*)	OA → PA	$G_\eta + Q_{\psi\psi} + P_{\psi\psi} \times$ $\times(I\cos\eta - J\sin\eta)^{-1} \times$ $\times(J\cos\eta - I\sin\eta)$
8	Same with limiting friction (adhesion, η = 0) (Fig. 5.6 *b*)	OA → PA	$G_\eta = G_0 = Q_{\psi\psi} + P_{\psi\psi} J$
9	The same, in the absence of friction (smooth boundary, η = π/4) (Fig. 5.6 *b*)	OA → PA	$G_\eta \to G_{\pi/4} = Q_{00} - P_{00} = T_0^{-1}$
10	Surface PO free from stress (Fig. 5.6*c*)	OA → PA	$F = R_\psi P_{\psi\psi}^{-1} \times$ $\times\left[R_0 (D+I)^{-1}(D-I) + Q_{00} R_0\right]$

Knowing the power series coefficients for the radii of curvature of the slip lines, it is possible to determine the coordinates of arbitrary points on them.

When finding the Cartesian *(x, y)* coordinates of point A (Fig. 5.6 *d*) initially, usually we define moving (*x*, *y*) coordinates, as they relate by simple relations to the radius of curvature corresponding to the slip line. If the series for the radius of curvature is given by

$$R(\phi) = \sum_{n=0}^{\infty} r_n \frac{\phi_n}{n!},$$

then the moving coordinates are given by

$$\bar{X}(\phi) = \sum_{n=0}^{\infty} t_n \frac{\phi_n}{n!},$$

$$\bar{Y}(\phi) = \pm \sum_{n=0}^{\infty} t_n \frac{\phi_{n+1}}{(n+1)!},$$

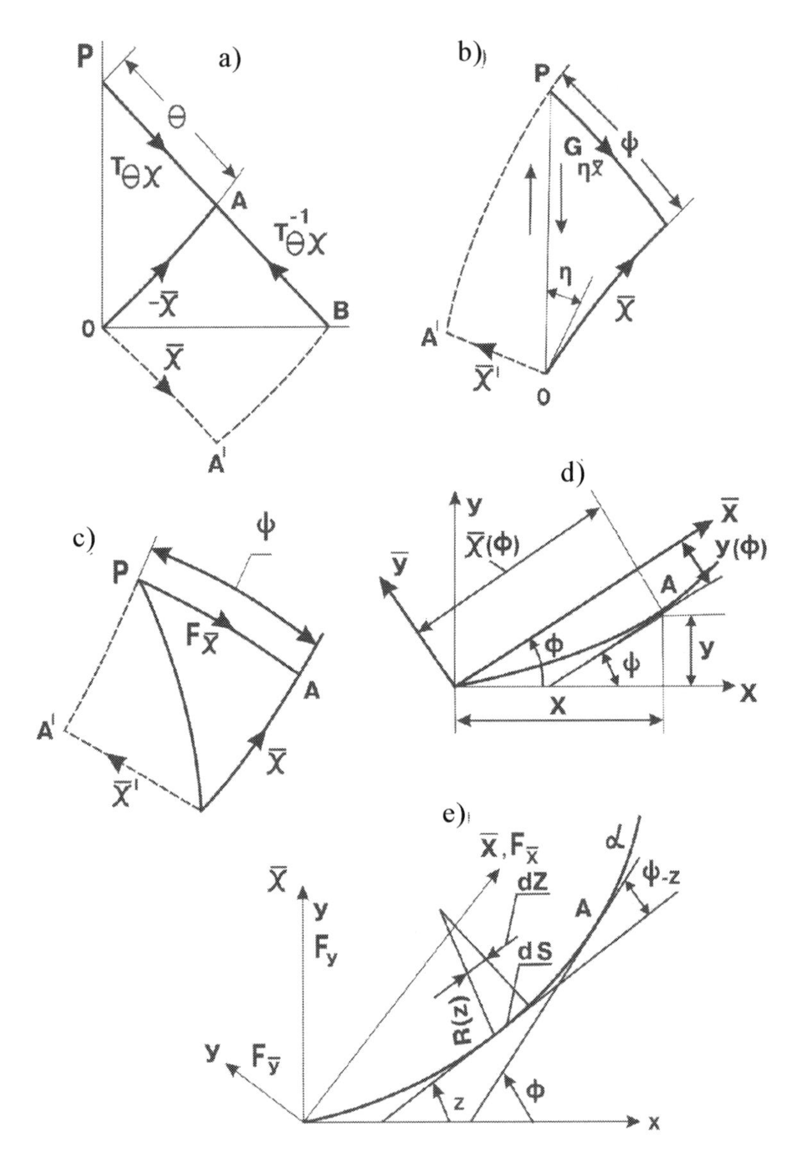

Fig. 5.6. Types of boundary conditions, the search for the coordinates and forces: a) there is no friction on the OB and OP boundary; b) the straight rough boundary OP with friction; c) stress-free boundary OP; d) Cartesian (xy) and mobile ($x'y'$) coordinates; e) the forces and moments.

wherein $t_{n+1} - t_{n-1} = |r_n||$, $t_0 = 0$, $t_1 = |r_0|$, and the signs "+" or "–" in the expression $\overline{Y}(\phi)$ are taken depending on whether the slip line has a positive or negative curvature. Directly from Fig. 5.6 *d* follow the expressions for the Cartesian coordinates of the point A

$$x = \bar{x}\cos\phi - \bar{y}\sin\phi;$$
$$y = \bar{x}\sin\phi - \bar{y}\cos\phi.$$

Table 5.2 Values of the basic operators

Operator	Expanded form ($\varphi_n = \varphi_n / n!$)
P_φ^*	$\begin{bmatrix} \varphi_0 & 0 & 0 & . \\ \varphi_1 & \varphi_0 & 0 & . \\ \varphi_2 & \varphi_1 & \varphi_0 & . \\ . & . & . & . \end{bmatrix}$
Q_φ^*	$\begin{bmatrix} \varphi_1 & \varphi_2 & \varphi_3 & . \\ \varphi_2 & \varphi_3 & \varphi_4 & . \\ \varphi_3 & \varphi_4 & \varphi_5 & . \\ . & . & . & . \end{bmatrix}$
S_φ	$\begin{matrix} \varphi_0 & \varphi_1 & \varphi_2 & . \\ 0 & \varphi_0 & \varphi_1 & . \\ 0 & 0 & \varphi_0 & . \\ . & . & . & . \end{matrix}\Bigg]$
R_φ	$-\begin{bmatrix} \varphi_0 & \varphi_1 & \varphi_2 & . \\ 0 & -\varphi_0 & -\varphi_1 & . \\ 0 & 0 & \varphi_0 & . \\ . & . & . & . \end{bmatrix}$
$P_{\varphi 0}$	$R_\varphi P_0^*$
	$R_0 Q_\varphi^*$
I	$\begin{matrix} 1 & 0 & 0 & . \\ 0 & 1 & 0 & . \\ 0 & 0 & 1 & . \\ . & . & . & . \end{matrix}\Bigg]$
J	$\begin{bmatrix} 0 & 0 & 0 & . \\ 1 & 0 & 0 & . \\ 0 & 1 & 0 & . \\ . & . & . & . \end{bmatrix}$

Calculation of forces and moments acting on the slip line can also be performed using the representation in the form of series. In Fig. 5.6 *e* the slip line with a radius of curvature $R(\alpha)$ is affected by the hydrostatic pressure p_0 at point 0. We calculate the force and moment acting on the concave side of the arc of the slip line. The normal and tangential forces acting on the portion ϕ of the α-slip line OA are obtained by summing the elementary forces

$$dF_{\overline{X}} = -k\cos(\phi - z)ds + p\sin(\phi - z)ds,$$
$$dF_{\overline{Y}} = k\cos(\phi - z)ds + p\cos(\phi - z)ds. \quad (5.53)$$

Normal pressure P acting on the element ds, can be expressed by the hydrostatic pressure p_0 at the reference point 0. For the α-line $p + 2k\alpha$ = const, so that $p = p_0 - 2kz$. Considering also that $ds = R(z)\,dz$, the expression (5.53) can be rewritten as

$$dF_{\overline{X}} = -k\cos(\phi - z)dz + p_0\sin(\phi - z)R(z)dz - 2kz\sin(\phi - z)R(z)dz;$$
$$dF_{\overline{Y}} = k\sin(\phi - z)R(z)dz + p_0\cos(\phi - z)R(z)dz - 2kz\cos(\phi - z)R(z)dz.$$

Integration of expressions (5.53) over the interval ϕ allows one to get the full value of the force $F_{\overline{X}}$ and $F_{\overline{Y}}$, which in the dimensionless variables are

$$\frac{F_{\overline{X}}}{k} = -\overline{X} + \frac{P_0}{k}\overline{Y} - 2\int_0^{\phi} zR(z)\sin(\phi - z)dz;$$
$$\frac{F_{\overline{Y}}}{k} = -\overline{Y} + \frac{P_0}{k}\overline{X} - 2\int_0^{\phi} zR(z)\cos(\phi - z)dz;$$
$$\overline{X} = \int_0^{\phi} R(z)\cos(\phi - z)dz;$$
$$\overline{Y} = \int_0^{\phi} R(z)\sin(\phi - z)dz.$$

The integrals in these formulas can be calculated by presenting them in the form of a series. As a result, we obtain

$$\int_0^{\phi} zR(z)\cos(\phi - z)dz = \sum_{n=0}^{\infty} c_n \frac{\phi^n}{n!},$$
$$\int_0^{\phi} zR(z)\sin(\phi - z)dz = \sum_{n=0}^{\infty} c_{n-1} \frac{\phi^n}{n!};$$

where $c_{n+1} + c_{n-1} = nr_{n-1}$, $c_0 = c_{-1} = 0$.

Turning to the Cartesian coordinates, we finally obtain

$$\frac{F_X}{k}=\frac{F_{\bar{X}}}{k}\cos\phi-\frac{F_{\bar{Y}}}{k}\sin\phi,$$
$$\frac{F_Y}{k}=\frac{F_{\bar{Y}}}{k}\cos\phi-\frac{F_{\bar{X}}}{k}\sin\phi.$$

5.2.6. Construction of the field of slip lines and the velocity hodograph

Using the matrix method for solving rolling problems is associated with the necessity of defining in advance the pattern of the field of slip lines and the corresponding velocity field. The only fundamental but important point of view for computing difference in comparison with the conventional variant of the method slip lines is the algorithmization of the process, allowing by the use of linear programming to avoid the need to resort to the method of trial and error. Due to the lack of a standard methodology for constructing the field of slip lines for the asymmetric process we will focus on this issue, based on the approach used in the construction of the field of slip lines and the velocity hodograph for symmetrical rolling.

The strip will be assumed to be thick and wide enough for the implementation of flat plastic flow conditions. The metal on the contact surface sticks to the rolls and plastic area on the axis of symmetry is not contracted to a single point, and has some length (Fig. 5.7). The compatibility conditions of the stress and velocity fields show that the velocity vectors of metal points on the contact surface are directed at a tangent to the surface and in the given conditions are equal in the vector modulus to the circumferential speed of the roll. The velocity vectors of the metal in the neighbourhood of point D (Fig. 5.7 *a*) in the rigid and plastic fields will have the same direction, but will differ in size: in the first region it will be equal to the velocity of metal at the outlet from the rolls (v_1), in the second – to the circumferential roll speed (v_b). The difference between the values v_l and v_b determines the amount of discontinuity of the tangential component of the velocity along a rigid–plastic boundary DC_1GC (the magnitude of velocity discontinuity along the slip line is constant).

On the velocity field (Fig 5.7 *b*) the direction parallel to the axis of the strip is chosen randomly, and from certain start (O_3) the velocity vectors v_1 and velocity v_b are produced (at this point they are directed parallel to the strip axis). The magnitude of the interval

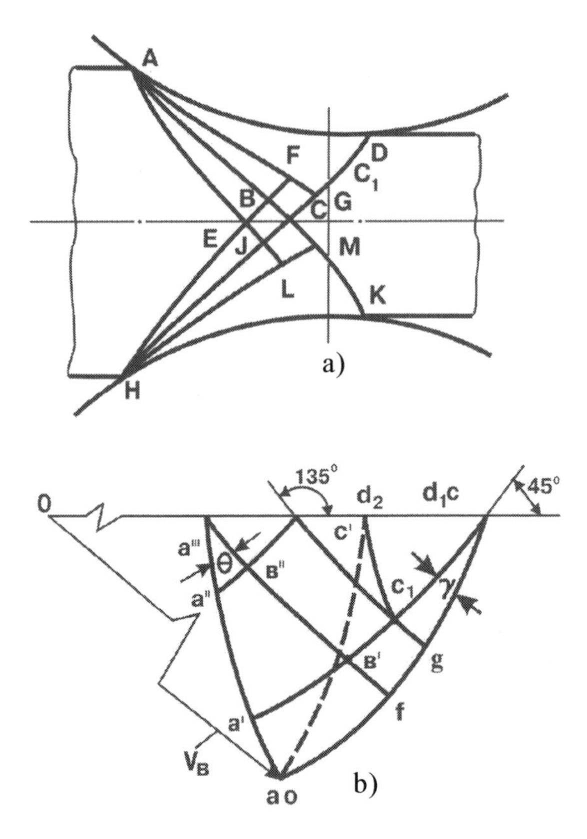

Fig. 5.7. Symmetric rolling: a – slip line; b – hodograph.

$d_2 d_1$ on the velocity field determines the difference $v_1 - v_b$ and the scale stress and velocity fields.

The rigid area at the exit from the rolls, having constant speed v_1, is displayed on the velocity field by a single point d_1. Due to constant values of the tangential velocity discontinuity along the rigid–plastic boundary DGC this line appears on the velocity field in the form of a circular arc centered at d_1 with a radius equal to the value of the discontinuity $v_1 - v_b$.

The slip line CBA is continuation of the rigid–plastic boundary of the lower half of deformation zone, a symmetrical line DC_1GO relative to the axis of the strip. Consequently, the tangential discontinuity velocity component propagates along the CBA. In symmetric deformation zone the magnitude of the velocity discontinuity along CBA is equal to the discontinuity along CGA.

Point C is displayed (when considering only the upper half of the deformation zone) by three points, C, C_1, C. Sections CC_1 and C_1C, equal to each other and determining the magnitude of the velocity discontinuity along the slip lines CGD and CBA, are orthogonal to the discontinuity lines at point C, and are inclined to the axis at angles respectively π/4 and 3π/4.

Further development of the velocity field is carried out by setting the lines $\overline{c_1 b}$ and $\overline{c_1 b}''$, reflecting the segment CB of the line CBA (lines $\overline{c_1 b}$ and $\overline{c_1 b}''$ are orthogonal at corresponding points of the segment CB). Having the line $\overline{c_1 b}''(\overline{c_1 b}')$ and randomly selecting angular parameters a and b, one can construct the velocity field.

The assumption of metal sticking to the rolls at the contact surface requires the presence of some volume (AFGD), moving with the rollers as a single unit. The velocity of the points of the AFGD area proportional to the distance of each point from the roll axis and is directed along the normal to the straight line passing through the axis of the roll and the given point. This means that the rotation of the d_2gfad_2 zone of the velocity field through an angle equal to π/2 clockwise forms the region DGFAD of the stress field (Fig. 5.7). After producing the CB slip line the CBE area if formed. The correctness of select the baseline $\overline{c_1 b}''(\overline{c_1 b}')$ to construct the velocity field and the corresponding field of slip lines is verified by the construction of the stress field CBE. If the line $\overline{c_1 b}''$ is chosen incorrectly, the slip line CB, EB will not provide construction of the curved triangle BAE – these lines do not converge at point A.

The constructed stress field should be consistent with the velocity field. The compatibility condition is based on the flow of the velocity vector across the line of contact being equal to zero or on the flows of the velocity vector through the rigid–plastic boundary being equal ($v_0h_0 = v_1h_1$). If this equality is not respected or not getting the right ratio h_1/h_0 is not obtained, it is necessary, setting other angles a and b and lines $\overline{c_1 b}''$, to build a new velocity field and the corresponding stress field.

The constructed stress field must also meet the equilibrium condition of the strip under the influence of normal and tangential forces on the contact surface and the longitudinal stresses applied to the strip. This condition is verified by the joint consideration of the conditions of equilibrium of rigid strip portions at the entry and exit from the deformation zone.

Thus, in symmetric rolling there are three parameters, which describe the disposing process. Typically we define angle θ_0 of the entry of strips into the rolls and this angle defines parameter b, so that there remain only two free parameters. Asymmetry of the process leads to the deformation of the grid of slip lines and the velocity hodograph diagram (Fig. 5.8). The asymmetry of the process may be caused by the strip entering the rolls at an angle,

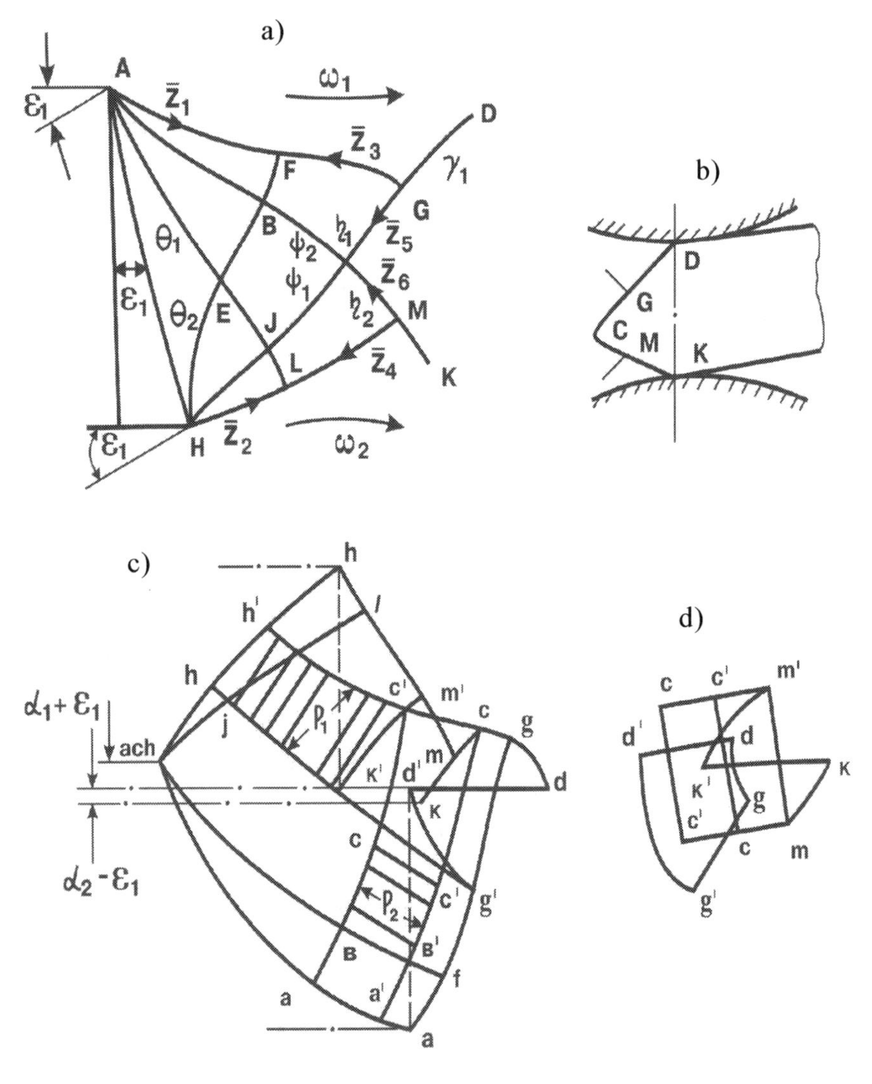

Fig. 5.8. Field of slip lines and hodograph under asymmetric rolling: a – field of slip lines; b – the impact of curvature of the strip to the upper roll on the slip line; c – hodograph; d – the influence of the curvature of the strip to the upper roll on the hodograph.

the angular velocity mismatch ω_1 and ω_2 of the rolls, the difference of the radii R_1, R_2 (index '1' and '2' refer respectively to the upper and lower rolls), uneven heating of the strip along the height and other causes. The strip material is continuously (plastically) deformed in the AGCMHEA region, on the lines HJCGD and ABCMK there occur breaks ρ_1 and ρ_2 of the tangential velocity components. As in the symmetric case, in the areas AGD and EMK the material is considered adhered to the rolls, but it is rotated, in general, with different angular velocities ω_1 and ω_2. Consequently, the strip speed at the exit from the rolls is

$$\begin{aligned}\Omega b_1 &= \omega_1 R_1 + \rho_1 \text{ at point } D;\\ \Omega b_2 &= \omega_2 R_2 + \rho_2 \text{ at point } K,\end{aligned} \tag{5.54}$$

where Ω – the angular velocity of rotation of the strip at the exit from the rolls, b_1 and b_2 – the radii of curvature of the outer (adjacent to the upper roll) and the inner (adjacent to the lower roll) layers of the strip at the exit from the rolls.

The state of the material in the deformation zone in the plane plastic flow of the metal is determined by the Hencky (5.29) and Geiringer (5.30) equations.. From these equations it is possible to find all unknown quantities σ, φ, υ_α, υ_β with the known slip lines, so that together with the equations (5.29), (5.30) it is necessary to consider the equations of the directions of the characteristics (5.26), (5.27)..

5.2.7. The matrix equation for the asymmetric process

The coupled system (5.31), (5.40), (5.28), (5.29) will be solved by the matrix–operator method. The field of slip lines at asymmetrical rolling is defined by six characteristics AF, HL, GF, ML, GC, MC and two lines *ab* and *aj* of the velocity hodograph (Fig. 5.8 *a*, *b*). For a complete description of the process it is also necessary to know be aware of the roll radii R_1 and R_2, the kinematic parameters ω_1, ρ_1, ρ_2, Ω, and finally, the mean pressure σ in at least one point (for example, *c*). We form the matrix equations for the fields of slip lines and velocity hodograph and then formulate the remaining conditions.

The matrix equations are derived using the relations given in the Tables 5.1 and 5.2. From these tables it follows that the segments EBF and EJL (Fig. 5.8 *a*) are vector representations (vectors $\overline{X}$ hereafter designated through $\overline{z}_i$):

$$\text{EBF} \mapsto Q_{Q_1\xi_1}\overline{z_1}; \quad \text{EJL} \mapsto Q_{Q_2\xi_2}\overline{z_2}, \tag{5.55}$$

where $\xi_i = \psi_i + \eta_i$ $(i = 1, 2)$; $\mapsto$ means "is represented as".

Considering the further parts of the field between two specified characteristics, we obtain

$$\bar{z}_\xi = Q_{\xi_1\psi_2} Q_{\theta_1\xi_2} \bar{z}_1 + P_{\psi_2\xi_2} Q_{\theta_2\xi_2} \bar{z}_2, \tag{5.56}$$

$$\bar{z}_4 = Q_{\xi_1\psi_2} Q_{\theta_1\xi_1} \bar{z}_1 + P_{\xi_2\psi_1} Q_{\theta_2\xi_2} \bar{z}_2, \tag{5.57}$$

$$\bar{z}_5 = Q_{\xi_1\psi_2} Q_{\theta_1\xi_1} \bar{z}_1 + P_{\psi_2\xi_1} Q_{\theta_2\xi_2} \bar{z}_2, \tag{5.58}$$

$$\bar{z}_6 = Q_{\xi_1\psi_2} Q_{\theta_1\xi_1} \bar{z}_1 + P_{\xi_2\psi_1} Q_{\theta_2\xi_2} \bar{z}_2 \tag{5.59}$$

Four of the remaining equations can be obtained by considering the velocity hodograph image (Fig. 5.8 *c*). The images of the slip lines, limiting a rigid rotating region AFGD, are curves, geometrically similar to these line rotated by an angle $\pi/2$ and having as the scale factor the corresponding angular velocity. Thus

$$gc \mapsto \Omega\bar{z}_5;\quad mc \mapsto \Omega\bar{z}_6;\quad gd \mapsto -\Omega r_1\bar{c};$$
$$mk \mapsto \Omega r_2\bar{c};\ af \mapsto \omega_1\bar{z}_1;\quad g'f \mapsto \omega_1\bar{z}_3;$$
$$hl \mapsto \omega_2\bar{z}_2;\ m'l \mapsto \omega_2\bar{z}_4, \tag{5.60}$$

where $\bar{c}$ is a circular arc of unit radius.

On the other hand, taking *eb* and *ej* as base lines, we obtain

$$ej \mapsto P_{\psi_1\psi_2}\bar{z}_7 - Q_{\psi_2\psi_1}\bar{z}_8;\quad cb \mapsto P_{\psi_2\psi_1}\bar{z}_8 - Q_{\psi_1\psi_2}\bar{z}_7. \tag{5.61}$$

Adding or subtracting the 'circular' vectors $\rho_1\bar{c}$ or $\rho_2\bar{c}$, we obtain

$$hj \mapsto Q_{\psi_2\theta_2}\bar{z}_8;\quad ab \mapsto Q_{\psi_1\theta_1}\bar{z}_7.$$

$$c'j' \mapsto P_{\psi_1\psi_2}\bar{z}_7 - Q_{\psi_2\psi_1}\bar{z}_8 + \rho_1\bar{c}, \tag{5.62}$$

$$c'b' \mapsto P_{\psi_2\psi_1}\bar{z}_8 - Q_{\psi_1\psi_2}\bar{z}_7 - \rho_2\bar{c}, \tag{5.63}$$

$$h'j' \mapsto Q_{\psi_2\theta 2}\bar{z}_8 - \rho_1\bar{c}, \tag{5.64}$$

$$a'b' \mapsto Q_{\psi_1\theta_1}\bar{z}_7 + \rho_2\bar{c}, \tag{5.65}$$

$$m'c' \mapsto -\Omega\bar{z}_0 + \rho_2\bar{c}, \tag{5.66}$$

$$g'c' \mapsto -\Omega\bar{z}_5 - \rho_1\bar{c}. \tag{5.67}$$

With *m'l* from (5.60) and *m'c'* from (5.66), as well as *g'f* from (5.60) and $g'c$ from (5.67) and guided by the rules of constructing the field using matrix operators (Table 5.1), we obtain alternative presentations for $c'j'$ and $c'b'$.

$$c'j' \mapsto \omega_2 P^*_{\eta_2} \bar{z}_4 + Q^*_{\eta_2}\left(-\Omega \bar{z}_6 + \rho_2 \bar{c}\right), \tag{5.68}$$

$$c'b' \mapsto \omega_1 P^*_{\eta_1} \bar{z}_3 + Q^*_{\eta_1}\left(\Omega \bar{z}_5 + \rho_1 \bar{c}\right), \tag{5.69}$$

and find

$$j'l \mapsto P_{\eta_2\psi_1}\left(-\Omega \bar{z}_6 + \rho_2 \bar{c}\right) + Q_{\psi_1\eta_2}\omega_2 \bar{z}_4, \tag{5.70}$$

$$b'f \mapsto P_{\eta_1\psi_2}\left(-\Omega \bar{z}_5 + \rho_1 \bar{c}\right) + Q_{\psi_2\eta_1}\omega_1 \bar{z}_5. \tag{5.71}$$

The last two expressions can obtain alternative representations of the hl and for af.

$$hl \mapsto P_{\theta_2\eta_2}\left(Q^*_{\psi_2}\bar{z}_8 + \rho_1 \bar{c}\right) + Q_{\eta_2\theta_2}\left(P_{\eta_2\psi_1}\left(-\Omega \bar{z}_6 + \rho_2 \bar{c}\right) + Q_{\psi_1\psi_2}\omega_2 \bar{z}_4\right); \tag{5.72}$$

$$af \mapsto P_{\theta_1\eta_1}\left(Q^*_{\psi_1}\bar{z}_4 + \rho_2 \bar{c}\right) + Q_{\eta_1\theta_1}\left(P_{\eta_1\psi_2}\left(-\Omega \bar{z}_5 + \rho_1 \bar{c}\right)\right) + Q_{\psi_2\eta_1}\omega_1 \bar{z}_3. \tag{5.73}$$

Comparing (5.62) and (5.68), (5.63) and (5.69), the expression for hl of (5.60) with (5.72), the expression for af of (5.60) with (5.73) we obtain four relations between the basis vectors

$$\omega_2 P^*_{\eta_2}\bar{z}_4 - \Omega Q^*_{\eta_2}\bar{z}_6 - P_{\psi_1\psi_2}\bar{z}_7 + Q_{\psi_2\psi_1}\bar{z}_8 = \left(\rho_1 - Q^*_{\eta_2}\rho_2\right)\bar{c}, \tag{5.74}$$

$$\omega_1 P^*_{\eta_1}\bar{z}_3 - \Omega Q^*_{\eta_1}\bar{z}_5 + Q_{\psi_1\psi_2}\bar{z}_7 - P_{\psi_2\psi_1}\bar{z}_8 = \left(-\rho_2 + Q^*_{\eta_1}\rho_1\right)\bar{c}, \tag{5.75}$$

$$\begin{aligned} &-\omega_1 \bar{z}_1 + \omega_1 Q_{\eta_1\theta_1} Q^*_{\psi_2\eta_1}\bar{z}_3 + \Omega Q_{\eta_1\theta_1} P_{\eta_1\psi_2}\bar{z}_5 + P_{\theta_1\eta_1} Q^*_{\psi_1}\bar{z}_7 = \\ &= \left(Q_{\eta_1\theta_1} P_{\eta_1\psi_2}\rho_1 + P_{\theta_1\eta_1}\rho_2\right)\bar{c}, \end{aligned} \tag{5.76}$$

$$\begin{aligned} &-\omega_2 \bar{z}_2 + \omega_2 Q_{\eta_2\theta_2} Q^*_{\psi_1\eta_2}\bar{z}_4 - \Omega Q_{\eta_2\theta_2} P_{\eta_2\psi_1}\bar{z}_6 + P_{\theta_2\eta_2} Q^*_{\psi_2}\bar{z}_8 = \\ &= -\left(P_{\theta_1\eta_1}\rho_1 + Q_{\eta_2\theta_2} P_{\eta_2\psi_2}\rho_2\right)\bar{c}. \end{aligned} \tag{5.77}$$

Expressions (5.56)–(5.59) and (5.74)–(5.77) give eight matrix equations linking eight vectors $\bar{z}_1 - \bar{z}_8$, defining slip baselines. Each of these matrix equations is equivalent to the system of equations whose order is determined by the dimension of the vector $\bar{z}_i$ (i = 1...8), or, in other words, the number of expansion terms of the basis vectors in the power series: with n members each of the matrices appearing in the equations has dimension $n*n$, each vector has n components, and each matrix equation is equivalent to n algebraic equations.

For ease of analysis and use these equations can be represented in the matrix form (5.78). Each matrix element (5.78) is a block matrix. While maintaining in the expansions the number of terms $n = 4$, the dimension of each block is 4*4 and the actual dimension of the matrix is 32*32. Vectors $\bar{z}_i, \bar{b}_i$, appearing in the matrix (5.78) are also block vectors, their true dimension is 4-m, so that vector $\bar{z} = \left[\bar{z}_1^{\mathrm{T}}, \bar{z}_2^{\mathrm{T}} \ldots \bar{z}_8^{\mathrm{T}} \right]^{\mathrm{T}}$ contains 32 elements, and the vector $\bar{b} = \left[0^{\mathrm{T}}_{\ldots} 0^{\mathrm{T}}, \bar{b}_5^{\mathrm{T}} . \bar{b}_8^{\mathrm{T}} \right]^{\mathrm{T}}$ also has the dimension 32, but contains only 16 zero elements.

$$\begin{bmatrix} Q_{\xi_1\psi_2} Q_{\theta_1\xi_1} & P_{\psi_2\xi_1} Q_{\theta_2\xi_2} & -I & 0 & 0 & 0 & 0 & 0 \\ P_{\psi_1\xi_2} Q_{\theta_1\xi_1} & P_{\psi_2\xi_1} Q_{\theta_2\xi_2} & 0 & -I & 0 & 0 & 0 & 0 \\ P_{\xi_1\psi_2} Q_{\theta_1\xi_1} & Q_{\xi_2\psi_1} Q_{\theta_2\xi_2} & 0 & 0 & -I & 0 & 0 & 0 \\ Q_{\psi_2\xi_2} Q_{\theta_1\xi_1} & P_{\xi_2\psi_1} Q_{\theta_2\xi_2} & 0 & 0 & 0 & -I & 0 & 0 \\ 0 & 0 & 0 & \omega_2 P^*_{\eta_2} & 0 & -\Omega Q^*_{\eta_2} & -P_{\psi_1\psi_2} & Q_{\psi_2\psi_1} \\ 0 & 0 & \omega_1 P_{\eta_1} & 0 & \Omega Q_{\eta_1} & 0 & Q_{\psi_1\psi_2} & -P_{\psi_2\psi_1} \\ 0 & -\omega_2 I & 0 & \omega_2 Q_{\eta_2\theta_2} Q_{\psi_2\eta_2} & 0 & -\Omega Q_{\eta_2\theta_2} P_{\eta_2\psi_1} & 0 & P_{\theta_2\mu_2} Q_{\psi_2} \\ -\omega_1 I & 0 & \omega_1 Q_{\eta_1\theta_1} Q_{\psi_2\eta_1} & 0 & \Omega Q_{\eta_1\theta_1} P_{\eta_1\psi_2} & 0 & P_{\theta_1\eta_1} Q^*_{\psi_1} & 0 \end{bmatrix} \begin{bmatrix} z_1 \\ z_2 \\ z_3 \\ z_4 \\ z_5 \\ z_6 \\ z_7 \\ z_8 \end{bmatrix} = \begin{bmatrix} 0 \\ 0 \\ 0 \\ 0 \\ \bar{b}_5 \\ \bar{b}_6 \\ \bar{b}_7 \\ \bar{b}_8 \end{bmatrix} \tag{5.78}$$

$$\bar{c}_5 = \left(\rho_1 - \rho_2 Q^*_{\eta_2}\right) \cdot \bar{C}, \ldots\ldots \bar{c}_7 = \left(\rho_1 - P_{\theta_2\eta_2} + \rho_2 Q_{\eta_2\theta_2} P_{\eta_1\psi_2}\right) \cdot \bar{C},$$

$$\bar{c}_6 = -\left(\rho_1 - \rho_1 Q^*_{\eta_1}\right) \cdot \bar{C}, \ldots\ldots \bar{c}_8 = \left(\rho_2 P_{\theta_1\eta_1} + \rho_1 Q_{\eta_1\theta_1} P_{\eta_1\theta_1} P_{\eta_1\psi_2}\right) \cdot \bar{C}.$$

Eight matrix equations, summarized in the matrix (5.78), exhaust all the possible relations between the basis vectors, and must ensure the construction of fields of slip lines and velocity hodograph. Consider this matrix, while assuming that the values of ρ_1, ρ_2, ω_1, ω_2, R_1, R_2, Ω, b_1, b_2 are given (to the question of their definition or determination we return below). Vectors $\bar{b}_i$ of the right side include operators that depend on the angular coordinates θ_1, θ_2.. η_1, η_2, given arbitrarily, based on preliminary considerations. In principle, this situation is the same as in symmetric rolling: there, having three (or two) parameters, the method of successive approximations is used to find a solution that satisfies both the physical conditions of the implementation of the process and technology requirements. In this case, the number of such parameters is much higher (32 factors), so that this method becomes unrealistic. Therefore, to solve the matrix equation (5.78) it is necessary to use methods of mathematical programming. Among the most effective is the method of sliding tolerance which will be used.

We now turn to issues related to specifying parameters and process conditions. As noted above, the asymmetric strip rolling process is characterized by 14 degrees of freedom (parameters) θ_1,

θ_2, ψ_1, ψ_2, ξ_1, ξ_2, γ_1, γ_2, ω_1, ω_2, ρ_1, ρ_2, Ω, p_0. The matrix equations (5.78) impose eight ties, the others are determined by the process conditions. In practice, in the process of direct calculations it is necessary to specify angular speeds ω_1, ω_2 of the upper and lower rolls, the magnitude of velocity discontinuity ρ_1 in the HJCGD line (the parameters ρ_1/ω_1 and ρ_1 are used as scale units on the diagrams of the slip lines and the hodograph), R_1 and R_2 are the radii of the upper and lower rolls (usually in the form of relations R_1/H, R_2/H, where H is the strip thickness at the entry), the relative compression $r_n = \varepsilon = 1-h/H$, h is the thickness of the strip at the exit and, finally, the value of pressure p_0 from the condition of the equality of the total force acting on the each of the rolls. Further conditions are introduced, based on the requirements of physical realization of the process and its special features. From these considerations, the values of the vertical component of the force and torque acting on the strip (two conditions) are specified at the exit. On entering the strip in the rolls three conditions can also be specified (e.g., angle of entry of the strip in the rolls, which is equivalent to the definition of two parameters – x, y – coordinates of the point of meeting of the strip with one of the rolls), and a horizontal force component. In this case, the vertical component and its associated moment (response on the roller table) are determined from the solution. Another two conditions follow from requiring that the translational displacement of the centres of the rolls relative to each other is equal to zero; this has no place in the actual process, but arises from the general solution.

5.2.8. Analysis of the results of calculation

A rolling program was compiled to solve the problems of asymmetric rolling, including blocks forming the matrix operators, operations with matrices, the formation of the matrix equation (5.78), and the solutions of this equation. The program of solving (5.78) is based on the method of moving tolerance, which allows to solve the nonlinear system of equations with constraints of the type of equality and inequality. The block form of the program allows us to solve the problem of calculating asymmetric processes using a PC [61].

The influence of the difference of the radii of the rolls on bending of the outgoing end of the rolled strip is shown in Fig. 5.9. The vertical axis is the ratio R_1/R_{av} where R_{av} is the radius of curvature of the midline of the outgoing end of the strip (the centre of curvature – in the side of the lower roll). In the calculations it was assumed

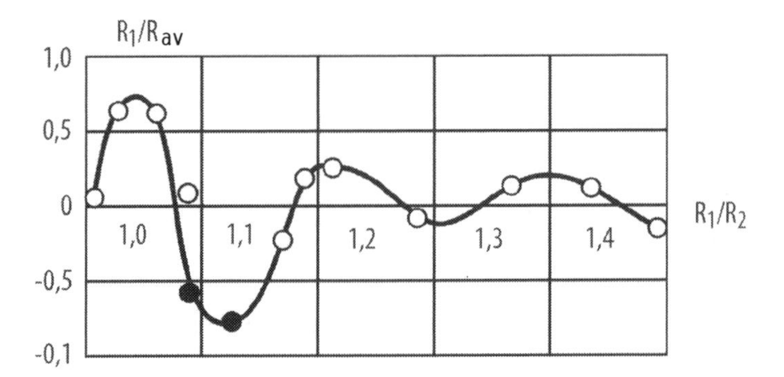

Fig. 5.9. Change in curvature of the front end of the band as a function of radius ratio of the rolls. Rolling conditions are mentioned in the text.

that the rotational speeds of the rolls are equal to 60 rpm, $R_1/H = 10$, $\varepsilon = 20\%$, the angle of inclination of the strip to the horizon at the entry is equal to zero. As seen in the figure, at a small difference of the radii of the rolls (9%), the curvature is positive, i.e. the end of the strip is bent towards the lower roll. When the difference of radii is between 9–16%, the strip is bent in the opposite direction. With further increase of the differences in the radii the pattern is repeated periodically, but at a decreasing (decaying) amplitude. Significant here is both the oscillatory nature of the distortions of the rolled strip and the fact that the greatest bending occurs at small differences of the roll radii ($R_1/R_2 \approx 1.05$). For values of $R_1/R_2 > 1.5$ the influence of this factor becomes insignificant. The moments on the upper and lower rolls are equal at points corresponding to the exit of the straight strip from the rolls. These conclusions are confirmed by the results of calculations of the asymmetric process given in other works, which are cited in [61].

The dependence of the curvature of the exiting end of the strip on the mismatch of the angular roll speed is shown in Fig. 5.10. The calculations were made for $R_1 = R_2 = 500$ mm, $R_1/H = 10$, $\varepsilon = 20\%$. For small mismatches of angular speeds (up to 8%) the curvature is positive, i.e. the exiting end of the strip is bent toward the lower roll rotating at a lower speed. For large values of mismatches (10% or more), the curvature changes sign and becomes permanent. This means that at high angular speed mismatches the strip can be 'wound' on the roll rotating faster.

The dependence of the curvature of the strip at the exit from the rolls on the relative reduction in asymmetric hot-rolling strips of

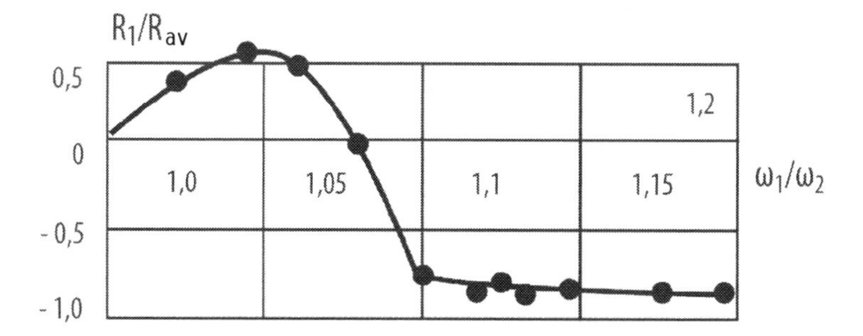

Fig. 5.10. Dependence of the curvature of the front end of the rolled strip on the roll speed ratio. The rolling conditions are given in the text.

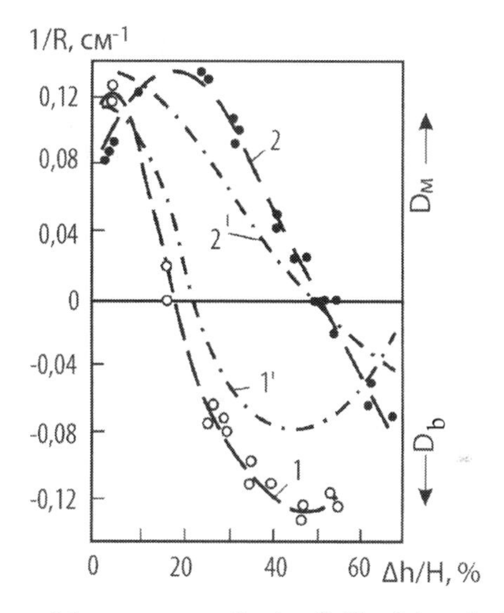

Fig. 5.11. Dependence of the curvature of strips ($1/R$) of the relative reduction($\Delta h/H$) in asymmetric ($\Delta D/D_{av}$ = 0.2) hot (t = 900°C) rolling of strips: 1, 1′ – H = 2.1 mm, the data of [56] and calculated by the matrix equation; 2, 2′– the same for H = 9.5 mm.

different thickness is shown in Fig. 5.11 [58]. Experimental data were obtained in rolling strips of steel St3 in the rolls with a large diameter D_l = 165 mm, smaller diameter D_{sm} = 135 mm, and with the same angular speed $\omega_l = \omega_{sm} = \omega$ = 40 rpm.

The dependence of the curvature of the strips exiting the rolls on the ratio of the initial strip thickness to the average diameter of the rolls is shown in Fig. 5.12. Experimental data were obtained at D_{av} = 150 mm and different thicknesses of rolled metal, the angular velocities of the rolls were 40 rpm. As can be seen the theoretical and the experimental results are in good agreement.

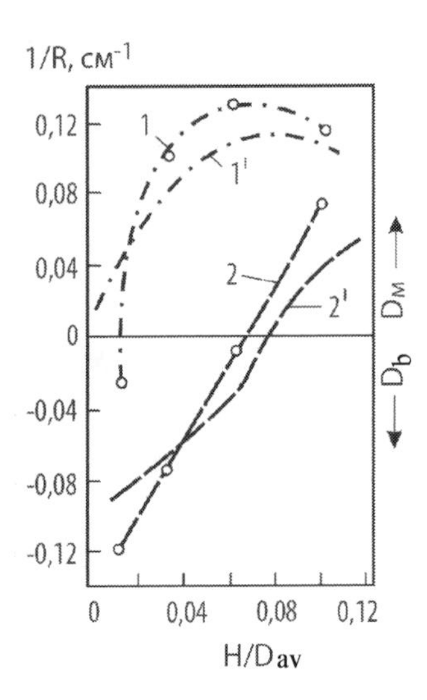

Fig. 5.12. Dependence of the curvature ($1/R$) of the strip exiting from the rolls on the ratio of the thickness to the average diameter of the rolls (H/D_{av}) at symmetrical hot rolling: 1, 1' – $\Delta h/H = 20\%$, the data [56] and calculated by the matrix equation; 2, 2'– when the same $\Delta h/H = 50\%$.

The dependence of curvature at the exit of the strip from the rolls on the relative compression for different thicknesses of the strip plate and different speeds of rotation of the rolls ($\omega_l = 40$ rpm, $\omega_{sm} = 35$ rev/min, $d_{av} = 150$ mm and $D_l = Ds_m = 150$ mm) is shown in Fig. 5.13.

The figure shows that the curves 1" and 2", calculated on the basis of the energy ratios, reflect less efficiently the real picture than curves the 1' and 2', obtained on the basis of the theory of slip lines. This, apparently, can be explained by the fact that the energy approach takes into account only the main power factors and does not take into account the change in stress throughout the deformation zone that is crucial for the kinematics of the outgoing end of the strip.

In the case of the strips entering the rolls at an angle (Fig. 5.14) it is necessary to link parameter ε_1 (or the coordinates of the meeting point) with the parameters of the field of slip lines. Figure 5.14 represents the different variants of entry of the strip with the latter inclined to the lower roll. It is very simple to establish a connection between the angle ε_1, the parameters of the field of slip lines and geometrical parameters in the variant (2), when the lower edge of

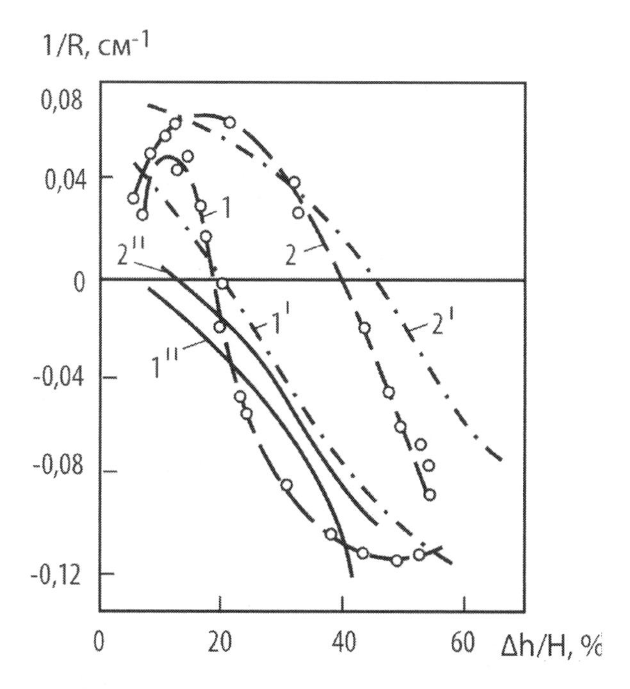

Fig. 5.13. Dependence of the curvature ($1/R$) of the strip on the relative reduction ($\Delta h/H$) in asymmetric hot rolling: $D_l = D_{sm} = 150$ mm, $\omega_l > \omega_{sm}$; 1, 1', 1'' – $H =$ 5 mm, the data of [56], the data calculated by the matrix equation and by the relations of the energy method [56]; 2, 2', 2'' – same as for $H = 9.8$ mm.

the strip is tangential to the roll at the entry point. In this case, on the one hand, directly from Fig. 5.14 we obtain

$$wH = H\sin\varepsilon;\quad Aw = H\cos\varepsilon. \tag{5.79}$$

On the other hand, these values can be found as the difference of the coordinates of the points A and H. In terms of the parameters of the slip lines, these differences between the coordinates can be represented as a vertical and a horizontal projection of the difference of the paths from A and H, respectively according to known slip lines to the common end point. Taking C as an endpoint point, we obtain

$$\begin{aligned} \mathrm{wH} &= (\mathrm{AF})_x + (\mathrm{FG})_x - (\mathrm{GC})_x - (\mathrm{HL})_x - (\mathrm{LM})_x - (\mathrm{MC})_x; \\ \mathrm{Aw} &= (\mathrm{AF})_y + (\mathrm{FG})_y + (\mathrm{GC})_y + (\mathrm{HL})_y + (\mathrm{LM})_y + (\mathrm{MC})_y. \end{aligned} \tag{5.80}$$

Here the indices x, y denote the projections of the slip lines on the horizontal and vertical axes of the Cartesian system X, Y of the coordinates with the origin at the point H. The projections of the

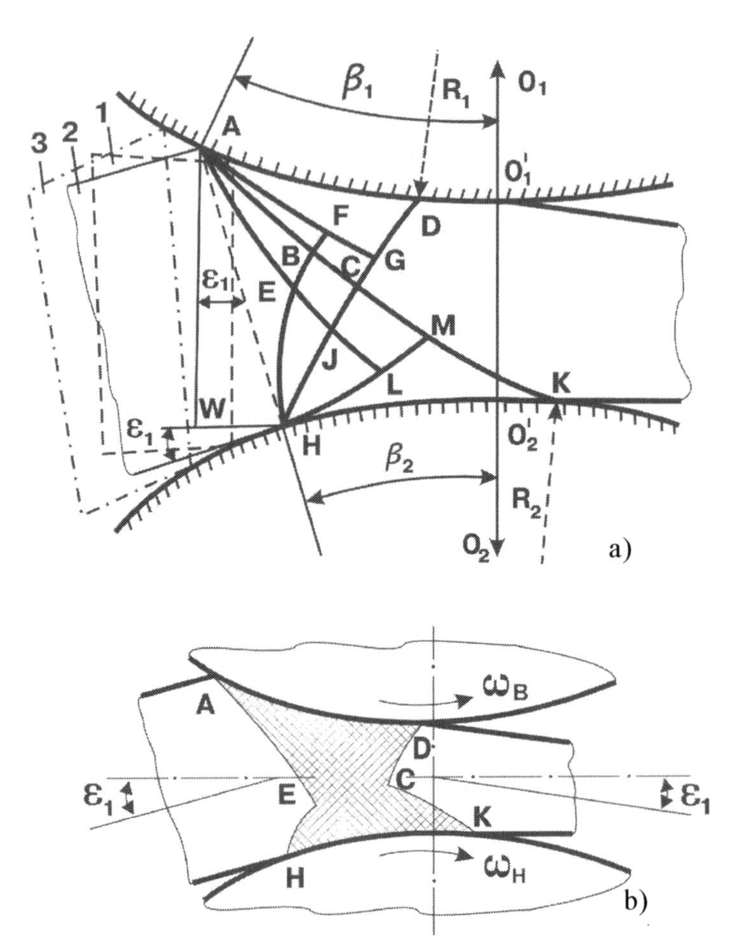

Fig. 5.14. Variants of entry of the strip into rolls (a) and the approximate form of the deformation zone at entry of the strip at an angle to the bottom roll (b): 1 – direct entry; 2 – at a tangent to the lower roll; 3 – with arbitrary angle at entry.

slip lines using matrix operators are as follows: through the radii of curvature of the basis lines we express the radii of curvature of the slip lines AF, FG, GC, HL, MC, determine the moving coordinates $\bar{x}$, $\bar{y}$ of the points F, G, C, M, L, H, and then the Cartesian x, y coordinates of these points. Corresponding dependences have the form

$$x=\bar{x}\cos\varphi+\bar{y}\sin\varphi; \quad y=-\bar{x}\sin\varphi+\bar{y}\cos\varphi;$$

$$\bar{x}=-\sum_{n=0}^{\infty}t_n\left(\frac{\varphi_n}{n!}\right); \; \bar{y}=-\sum_{n=0}^{\infty}t_n\left(\frac{\varphi_n}{(n+1)!}\right),$$

where φ is the angle of the tangent (movable axis $\bar{x}$) to the Cartesian axis at the examined point. The $\overline{y}$ axis is rotated by an angle $\pi/2$ counterclockwise with respect to the axis $\overline{x}$, t_n are factors related to the power series coefficients r_n for the radius of curvature $R(\varphi)=\sum_{n=0}^{\infty} r_n\left(\frac{\varphi^n}{n!}\right)$ by the recurrent relationships

$$t_{-1}=0,\ t_0=0,\ t_1=r_0,\ t_{n+1}=-t_{n-1}-r_n.$$

The values r_n are determined using matrix operators.

Comparison of right-hand sides of the expressions (5.79) and (5.80) gives the required conditions relating the angle ε_1 with the parameters of the field of slip lines

$$\mathrm{H}\sin\varepsilon_1=(\mathrm{AF})_x+(\mathrm{FG})_x-(\mathrm{GC})_x-(\mathrm{HL})_x-(\mathrm{LM})_x+(\mathrm{MC})_x=\Delta_x,$$

$$\mathrm{H}\cos\varepsilon_1=(\mathrm{AF})_y+(\mathrm{FG})_y+(\mathrm{GC})_y+(\mathrm{HL})_y+(\mathrm{LM})_y+(\mathrm{MC})_y=\Delta_y.$$

In other strip entry variants the schematic diagram of the arguments are preserved.

Since the strip is rigid behind the exit points D and K, then, as before, it is necessary to bear in mind the relations (5.54), given that $b_1 = R_{av} + h/2$, $b_2 = R_{av} - h/2$. Furthermore, the following condition must be satisfied for the accepted variant of entry of the strip into the rolls:

$$(R_2+h_0)\cos\beta_2+R_1\cos\beta_1=L, \tag{5.81}$$

where β_1, $\beta_2 = \varepsilon_1$ are the angles of nip of the strips in the upper and lower rolls, L is the distance between the centres of the roll pair.

The angles should not exceed the limit values due to technological requirements. For given values of R_1, R_2, h, L expression (5.81) uniquely determines the angle β_1. On the other hand, when formulating the problem of process control the expression (5.81) acts as a constraint equation.

These relations, together with the matrix equations (5.78), were used to determine the effect of the entry angle on the output parameters of the strip during rolling and to assess the possibility of controlling these parameters. The calculations described above were carried out used with the addition to the software of blocks reflecting the specific task.

The relationship between the entry angle ε_1 of the strip and the curvature ρ of its forward end is shown in Fig. 5.15. The strip steel had the ultimate shear strength k = 210 N/mm^2. The radii of the rolls R_1 = R_2 = 400 mm, the angular speed ω_1 = ω_2 = 60 rpm.

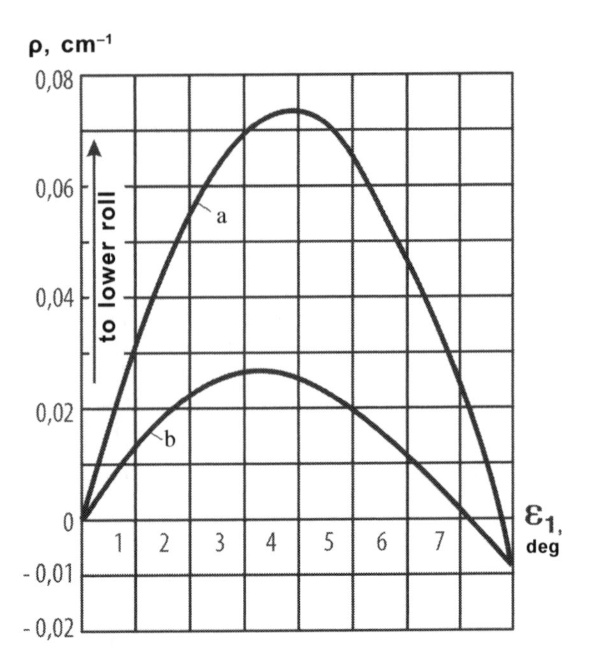

Fig. 5.15. Dependence of curvature (ρ, cm^{-1}) of the front end of the rolled strip on the entrance angle (ε_1, deg) of the strip in rolls when it is tilted to the lower roll: a – rolling without back tension; b – rolling with back tension equal to $0.1k$. Steel 16GS; $k = 210$ N/mm^2; $R_1 = R_2 = 400$ mm; $\omega_1 = \omega_2 = 60$ rpm; original thickness of the strip 10 mm; 40% reduction.

The thickness of the strip at entry was 10 mm, reduction $\varepsilon = 0.40$. Tilting of the strip at entry was assumed to the lower roll. Curve (a) corresponds to the absence of back tension, curve (b) – to back tension, leading to the stress in the strip equal to $0.1k$. According to the results shown in Fig. 5.15, in the range of the inclination angle $0° \leq \varepsilon_1 \leq 7°$ the strip is bent towards the lower roll, both in the absence and in the presence of back tension, and in this range the curvature increases with increasing angle ε_1. For the values ε_1, close to 3–5°, increase the curvature of the front end is slowing and at $\varepsilon_1 \approx 7°$ the direction of curvature is reversed. In the range $0° \leq \varepsilon_1 \leq 7°$ the presence of back tension reduces the distortion at the exit, but does not change its sign. At the same time, the effect of back tension on the curvature of the front end of the strip depends essentially on reduction: the curvature of the rolled strip increases with decreasing reduction, and a sufficiently large value of the ratio of the magnitude of back tension to the reduction the sign of the

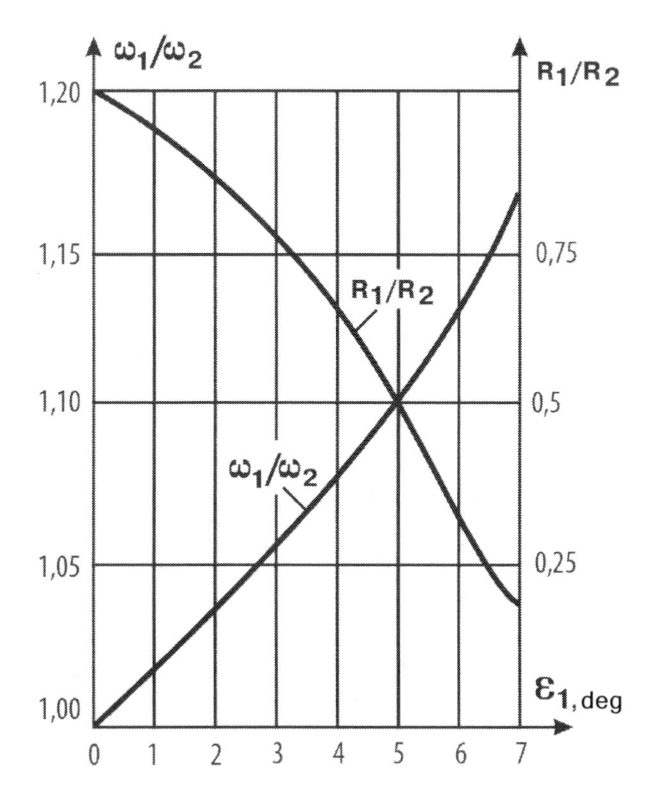

Fig. 5.16. Ratios ω_1/ω_2 of angular velocities and R_1/R_2 of the radii of the rolls, ensuring rolling of a flat strip when entering the rolls at an angle ε_1, deg. Rolling conditions are the same as in Fig. 5.15.

curvature of the strip at the exit from the rolls can be changed to opposite (more on this, see [57, 62]).

The ratios of the angular velocities ω_1/ω_2 and the radii R_1/R_2 of the rolls, which ensure the output of the straight strip from the rolls when it is inclined into the rolls, are shown in Fig. 5.16. As can be seen, at small angles of inclination of the strip in the roll, the output of the straight strip from the rolls is achieved due to a slight deviation of the angular velocities and a small difference in the radii of the rolls. For example, at $\varepsilon_1 = 1°$, the output of the straight (flat) front end of the strip is provided at $\omega_1/\omega_2 = 1.03$ and $R_1/R_2 = 0.94$. At significant tilt angles of the strip exceeding 3°, process control requires the use of rolls with greatly different radii. It should be noted that this situation is quite understandable from the point of view of the results obtained above, which show the influence of the difference in the radii of the rolls and the misalignment of their angular velocities on the curvature of the leading end of the strip: in the first case, there is a periodic dependence the amplitude decreasing with an increase in the ratio R_1/R_2 (Fig. 5.9); in the second – the constant value of the curvature, beginning with a certain amount of misalignment (Fig. 5.10). So, controlling the process by changing the parameters R_1/R_2 and ω_1/ω_2 with increasing their values becomes less effective.

How important it is in production practice to use the asymmetry effect of the rolling process to create favourable technological conditions is shown in Ref. [62] using the example of the 1680 wide-strip hot rolling mill of the Zaporozhstal' metallurgical plant. The peculiarities of the deformation of the strips in this mill are that in the first roughing mill rolling takes without the rear tension of the rolls, since there are no vertical rolls in the stand. The upper layers of the rolls have a higher temperature than the lower ones. The lower work rolls wear out more than the upper ones. The strip plate at the entrance to the rolling stand is deflected by 80–100 mm to the lower roll.

Keeping in the text and in Fig. 5.17 the notations adopted in Ref. [62] it is assume that a strip plate having sufficient stiffness enters the rolls at an angle θ, and the bulk of it is supported by the rolls of the roller table in front of the stand. Then the lengths of the arcs

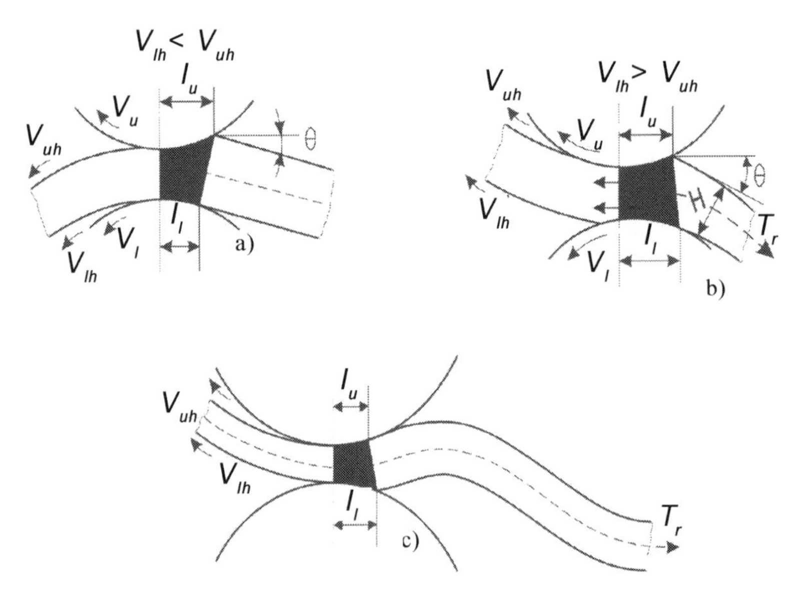

Fig. 5.17. Scheme of gripping the strip by the rolls and the formation of the front end in rolling without rear tension (a), and with rear tension (b, c) [62].

of contact on the side of the upper ℓ_u and the lower rolls ℓ_l will be different:

$$\ell_u = R_u(\alpha+\theta); \ \ \ell_l = R_l(\alpha-\theta);$$

where R_u and R_l – radii of the upper and lower rolls; α – angle of nip.

When $R_u = R_l$, we get $\ell_u > \ell_l$. The reduction from the upper roll side will be larger and, consequently, at the same circumferential velocities of the upper and lower rolls, the metal flow velocity on the upper surface of the strip will be greater than at the lower side, i.e. $V_{uh} > V_{lh}$. In such conditions, the front end of the strip plate leaving the rolls will be bent towards the lower roll (Fig. 5.17 *a*). This is facilitated by a higher temperature of the upper layers of the strip plate. These arguments are confirmed by the production experience – almost all the strip plates in similar conditions come out from the first stand with a bend to the lower roll. To avoid distortion of the front ends of the strip plate, it was recommended to use a lower roll of a larger diameter in the mill stand 1.

In the stands 2–4 of the 1680 WSHRM, the strip plates are rolled with a back tension, which is created by vertical rolls, and with an inclination to the lower roll by an angle $\theta \leq 5°$ (Fig. 5.17 *b*). Under these conditions, when entering the stand, the strip plates are pressed against the lower roll, as a result of which the length of the gripping arc and the reduction from the lower roll side increases, the velocity of the lower layers of the metal increases and at $\upsilon_{uh} > \upsilon_{lh}$ the front end of the rolled metal curves onto the upper roll. As the thickness of the strip plate decreases, the influence of tension increases. Therefore, to obtain strip plates with straight front ends in the stands 2–4 of the 1680 WSHRM, it was recommended [62] to reduce the speed or diameter of the lower roll, which gave positive results.

The recommendations made are not unambiguous for all wide-strip mills, since the process of asymmetric rolling is very sensitive to the slightest changes in its parameters. That is why, in order to make the right technical decisions and instructions on the technology of this process, in each specific case it is essential to carry out detailed investigation with the help of the mathematical models, algorithms and methods of numerical analysis proposed above. The above calculation results suggest the following conclusions.

1. The curvature of the leading end of the strip due to the difference in the radii of the rolls has an oscillatory character with a rapidly decreasing amplitude of successive deviations of the strip to the upper and lower rolls. Initially (with a slight difference in the

radii of the rolls), the strip deviates towards a roll with a smaller radius, and then to a roll with a large radius and so on. A similar (oscillating, with decreasing amplitude) form has the dependence of the torque on the rolls on the ratio of their radii.

2. The dependence of the curvature of the leading end of the strip on the misalignment of the angular velocities of the rolls (the ratio ω_1/ω_2) is different: after curving to a slower rotating roll for small values of the ratio ω_1/ω_2, the strip then curves in the opposite direction and with further increase in the discrepancy of the roll speeds this curvature increases.

3. The results of calculations related to the determination of the effect of the angle of inclination of the strip at the entrance to the rolls on the curvature of its front end show that at low angles of inclination (up to 2°–3°), it is possible to effectively control the process (to provide an exit from the rolls of the flat front end of the strip) due to the difference in the radii of the upper and lower rolls and the misalignment of their angular velocities. At angles of inclination of the strip entering the roll exceeding the indicated values, such process control is inefficient or impossible. This is due to the above-noted peculiarities of the dependence of the curvature of the leading edge of the strip on the ratio R_1/R_2 (sufficiently rapid decrease in the oscillation amplitude) and ω_1/ω_2 (constant curvature towards a faster roll after one oscillation period).

4. The use of the method of slip lines in matrix–operator variants and in combination with the method of non-linear programming allows with sufficient accuracy for practical use to determine the kinematic and power parameters of asymmetric hot rolling, including those which are not detected by other methods or determined not precise enough. The proposed approach also provides an opportunity by imposing restrictions on the range of variation of the individual parameters and appropriate objective functions to manage the process of asymmetric rolling.

5.3. Effects of process asymmetry in cold strip rolling

Cold rolling of sheet steel is carried out, as a rule, using technological lubrication. Any asymmetry of the process results in a non-uniform supply of lubricant to the deformation zone from the upper and lower roll side. As a result of this, the thickness of the lubricating film and, as a consequence, the friction coefficient values are different in the contact areas of the upper and lower rolls with the deformable metal. There is an additional asymmetry of the rolling process, due to the already mentioned difference in the thicknesses of the lubricant layer between the upper and lower surfaces of the strip and the corresponding rolling rolls.

Let's consider this issue in more detail.

The theory of the rolling process with the use of process lubrication is described in detail in monographs [63, 64, etc.]. In the same place, formulas are given for calculating the thickness of the lubricant layer in the deformation zone for symmetrical rolling of strips. The experimental verification showed their suitability for practical use. To calculate the thickness of the lubricating film in the input section of the deformation zone, it is necessary to substitute the equations describing the motion of the lubricant layer between the surfaces of the strip and rolls, which determine the geometry of the input zone of the deformation centre, and also set the required boundary conditions [63]. Using this approach, an equation was obtained [65] for finding the thickness of the lubricant layer ξ_{ent} in the input section of the deformation zone for asymmetric rolling conditions, when the strip is sent with an inclination (at an angle) to one of the rolls (Fig. 5.18):

$$\frac{1-\exp\left(-\theta p_{ent}\right)}{6\theta\mu_0\left(V_R+V_{ent}\right)}=\int_{-\infty}^{0}\frac{-(\alpha\pm\Psi)\cdot x+0.5R^{-1}\cdot x^2}{\left[\xi_{ent}-(\alpha\pm\Psi)\cdot x+0.5R^{-1}\cdot x^2\right]^3}dx,$$

where α is the angle of nip; μ_0 is the dynamic viscosity of the lubricant at atmospheric pressure and room temperature; θ is the piezocoefficient of viscosity; p_{ent} is the pressure in the lubricant layer in the inlet section of the plastic zone of the deformation region; V_R is the circumferential speed of the roll; V_{ent} is the speed of the strip at entry to the deformation region; R is the radius of the roll; Ψ is the angle of inclination of the surface of the strip to the rolling plane (the plus sign before Ψ is accepted when the angle of entry of the surface of the strip into the deformation region is greater than the angle of nip α by an amount Ψ, i.e., from the surface of the strip deviated from the roll and from the horizontal axis by the value of angle Ψ, the minus sign when the angle of entry of the surface of the strip into the deformation region is less than the angle of nip, i.e., from the side of the surface of the strip deflected to the roll), x is the coordinate in the direction of motion of the rolls and strip; y – in a direction parallel to the line connecting the centres of the

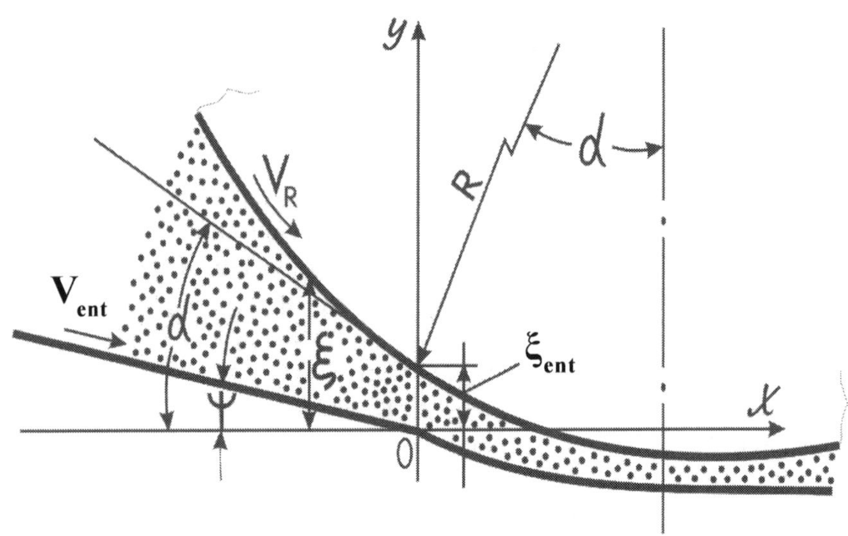

Fig. 5.18. Scheme of the deformation zone at asymmetric rolling of strips with lubrication.

upper and lower rolls; ξ is the current thickness of the lubricant layer; ξ_{ent} is the thickness of the lubricant layer in the entry section of the deformation region.

According to the scheme shown in Fig. 5.18, the lubrication layer ξ_{ent} in the entry section of the deformation region from the side of the surface of the strip deflected to the upper roll will be thicker than the lubricant layer from the lower surface side of the strip. The reason for this is that, on the upper roll side, the angle of entry of the lubricant into the deformation zone is smaller (oil wedge). Accordingly, the friction coefficient will be lower on this side than from the lower rolling roll side. The calculations of ξ_{ent} on the side of both the upper and lower rolls can be performed by solving the above equation with respect to ξ in terms of the magnitude and sign of the angle ψ of the deviation of the strip from the horizontal axis at the entrance to the deformation region. The method of solution, as well as the results of experimental and theoretical studies of this problem are given in [63, 65]. Note that for $\Psi = 0$, i.e. for symmetric rolling conditions, the above expression takes the form of solutions known earlier [63].

Thus, when analyzing the process of asymmetric rolling of strips with the use of process lubrication, it should always be borne in mind that any asymmetry (due to different diameters or speeds of rotation of the upper and lower rolls, the deviation of the entry angle of the strip into the deformation region from the rolling direction, the temperature asymmetry in the thickness of the strip etc.) is accompanied by an additional asymmetry of frictional conditions on the upper and lower contact surfaces due to the different amount of lubricant entering the deformation zone from the sides of the upper and lower rolls. This means that the initial (primary) asymmetry of the parameters of the rolling process generates a secondary effect of the asymmetry of the friction conditions in the deformation zone, which can either enhance or weaken the effect of primary asymmetry on the process. There is a kind of causal relationship. The total result from the effect of both factors in each particular case should be determined by appropriate calculations using the above methodology.

The considered effects can be used in control systems for the rolling of sheets and strips on mills of various designs. For example, in rolling stands with rolls driven from one motor, the mismatch of the moments on the spindles caused by differences in the friction conditions (coefficients) on the contact surfaces of the rolled metal with the upper and lower rolls can reach 30–40%, which adversely affects the operation of the elements of the main lines of the mill. In order to improve the quality of rolled metal, to increase the speed of load equalization in the stands with individual drive of rolls and to ensure the alignment of the moments on the spindles in rolls with a group drive of rolls, the Institute of Ferrous Metallurgy of the National Academy of Sciences of Ukraine proposed (V.L. Mazur, V.A. Trigub and E.A. Parsenyuk) to equalize the moments by changing the entry angle of the strip to the deformation region as a function of the difference of the moments on the spindles.

The alignment of loads on the spindles can be achieved by changing the frictional conditions between the strip and the working rolls. The coefficient of contact friction between the surfaces of the rolls and the deformed metal depends on the thickness of the layer (amount) of lubricant in the deformation zone. In turn, the thickness of the lubricating film on the side of the upper and lower rolls, as shown above, depends on the entry angle of the strip in the deformation zone. Consequently, by varying the angle of inclination of the strip entering the contact zone, it is possible to influence the values of the lubricant layer and the coefficient of friction on the side of the upper and lower rolls, and as a result, the moments of rolling on each roll. That is why by changing (adjusting) the entry angle of the strip into the deformation region as a function of the torque difference on the spindles, it is possible to equalize (regulate) the loads on the motors in the stands with individual drive of the rolls or to equalize the moments on the spindles in the rolls with the group drive of the rolls.

Rolling in rolls with different surface roughness is one of the special cases of the process under asymmetric conditions with different coefficients of friction on the contact surfaces of the upper and lower rolls. The different roughness of the upper and lower rolls in a number of cases increases the efficiency of the rolling process and, as will be shown below, positively affects the physical and mechanical properties of the rolled metal.

When rolling in the uneven rolls strips without tension, the forward creep of the metal from the upper and lower rolls is different. As a result, the rate of metal exit along the height of the deformation region is different and the strip bends. In the process of rolling with tension, which is asymmetrical in friction, in a steady state, the strip leaving the rolls is straight (there is no bending in the vertical plane). Consequently, the rate of metal exit from the deformation region at the points of contact with the upper and lower rolls is the same. However, as a rule, the layers of rolled metal adjacent to the roll with a higher roughness tend to leave at a higher speed and entrain the metal layers behind them from the side of the less rough roll (it is also possible to reverse the layers of metal at higher speeds, adjacent to the smooth roll, but in practice this situation is observed less often). As a result, transverse shear regions arise over the thickness of the strip. The additional tangential forces on the side of the upper and lower rolls are equal in magnitude and opposite in direction. The action of these additional tensile stresses in the strip on the side of the rough roll is equivalent to the action of an additional rear tension, and on the side of the relatively smooth roll, an additional front tension. These additional tensile stresses lead to a reduction in the rolling force and an increase in the efficiency of this process.

In an approximate analysis, following Ref. [66], when cold rolling thin strips with tension in rolls with different surface roughness, the deformation will be assumed to be two-dimensional. A diagram of the forces acting on the element of the strip, separated by vertical sections in the deformation region during rolling in differently rough rolls, is shown in Fig. 5.19. In each section of the deformation region the stresses σ_x, σ_y are assumed to be constant and equal to their average values for the strip thickness. The quantities related to

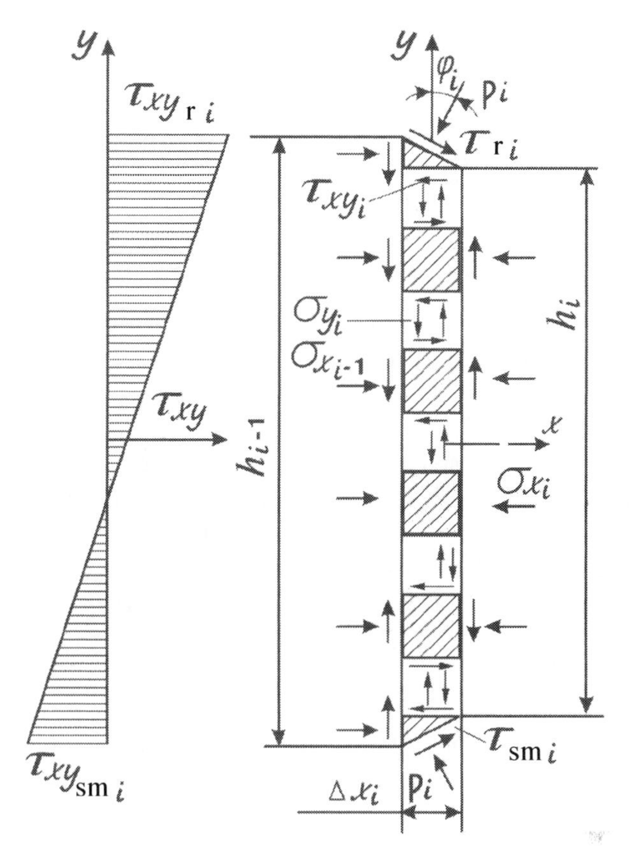

Fig. 5.19. Tangential stresses in the *i*-th section of the deformation zone during rolling in rolls with different surface roughness.

the rough roll are denoted by the 'r' index, and by the 'sm' index to the smooth one.

The equilibrium equation of *i*-th element (Fig. 5.19) has the form

$$\sigma_{x_{i-1}} h_{i-1} - \sigma_{x_i} h_i - 2 p_i \Delta x_i \operatorname{tg} \varphi \pm \left(\tau_{ri} + \tau_{smi} \right) \Delta x_i = 0 \tag{5.82}$$

where p_i are the normal contact stresses; $\sigma_{x_{i-1}}$, σ_{x_i}, are the stress averaged over the cross section, acting in the direction of the x-axis; τ_{ri} and τ_{smi} are the contact tangential stresses on the side of the rough and smooth rolls; h_{i-1} and h_i is thickness of the strip in the (i–1)-th and i-th sections of the deformation region; φ_i is the angular coordinate of the element in question (current angle); Δx_i is the horizontal projection of the i-th portion of the contact arc.

The plus sign before the last term refers to the backward creep zone, and the minus to the forward creep zone.

For the i-th section of the deformation region, the plasticity conditions in stresses averaged over the height of the rolled strip have the form [66, 67, p.59]

$$\sigma_{y_i} - \sigma_{x_i} = 2k_i\psi_i; \quad \psi_i = \frac{1}{h_i}\int_{-h_i/2}^{+h_i/2}\sqrt{1-\left(\frac{\tau_{xy_i}}{k_i}\right)^2}\,dy, \tag{5.83}$$

where k_i is the plastic shear resistance of the rolled metal; τ_{xy_i} are the shear stresses (Fig. 5.19); $\sigma_{y_i} = p_i \cos\varphi_i$, for the case of cold rolling thin strips it can be taken $\sigma_{y_i} \approx p_i$.

In general, ψ_i is a function of the argument x and depends on the magnitude and distribution of shear stresses in the height of the deformation region. The linear dependence is used in the present analysis to describe the shear stresses within the entire deformation region. Tangential stresses in the near-contact layers on areas parallel to the planes of coordinates near the rough $\tau_{xy_{ri}}$ and smooth $\tau_{xy_{smi}}$ rolls have different signs (Fig. 5.19) and are

$$\tau_{xy_{ri}} = \tau_{ri}\cos\varphi_i; \quad \tau_{xy_{smi}} = -\tau_{smi}\cos\varphi_i. \tag{5.84}$$

using expressions (5.84), we write

$$\tau_{xy_i} = \left(\tau_{xy_{ri}} - \tau_{xy_{smi}}\right)y_i / h_i + \left(\tau_{xy_{ri}} + \tau_{xy_{smi}}\right)/2 \tag{5.85}$$

or

$$\tau_{xy_i} = \left[\left(\tau_{ri} + \tau_{smi}\right)y_i / h_i + \left(\tau_{ri} - \tau_{smi}\right)/2\right]\cos\varphi_i.$$

Note that in accordance with the dependences $\tau_{xyi} = 0$ at $y = \frac{\tau_{ri} - \tau_{smi}}{\tau_{ri} + \tau_{smi}} \cdot \frac{h_i}{2}$ and $\tau_{ri} > \tau_{smi}$.

Substituting (5.85) in the second equation (5.83) and integrating with respect to τ_{xy_i}, we obtain

$$\psi_i = \frac{1}{\tau_{xy_{ri}} - \tau_{xy_{smi}}}\int_{\tau_{xy_{smi}}}^{\tau_{xy_{ri}}}\sqrt{1-\left(\frac{\tau_{xy_i}}{k_i}\right)^2}\,d\tau_{xy}. \tag{5.86}$$

After integrating the expression (5.86) and the substitution of (5.84) we find

$$\psi_i = \frac{\cos\varphi\left(\tau_{ri}\sqrt{k_i^2-\tau_{ri}^2\cos^2\varphi_i}+\tau_{smi}\sqrt{k_i^2-\tau_{smi}^2\cos^2\varphi_i}\right)+}{2k_i\left(\tau_{ri}+\tau_{smi}\right)\cos\varphi_i}\rightarrow$$
$$\leftarrow\frac{+k_i^2\left(\arcsin\dfrac{\tau_{ri}\cos\varphi_i}{k_i}+\arcsin\dfrac{\tau_{smi}\cos\varphi_i}{k_i}\right)}{2k_i\left(\tau_{ri}+\tau_{smi}\right)\cos\varphi_i} \quad (5.87)$$

Contact shear stresses (friction force) are expressed in the form of the law of dry friction with the restriction on the upper limit:

$$\left.\begin{array}{l}\tau_{ri}=f_r p_i \text{ at } f_w p_i<k_i;\ \tau_{ri}=k_i \text{ at } f_w p_i\geq k_i;\\ \tau_{smi}=f_{sm} p_i \text{ at } f_{sm} p_i<k_i;\ \tau_{smi}=k_i \text{ at } f_{sm} p_i\geq k_i;\end{array}\right\} \quad (5.88)$$

where f_r and f_{sm} are the friction coefficients from the rough and smooth rolls; within the deformation zone the f_r and f_{sm} values may be taken constant or vary as a function of the coordinate x.

The system of equations (5.82)–(5.83), (5.87)–(5.88) can be used to determine all parameters of the process of rolling strips in rolls with different roughness. Distribution of p along the length of the deformation zone is determined by the method of successive calculations using the algorithm described in [68, 69]. In the same paper [69], the expression of the variables φ_i, h_i, Δx_i included in the equations are presented. A special feature of this solution is only the need for additional iterations to refine at each step the values φ_i and p_i.

The initial conditions for the calculation of normal contact stresses are dependences:

$$\left.\begin{array}{l}p_0=\left(2k_0\psi_0-\sigma_0\right)/\cos\varphi\approx 2k_0\psi_0-\sigma_0 \text{ for the backward creep zone;}\\ p_n=2k_n\psi_n-\sigma_n \text{ for the forward creep zone.}\end{array}\right\} \quad (5.88)$$

The indices 0 and n are the input and output sections of the deformation zone. To reduce the amount of calculations we denote

$$A_i=p_{i-1}h_{i-1}\cos\varphi_{i-1}-2k_{i-1}\psi_{i-1}h_{i-1}+2k_i\psi_i h_i;$$
$$B_i=h_i\cos\varphi_i+2\Delta x_i\,\mathrm{tg}\,\varphi_i;\ C_i=h_i\cos\varphi_i;$$
$$D_i=p_{i+1}\left(h_{i+1}\cos\varphi_{i+1}+2\Delta x_{i+1}\,\mathrm{tg}\,\varphi_{i+1}\right)+2k_i\psi_i h_i-2k_{i+1}h_{i+1}\psi_{i+1}.$$

Now, substituting the expressions (5.83) and (5.88) in the difference equation (5.82) and solving it with respect to p_i, taking into account the adopted notation we obtain:

For the backward creep zone

$$p_i = \frac{A_i}{B_i - (f_r + f_{sm})\Delta x_i} \text{ at } f_r p_i < k_i \text{ and } f_{sm} p_i < k_i;$$

$$\left.p_i = \frac{A_i + k_i \Delta x_i}{B_i - f_{sm}\Delta x_i} \text{ at } f_r p_i \geq k_i, \ (\tau_{ri} = k_i) \text{and } f_r p_i < k_i;\right\} \quad (5.90)$$

$$p_i = \frac{A_i + 2k_i \Delta x_i}{B_i} \text{ at } f_r p_i \geq k_i, (\tau_{ri} = k_i) \text{and } f_{sm} p_i \geq k_i (\tau_{smi} = k_i);$$

and for the forward creep zone

$$p_i = \frac{D_i + (f_r + f_{sm}) p_{i+1} \Delta x_{i+1}}{C_i} \text{ at } f_r p_{i+1} < k_{i+1} \text{ and } f_{sm} p_{i+1} < k_{i+1};$$

$$\left.\begin{aligned} p_i &= \frac{D_i + f_{sm} p_{i+1} \Delta x_{i+1} + k_{i+1} \Delta x_{i+1}}{C_i} \text{ at } f_r p_{i+1} \geq k_{i+1}, (\tau_{ri} = k_{i+1}) \text{ and } f_{sm} p_{i+1} < k_{i+1}; \\ p_i &= \frac{D_i + 2k_{i+1} \Delta x_{i+1}}{C_i} \text{ at } f_r p_{i+1} \geq k_{i+1}, (\tau_{ri+1} = k_{i+1}) \text{ and } f_{sm} p_{i+1} \geq k_{i+1} (\tau_{smi+1} = k_{i+1}). \end{aligned}\right\} \quad (5.91)$$

Note that the right-hand side of the given dependences also contains the variable p_i, since it implicitly, through (5.88) for τ_{ri} and τ_{smi}, enters the function ψ_i according to the formula (5.87). That is why at each step of the calculation iterative cycles are needed to refine the values of this variable.

When implementing the above algorithm, initial data are input and the angle of nip α, the yield strength of the rolled metal σ_{Ti}, φ_i, h_i, k_i, Δx_i are calculated. Next, contact stresses are calculated in the backward creep zone. For the first section of the deformation region, we take $\psi_1 = 1$ as the initial approximation, determine p_i with the help of condition (5.89), calculate τ_{r1} and τ_{sm1}, and use the formula (5.87) to find the refined value ψ_1. After this, the cycle is repeated until the relative error of computing ψ_1 becomes less than the specified value, for example, 0.01. Then we turn to the calculation of p_i, τ_{ri}, τ_{smi}, ψ_1 for the subsequent sections of the deformation zone. And for the initial approximation for p_i and ψ_1 we take the values p_{i-1} and ψ_{i-1}. The question of which of the three formulas (5.90) should be used for the calculation of p_i is solved by a logical check of the equality τ_{r1} and τ_{sm1} to the value k. The value of ψ_i is refined in the iteration cycle.

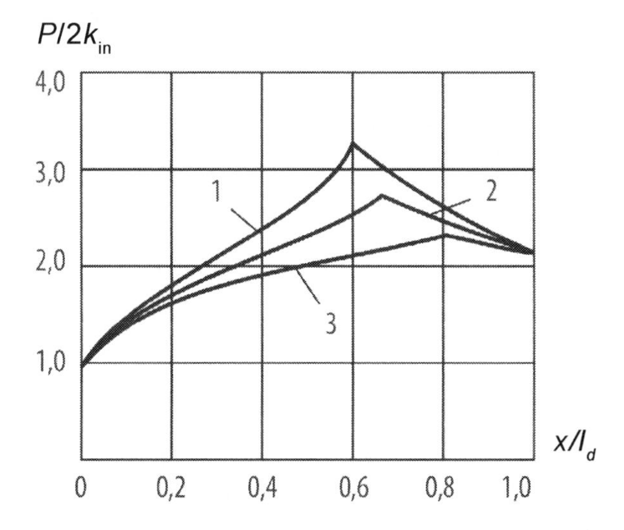

Fig. 5.20. The change in the specific pressure along the length of the deformation zone during the rolling of strips in equally rough (curves 1 and 3) and differently rough (2) rolls. Rolling conditions: h_{in} = 2.5 mm, h_{out} = 2.0 mm, R = 250 mm; $\sigma_{T_{ini}}$ = 200 N/mm²; b = 43 N/mm²; n_y = 0.61; σ_{in} = 20 N/mm²; σ_{out} = 40 N/mm². The hardening of the metal was calculated from the formula $\sigma_T(\varepsilon) = \sigma_{T_{init}} + b\varepsilon^{n_y}$ where ε is the degree of deformation,%. Curves: 1 – $f_r = f_{sm}$ = 0.1; 2 – f_r = 0.1, f_{sm} = 0.03; 3 – $f_r = f_{sm}$ = 0.03. Indices: in, out denote the input and output cross-sections of the deformation region; r, sm – rough and smooth surfaces of the roll; ini is the initial state of the metal.

Similarly, contact stresses are calculated in the forward creep zone. The calculation is carried out from the output section of the deformation region to the intersection of the branches p_i for the backward creep and forward creep zones.

Curves of contact pressure in the deformation zone during rolling of strip of 08 low-carbon steel in coarse-rough, smooth and variously rough rolls, calculated using the technique proposed in our paper [68], are shown in Fig. 5.20. For the selected example, the total rolling force and the mean contact pressure were: 5923.6 kN and 529.82 N/mm² for $f_r = f_{sm}$ = 0.1; 5387.7 kN and 481.89 N/mm² for f_r = 0.1 and f_{sm} = 0.03; 4878.2 kN and 436.32 N/mm² at $f_r = f_{sm}$ = 0.03. That is, the rolling force of the strips in the uneven rolls takes an intermediate value between the rolling forces in the rolls with coarse and smooth surfaces.

Curves of plastic deformation of a strip in rolls with different surface states are shown in Fig. 5.21. When rolling strips of thickness h_1 from a strip plate of thickness h_0 in two coarse-rough rolls, the rolling force will be P_r, and in two smooth rolls P_{sm}. If, during rolling in two rough rolls a back tension is applied to the strip, equal in magnitude to the additional tangential stress in the deformation region, which occurs when rolling in differently rough rolls, the plastic curve 1′ will pass below curve 1 and the rolling force P_r will be less than P'_r. Accordingly, in the case of rolling in both smooth rolls, we obtain $P'_r < P'_{sm}$. The rolling force in the rolls with different roughness $P_{r.sm}$ is approximately equal to the half-sum of the quantities P'_{sm} and P'_r. Therefore, when the strips are rolled in differently rough rolls, when the additional tensile stresses in the deformation zone are small and weakly influence the force, the value of $P_{r.sm}$ will be less than P_r, but larger than P_{sm} (Fig. 5.21 *a*). An example of such a rolling procedure is shown in Fig. 5.20. Under the conditions when the role of additional tensile stresses arising in the deformation region due to the difference in the roughness of the upper and lower rolls is manifested strongly, it is found that $P_{r.sm} < P_{sm} < P_r$ (Fig. 5.21 *b*).

The results of an experimental study of the rolling of strips in variously rough rolls at the five-stand 1700 mill of the Karaganda

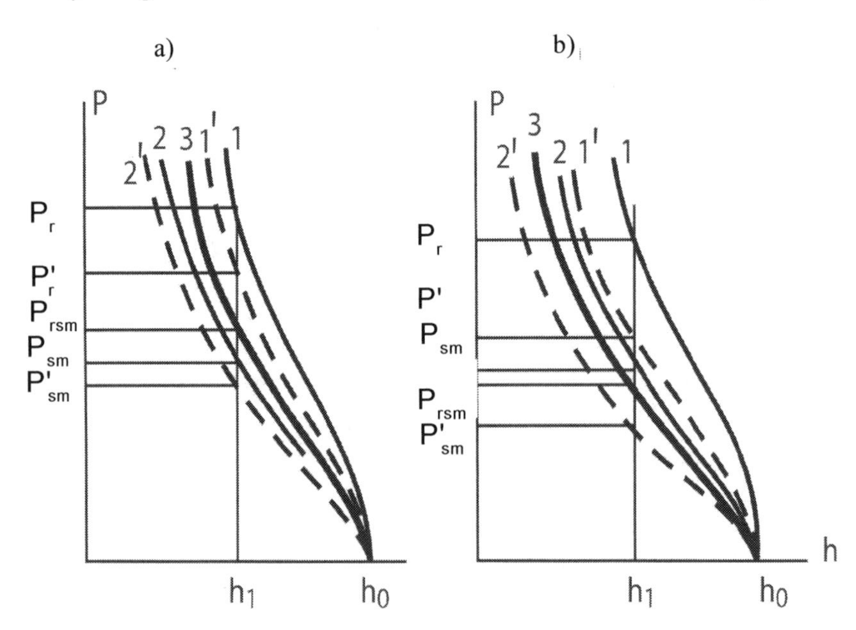

Fig. 5.21. Force P of rolling in rolls with different surface roughness: 1 – rolling in two rough (r) rolls; 1'– the same, with tension; 2 – rolling in two smooth (sm) rolls; 2'– the same, with tension; 3 – rolling in rolls with different roughness (rsm); h_0 and h_n – thickness of the strip before and after rolling. a) $P_{sm} < P_{rsm} < P_r$; b) $P_{rsm} < P_{sm} < P_r$.

Metallurgical Combine showed that when rolling strips of steel with a thickness of 0.5 mm from a strip plate of thickness 2.5 mm and a width of 1015 mm, the use in the last stand of rolls with different surface roughness (the upper one Ra = 4–5 μm, the lower one Ra = 0.4–0.6 μm) reduces the rolling force by 5–8% compared to the rolling force in two coarse-rough rolls [70]. When rolling strips in two equally coarse-rough rolls (Ra = 4–5 μm), the torque developed by the fifth stand engine, after gripping the front end of the strip by rolls, was M_5 = 61 kN · m. Since the total tension between the fourth and fifth stands T_{45} was 180 kN, and the working roll radius was 0.3 m, the strain moment was equal to M_d = 56.6 kN · m.

When rolling strips in rolls with various surface roughness M_5 = 58 kN · m, the total tension between the fourth and fifth stands T_{45} = 182 kN and M_d = 52.6 kN · m. Consequently, the use of work rolls with different surface roughness in the rolling conditions considered provided a reduction in the deformation moment by 8.6%.

One of the basic problems in the theory and technology of the rolling process in miscellaneous rough rolls is to define the direction of bending of the rolled strip. Theoretically, using the method of slip lines, this issue was discussed above. In the experiments performed on the 200 duo-quarto mill of the Institute of Ferrous Metallurgy in rolls with a diameter of 55 mm 1.92 mm thick strips of steel E0300 and 2.12 mm thick steel E0100 strips were rolled in five passes to a thickness 0.46–0.54 mm. The surface roughness of the rolls varied significantly. One roll was polished Ra = 0.3 μm, the other was processed by the electric spark method to Ra = 15–20 μm. At the first pass and the absolute reduction, approximately 0.5–0.6 mm, the samples slightly bent towards the coarse-rough roll. In the second and third passes, when rolling was carried out from a thickness of 1.45–1.60 mm to 1.2–1.3 mm and further to 1.0–1.1 mm, the samples came out almost flat from the rolls. In the fourth and fifth passes, when leaving the deformation zone, the strips bent toward the smooth roll. In the last pass the bend was stronger. Similar experimental results were obtained by V.A. Nikolayev and his co-workers [61].

The physical nature of the noted patterns of the bending of the strip towards the upper or lower rolls is that, in the asymmetric rolling process, the metal forward creep on the side of each of the rolls varies depending on the thickness of the strip, the reduction, the geometry of the deformation zone, the properties of rolled metal, its hardening, coefficients of friction on the contact surface with the upper and lower rolls. In accordance with the scheme in Fig.

5.22 for small reductions, when $\alpha < \alpha_1$, the value of the angle of the neutral section γ and the forward creep S on the side of the smooth roll is greater than on the side of the rough roll $\gamma_{sm} > \gamma_r$ side and the strip is bent onto the rough roll when leaving the deformation zone. Since the difference in the values of forward creep S_{sm} and S_r is small, the curvature of the strip after rolling is also small. At a certain value of the angle of nip ($\alpha = \alpha_1$), the values of the neutral angles and forward creep from each side are equal ($\gamma_{sm} = \gamma_r$) and, as a result, the strip leaves the rolls flat. In the cases $\alpha_1 < \alpha \leq \alpha_{sm}$, the strip is bent already on the smooth roll, since $\gamma_r > \gamma_{sm}$. In this range of reduction, the difference between the values of γ_r and γ_{sm} is already large, so that the rolled strip will be strongly curved. At $\alpha_{sm} < \alpha \leq \alpha_r$, the surface of the rolled metal slips relatively to the smooth roll. The strip strongly bends to the smooth roll. Slippage of the rolling surface over a smooth roll causes vibrations to occur. It should be noted that the nature of the dependences $\gamma_{sm}(\alpha)$ and $\gamma_r(\alpha)$ is rather complicated, since with increasing reduction and the angle of nip α cause changes of the rolling force, the imprint of the roughness of the rolls on the strip, the thickness of the lubricating film during rolling with lubrication, and, as a result, the coefficients friction from the smooth and rough rolls.

The scheme depicted in Fig. 5.22 shows the position when the curves $\gamma_{sm}(\alpha)$ and $\gamma_r(\alpha)$ have a common point of intersection. There is also the case when there is no such point and the curve $\gamma_{sm}(\alpha)$ passes below the curve $\gamma_r(\alpha)$ for all values of α. In such conditions,

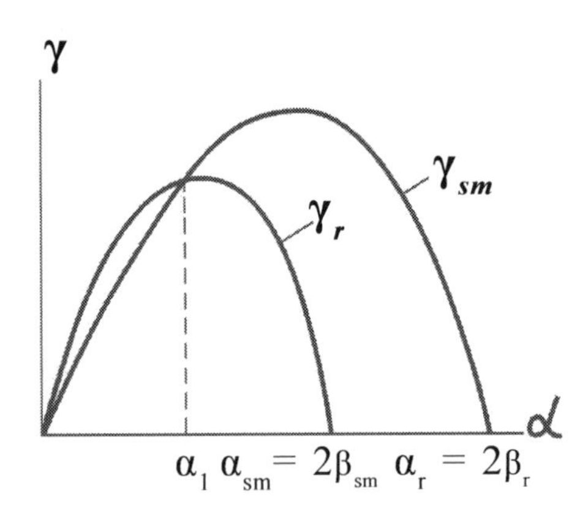

Fig. 5.22. Scheme for determining the direction of bending of the strip in rolling in rolls with different surface roughness; α, β, γ – angles of nip, friction and position of the neutral section. Indexes: sm – by smooth roll; r – rough.

the strip at the exit from the deformation zone will always be bent towards the smooth roll.

The performed theoretical analysis and experimental studies have shown that the use of the work rolls with different roughness allows solving the following technological tasks during rolling.

1. In order to obtain the greatest efficiency from the energy point of view of the process of asymmetric rolling in the rolls of different surface roughness, it is advisable to give greater roughness to the roll having a larger diameter. Such a technology is rationally applied to those mills where sheets and strips of hard-to-deform steels and alloys are rolled and rolling is carried out at a low speed.

2. In high-speed mills where the danger of the damage in the rolls is high due to the bending of the strip during breaks, to balance the rate of metal exit from the deformation zone along the height of the strip and reduce its bending, the increased roughness should be applied to rolls of smaller diameter. That is, the asymmetry of the rolling process, due to the difference in the diameter of the rolls, is compensated by creating an asymmetry of the friction conditions.

3. In sheet rolling and tempering mills, in which the separation the power flow of the drive motor to the two working rolls is carried out through the gear stand with a gear ratio equal to one (as in the 1700 temper mill of the Karaganda Metallurgical Combine), to equalize the forces in the branches of the closed loop of the drive, to reduce overloads and vibrations in the spindles, it is necessary that the roll of smaller diameter was more rough.

Because of uneven wear and tolerances during grinding, the diameters of the upper and lower working rolls in the stand are, as a rule, somewhat different. In mills with a drive of this type, when the strip enters the stand between working rolls, a force interaction arises and a closed kinematic chain is formed, consisting of gear rolls, spindles, work rolls and rolled metal. Moreover, if the gear ratio between the gears is equal to one, then, although insignificantly, this ratio differs from the unity due to the difference in their diameters. Due to the inequality of the transmission ratios the drive branches are twisting, and considerable forces are exerted in the closed kinematic chain. Under their action, self-regulation processes occur in the deformation zone, which stop the twisting of the closed system, and the magnitude of the forces in it stabilize. The inequality of the friction coefficients on the upper and lower rolls, due to the difference in their roughness, exerts the same effect on the nature of the loads as the inequality of the circumferential velocities. Therefore, by choosing the appropriate difference in the roughness of the roll surfaces in the pair, it is possible to largely compensate for the difference in their diameters and substantially equalize the loads in the branches of the closed drive circuit. For this, as in the case of preventing the winding of the strip on the rolls thus damaging the latter, it is necessary that the roll of smaller diameter be rougher. Other ways of reducing overloads and vibrations in the spindles of tempering mills with a group drive are discussed in detail in the chapter on the tempering process, and in our work [71].

4. When rolling or skin pass rolling the strips with rollers of the same diameter in the mills with a group drive (through the gear stand from one motor), as, for example, in the 1700 tempering mill of the Karaganda Metallurgical Combine, in order to avoid shocks in the gear stand, the roll with a lower roughness should be placed on the same shaft with the driving gear roll. In this case, the lateral clearance in the meshing of the gear wheels will always be closed, and, therefore, the prerequisites for formation of impacts in the meshing will be eliminated.

Thus, the decision on which of the work rolls to impart an increased roughness should be determined by specific conditions, taking into account the above recommendations.

When rolling strips in rolls of different surface roughness the effect of increasing the efficiency of the process increases with increase of the difference in the values of the roughness of one and the other rolls. Therefore, one roll is given a very high roughness (Ra = 4–6 μm), and the other is as smooth as possible (Ra = 0.2–0.4 μm). Such roughness, however, is often short-lived. The roughness wears out more intensively, the more the initial surface finish differs from the equilibrium surface, which corresponds to the temperature–force conditions of the rolling process. The improved rolling method in the uneven rolls is that the orientation of the microrelief of one and the other work rolls in the set is performed with a different orientation with respect to the roll axis. That is, on the surfaces of the upper and lower rolls of one set, a roughness is created, differing in the direction of microroughness. The directional microrelief can be produced most efficiently in the form of rectilinear scratches. In this case, the risks on the surface of one roll must be oriented perpendicular to the scratched on the surface of another roll. In one roll the roughness must be parallel, and in the other perpendicular. Since the microrelief of the parallel and perpendicular type 'captures' different amounts of lubricant, the friction coefficients on the side of the upper and lower rolls will be different even at the same microroughness height. As a result, the deformation zone becomes asymmetrical in the friction conditions.

When rolling with the rolls of the same magnitude but different directions of microroughness, the rolled metal receives the same roughness on both sides, and the requirements of the standards for the surface state are very simply satisfied. At the same time, the effect of increasing the efficiency of the rolling process due to the asymmetry of external friction persists.

Studies of the process of rolling strips, asymmetric in friction, made it also possible to proposed a method for processing the welded joint of strips before subsequent cold rolling, according to which the roughness of the welded seam is different from the top and bottom sides. Moreover, the roughness can be different both in magnitude and in the direction of microroughness. The energy–power parameters of the rolling of the welded section with the specified treatment are reduced, which positively affects its reliability.

The stability of the rolling process of sections of the strip with welded seams can also be increased by creating its asymmetry can by deflecting the strip to one of the rollers. By adjusting the angle of the strip entry into the rolls during the passage of the weld seam through the deformation zone, it is possible to weaken the increment of the rolling force due to the lower ductility of the weld as compared to the ends of the welded strips.

In the technical literature, a large number of original ways of influencing the rolling process are presented by using the effect of the difference in the surface roughness of the upper and lower rolls for solving many technological problems. Thus, in the Zaporozh'e Industrial Institute, a method of cold rolling of strips on a continuous mill was developed, providing for the use of working rolls in different mill stands with different surface roughness. In the first stand of the continuous mill rolling is carried out in the rough work rolls with a roughness $Ra = 4 \div 5$ µm. In the second and third stands of the five-stand mill in ground ($Ra \approx 0.8$ µm) working rolls. In the fourth stand (pre-finishing), a lower roll with a roughness $Ra = 0.8$ µm is used, and the upper one, to which the strip is deflected by the rolls of strain gauges, with a coarser surface ($Ra = 2$–3 µm). In the last fifth stand (finishing), very rough work rolls ($Ra \approx 4$ µm) are used. That is, in the prefinishing stand the strip during rolling is inclined to the roll whose surface roughness is 2.5–3.75 times greater than that in the other roll in this stand, and 1.3–2 times less than the surface roughness of the work rolls in the subsequent (finishing) of the mill stand. According to the authors of this method, its application will increase the durability of he microprofile of the roll surface. As a result, the risk of inter-turn welding of strips in coils when annealing is reduced and the of appearance defects caused by this phenomenon is prevented.

Considering the topic of rolling strips and sheets in asymmetrical conditions, it is probably necessary to pay attention also to technical solutions in which asymmetry is used in relation to rolling rolls or roll bundle assemblies. Most often, these decisions concern asymmetric profiling of working rolls both along the length of their barrels and in the circumferential direction (along the perimeter). For example, in order to improve the operational stability of work rolls in wide-strip hot rolling mills and to improve the transverse profile of rolled strips, it has been proposed[4] to make the generator of the roll drum asymmetrical along its length, with one half of the barrel being cylindrical and the other half curved with a decrease in diameter from the middle to the edge of the barrel.

The idea of this solution is that the rolls of wide-strip rolling mills wear out unevenly along the length of the barrel. Because of this, the thickness of the edges varies and the rolling strips become wedge-shaped. When rolls with a symmetrical barrel profiling are used by the end of the working roll operation campaign (before changing the rolls), which lasts, as a rule, 5–7 hours, the difference in wear of the edge portions of the roll and the thicknesses of the right and left edges of the strip reaches 0.07–0.15 mm. At occurrence of such uneven wear, the set of the work rolls is changed. To compensate for this negative phenomenon, the work rolls are shaped asymmetrically along the length of the barrel. This edge of the roll, where wear is expected to be large, is assumed to show the increase in the diameter of the barrel.

In many hot and cold rolling mills, in order to increase the accuracy of the geometric dimensions and to ensure the required profile of the rolled metal, the working rolls are axially moved. In this case, the roll bundle of the rolled stand includes the work rolls formed with a curved asymmetric barrel profiling comprising a concave portion with a start at the edge of the barrel. The upper and lower rollers with such profiling are deployed 180° apart. Moreover, the concave portion of the drum of the work roll is asymmetric with respect to its centre. In this case, when the work rolls are displaced axially during the rolling of the strips, the gap between

[4]V.A. Nikolaev, V.L. Mazur, S.S. Pilipenko, O.N. Soskovets, V.P. Sosulin

their generatrices can be convex or concave. Due to this, operative regulation of the profile and shape of the rolled strips is achieved.

Let's consider one more rarely used variant of the process of cold rolling of strips in the asymmetric conditions and a rolling roll for its realization.

During cold rolling of thin strips (tinplate) on continuous mills, as well as during skin pass rolling, the roughness of the surface of the rolls is imprinted on the surface of the rolled metal. This question is considered in detail in the chapter devoted to the process of skin pass rolling thin sheet steel. By varying the roughness on the surface of the rolling or skin pass rolling rolls along the length of their barrel, one can influence the distribution of the surface roughness value across the axis of the strip (in a direction perpendicular to the rolling axis).

The technology of hot tinning sheet production involves rolling a tinplate on a continuous mill, skin pass rolling, cutting strips into cards and then coating these cards in hot tinning units. In this case, the direction of movement of tinplate cards in hot tinning units is perpendicular to the rolling axis of these cards. For example, a strip of width 712 mm after cold rolling on a transverse cutting unit is cut to measuring lengths of 512 mm. At the subsequent hot tinning, the received tinplate cards of the tinplate are set in the unit in a direction parallel to the long side (712 mm), i.e., if at rolling a size of 712 mm represents the width of the strip, then when tinning it represented the length of the tinplate card.

Features of the aggregates and technology of hot tinning of tinplate determine significant differences in thickness of the coating in the direction of tinning. At the front end of the tinplate during the tinning process the thickness of the tin coating is usually less than at the rear end. It is established that the thickness of the coating during tinning is directly proportional to the roughness of the metal before tinning. In order to compensate for (prevent) systematic thickening of the coating from the front end to the back end of the card by changing the roughness of the tinplate surface before tinning (cold rolled tin), it is necessary that the surface roughness of the card decreases from the front end to the rear end. Accordingly, the roughness of the cold-rolled strip surface must be variable in width. Moreover, a larger amount of roughness should be on the surface from the side, which during subsequent tinning enters the tinning aggregate. Consequently, the roughness of the roll barrel surface must change uniformly (monotonically) from one edge to the other[5].

The uniform increase in surface roughness from one edge of the roll barrel to the other is achieved in the process of grinding, polishing the rolls or shot blasting.

The above technical proposals are certainly not cardinal and complex for wide practical use on industrial mills. Sometimes they can cause side negative effects. Therefore, their use is advisable only in cases where simpler in implementation, more accessible and effective technological methods still do not solve the tasks. Nevertheless, these similar proposals considered below in the subsequent sections of the book well illustrate the multifaceted possibilities of influencing the rolling process by the asymmetry of its input parameters.

5.4. Influence of asymmetry on the rolling process on the texture of steel sheet

One very significant, but poorly understood effects of the asymmetric rolling process is the effect of the asymmetry of the process on the structure, texture and mechanical properties of rolled metal in such circumstances.

It is known [42] that the ability of a steel sheet to deep drawing substantially depends on the coefficient of normal plastic anisotropy which characterizes the differences in the properties of the metal, measured in the normal and parallel directions to the plane of the sheet and is determined by the ratio of the strains in the width and thickness of the sample under tension:

$$r = \frac{\ell n(b_O / b)}{\ell n(h_O / h)},$$

where b_o and h_o are the initial width and thickness of the sample; b and h are width and thickness after stretching by 15–20%.

The value r varies depending on the orientation of the sample relative to the rolling direction. Therefore, the normal plastic anisotropy of the sheet steel should be estimated by the average value $\bar{r} = 1/4(r_{0°} + 2r_{45°} + r_{90°})$, where 0°, 45°, 90° are the orientation angles of the test sample to the rolling direction. The difference

[5]Proposed by V.L. Mazur, A.I. Dobronpravov, V.A. Kuvshinov.

of the values of r in different directions in the plane of the sheet characterizes the coefficient of plane anisotropy

$$\Delta r = \frac{1}{2}\left(r_{0^\circ} + r_{90^\circ}\right) - r_{45^\circ}.$$

The high-quality sheet steel, which meets the requirements of various types of deformation of punching operations, must have high values of $\bar{r}$ with a slight planar anisotropy Δr. Since the normal plastic anisotropy of the metal is determined by its crystallographic texture, the conditions for the production technology of sheet steel, which affect its texture, should be the subject of increased attention of metallurgists. What has been said fully concerns the effect of the asymmetry of the rolling process on the texture and the normal plastic anisotropy of the steel.

The texture of the annealed cold-rolled sheet is described by the following ideal orientations: an axial texture with [111] axis perpendicular to the rolling plane; (111) <112>; (111) <110>; (001) [110]; (112) <110>. Favourable for stamping are orientations (111) <*uvw*>, unfavourable (001) [110]. Texture (112) <110> causes an increased planar anisotropy of sheet steel.

The Institute of Ferrous Metallurgy investigated the influence of the asymmetry of the rolling process on the texture of cold rolled steel due to the different roughness of the upper and lower working rolls, and, of course, the unequal frictional conditions on the contact surfaces of the rolled metal in the deformation zone on the side of one and the other rolls.

To determine the influence of surface roughness on the texture the 08Yu sheet steel 2.2 and 3.6 mm thick was rolled in several passes with a total reduction of 30; 50 and 70% in rolls with a diameter of 200 mm and different surface roughness in the following conditions: I – the surfaces of both rolls had the same roughness value of 0.8 μm *Ra*; II – both rolls had the same roughness of 5.0 μm *Ra*; III – both rolls had the same roughness of 0.3 μm *Ra*; IV – the surface of one roll had a roughness of 0.3 μm *Ra*, the other 8.0 μm *Ra*.

The cold-rolled steel was subjected to recrystallization annealing in hood-type furnaces using the technology adopted for the metal (particularly complex drawability): ageing temperature 680°C, holding time 12 hours.

The texture interference curves (110) were plotted according to the Schulz method from the sample plane of sheet steel. In the initial position, the sample was aligned by Bragg–Brentano focusing. Then it was slowly tilted around an axis passing through the plane of the primary and reflected rays, and simultaneously rotated around the normal to its surface.

The results of these investigations are detailed in our papers [72, 73]. Here we dwell on the main results and conclusions.

The main feature of the deformation texture of sheet steel rolled with a reduction of 30% under asymmetric conditions with different roughness of the upper and lower rolls (0.3 and 8.0 μm *Ra*) is the strong slope of the symmetry plane of the pole figure to the rolling plane in the layers adjacent to the roll surface. Changing the friction conditions on one of the rolls (increasing the roughness) leads to a sharp change in the texture in the layers adjacent to the smooth roll, as compared to rolling in both smooth rolls. The degree of texture is higher than with rolling by two smooth rolls, especially for $0.25h$; $0.375h$ and $0.5h$ layers. The nature of the change in the texture along the depth from the side of the rough roll is very similar to that observed in the case of rolling on two rolls with a rough roughness of the surface. The surface layer has an almost random orientation of the crystallites.

When rolling in rolls with different surface roughness with a reduction of 50%, a large slope of the plane of symmetry of the pole figure to the rolling plane also forms. The texture distribution along the thickness of the sheet on the side of the smooth roll is very different from that observed for rolling in both smooth rolls with the same reduction. The difference is manifested not only in the inclination of the orientation (112)<110> to the rolling plane along the thickness of the sample and the features of the change in the angle γ, but also in the change in the degree of texture of the steel, which is close to that found for the two rough rollers. On the side of the rough roll as well as at the reduction of 30%, the texture distribution along the thickness is similar to that observed for a metal rolled in symmetrical conditions by two rough rolls. The texture of the surface layer is weakly expressed and sharply enhanced when removed from the surface.

Rolling in the rolls with different surface roughness with a reduction of 70% leads to a texture distribution that almost repeats on the side of the smooth roll the texture change in thickness, characteristic for rolling in two smooth rolls, and on side of the rough roll – in two rolls with high roughness.

The annealing of the metal rolled under asymmetric conditions in the upper and lower rolls with different surface roughness (0.3 and 8.0 μm *Ra*) and with a reduction of 30 and 50% leads to the formation of an orientation of the crystallites, which is close to random throughout the section of the sheet. A characteristic feature of the texture of these samples is its greater scattering in comparison with the annealed metal, which was rolled under symmetrical conditions with two equally rough rolls. This is particularly pronounced in the middle layers of the metal, rolled with a reduction of 30 and 50%. Probably, the inclined texture during the recrystallization of cold-rolled steel during the annealing process passes mainly to a random component. This assumption is supported by the fact that in the case of asymmetric rolling of sheet steel with differently rough rolls with a reduction of 70%, the deformation texture is better inherited by recrystallization.

In industrial practice, in particular in the production of thin-sheet metal for deep drawing, the presence of a random component in the texture of cold-rolled steel is undesirable. Here the problem consists in the opposite: strengthening of the axial texture favourable for ensuring high values of normal plastic anisotropy with the {111} plane orientation in the rolling plane. Therefore, it is necessary to use the advantages of asymmetric deformation on the one hand to increase the efficiency of the rolling process, and on the other hand, to improve the texture of cold-rolled annealed steel. This problem can be solved by the fact that during the process of asymmetric deformation of the metal in the multi-stand mill due to the different roughness of the upper and lower rolls, the ratio of the roughness values of the upper and lower rolls in each subsequent mill stand is reversed.

It has been shown above that the microrelief of the surface of the rolls and the strip affects the texture of the sheet steel and the normal plastic anisotropy caused by it. The greatest slope of the symmetry plane of the pole figure to the plane of the rolled metal is observed when rolling in rolls with significantly different surface roughness: one roller is smooth (for example, Ra = 0.3 μm), another rough (e.g. Ra = 8 μm). The annealing of the metal rolled in differently rough rolls leads to the formation of an orientation of the crystallites which is close to random in the entire section of the sheet. The texture of recrystallized steel, in which crystallites predominate with erratic orientation, is undesirable from the point of view of metal stamping.

Consequently, the rolling of strips in rolls with different roughness should be carried out so that after the annealing of the sheet steel the character of the texture distribution along the sheet section remains approximately the same as in the cold-deformed metal.

When rolling in uneven rolls, as in other ways of creating the asymmetry of this process, the strip is bent to the upper or lower roll. Because of this, during the rolling of strips with tension, the distribution of tensile stresses is asymmetric in the thickness of the strip. Alternating the location of the smooth and rough rolls in each subsequent stand of the multi-stand mill increases the marked uneven distribution of the tensile stresses from the tension along the thickness of the strips. On the asymmetry of the rolling process, because of the difference in the roughness of the surfaces of the upper and lower rolls, the asymmetry is imposed, caused by the uneven distribution of the tensile stresses along the thickness of the strip. As a result, due to the appearance of additional shear deformations, the efficiency of the rolling process increases.

The improvement in the texture of the metal is due to the fact that the rolling of the strips with alternation along the paths of the arrangement of the rough and smooth rolls (when the roughness of the upper and lower rolls changes in each subsequent stand) from the point of view of the effect on the texture of the metal is almost equivalent to rolling in two equally rough rolls in all stands. So, if after rolling in the first stand with a high roughness of the upper and low roughness of the lower rolls on a relatively smooth surface of the strip, a steep slope of the plane of symmetry of the pole figure describing the texture of the metal, then after rolling in the next stand, where the upper roll is already smooth and the lower one rough, a slope of the texture is formed also on the opposite side of the rolled metal. At the same time, the number of crystallites with orientations (112) <110> is equalized; (001) [110]; (111) <112>, as well as the proportion of randomly oriented crystallites from the upper and lower surfaces of the rolled metal. As a result, the angle of inclination of the largest axis of elongated grains to the plane of the sheet is about the same as after rolling under symmetric conditions in two equally rough rollers.

After annealing, the sheet steel is no longer texture-free as after rolling in the rolls with different roughness without changing the ratio of the roughness of the upper and lower rolls in the passes, and similar in texture to the rolled metal under symmetrical conditions. That is, the rolling in each pass is carried out in an asymmetric regime as regards the roughness of the upper and lower rolls, using all the advantages of this process, and the texture of the annealed metal is the same as after rolling under symmetrical conditions (a texture favourable for subsequent stamping).

In the general case, it is possible to change to the opposite not only the ratio of the roughness values of the upper and lower rolls, but also to alternate in the stands the direction (type) of the microrelief of the rolls. For example, in the first stand the upper roll can have a roughness of a parallel type, and the lower one – a roughness of the perpendicular type, in the second stand, on the contrary. Differences in the orientation of the microrelief of the upper and lower rolls, because of the different amounts of the lubricant taken up by the roughness, create an asymmetry of the friction conditions in the deformation zone during rolling. However, the effect of the asymmetry of the process is weaker here than with an appreciable difference in the roughness values of the upper and lower rolls.

The asymmetry of the rolling process in each successive pass can be reversed by other methods, for example, in each subsequent pass, by alternating the arrangement of the upper rolls with different diameters or rotational speeds, changing the temperature asymmetry of the rolling process in each pass or the asymmetry in the lubrication conditions. However, these ways of implementing the process asymmetry are more complex and less effective.

In the continuous mills, the arrangement of rough and smooth rolls in the last stand of the mill is expediently set so that the direction of bending of the strip emerging from the last stand coincides with the direction of bending of the strip on the drum of the coiler.

The disordered component in the texture adversely affects the coefficient of normal plastic anisotropy and the ability of annealed cold-rolled steel to deep drawing. However, for a metal of a different purpose, for example, certain types of electrical (dynamo) steel, cold-rolled low-carbon steel intended for subsequent enameling, such 'non-texture' is desirable.

Thus, experts[6] at the Karaganda Metallurgical Combine proposed methods for the production of cold-rolled isotropic (dynamo) electrical steel, which involve asymmetrical rolling of strips in one or several stands of a finishing group of the 1700 wide-strip hot rolling mill, that is, at the first rolling production stage. In the first case, finishing rolling in one or several match is carried out with a mismatch in the speeds of the rolls, depending on the temperature

[6]V.N. Kulikov, V.I. Sidor'kin, T.S. Seisimbinov, A.G. Svichinskii, B.V. Balandin

difference on the upper and lower surfaces of the strip plate. The rolls with a greater speed impact on the surface of the roll with a lower temperature. Conversely, a roll with a lower speed – on a surface with a higher temperature. In another case, the asymmetry of the process of hot rolling strips in at least one stand is created by using one drive and a second non-drive roll. The drive roll is a roll of smaller diameter. In the opinion of the authors of these methods, their use makes it possible to significantly improve the magnetic properties of cold-rolled steel with a silicon content of ~0.3%.

As an example, let us also turn to the technology of producing cold-rolled low-carbon steel of the 08Yu type with an aluminum content of 0.01–0.07%, which is used for the production of mainly dishes by the method of cold stamping with subsequent enameling of stamped products. The thin-sheet low-carbon steel for this purpose must have the ability to be processed by drawing, and its surface must have a roughness which ensures improved grip of the lubricant during the drawing of the products, and an improvement in the metal stamping. At the same time, the surface roughness should not be too high, so that the smoothness of the surface of the enamel coating applied to the stamped products is not affected. These requirements are satisfied with roughness of the surface of cold-rolled steel Ra = 0.8–2 μm.

Experiments show that the tendency to the formation of 'fish scales' defects on the surface of enameled ware (articles) is significantly influenced by the crystallographic texture of steel: the weaker the texture, the lower is the tendency to form 'fish scales'. However, cold-rolled, low-carbon aluminum-deoxidized steel sheets have a rather strong texture and a high coefficient of normal plastic anisotropy. There are no special possibilities to reduce the texture of the steel due to the influence on the modes of hot and cold rolling of strips and heat treatment of cold-rolled steel, since all technological regimes are chosen to ensure good metal stamping. At the same time, it is possible to influence the texture of the steel by cold rolling the strips in differently rough rolls

It was shown above that the main feature of the deformation texture of sheet steel rolled with different roughness of the upper and lower rolls is a steep slope of the symmetry plane of the pole figure to the rolling plane in the layers adjacent to the surface of the rolls. If, after annealing the sheet rolled under the conditions where both rollers have the same roughness, the pattern of the texture distribution along the metal thickness remains approximately the same as in the cold-worked steel, but the recrystallization texture is 2–2.5 times weaker than the deformation texture, then the annealing of the metal, rolled in differently rough rolls, leads to the formation of an orientation of the crystallites which is close to random throughout the section of the sheet. The result is almost texture-free steel. Therefore, rolling in the rolls of different roughness can be used to suppress the texture in cold-rolled annealed steel and reduce its propensity to form defects of the 'fish scales' type with subsequent enameling.

It has been experimentally established[7] that the effect of different roughness of the surfaces of the upper and lower work rolls of the rolling mill, and consequently of the roughness of the surface of the upper and lower sides of the cold-rolled strip on the degree of texture and the propensity to form 'fish scales' of finished rolled products, is manifested in the ratio of the roughness of the upper and lower surfaces not less than 1.1–1.2.

Rolling of strips, creating in all cases the maximum different roughness of the surface of the sides of cold-rolled steel, is inexpedient for several reasons. First, the more the difference in the roughness of the strip surfaces is required, the greater the difference in the roughness of the working surfaces of the upper and lower rolls, and this is undesirable, since the process becomes highly asymmetric with all resulting from this consequences (curvature of the front end of the strip, uneven loading of spindles and drive motors of the upper and lower rolls, etc.). Secondly, a great difference in the roughness of finished products is undesirable from the viewpoint of ensuring the uniformity of the thickness of the coating of the sides and the degree of its adhesion to the surface of the steel substrate. The different roughness of the surfaces of cold-rolled steel for enamelling should be used only depending on the aluminum content in it. The degree of roughness of the surfaces of the upper and lower sides of the strip should be the stronger, the higher the aluminum content of steel.

The main conclusion resulting from the above examples is that the asymmetry of the rolling process, created not only by the difference in the frictional conditions on the contact surfaces in the deformation zone due to the different roughness of the working surfaces of the upper and lower rolls, but also by other conditions (difference in roll diameter, speed of their rotation, the difference in temperature along the thickness of the rolled strips, etc.) is able to significantly change the direction, orientation of the structural elements and the

[7]Experiments were carried out by D.D. Khizhnyak, B.P. Kolesnichenko, V.L. Mazur and I.V. Garashchuk in co-operation with experts of the Zaporozhstal' Metallurgical Combine.

texture of the metal. This effect is manifested not only when rolling steel, but also as a result of asymmetric rolling conditions of porous and powder materials, which is very important in the technology of their production.

Thus, the asymmetry of the rolling process significantly changes the stress distribution along the thickness of the strip, which leads to an increase in the stress gradient and affects the structure and texture of the rolled metal. Therefore, to ensure the required quality of sheet rolling products, accurate and theoretically justified calculation methods are needed. As was shown above, these requirements are best satisfied by the method of slip lines, the use of which in the matrix operator version with the use of nonlinear programming for numerical field construction makes it possible to obtain good results.

5.5. Using process asymmetry to determine the friction coefficient in rolling

Specific features of the process of rolling strips in asymmetric conditions are effectively used not only in production practice fo improvement of technological regimes but also in scientific-research experiments, for example, in the measurement of the force and the friction coefficient in the deformation zone during rolling. Let's give concrete examples.

Methods for determining the friction coefficient in a deformation region when rolling in accordance with the experimental diagrams of the specific forces of friction and pressure, the maximum angle of nip, using torque measurements, limiting reduction, forward creep, total metal pressure on rolls, etc. A method is widely known for determining the force and friction coefficient in a deformation zone during rolling, including rolling the strip in rollers and applying to the braking force to the strip. When unambiguous slipping is achieved in the deformation zone, the metal pressure on the rolls is measured, the rolling speed, the braking force of the strip and the circumferential velocity of the rolls, the results of which calculate the force and friction coefficient. This classic method of determining the coefficient of friction in a deformation zone during rolling was developed more than 50 years ago by I.M. Pavlov, and was used widely by A.P. Grudev and many other scientists. A method is also known for determining the force and the friction coefficient in a deformation zone, which includes rolling the strip in rolls with a continuous unambiguous slip on its surface and measuring the total pressure on the rollers and the torque on the rolls. The results of these measurements are used to calculate the force and friction coefficient.

These and other common methods have drawbacks, the main of which are the methodological complexity and laboriousness of the process of measuring the friction coefficient and frictional force, and the insufficient accuracy in measuring calculations and results. For example, in determining the friction coefficient in hot rolling, chilling of the heated sample during its fixing in the braking device, causes an unaccountable temperature decrease. As a result, the accuracy of the results deteriorates. Devices for the implementation of these methods are not suitable for rolling long samples, rolling with the application of not only the rear but also the front tension.

In order to simplify the process and increase the accuracy of measuring the force and friction coefficient in the deformation zone during rolling, the experts of the Institute of Ferrous Metallurgy of the National Academy of Sciences of Ukraine proposed[8] after the strip gripping by the rolls to brake one of the rolls before the slip of the strip relative to the rolls, and at the same time register the change in forces and torques on the rollers, the peripheral speed of the rolls and the speed of the rolled strip. From the results of measurements, it is easy to calculate the values of the frictional force and the coefficient of friction.

The full friction force on each roll T can be calculated by the formula $T = \frac{M}{P}$, and the friction coefficient using the formula $f = \frac{M}{RP}$, where M – torque on the roll; R – radius of the roll; P – rolling force.

Other formulas can also be used.

In the implementation of this method, a sheet steel sample is set into rolls having an individual drive. After the sample is gripped and the transition to the steady rolling process takes place, one of the rolls starts to brake with the field of the motor or applying a torque resistance to rotation by means of friction devices installed on the shaft. The value of the moment of resistance can be adjusted by strengthening or weakening the field or by regulating the force in the friction devices. Before the commencement of braking, the circumferential speeds of the rolls are equal and, due to the symmetry of the deformation zone, the frictional forces in the forward creep and backward creep zones on the upper and lower rolls are equal.

[8]V.L. Mazur, V.A. Trigub.

When one of the rolls is braked relative to another, the deformation centre becomes asymmetric. The extent of the forward creep zone on the contact surface of the rolled metal with a roll rotating at a higher speed the will gradually decrease. As a result, the forward creep zone will disappear and at all points of the contact arc the metal speed will be less than the speed of the roll. There will be a state of one-directional slip – the entire arc of contact will be a backward creep zone.

On the contact surface of the braked roll with the rolled metal, the backward creep zone will wedge out. Here, too, there will be a state of one-directional slip, but of another sign, – at all points the metal will outstrip the roll. Thus, the frictional forces on opposite contact surfaces of the deformation zone will be directed in opposite directions. Further, as the torque of the roll is increased by rotation, the braking torque will exceed the moment of adhesion of the surfaces of the roll and the rolled metal, and a breakdown will occur – the slippage of the roll on the strip. During the entire braking time, changes in the metal pressure on the rolls, the moments on the rolls, the speed of the rolls and the rolled strip are recorded. Analysis of these data at different periods of braking of the rolling process allows us to determine the frictional forces (contact stresses) in different areas of the deformation zone.

The frictional force and friction coefficient can be calculated most simply by the above formulas, substituting in them the values of the metal pressure on the rolls P and the moments M on the rolls at that instant (instant of time) when a one-directional but different sign of slip occurred on the contact surfaces of the strip and the upper and lower rolls (on one roll only forward creep, on the other – only delay), and deformation still continues, i.e. In the instant preceding the occurrence of slippage.

The method for measuring the force and the friction coefficient in the deformation zone during rolling simplifies the procedure and reduces its laboriousness. To determine the force and friction coefficient by this method, it is not necessary to install the device for gripping and braking the trailing end of the strip. With the continuous registration of rolling parameters from the onset of braking to the onset of slippage, one can analyze the change in frictional forces (contact stresses) during different periods of the braking process. This makes it possible to study the rolling process with different combinations of the lengths of the forward creep, sticking, and backward creep zones. The force and friction coefficient during rolling with tension can be measured.

Thus, the distinctive feature of the method considered is that the friction coefficient in the deformation zone is determined in the case of an asymmetric rolling process. In this case, asymmetry is created by braking one of the rolls.

As a remark, it can be noted that in the considered method an *apriori* assumption is made about the equality of the friction coefficients on the contact surfaces of the rolled strip with the upper and lower rolls, although, strictly speaking, some difference is possible. To implement the method, it is again necessary to equip the mill with a set of control, measuring and recording equipment, a system of affecting the drive or a device for braking the roll.

The reduction in labour intensity and simplification of the process of measuring the friction coefficient during rolling can be achieved by the fact that in asymmetric rolling, the reduction, the angle of contact between the strip and rolls, the radius of curvature of the strip after its exit from the rolls and the magnitude of its calculation determine the friction coefficient.

In the case of an asymmetric rolling process in two rolls, the force action scheme on the rolled strip, the conditions at the contact, the stress–strain state or the speed conditions in the reduction zones related to each roll are different. An external feature of the process of asymmetric rolling is the bending of the strip before entering the rolls and after it leaves the rolls. Other conditions being equal, the bending of the strip depends on the relative reduction, the thickness of the rolled strip and the friction coefficient in the deformation zone. If the rolling is carried out in rolls with different diameters, then for small reductions the strip bends towards the roll of smaller diameter, and at large reductions towards the roll of a larger diameter having a large circumferential speed. Knowing the amount of reduction of the strip and the parameters defining the asymmetry of the rolling process (the difference in the diameters of rolls, the difference in circumferential speed of the roll surface) and measuring the curvature of the strips coming out of the rolls, one can calculate the friction coefficient in the deformation zone[9].

The rolling scheme of samples under asymmetric conditions, for example, rolls of different diameters rotating at the same angular velocity, is shown in Fig. 5.23.

[9]The method was developed by V.L. Mazur and I.I. Leepa.

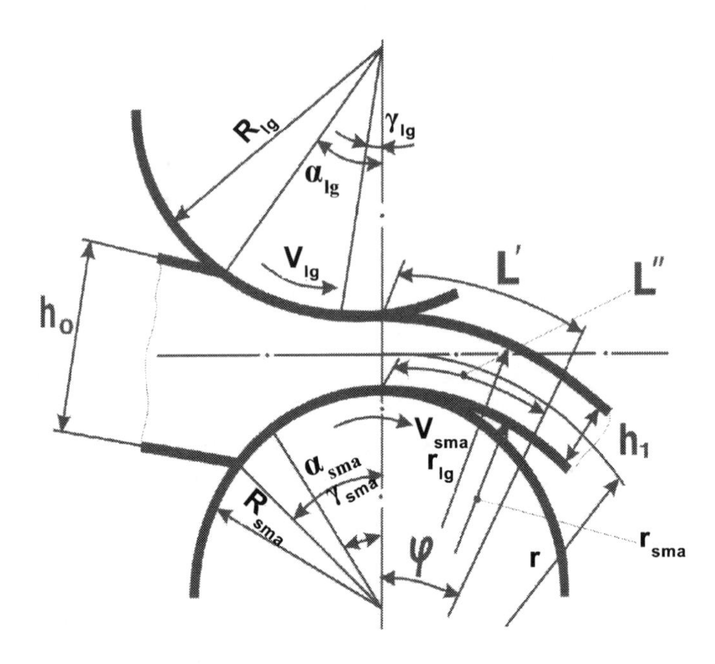

Fig. 5.23. Scheme of rolling strips in rolls of different diameters. The designations are given in the text.

According to the provisions of the known theory of rolling the forward creep and neutral angle relative to the rolls of larger and smaller diameters are expressed by the following Eckelund–Pavlov and Golovin–Dresden equations

$$S_{\mathrm{lg}} = \frac{R_{\mathrm{lg}}}{h_1} \cdot \gamma_{\mathrm{lg}}^2;\ \gamma_{\mathrm{lg}} = \frac{\alpha_{\mathrm{lg}}}{2}\left(1 - \frac{\alpha_{\mathrm{lg}}}{2f}\right);\ S_{\mathrm{sma}} = \frac{R_{\mathrm{sma}}}{h_1} \cdot \gamma_{\mathrm{sma}}^2; \tag{5.92}$$

$$\gamma_{\mathrm{sma}} = \frac{\alpha_{\mathrm{sma}}}{2}\left(1 - \frac{\alpha_{\mathrm{sma}}}{2f}\right).$$

Here, h_1 – the thickness of the strip after rolling; f – friction coefficient; R – radius of the roll; S – forward creep; γ – neutral angle; α – angle of nip. Subscripts 'lg' and 'sma' denote the parameter value on the size of the larger roll ('lg') or smaller ('sma') diameters.

When rolling in the asymmetric conditions caused by the difference in the diameters of the upper and lower rolls, respectively, the angles of nip for the large and small diameter rolls are calculated from the condition of equality of the total pressures on the top and bottom rolls using the expression

$$\alpha_{lg}=\sqrt{\frac{\Delta_{lg}h}{R_{lg}}};\ \Delta_{lg}h=\frac{\frac{R_{sma}}{R_{lg}}}{1+\frac{R_{sma}}{R_b}}\cdot\Delta h;\ \alpha_{sma}=\sqrt{\frac{\Delta_{sma}h}{R_{sma}}};\ \Delta_{sma}h=\frac{1}{1+\frac{R_{sma}}{R_{lg}}}\cdot\Delta h, \quad (5.93)$$

where $\Delta_{lg}h$ and $\Delta_{sma}h$ – absolute reduction of the strip by rolls of larger and smaller diameters; $\Delta h = \Delta_{lg}h + \Delta_{sma}h$ – total absolute reduction of the strip equal to the difference of the strip thickness at the inlet and outlet of the rolls $\Delta h = h_0 - h_1$, where h_0 – initial thickness of the strip.

Considering the section of the strip leaving the rolls, highlighted by two radial sections with spacing φ, we can write

$$l'=V't';\ \ l''=V''t;\ \ \frac{l'}{l''}=\frac{V'}{V''}, \quad (5.94)$$

where t – duration of strip rolling area under consideration; V' and V'' – the strip exit speed from the side of the rolls of larger and smaller diameters, respectively; l' and l'' – the length of the considered area of the strip on the surfaces corresponding to the rolls of larger and smaller diameter.

On the other hand, it is obvious that

$$l'=r_{lg}\cdot\varphi;\ l''=r_{sma}\cdot\varphi;\ \frac{l'}{l''}=\frac{r_{lg}}{r_{sma}}, \quad (5.95)$$

where r_{lg} and r_{sma} – the radii of curvature of the rolled strip adjacent (relevant) to the rolls of larger and smaller diameters; φ – the length of the considered area (arc) of the rolled strip, expressed in radians. According to the adopted scheme

$$r_{lg}=r+\frac{h_1}{2};\ \ r_{sma}=r-\frac{h_1}{2}, \quad (5.96)$$

where r is the radius of bending of the rolled strip along the axial line (radius of curvature).

Comparing the expressions (5.94), (5.95), (5.96), we write

$$\frac{V'}{V''}=\frac{r_{lg}}{r_{sma}}=\frac{r+\frac{h_1}{2}}{r-\frac{h_1}{2}}. \quad (5.97)$$

It is known that

$$V'=V_{lg}\left(1+S_{lg}\right);\ V''=V_{sma}\left(1+S_{sma}\right), \quad (5.98)$$

where V_{lg} and V_{sma} are the circumferential speeds of the rolls of larger and smaller diameters, respectively.

At the same angular velocity ω of rotation of the two rolls

$$V_{\mathrm{lg}}=\omega R_{\mathrm{lg}};\quad V_{\mathrm{smc}}=\omega R_{\mathrm{smc}}. \tag{5.99}$$

Using equation (5.97) according to (5.98) and (5.99), we obtain

$$\frac{1}{i}\cdot\frac{1+S_{\mathrm{lg}}}{1+S_{\mathrm{sma}}}=\frac{r+\frac{h_1}{2}}{r-\frac{h_1}{2}},\ \text{where } i=\frac{R_{\mathrm{sma}}}{R_{\mathrm{lg}}}. \tag{5.100}$$

Solving the equation (5.100) with respect to the radius of curvature, we obtain

$$r=\frac{1+S_{\mathrm{lg}}+i\left(1+S_{\mathrm{sma}}\right)}{1+S_{\mathrm{lg}}-i\left(1+S_{\mathrm{sma}}\right)}\cdot\frac{h_1}{2}. \tag{5.101}$$

Thus, rolling the strips under asymmetric conditions, in particular in rolls of different diameters, by measuring the amount of reduction and the radius of curvature of the strip leaving the rolls (5.92), (5.93), and (5.101), we can calculate the friction coefficient. If we substitute the value of S_{lg} and S_{sma} in expression (5.101) according to (5.92) and (5.93), then it is not possible to solve it with respect to the friction coefficient f without simplifications. Therefore, it is much more convenient to find f with the help of formulas (5.92), (5.93), (5.101) by the inverse conversion method, for example, by selection. In this case, by successively setting the different values of f in formulas (5.92), (5.93), (5.101), the computation is carried out until the calculated value of the radius of curvature of the strip leaving the rolls coincides with the radius of curvature measured after rolling.

If the asymmetry of the rolling process is created due to the difference in the speeds of rotation of the upper and lower rolls of the same radius, then in expression (5.101) we must take $i = V_{\mathrm{sma}}/V_{\mathrm{lg}}$.

When calculating the friction coefficient, the curvature should be considered positive and the measured radius of curvature of the strip should be substituted into the formula (5.101) with the plus sign in those cases when the strip leaving the rolls is bent towards the roll of smaller diameter. Conversely, the value of r is taken with the minus sign when the rolled strip is bent towards the roll of a larger diameter.

This method can be applied in combination with other known methods for measuring the friction coefficient. For example, in the process of rolling the samples, in addition to the bending radius of the strip emerging from the rolls, it is possible to measure the pressure on the rolls, the torque, the forward creep on the side of both rolls, and other rolling parameters, and calculate the friction coefficient from the results of these measurements. As a result, the value of the friction coefficient determined simultaneously by different methods will be more accurate. The reliability and accuracy of the information obtained about the friction coefficient in rolling will increase.

A variation of the proposed method are methods for measuring the friction coefficient in rolling under asymmetric conditions, in which the curvature (deviation from the horizontal plane) of the trailing end of a rolled sample is measured, or the direction of the resultant forces in asymmetric rolling, and the friction coefficient is calculated from their values. However, these options are not convenient to implement.

Advantages of the method are that, when applied, the methodology is simplified and the labour costs for determining the friction coefficient are reduced. To implement the proposed method, complex measurement equipment is not required, the friction coefficient measurement can be performed at any mill. In this way, it is possible to measure the friction coefficient in the deformation zone in asymmetric rolling with the application of back tension. As can be seen, in the above calculation, according to Eqs. (5.92), the friction coefficient is assumed to be the same on the contact surfaces on the side of the upper and lower rolls. It is also clear that the dependence of the curvature of the strip rolled in asymmetric conditions on the friction coefficient in the deformation zone is approximate. More precisely, this dependence is determined by the method of slip lines, as described in our publications [57–61]. Therefore, from the experimentally determined radius of curvature of the rolled strip, it is possible to calculate the friction coefficient more accurately using a solution based on the method of slip lines.

6

Mathematical model of the process of cold rolling of strips in continuous mills

6.1. The model of the stationary process

The continuous rolling mill (CRM) is a complex system and its working condition is characterised by a large number of parameters. The parameters of the rolling process in each separate stand of the mill are inter-linked through the strip being rolled. This circumstance determines the need for using a complex approach to the development of the mathematical model of the continuous rolling process and, in particular, for the mill as a whole, and not for its individual stands, sections and aggregates.

The structural elements of the generalised mathematical model of the continuous cold rolling mill is the model of the deformation zone (see chapter 1) used for calculating the energy–force, kinematic and geometrical parameters of the rolling process in every stand of the mill; the model of the variation of the size of the roll gap; the model for calculating the temperature of the rolled strip; the dynamic model of the stand which includes the equation of motion of the masses of the stand and the electric drive of the rolls (Fig. 6.1).

6.1.1. Selection of the method for calculating the deformation resistance

In advanced mathematical models of continuous cold rolling the deformation resistance of the metal is described taking into account not only the effect of the degree of deformation but also the temperature

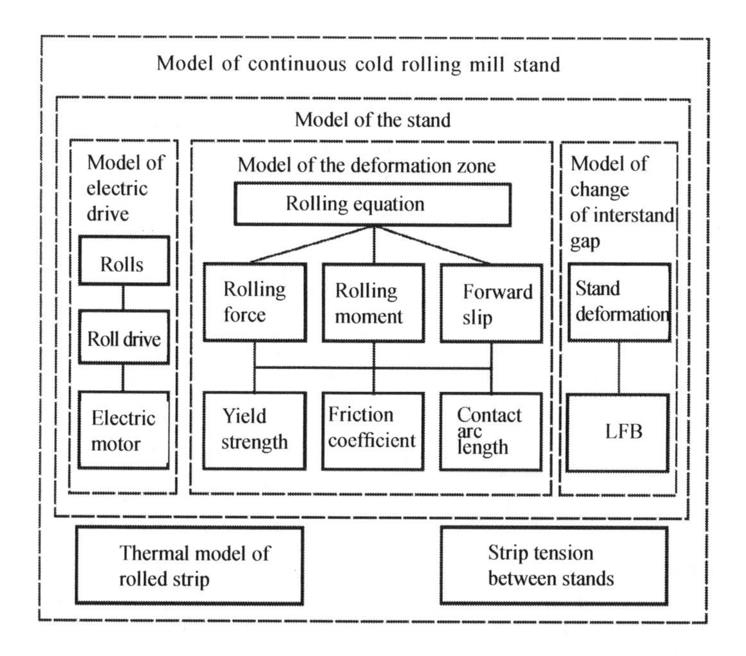

Fig. 6.1. Structural diagram of the mathematical model of the continuous cold rolling mill.

and rate conditions of deformation. Usually, the deformation resistance is represented in the form of the sum of the static and dynamic components:

$$\sigma_{T} = \sigma_{T\,st} + \sigma_{T\,dyn}. \tag{6.1}$$

The static component $\sigma_{T\,st}$ (yield strength) depends on the total relative deformation of the metal and on temperature. It is determined by standard tensile tests of sheet steel samples with the strain rate $\dot{\varepsilon} = 10^{-4}$ 1/s.

To describe the yield strength of the steel at room temperature (20°C), we use the resultant definition and the already mentioned dependence of the type:

$$\sigma_{T^{20}st} = \sigma_{T\,in} + a\varepsilon^{b}, \tag{6.2}$$

here $\sigma_{T\,in}$ is the yield strength of the metal prior to cold deformation; a, b are the hardening constants.

Hardening of the steel can be investigated on samples taken during cold rolling of the strips in the continuous rolling mill. For this purpose, samples are taken in the gaps between the stands and

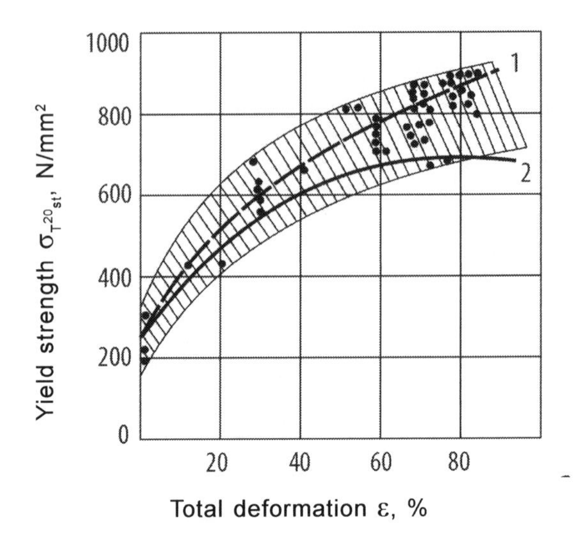

Fig. 6.2. Dependence of the yield strength of 08kp steel on the degree of cold deformation. The numbers at the curves: 1 – experimental curve; 2 – the curve calculated using the equation proposed by A.V. Tret'yakov [74]: $\sigma_{T^{20}_{st}} = \sigma_{T_{in}} + 34\varepsilon^{0.6}$. The crosshatched area is the field of the experimental points.

after the final stand of the mill and subsequently subjected to the mechanical tensile tests. The dependence $\sigma_{T^{20}st}(\varepsilon)$ obtained in this manner for low-carbon steel (0.08% C) is shown in Fig. 6.2. The yield strength of 08kp steel can be described most conveniently using the coefficients a = 43 and b = 0.61.

The effect of temperature on the yield strength of the metal in the quasi-static test conditions according to the data in [75, 76] can be taken into account using the temperature coefficient n_t:

$$\sigma_{T^{20}st} = n_t \sigma_{T^{20}st} \tag{6.3}$$

here n_t is the empirical coefficient which depends on the chemical composition of the metal.

At present, several relationships are available for determining the dynamic component of the deformation resistance $\sigma_{T\,dyn}$ [77–80]. When constructing the mathematical model it is necessary to select the best relation describing most accurately the effect of the main factors on deformation resistance.

When selecting the equation for calculating $\sigma_{T\,dyn}$ the following circumstances are taken into account. Firstly, the deformation resistance of low-carbon steel not subjected to cold working depend strongly on the strain rate. For example, according to the data in [74, 80] the deformation resistance at the rate $\dot{\varepsilon}$ = 1000 1/s is 100% or more greater than the yield strength of the steel in quasi-static tests ($\dot{\varepsilon} = 10^{-4}$ 1/s). The effect of the strain rate was taken into account in the equations derived by many authors [77–82]. For example, A.V. Tret'yakov et al [79] proposed to calculate the yield strength from the following dependence:

$$\sigma_T = \sigma_{T\,st}\,\alpha\,[1-\beta(\dot{\varepsilon}-1)], \tag{6.4}$$

where α and β are the coefficients equal to 1.12 and 0.00202, respectively, for the low-carbon steel.

Secondly, the dependence $\sigma_T(\dot{\varepsilon})$ is strongly affected by the degree of preliminary deformation. According to Roberts [77] the gradient of the function $\sigma_T(\dot{\varepsilon})$ for the cold-worked low-carbon steel is approximately 2.2 times smaller than for the steel without cold working. If the yield strength of the steel not subjected to deformation at a strain rate $\dot{\varepsilon}$ = 1000 1/s increases by almost 100%, then for the steel subjected to preliminary deformation of 68.1% the deformation resistance increases by only 20% in comparison with the value in the quasi-static test conditions. This nature of the effect of the degree of deformation on the dynamic component of deformation resistance is taken into account in equations derived by, for example, V.J. Roberts [77]. V.J. Edwards and N.A. Fuller [78], S. Hennig and K.-H. Weber [80].

Thirdly, the results obtained in the investigations by A.P. Grudev and Yu.B. Sigalov [82] show that the dynamic component of the deformation resistance is greatly affected by the temperature of the deformed metal.

The interaction of the three previously mentioned factors in describing the dynamic component of the deformation resistance is described more sufficiently by the equation derived by Japanese investigators [81]:

$$\sigma_{T_{dyn}} = \exp(6.5)\left[\frac{\dot{\varepsilon}}{\dot{\varepsilon}(\bar{\varepsilon})}\right]^{\frac{\kappa T}{0.14}}, \tag{6.11}$$

here $\dot{\varepsilon}(\bar{\varepsilon}) = 5 \cdot 10^{11} \cdot 60.842^{\bar{\varepsilon}}$; $\bar{\varepsilon}$ is the logarithmic strain; $\bar{\varepsilon} = h_0/h$, where h_0 and h is the strip thickness prior to and after deformation; κ is the Boltzmann constant ($0.8625 \cdot 10^{-4}$, eV/K); T is temperature, K.

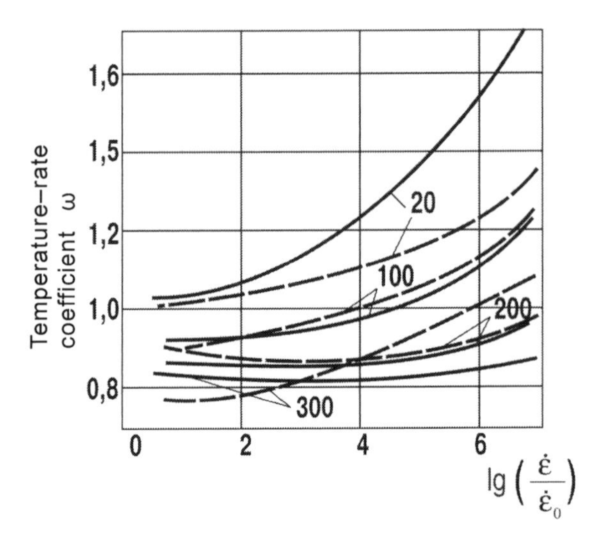

Fig. 6.3. Dependence of the ratio $\omega = \frac{\sigma_T}{\sigma_{T_{st}}^{20}}$ on the strain rate and deformation temperature of low-carbon steel. Solid lines – calculated dependences; dashed lines – experimental data by A.P. Grudev and Yu.B. Sigalov [82]. The numbers at the curves – deformation temperature, °C; $\dot{\varepsilon}$ – strain rate in the quasi-static tests.

To verify the adequacy of equation (6.11), the calculated values of the deformation resistance were compared with the experimental data in [77, 82]. The values of the temperature–rate coefficient of the deformation resistance were calculated for this purpose:

$$\omega = \frac{\sigma_T}{\sigma_{T_{st}}^{20}}.$$

The results of the calculations ω, compared with the experimental data in [82], are presented in Fig. 6.3. It is important to note the good agreement between the calculated and experimental values at deformation temperatures of 100 and 200°C. For example, at 200°C the curves coincides almost completely. The results of the fat only at 20°C. Comparison of the calculated values of deformation resistance with the experimental data [74] at this temperature shows that they are in good agreement.

The results of calculation of the deformation resistance using the equation (6.11) in comparison with the experimental data [77] (Fig. 6.4) show that equation (6.11) takes into account with sufficient accuracy the effect of preliminary hardening of the steel on the dynamic component of yield strength. The difference in the results usually does not exceed 10%. The quantitative difference of the

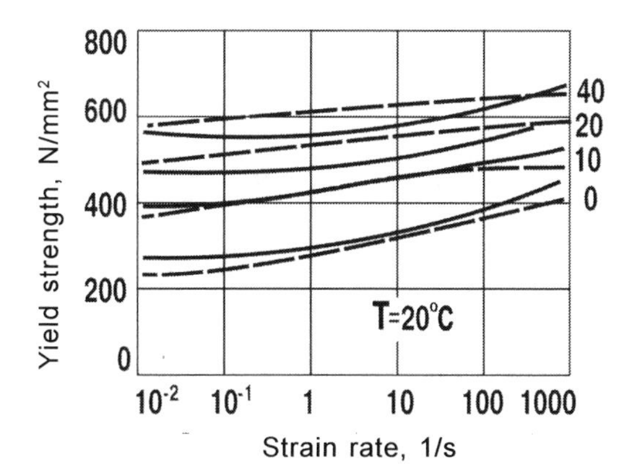

Fig. 6.4. Effect of the strain rate on the deformation resistance of low-carbon steel with preliminary cold working: solid lines – calculated dependence, dashed lines – experimental data by V.J. Roberts [77]. The numbers of the curves are the degrees of preliminary deformation, %.

results of the calculations and the experimental data [76] in quasi-static loading in the temperature range 0–200°C is slightly greater, 13%.

The investigations confirm that the dynamic component of the deformation resistance in cold rolling is described more accurately and completely by equation (6.11). Combining this dependence with the expressions (6.2), (6.3) yielded the generalised dependence for calculating the deformation resistance of low carbon steel in cold rolling:

$$\sigma_T = n_t\left(\sigma_{T_{in}} + a\varepsilon^b\right) + \exp(6.5)\left[\frac{\dot{\varepsilon}}{\dot{\varepsilon}_0(\varepsilon)}\right]^{\frac{\kappa T}{0.14}}. \tag{6.12}$$

6.1.2. Investigation of the deformation resistance of steel in the deformation zone

The initial data for calculating the deformation resistance in any cross-section of the deformation zone using equation (6.12) are the preliminary degree of deformation, the strain rate and the temperature of the metal in this cross section. The increase of temperature of the metal in the *i*-th cross-section of the deformation zone was calculated using the following dependence [67]:

$$\Delta T = k \cdot m(p+\sigma)\ln\frac{h_0}{h_1}. \tag{6.13}$$

Here k is the dimensionless coefficient of heat exchange between the rolled strip and the rolls; $m = 1/(A \cdot c \cdot \rho)$; A is the mechanical equivalent of heat; c is the specific heat of the metal; ρ is the density of metal; p is the specific pressure in the deformation zone; σ is the rear or front specific tension depending on the deformation zone in which temperature is calculated.

The amount of heat, transferred to the rolls by the strip was calculated by the authors of [83, 84] using the dimensional coefficient of contact heat exchange α_{ch}. For example, A.V. Tret'yakov et al [84] determined that the mean value α_{ch} is equal to $4.1 \cdot 10^{-2}$ kcal/m^2 · h · °C. However, in the calculations of the metal temperature in the deformation zone it is more convenient to use the dimensionless heat exchange coefficient k. V.J. Roberts [77] assumes that the working rolls receive half the heat generated as a result of the work of the friction forces. In this case, k changes in the range 1.0–0.8. The authors of [81] assumed that $k = 0.9$. N. Funke and K. Kottman [85] obtained $k = 0.875$ on the basis of the experimental data for the five-stand cold rolling mill. Since the values of the coefficient k recommended in various sources differ only slightly, in this work it is assumed that this coefficient is equal to 0.9.

The dependences for calculating the reduction, the rate and temperature in the i-th section of the deformation zone will be represented in the form suitable for use in the discrete model of the deformation zone (chapter 1):

$$\begin{aligned} &\varepsilon = \frac{h_0 - h_1}{h_0}; \\ &\bar{\varepsilon}_i = \ln\frac{h_0}{h_i}; \\ &\dot{\varepsilon}_i = 2V_r \gamma \cos\gamma \frac{\operatorname{tg}\varphi_i}{h_i}; \\ &T_i = T_{i-1} + k \cdot m \cdot \left(p_{i-1} + \sigma\right)\ln\frac{h_{i-1}}{h_i}; \\ &i = 1, \ldots, N+1, \end{aligned} \tag{6.14}$$

here γ is the angle of the central cross-section in the deformation zone.

The introduction of specific pressure p into equation (6.13) complicates temperature calculations because to determine p_i in the i-th cross-section of the deformation zone we do not have the value of the deformation resistance σ_{Ti} which in turn cannot be determined

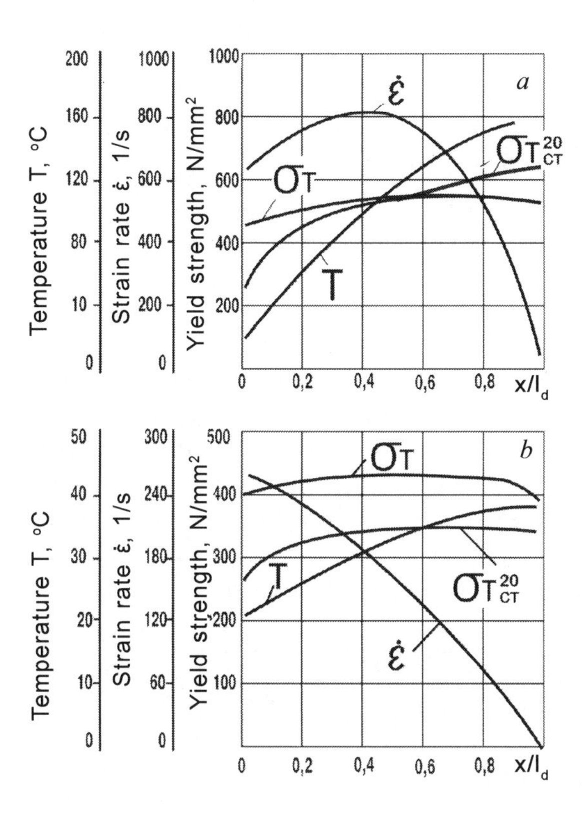

Fig. 6.5. Variation of the yield strength of 08kp annealed steel along the length of the deformation zone in rolling 1 mm strips at a speed of 5 m/s in the rolls with a diameter of 200 mm. *a*) reduction 50%; *b*) reduction 10%.

if the temperature of the metal in this cross section is not available. This problem was solved by the iteration process for each cross-section of the deformation zone. However, the algorithm in this form greatly increases the calculation time. Therefore, the calculations of the metal temperature in the *i*-th cross-section of the deformation zone were carried out using the values p_{i-1} in the equation (6.14). As shown by the verification calculations, at a sufficiently large number of divisions of the deformation zone, for example, $N = 50$, this assumption results in the error smaller than 1%.

The discrete model of the deformation zone (chapter 1) can be used to investigate the change of the deformation resistance along the length of the zone. As an example, the parameters of the process of rolling the annealed and cold-worked 08kp steel with a thickness of 1 mm in the rolls with a diameter of 200 mm were simulated. The rolling speed was equal to 5 m/s. The curves of the variation of the strain rate, temperature and deformation resistance of the metal along the length of the deformation zone in rolling the annealed steel with the reduction in parts of 50 and 10% are shown in Fig. 6.5.

The results of similar investigations conducted on the cold-worked steel (preliminary reduction 30%) are shown in Fig. 6.6.

According to the calculated data, the largest difference in the results of calculating the deformation resistance with the temperature and strain rate taken and not taken into account is obtained in rolling the metal with no cold working. For example, in rolling with 10% reduction (Fig. 6.5 *b*) the deformation resistance in the entry cross-section of the deformation zones, calculated taking into account the effect of $\dot{\varepsilon}$ and T, is 52% greater than the yield strength calculated without considering the temperature–rate factor. Closer to the output cross-section of the deformation zone this difference decreases to 18%.

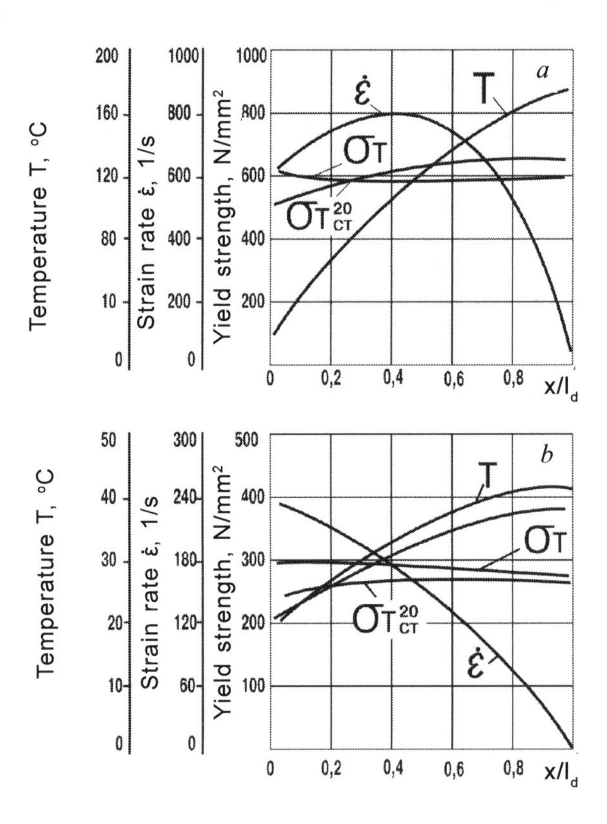

Fig. 6.6. Change of the yield strength of the hardened 08kp steel (preliminary reduction 30%) along the length of the deformation zone. The rolling conditions and notations are the same as in Fig. 6.5.

The situation greatly changes with increase of the reduction. For example, in rolling with 50% reduction (Fig. 6.5 *a*) the values of σ_T becomes smaller than $\sigma^{20}_{T_{st}}$, starting in the centre of the deformation zone. The results can be explained by the fact that in the first case (ε = 10%) σ_T is controlled by the strain rate and taking the strain rate into account increases the deformation resistance. In rolling with 50% reduction the temperature of the metal greatly increased (from 38 to 160°C) together with the increase of the static component of the deformation resistance (from 34 to 60 N/mm^2) and this weakens the effect of the rate factor. However, starting in the centre of the deformation zone, the temperature factor already has a stronger effect on the rate factor (Figure 6.5 *a*).

In rolling the cold worked steel (Fig. 6.6), the reduction per pass has no longer such a strong effect on deformation resistance. Although, as in the case of the rolling of annealed steel at ε = 10% the value σ_T is greater than $\sigma^{20}_{T_{st}}$ almost along the entire length of the deformation zone (Fig. 6.6 *b*), and on average σ_T 10% greater than $\sigma^{20}_{T_{st}}$. In rolling with a larger reduction (50%) σ_T is smaller than $\sigma^{20}_{T_{st}}$, starting at one third of the length of the deformation zone (Fig. 6.6 *a*). On average, σ_T is only 1–2% greater than $\sigma^{20}_{T_{st}}$ in the deformation zone.

Thus, the results of the combined effect of the temperature, degree of deformation and strain rate in rolling on the deformation resistance of the steel without cold working and cold worked steel differs.

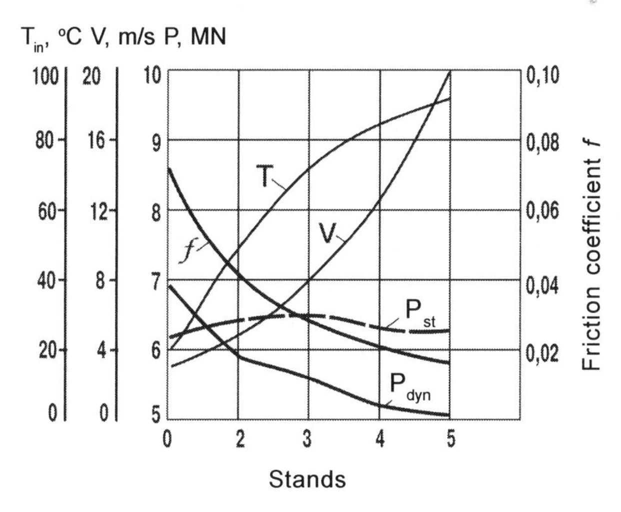

Fig. 6.7. The rolling force of 2.2/0.25 × 730 mm plate in the 1200 rolling mill of the Magnitogorsk Metallurgical Concern, calculated taking into account (P_{dyn}) and disregarding (P_{st}) the effect of temperature and the strain rate on the yield strength of the steel.

This fact must be taken into account when modelling the process of continuous cold rolling during which the temperature of the strip and the strain and strain rate and also the degree of preliminary hardening of the metal change in wide ranges.

To evaluate the effect of the dynamic component of the yield strength on the parameters of continuous cold rolling, calculations were carried out of the force in the stands of the 1200 continuous five-stand rolling mill (Fig. 6.7). The results show that the rolling force P_{dyn}, calculated taking into account the effect of temperature, the degree of deformation and strain rate on the deformation resistance in the first stand of the mill is greater and the rolling force P_{st}, calculated disregarding the temperature–rate factor. This result can be explained by the fact that the first stand is used to roll the metal with no working and the deformation resistance of this metal depend strongly on the strain rate. The magnitude of reduction in the first stand was 25–30%. These reductions result in a relatively small increase of the metal temperature in the deformation zone (~70°C) and, therefore, the effect of the rate factor is strong in contrast to the effect of the temperature factor.

In the next stands of the rolling mill regardless of the fact that the rolling speed increases, the rate factor has no longer any significant effect on the deformation resistance, and consequently, P_{dyn} becomes smaller than P_{st}. This is explained by the fact that in the final stands of the rolling mill the metal is already cold-worked and its deformation resistance depends only slightly on the strain rate. The simulation of the process of rolling the strip with different speeds showed that the $P_{dyn}(V)$, $P_{st}(E)$ curves for the fifth stand are equidistant and the difference in the values of P_{dyn} and P_{st} is caused mostly by the effect of the temperature of the rolled strip on the deformation resistance.

Thus, the simulation of the process of continuous cold rolling shows that if the effect of temperature and strain rate on the deformation resistance of the metal is not taken into account, the calculated values of the forces and other energy–force parameters of rolling for the final stands (starting with the second stand) are too high. The increase of the rolling force in, for example, the fifth stand of the 1200 mill at a rolling speed of 20 m/s may reach 23%.

6.1.3. Calculation of the friction coefficient in the deformation zone

In the cold rolling thin sheet steel the energy and force parameters of the process are influenced mostly by the friction coefficient f. It suffice to say that the variation of the friction coefficient from 0.04 to 0.06 in the final stand of the 1700 continuous five-stand cold rolling mill increases the rolling force from 12 to 21.8 MN.

The fundamental studies of external friction in rolling, the general physical relationships of the effect of the parameters of the cold rolling process on the friction coefficient were carried out by A.P. Grudev et al and published in, for example, [86]. However, the equations for calculating the friction coefficient in rolling were obtained in most cases on the basis of laboratory experiments carried out in the conditions differing from the rolling conditions of the strips in the industrial stands. Consequently, these equations cannot be used for investigating the production processes.

To derive the equation for calculating f it is more efficient to consider the method of reversed conversion of the friction coefficient using the accepted mathematical model with the measured parameters of the rolling process. The values of the friction coefficient obtained by this procedure also include the error of the accepted model, i.e., play the role of the correction coefficient and, therefore, cannot be used for the theoretical analysis of the friction forces in the deformation zone. However, the advantage of the method of reversed conversion is that it greatly improves the agreement between the calculated and experimental values of the energy and force parameters of the rolling process. Therefore, the method is used quite widely in simulation of the process of cold rolling thin metal sheets.

The authors of this book investigated the friction coefficient in cold rolling of strips of low carbon steel in the conditions of the continuous 1700 five-stand in rolling mill (the recorded parameters of the process of rolling of two strips with a thickness of 0.5 mm as an example are presented in Table 6.1).

When using the reverse conversion methods, the numerical value of the friction coefficient is influenced my many parameters, included in the accepted mathematical model of the rolling process: geometrical, kinematic, and also the parameters characterising the mechanical properties of the rolled metal and the condition of the strip–roll friction pair. The kinematic criterion for taking these parameters into account is represented by the rolling speed V of the metal in the stand. The

Table 6.1. Recorded (typical) parameters of the process of rolling 2.5 → 0.55×10.15 mm strips of 08kp steel in the 1700 continuous cold rolling mill

Stand No.	Relative reduction, %	Tension, kN		Rolling speed, m/s	Force, MN
		Rear	Front		
1	20.0/21.0	4/4	326/334	2.79/2.70	11.6/11.8
2	25.5/36.5	326/334	223/240	4.35/4/25	12.3/9/4
3	24.5/27.5	223/240	146/146	5.76/5.8	17.2/15.4
4	26.0/23.5	146/146	80/103	7.80/7.60	14.6/15/6
5	23.0/21.5	80/103	20/20	10.1/9/7	13.3/15/3

parameter characterising the mechanical properties of the rolled metal is the yield strength of the metal σ_T in front of the stand.

The dependences of the friction coefficient on the selected similarity criteria correspond to the family of the curves, described by the gamma distribution law in which the independent variable is the relative reduction in the stand:

$$f = a_0 \varepsilon^{(a_1 + a_2 h_0 / R + a_3 \sigma_T + a_4 V)} \cdot \exp(a_5 \varepsilon). \tag{6.15}$$

The coefficients a_0–a_5 are determined by the multiple regression method. For this purpose, the logarithm of equation (6.15) was prepared and the following linear form obtained:

$$\ln f = \ln a_0 + \left(a_1 + a_2 h_0 / R + a_3 \sigma_T + a_4 V\right) \ln \varepsilon + a_5 \varepsilon.$$

As a result of the processing of the experimental data, the rolling force in the individual stands of the mill was as follows: a_0 = 0.185, a_1 = 0.393, a_2 = –99.8, a_3 = 0.00007; a_4 = 0.0193; a_5 = –3.55. Correlation analysis shows that the closest relationship with the friction coefficient is observed in the case of the geometrical parameters h_0/R and ε. The coefficient of paired correlation between ln$f$ and ε is equal to 0.75, between ln$f$ and the complex ε, h_0/R is equal to 0.92. The multiple correlation coefficient with all the parameters included in the equation (6.15) is equal to 0.94. The mean calculation error of f in calculations using equation (6.15) did not exceed 21%.

The effect of the surface roughness of the working rolls on the friction coefficient was taken into account by the dimensionless coefficient K_r. In rolling the strips in the ground rolls with the surface roughness of 0.5–1.2 μm Ra the coefficient was assumed to be equal to unity. To determine the values of K_r in rolling in the rough rolls (surface roughness after shot blasting 4–5 μm Ra) special investigations were carried out during which the surface roughness of the rolls of the fifth stand of the 1700 mill was varied. In particular, one half of the melt of 08kp steel was rolled to a thickness of 0.55 mm in the ground rolls of the fifth stand (Ra = 0.85 μm), the other half in the rough rolls (Ra = 3.0 μm). Since the dependence of the friction coefficient on the surface roughness of the rolls Ra was similar to the linear dependence [86], the expression for K_r was used in the form $K_r = 1+0.25\ (Ra_r - 0.85)$.

The surface roughness of the rough rolls in service decreases in dependence on the amount of the rolled metal G:

$$Ra_r = 0.85 + 3.6e^{0.0026G} \qquad (6.16)$$

Taking into account (6.16) coefficient K_r for the case of rolling the strips in the knurled rolls was selected equal to

$$K_r = 1 + 0.9e^{0.0026G} \qquad (6.17)$$

where G is the mass of rolled metal, tonnes.

The effect of the type of lubricant used in the rolling on the friction coefficient was taken into account by the coefficient K_{lub}. In rolling with emulsions of mineral oils K_{lub} = 1.0. For a water emulsion of palm oil it is usually assumed that K_{lub} = 1.4. To take into account the surface roughness of the working rolls and the type of lubricant, coefficient a_0 in the expression (6.15) has the following form $a_0 = 0.185\ K_r/K_{\text{lub}}$.

The calculated values of the friction coefficient for different stands of the cold rolling mill are presented in Fig. 6.8.

The maximum values of f were obtained in the first stand of the mill. This may be explained by the fact that the pickled strip without lubrication is rolled in the first stand of the mill and the surface of the strip contains a large amount of contaminants (sludge, oxides, metallic particles); rolling is carried out in the shot blast rolls with a high surface roughness Ra = 4.5 μm. Consequently, the friction coefficient in this stand is maximum. In the first stand and subsequent interstand gap the contaminants are washed away from the surface of the strip by the emulsion. The strip entering the second stand is relatively clean but rough (Ra = 2.0–2.5 μm). It is well-known [87] that in rolling rough strips in smooth ground rolls

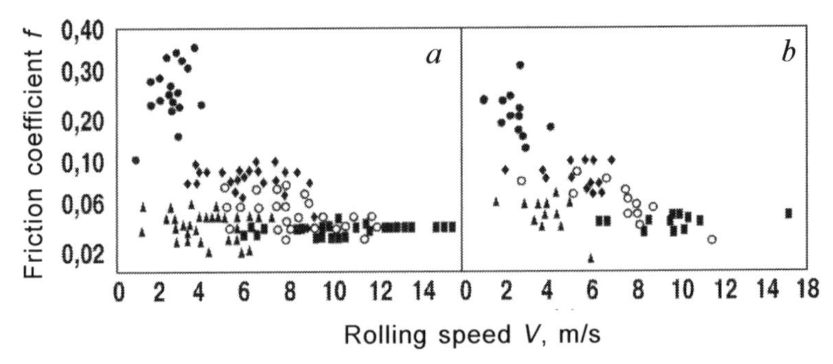

Fig. 6.8. The values of the friction coefficient in cold rolling of the 2.0/0.55×1015 mm strips in the 1700 mill calculated disregarding (*a*) and taking into account (*b*) the effect of temperature and strain rate. Notations: ● – stand 1; ▲– stand 2; ♦– stand 3;○– stand 4; ■ – stand 5.

(Ra_r = 0.5–1.2 μm) the friction coefficient is always lower than when rolling the smooth surface of the strip in the rough rolls. Therefore, rolling in the second stand takes place at relatively low values of the friction coefficient.

The data presented in Fig. 6.8 show that the scatter of the values of the friction coefficient decreases from the first to the last stand. This is the consequence of the decrease in each consecutive stand of the non-constancy of the parameters of the rolling process – thickness, strip temperature, yield strength, reduction, and others. Comparing the results of the calculation of *f* with taking into account the effect of the temperature–rate conditions of deformation on σ_T of the rolled metal and without taking these effects into account, it may be noted that for the first stand of the mill the friction coefficient is 20–25% lower than if the above affect is not taken into account. For other stands the results are virtually very similar. The effect of the temperature of the rolled strip on the friction coefficient is manifested evidently not as much by the deformation resistance of the rolled metal as by its effect on the viscosity and other properties of the lubricant.

Many investigators, using the method of reverse calculations of the friction coefficient from the rolling force, explain the decrease of the friction coefficient from the first stand of the continuous cold rolling mill to the final stand by the effect of the rolling speed. Our investigations (measurements of the rolling parameters) were carried out at a variable rolling speed. The rolling speed in the final stand of the mill changed in the range from 5 to 16 m/s. The investigations did not show the strong dependence of the friction coefficient on the rolling speed in a separately consider stand which is attributed to the change of the value of the calculated friction coefficient from the first to last stand of the rolling mill.

6.1.4. Calculation of the strip temperature in the line of the rolling mill

To determine the deformation resistance of the metal using equation (6.4), it is necessary to know the temperature of the strip prior to entry into each stand of the mill.

In the calculations of the change of the strip temperature in the space between the stands the majority of the authors used the equation of heat transfer of a thin flat strip with one- or two-sided heat transfer:

$$Vc\rho h\left(-\frac{\partial T}{\partial X}\right)dx = n\alpha\left(T - T_e\right)dx, \tag{6.18}$$

where V is the speed of movement of the strip; c is the specific heat of the strip metal; ρ is the specific weight of the strip metal; h is the thickness of the strip; T is the strip temperature; x is the distance of the investigated section of the strip from the previous stand; n is a coefficient which takes into account the cooling method (for the two-sided heat transfer $n = 2$, for one-sided heat transfer $n = 1$); α if the heat transfer coefficient; T_e is the temperature of the cooling emulsion.

After integrating equation (6.18) taking into account the temperature of the strip at exit from the stand T_{exit}, the following equation was obtained

$$T = T_e + \left(T_{\text{exit}} - T_e\right)\exp\left(-\frac{n\alpha x}{Vc\rho K_1}\right), \tag{6.19}$$

here K_1 is the correction coefficient characterising the fraction of the strip length on which the liquid falls in the gap between the stands.

To calculate the strip temperature at entry into the next stand (T_{in}), the value x in equation (6.19) should be replaced by the distance between the stands. The temperature T_{in} for the first stand is equal to the strip plate temperature. To calculate the strip temperature in coiling, the value x in equation (6.19) should be replaced by the distance from the last stand to the coiler.

The heat transfer coefficient α, included in equation (6.19), is difficult to determine in practice because its value depends on the method of supply and amount of the cooling liquid, its pressure, angle of incidence, etc [85]. According to the data [75], in forced feeding of the water–oil mixture on the strip the heat transfer coefficient for the five-stand 1200 rolling mill changes in the range 2500–3200 kcal/m^2 · h · °C. The authors of [85] proposed to use α = 3000 kcal/m^2 · h · °C for a similar stand with water–oil cooling. According to the data in [75], in cooling the strip with the emulsion the heat transfer coefficient changes in the range 1730–2340 kcal/m^2 · h · °C.

In the development of the thermal model of the stand the basic values of the heat transfer coefficient were as follows: in cooling the strip between the stands with the water–oil mixture α = 3000 kcal/m^2 · h · °C, in cooling with the emulsion α = 2000 kcal/m^2 · h · °C, and in air cooling 150 kcal/m^2 · h · °C.

The correction coefficient K_1 in equation (6.19) for a specific stand can be determined by comparing the experimental and calculated values. In experiments, the coefficient can be determined by measuring the strip temperature at the start and end of the gap between the stands. However, these measurements cannot be carried out with sufficient accuracy. Therefore, the only method of calculating K_1 is the comparison of the calculated and actual strip temperature in coiling at the known strip plate temperature.

The results of calculations of the strip temperature made of 08kp steel in rolling in the 1200 stand the mill are presented in Figure s6.9–6.11 and in the following table.

The variation of the strip temperature in the deformation zone of each stand was determined by the procedure described previously in section 6.1.1. The cooling of the strip in the gaps between the stands was calculated using equation (6.19). The calculations were carried out for the following rolling conditions: the temperature of the cooling liquid T_e = 35°C, the strip plate temperature 45°C, the thickness of strip plate 2.0 mm, rolling speed 20 m/s. The specific tension of the strip between the stands was 120 N/mm^2. The total relative reduction ε_Σ = 89%.

The strip temperature was calculated for three reduction modes (%): with reduced reduction in the final stand of the mill (mode A), with uniform distribution of the reduction in the stands of the mill (mode B) and with higher relative reduction in the last stand of the mill (mode C), see the table overleaf.

In the distribution of the reductions in the mode A, the strip temperature after the fourth and fifth stand of the mill was lower than in the two other cases (Fig. 6.9). This circumstance must be taken into

Stand	1	2	3	4	5
Mode A	37.0	46.0	36.0	29.0	26.5
Mode B	28.8	35.8	36.5	37.8	37.0
Mode C	20.0	24.5	39.0	43.8	44.5

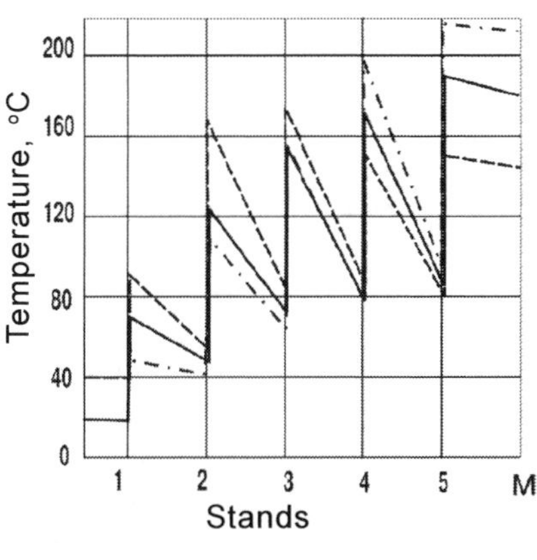

Fig. 6.9. Effect of the reduction mode on the strip temperature in the continuous cold rolling mill: A, B, C – the reduction modes (see the text): A – dotted lines, B – solid lines, C –dot-and-dash lines, M – coiler.

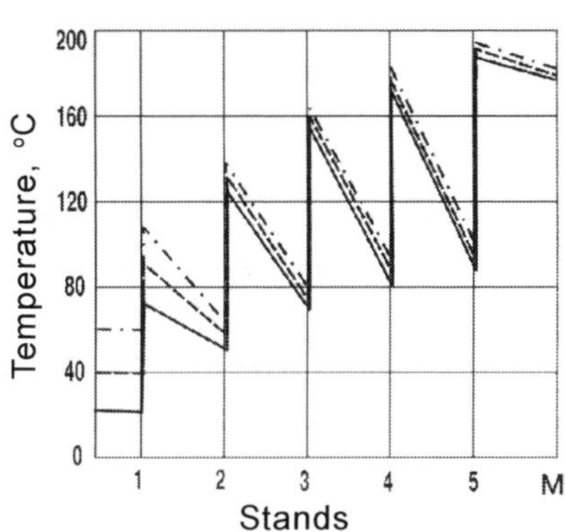

Fig. 6.10. Effect of strip plate temperature on the strip temperature in the continuous mill. The notations are the same as in Fig. 6.9.

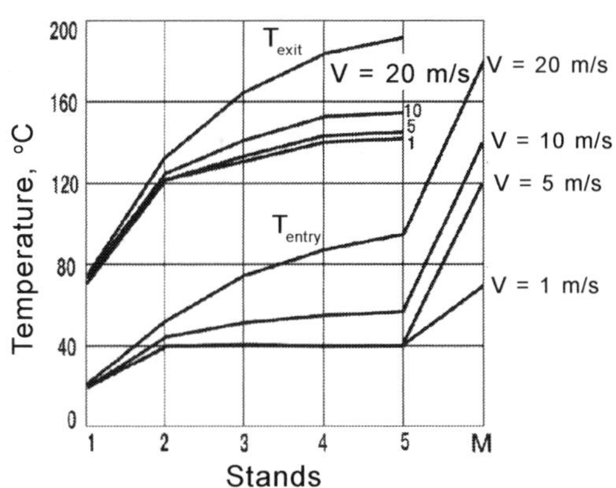

Fig. 6.11. Dependence of the strip temperature at entry T_{entry} into and exit T_{exit} from the stands of the continuous rolling mill on the rolling temperature V. The numbers are the curves indicate the rolling speed, m/s.

account in selecting the reduction modes because the coiling of the hot strip as often a beneficial effect on the flatness of the strip [88]. The variation of the strips plate temperature in front of the stand in cold rolling has almost no effect on the temperature of the strip coiled on the drum of the coiler (Fig. 6.10). The effect of the rolling speed on the distribution of the strip temperature in different sections of the mill is shown in Fig. 6.11. It should be noted that the increase of the rolling speed from 10 to 20 m/s with other conditions being equal increases the strip temperature in coiling by 40°C.

6.2. The model of the non-stationary process

6.2.1. The equation of the stand–drive–strip dynamic system

The continuous cold rolling mill is a complicated electromechanical system where the main members are in constant interaction. The complete model of the system, taking into account all inertia elements, all the relationships between them and the acting loads, is also very complicated, and the analysis of the dynamic processes in such a system is associated with considerable difficulties. Therefore, the investigated object is represented in the form of an equivalent (as regards the energy) calculation system which reflects the most important dynamic properties. In turn, the calculation scheme, consisting of the reduced masses, inertia moments and elastic imponderable relationships is simplified.

In the proposed model, the system of the working stand is represented by a single-mass calculation procedure with an elastic relation (Fig. 6.12), as proposed in [89,90].

The movement of this system can be described by the following equation:

$$m\ddot{x} + C_k x = P(t) - \text{sign}(\dot{x}) F(t), \quad (6.20)$$

where m is the reduced mass of the roll section; C_k is the stifness of the stand; x is the displacement of the roll section; $P(t)$ is the rolling force; $F(t)$ is the friction force of the roll carriage on the columns of the stand.

If the coordinate x is expressed as the sum

$$x = x_0 + \Delta x,$$

where x_0 is the elastic deformation of the stand in the steady rolling

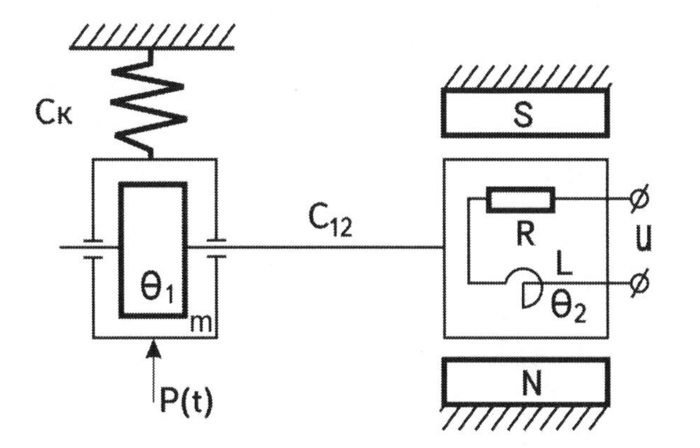

Fig. 6.12. Calculation scheme of the electromechanical systems of the drive and the stand.

process; Δx is the increment of deformation of the stand in the transition process, the following differential equation of oscillations of the stand in increments is obtained:

$$m\Delta x + C_k\Delta x = \Delta P(t) - \text{sign}(\Delta x)F(t). \qquad (6.21)$$

Analysis of the frequency characteristics of the main drive lines of the stands of the thin sheet continuous rolling mills, and also the results of experimental investigations, carried out in [89–91] show that the two-mass calculation scheme of the main line of the stand (Fig. 6.12) expresses the actual transition processes with the accuracy sufficient for practice. The movement of the reduced masses in relation to the equilibrium state is described by the system of differential equations:

$$\Theta_1\ddot{\varphi}_1 = \Delta M_{\text{tor}} - 2\bar{n}(\dot{\varphi}_1 - \dot{\varphi}_2) - C_{12}(\varphi_1 - \varphi_2), \qquad (6.22)$$

$$\Theta_2\ddot{\varphi}_2 = \Delta M_{\text{dr}} + C_{12}(\varphi_1 - \varphi_2), \qquad (6.23)$$

where ΔM_{tor}, ΔM_{dr} are the increments of the torque in the rolls and the momentum of the drive; Θ_1, Θ_2 are the moments of inertia of the roll section and the armature of the drive reduced to the shaft; C_{12} is the reduced rigidity of the drive shaft; φ_1, φ_2 are the angles of rotation of the masses Θ_1 and Θ_2; $\bar{n}$ is the damping factor.

The movement of the electric drive is described with the sufficient accuracy by the following equation [89]

$$\Delta U = L\Delta I + R_a\Delta I + C_e\dot{\varphi}_2, \qquad (6.24)$$

where R_a, L is the active resistance and inductance of the armature circuit of the drive; U is the voltage of the armature winding of the drive; I is the intensity of current in the circuit of the drive armature.

The interstand tension in rolling in the continuous rolling mill was analysed using the differential equation derived by D.P. Morozov:

$$\frac{dT_{k,k+1}}{dt} = \frac{E_k Q_k}{l_{k,k+1}} + \left(V'_{k+1} - V_k\right), \tag{6.25}$$

Here $T_{k,k+1}$ is the tension of the strip between the k-th and k+1-th stands of the continuous mill; E_k is the modulus of elasticity of the strip; Q_k is the cross-section of the strip; $l_{k,k+1}$ is the distance between the stands; V'_{k+1} is the speed of entry of the strip into the k+1-th stand; V_k is the speed of exit of the metal from the k-th stand.

The speed of exit of the strip from the k-th stand Vk in the stationary mode is determined by the circumferential velocity of the working rolls V^0_{wr} and by the forward slip of the metal in the deformation zone S^0_k:

$$V_k^0 = V_{\text{wr}}^0\left(1 + S_k^0\right). \tag{6.26}$$

At any deviation of the deformation conditions in the gap between the rolls the circumferential velocity of the roll increases by ΔV_{wr}, and the forward slip of the metal in the deformation zone changes by the value ΔS_k. Consequently, the speed of the strip at exit from the stand will be:

$$V_k = \left(V_{\text{wt}}^0 + \Delta V_{\text{wt}}\right)\left[1 + \left(S_k^0 + \Delta S_k\right)\right]. \tag{6.27}$$

The increment of the circumferential velocity of the roll is determined by its angular velocity in relation to the equilibrium state:

$$\Delta V_{\text{wt}} = R\dot{\varphi}_{1k} \tag{6.28}$$

which in turn is determined by solving the system of equations (6.21)–(6.25). The magnitude of forward slip at every moment of the transition process is calculated by integration of the differential rolling equation.

The speed of entry of the strip into the deformation zone of the k-th stand is determined by the condition of strong continuity of the metal:

$$V_k' = V_k\left(1 - \varepsilon_k\right), \tag{6.29}$$

here $\varepsilon_k = (h_{0k} - h_{1k})/h_{0k}$.

In this form, the system of the equations (6.21)–(6.29) does not yet reflect completely the relationships and dependences which show how the interaction of the elastic systems of the drive, the stand and the strip takes place. For this purpose, it is necessary to develop an algorithm for calculating the force and torque at deviations of the rolling conditions from the stationary mode. The rolling force P and torque M_t are the perturbing effects which displace the system from the steady-state. A suitable example of the cold rolling process in the non-stationary conditions is the rolling of sections of welded joints in the strip. This case will be examined in greater detail.

6.2.2. Mathematical model of contact stresses in the deformation zone in rolling welded joints

The construction of the mathematical model of rolling welded joints is complicated by the fact that the main differential equation of the contact stresses includes the deformation resistance which is expressed by a complicated dependence which changes with time along the length of the deformation zone.

The contact stresses were calculated using the discrete analogue (1.11)–(1.12) of the differential rolling equation. The algorithm of calculating the contact stresses in the deformation zone in rolling a welded joint is presented in the form of a block diagram in Fig. 6.13.

When defining the initial data, we select the number of divisions of the deformation zone which is determined by the error of approximation of the exact solution of the equation by solving the difference problem. Previously, it was shown that when the zone is divided into 50 parts, the error does not exceed 1.0%. Other parameters, in particular, strip thickness, yield strength, the strain hardening law, and others, are determined by the properties of rolled metal, the rolling conditions and the characteristic of rolling equipment (block 1).

In the first stage of the calculations (block 2) we determine the contact stresses in the steady rolling process of the strip to the section of the welded joint. The position of the welded joint is given by the coordinates of its boundaries. The front boundary of the welded joint is given the coordinate 0, the coordinate of the rear boundary of the welded joint is determined by the length of the welded joint in the units of division of the deformation zone:

$$L_w = \mathrm{E}\left(\frac{l_w}{l}n + 0.5\right),$$

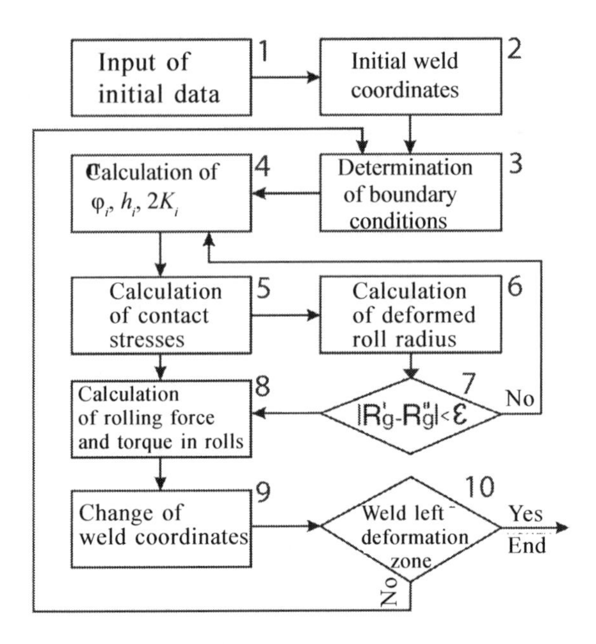

Fig. 6.13. The block diagram for calculating the contact stresses in the deformation zone in rolling a welded joint.

where L_w is the length of the welded joint in the units of division of the deformation zone; E is the integer part of the number; l is the length of the contact arc; l_w is the length of the welded joint; n is the number of divisions of the deformation zone.

The values of the variables included in the right-hand part of the equations (1.11) and (1.12) are determined in the block 3. The static component of the deformation resistance is calculated using the empirical relationships examined previously:

$$2k_i = \frac{2}{\sqrt{3}}\left(\sigma^j_{T_i} + a^j_i \varepsilon_i^{b^j_i}\right),$$

were $\sigma^j_{T_i}$ is the initial yield strength in the i-th cross-section of the deformation zone, the index j denotes the number of the cross-section of the welded joint prior to rolling (Fig. 6.14; a^j_i, b^j_i are the hardening constants of the steel; ε^j_i is the total reduction of the strip at the i-th section of the deformation zone.

Investigation of the mechanical properties of the metal in the zone of the welded joint showed that the variation of the yield strength of the welded joint in the direction of the strip length can be described by the sinusoidal function of the following form:

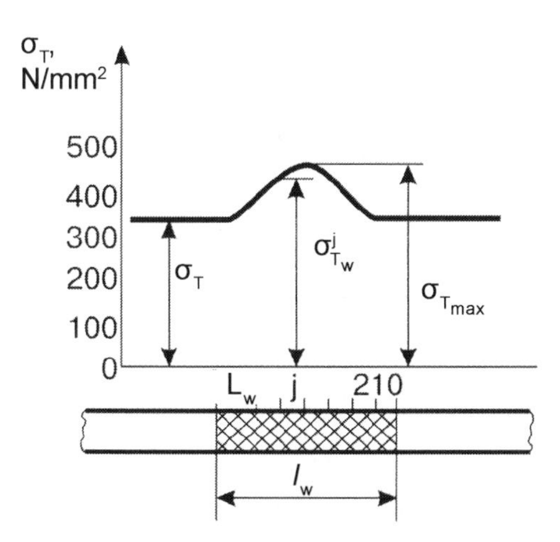

Fig. 6.14. Distribution of yield strength of the metal in the cross-section of the welded joint.

$$\sigma'_{T_w} = \sigma_{T_{av}}\left[\left(1+\frac{a_1}{2}\right)+\frac{a_1}{2}\sin\left(x-\pi\right)\right], \tag{6.30}$$

where $\sigma_{T_{av}} = ½\,(\sigma_{T_1} + \sigma_{T_2})$ is the average yield strength of the contacting planes; $x = \pi\,(2j/Lw-1)$ is the coordinate along the length of the welded joint; a_1 is the coefficient indicating by how much the yield strength of the weld metal is greater than the yield strength of the metal of the strip: $a_1 = \dfrac{\sigma_{T_w} - \sigma_{T_{av}}}{\sigma_{T_{av}}}$.

The boundary conditions have the form; $\sigma_0 = -\sigma_{rear}$; $\sigma_n = -\sigma_{front}$, where σ_{front} and σ_{rear} are the front and rear specific tensions.

In passage through the deformation zone the welded joint gradually occupies different positions: the welded joint in front of the deformation zone, the welded joint is partially located in the deformation zone or completely fills the zone if the length of the welded joint is greater than the length of the deformation zone. Depending on the position of the welded joint, the boundary conditions in the block 4 change.

The difference problem is solved in the block 5. In the direction from entry to exit of the deformation zone the equations (1.11) were used to calculate the longitudinal and then normal contact stresses in the backward creep zone. Using the equations (1.12) the longitudinal and normal stresses in the forward creep zone are calculated. The point of intersection of the resultant curves determines the position

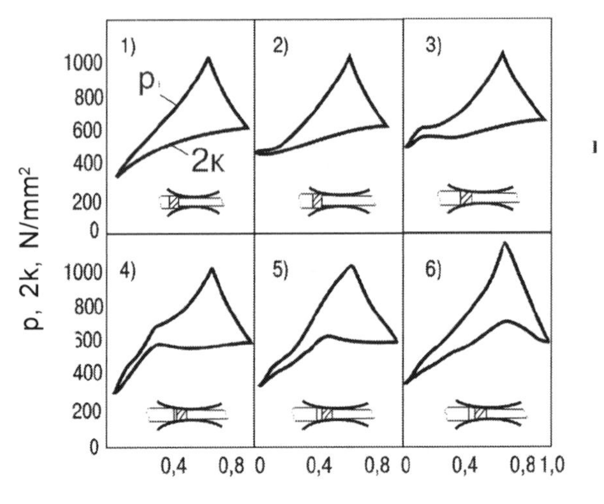

Fig. 6.15. Variation of the deformation resistance $2k$ and normal stresses p in the deformation zone in rolling the welded joint. The numbers on the figure indicate the position of the section of the welded joint. The rolling conditions: $h_0 = 3.0$ mm; $\sigma_T = 320$ N/mm²; $l_w = 10$ mm; $h_1 = 1.8$ mm; $\sigma = 430$ N/mm²; $R = 250$ mm; $f = 0.08$; $\sigma_{front} = \sigma_{rear} = 0$.

of the neutral section in the deformation zone. The accuracy of calculating the normal contact stresses is improved taking into account the elastic deformation of the roll (blocks 6 and 7). In block 8 calculations are carried out to determine the rolling force and the torque on the rolls for every position of the welded joint in the deformation zone. In block 9 the welded joint moves along the deformation zone. The movement of the welded joint is simulated by changing the coordinates of its boundaries.

As an example, Fig. 6.15 shows the curves of deformation resistance and normal stresses in the deformation zone in rolling a welded joint. The thickness of the strip in the section of the welded joint was assumed to be constant in this case.

The variation of the rolling force and the torque is shown in Fig. 6.16.

In the selected example, the maximum rolling force is obtained when the welded joint is between the positions 3 and 4, shown in Fig. 6.15.

The proposed algorithm for calculating the force P and torque M_t in rolling a welded joint can be used not only for the qualitative

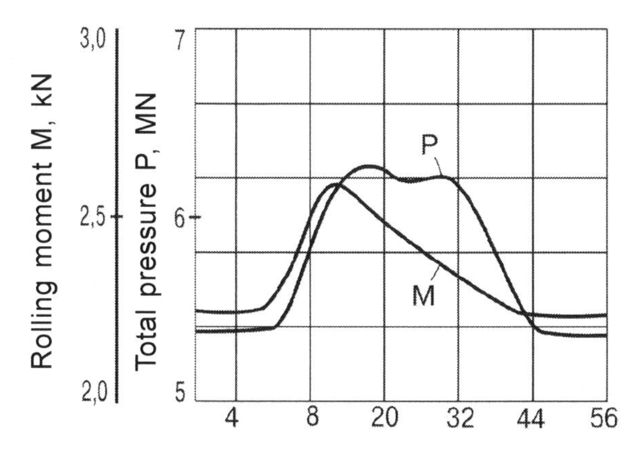

Fig. 6.16. Variation of the force P and torque M in rolling the welded joint. The rolling conditions at the same as in Fig. 6.15.

analysis of the nature of loading the stands of the rolling mill but also carry out, using the equations of motion, the quantitative evaluation of the dynamic loads in the electromechanical system of the drive and the stand in rolling the welded joint. In addition to this, it is also possible to solve the problem of the selection of rolling conditions which would ensure the maximum decrease of the dynamic loading the strip during rolling butt welded strips. These questions are investigated in greater detail in a previous work by the authors of this book [92].

6.2.3. Methods for solving the dynamic problem

The system of differential equations of movement of the stand–drive–strip mechanical system (6.21)–(6.25) was solved by the finite difference method.

The duration of the perturbing effect of the welded joint on the deformation zone was divided into m equal periods Δt. It was assumed that in each time period (t_i, t_{i+1}) the parameters of the transition process remain unchanged. This assumption results in a small error if the step Δt is selected sufficiently small. It was assumed that the time period Δt corresponded to 1/10 of the period of intrinsic oscillations of the stand taking into account the elastic connection on the side of the deformed metal and approximately 1/50 of the period of the intrinsic oscillations of the drive. The calculation show that a further decrease of the time period Δt (halving this period) results in a change of the results by no more than 2%. Consequently, the selected time step ensures the sufficiently high accuracy of solving the system of equations (6.21)–(6.25).

The continuous problem (6.21) can be determined by the difference analogue:

$$\Delta x_{i+1} = \left(2 - \frac{C_k}{m}\Delta t^2\right)\Delta x_i - \Delta x_{i-1} + \frac{\Delta P(t_i)}{m}\Delta t^2 - \text{sign}\left[\frac{\Delta x_i - \Delta_{x-1}}{\Delta t}\right]F(t_i)\Delta t^2. \quad (6.31)$$

$\Delta x_0 = 0$; $\Delta x_1 = 0$.

The initial conditions were defined assuming that at the initial moment of time the increment of the elastic deformation of the stand and the forward speed of the roll section are equal to 0 (stationary process).

The difference analogue of the differential equations (6.22)–(6.24) is described by the following system of algebraic equations:

$$\begin{aligned}
\varphi_{1_{i+1}} &= \left(2 - 2n\Delta t - \frac{C_{12}}{\Theta_1}\Delta t^2\right)\varphi_{1_i} - (1 - 2n\Delta t)\varphi_{1_{i-1}} + \\
&+ \left(\frac{C_{12}}{\Theta_1}\Delta t^2 + 2n\Delta t\right)\varphi_{2_i} + \varphi_{2_{i-1}}\Delta t + \frac{\Delta M_{tor_i}}{\Theta_1}\Delta t^2; \\
\varphi_{2_{i+1}} &= \left(2 - \frac{C_{12}}{\Theta_2}\Delta t^2\right)\varphi_{2_i} - \varphi_{2_i-1} + \frac{C_{12}}{\Theta_2}\varphi_{1_i}\Delta t^2 + \frac{\Delta M_{\text{dr}_i}}{\Theta_2}\Delta t^2; \\
\Delta I_{j+1} &= \left(1 - \frac{R_a}{L}\Delta t\right)\Delta I_i - \frac{C_e}{L}\left(\varphi_{2_{i+1}} - \varphi_{2_i}\right); \\
\Delta M_{\text{dr}_{i+1}} &= \frac{C_m}{\Theta_2}\Delta I_{i+1};\ \varphi_{10} = \varphi_{11} = \varphi_{20} = \varphi_{21} = 0;\ \Delta I_1 = 0.
\end{aligned} \quad (6.32)$$

The values of the perturbing parameters ΔP and ΔM_t, included in the equations (6.31) and (6.32) are calculated for each time period $[t_i, t_{i+1}]$. The angular velocity of the roll is expressed by the first derivative of the angle of rotation with respect to time $\omega_{1_i} = \dot{\varphi}_{1_i}$ In the different form it is:

$$\omega_1 = \frac{\varphi_{1_{i+1}} - \varphi_{1_i}}{\Delta t}. \quad (6.33)$$

The circumferential speed of the roll at the time t_{i+1} is equal to

$$V_{r_{i+1}} = V_{r_i} + R\omega_{1_{i+1}}. \quad (6.34)$$

The difference analogues of the differential equation of the strip tension in the interstand gaps (6.25) will be written. For the front tension in the k-th stand we have

$$T_{k,k+1}^{i+1} = T_{k,k+1}^{i} + \frac{E_k \Theta_k}{l_{k,k+1}} \left(V_{k+1}^{\prime i} - V_k^i \right) \Delta t. \tag{6.35}$$

The rear tension is calculated from the equation:

$$T_{k-1,k}^{i+1} = T_{k-1,k}^{i} + \frac{E_{k-1} Q_{k-1}}{l_{k-1,j}} \left(V_k^{\prime i} - V_{k-1}^i \right) \Delta t. \tag{6.36}$$

The initial values of the front and rear tensions are represented by the tensions in the steady process of rolling the rear end of the strip. The recurrent relationships (6.35), (6.36) complete the construction of the different scheme for calculating the parameters of the transition process in the stand which has the form of a system of algebraic equations. It can be solved by the iteration method. Calculations in each time period are carried out in the same order as the order in which the system (6.31)–(6.36) is written, and are completed when the relative difference of the values of the arbitrary selected dynamic parameter is not smaller than some value is defined in advance.

6.2.4. Simulation of the transition process in rolling a welded joint

The computational experiments carried out to investigate the transition process in rolling the welded joint were conducted using the conditions of cold rolling of 0.5 × 1015 mm strips made of steel 08kp in the five-stand 1700 rolling mill. The strips of this grade were selected because they often represent a large fraction in the volume of production of thin sheet rolling mills, and the rolling of the strips is characterised by the highest labour content because of problems with the rolling of welded joints.

The parameters of the main lines of the stands of the rolling mill are presented in Table 6.2. Since the majority of breaks in the welded joints takes place in the final interstand gaps – between the fourth and fifth stand – the investigations were carried out for the penultimate (fourth) stand of the rolling mill.

The result of the solutions of the system of equations (6.21) – (6.29) yielded the curves of variation of the torque in the drive, the rolling force, strip tension, thickness of the strip at exit from the

Table 6.2. Initial data for solving the system of dynamic equations of the transition process in the five-stand 1700 rolling mill

Process parameters	Stand 1	Stands 2–5
Weight of working roll with chock, kg	960	960
Weight of support roll with chock, kg	10500	10500
Stifness of stand, MN/mm	4500	4500
Inductance of armature of drive L, H	7 × 10	32.24 ×10
Resistance of armature circuirt, R_a, ohm	0.0075	0.0055
Electromechanical time constant of drive, s	0.043	0.081–0.115
Drive constant C_e, V·s	0.628	0.293
Drive constant C_m, kgf · m/a	0.61	0.29
Reduced moment of inertia of drive armature Θ_a, kgf · m · s^2	765	765
Reduced moment of inertia of rolls Θ_r, kgf · m.s^2	186.2	159.7
Reduced stiffness of shaft line C_{12}, kgf · m (upper/lower roll)	$4.15 \cdot 10^5/9.35 \cdot 10^5$	$5.2 \cdot 10^5/6.2 \cdot 10^5$

stand, etc. The calculated values of the parameters of the transition process are presented in Fig. 6.17.

Analysis of the transition process in the drive–stand–strip system shows that during the passage of the welded joint to the deformation zone the forward slip S of the metal in the deformation zone, force P and the rolling momentum M_r. The first circumstance results in a change of the speed of exit of the strip from the interstand gap, the second and third –in the gradual vertical and the rotational oscillations of the rails. The calculation show that the variation of the front tension of the strip in rolling the welded joint of different thickness consists of several components: change of tension as a result of the progressive oscillations of the roll section (high-frequency component); torsional oscillations of the drive–roll system (low-frequency component), and also the variation of the forward slip of the metal in the deformation zone and the change of the speed of rotation of the armature of the drive.

In rolling a welded joint with a typical configuration (after the thicker end the thin end is rolled, and the welded joint has a thicker area) the front tension increases most markedly. This is caused by the fact that during the time when the welded joint is in the gap between the rolls the rolling force and the thickness of the strip at exit from the stand increase. After rolling the welded joint, the strip of smaller

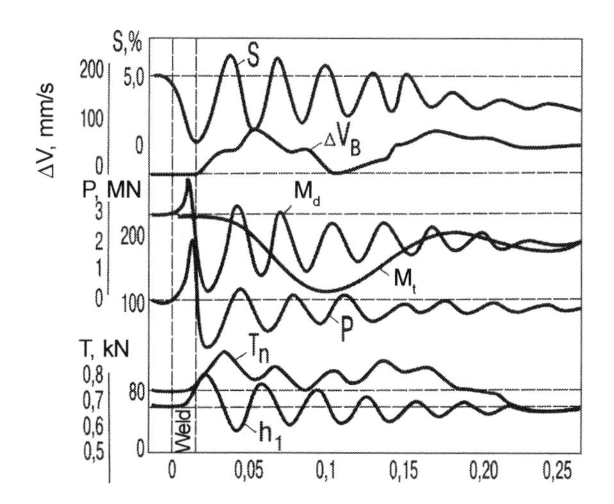

Fig. 6.17. Results of simulation of the transition process in the fourth stand of the 1700 mill in rolling a welded joint. The rolling conditions: h_1 = 0.9 mm; T_0 = 146 N; B = 1015 mm; h_2 = 0.8 mm; T_1 = 146 N; l_w = 30.0 mm; h_w = 1.0 mm; V = 5.0 m/s; ε = 25%.

thickness is rolled and this is accompanied by an additional decrease of the reduction. This results in a decrease of the forward slip and this increases the front tension. For example, in rolling a welded joint with the dimensions h_1 = 0.9 mm; h_w = 1.0 mm; h_2 = 0.8 mm; l_w = 30 mm with a speed of 5.0 m/s, the front tension increases by 60%. It is characteristic that the maximum of the total front tension corresponds to the minimum thickness of the strip at exit from the stand. This results in an additional increase of the specific tension in this cross-section (up to 100%). The simulation of the transition process at a stepped variation of the thickness of the rolled strip and the constant properties and geometry of the welded joints shows that the change of strip tension in this case is considerably smaller, only 20%.

Thus, the results obtained using the discrete model of the rolling process show that the combined effect of the differences in the thickness of the strips and increase of the thickness of the welded joint resulted in a large increase of the tension in the passage of the welded joint through the deformation zone which may cause fracturing of the strip. More detailed analysis of the process of rolling sections of the strips with welded joints is presented in chapter 7 of this book and also in a previous work by the authors of this book [92].

7

Optimisation of the technological conditions of continuous cold rolling of strips

7.1. Selection of the criterion and optimisation method

The adherence to the technological and design restrictions of the cold rolling process does not make it possible to distribute uniformly the reductions in the stands of the continuous rolling mill because the same restrictions are satisfied by a large number of greatly different conditions. Practical experience shows that in many cases when using the same rolling mill for the identical range the accepted deformation –rate rolling conditions differ. Therefore, in any case, the selection of the rolling conditions for each specific case is in fact a solution of the optimisation problem. The main stages of solving this problem consist of the selection of the criterion (strategy) of optimisation, the determination of the restrictions for the technological conditions and design parameters of the rolling mill, the selection of the optimisation method, and the development of the appropriate algorithm and methods of realising this algorithm.

The following strategies (criteria) in optimisation can be used:

a) achieving the maximum degree of deformation of the strip:

$$\mu_c = \prod_{i=1}^{n} \frac{h_{01}}{h_{1i}} \to \max. \tag{7.1}$$

Here h_{0i}, h_{1i} is the thickness of the strip respectively at entry and exit for each i-th stand; n is the number of stands;

b) achieving the maximum productivity of the rolling mill

$$\Pi = \sum_{i=1}^{n} (A_i V_i) \to \max, \tag{7.2}$$

where A_i, V_i is respectively the mass of 1 unit meter of the strip and the rolling speed in the i-th stand;

c) achieving the minimum specific power capacity of the rolling process

$$E_\Sigma \sum_{i=1}^{n} E_i \to \min, \tag{7.3}$$

where E_i is the specific power capacity of the process of rolling the strip in the i-th stand;

d) achieving the minimum longitudinal thickness difference of the strip;

$$K_b = \prod_{i=1}^{n} K_i \to \max, \tag{7.4}$$

here K_i is the coefficient of equalisation of the longitudinal thickness difference in the i-th stand;

e) ensuring the flatness of the strip:

$$\frac{\Delta h_i}{h_i} = \text{const}, \tag{7.5}$$

where Δh_i, h_i is respectively the transverse thickness difference and the thickness of the strip at exit from the i-th stand.

These optimisation criteria depend directly or indirectly on: the nature of loading the stands of the rolling mill as regards the force or rolling power. The mathematical expression of the optimality criterion, reflecting the given distribution of the load in the stands of the rolling mill, was obtained by solving the well-known isoperimetric problem [93]:

$$F = \prod_{i=1}^{n} \left(k_i \frac{Q_c}{Q_i} \right)^{\frac{1}{n}}, \quad \text{F} \to \min = n, \tag{7.6}$$

here $Q_\Sigma = \sum_{i=1}^{n} \frac{Q_i}{k_i}$ is the conventional total load on the stand; n is the number of stands in the rolling mill; k_i are the numbers defining the ratio of the loads in the stands of the mill or the loading coefficients; Q_i is the load in the i-th stand. This optimality criterion should tend to a minimum equal to the number of stands in the rolling mill.

Another optimality criterion can also be used. This criterion has the form of the sum of the absolute values of the load differences in the i-th and $(i+1)$-th stands which for the given distribution of the loads k_i in the individual stands of the mill tends to 0

$$F = \sum_{i=2}^{n} \left| \frac{Q_i}{k_i} - \frac{Q_{i-1}}{k_{i-1}} \right|, \quad \mathrm{F} \to \min = 0. \tag{7.7}$$

The required reduction schedule should ensure the equality:

$$\frac{Q_1}{k_1} = \frac{Q_2}{k_2} = \ldots = \frac{Q_n}{k_n}. \tag{7.8}$$

We examine the algorithm of the iteration method of optimisation of the reduction schedule according to the distribution of the loads in the stands of the rolling mill defined by the coefficients k_i [94]. In a general case, the loads in the i-th step of the iteration process Q'_i, obtained for the initial distribution of the reductions ε'_i, do not satisfy the relationship (7.8). It is therefore necessary to correct the reductions in the stands of the rolling mill. This can be carried out efficiently if we assume the directly proportional dependence of load on reduction:

$$\frac{\varepsilon'_i}{\varepsilon''_i} = \frac{Q'_i}{Q''_i}, \tag{7.9}$$

Here ε'_i, ε''_i are the relative reductions in the i-th stand in the first and second iteration; Q'_i, Q''_i are the loads in the i-th stand obtained in the first and second iteration.

The reduction correction coefficient in the i-th stand, starting from the second step of the iteration process, is determined as the ratio of the conventional mean load per mill to the load in the investigated stand taking into account the accepted load factor:

$$a'' = Q'_{\Sigma} \frac{k_i}{nQ'_i}. \tag{7.10}$$

The conventional total load per mill Q'_{Σ} is calculated in each iteration step. The load in the next iteration step Q''_i should tend to the relation $Q'_{\Sigma}\, k_i/n$.

The new approximate distribution of the reductions ε'_{bi} is calculated from the equation

$$\varepsilon''_{bi} = a''_i \, \varepsilon'_{bi}. \tag{7.11}$$

The rate of convergence can be improved by introducing the relaxation coefficient k_p, and to take into account the anomalous effect of the reduction on the load it is possible to take into account the change of the sign of the increment of the degree of deformation according to the results of the previous iteration step. The sign of the increment of the reduction in each stand corresponds to the sign of the product of the increment of the load and the reduction in the previous step:

$$k''_i = \text{sign}\left\{(Q''_i - Q'_i)(\varepsilon''_i - \varepsilon'_i)\right\}; \tag{7.12}$$

$$\varepsilon''_i = \varepsilon'_i \left[1 + k''_i \, k_p \left(\frac{Q_{cp} k_i}{Q_i} - 1\right)\right], \tag{7.13}$$

here k_p is the relaxation coefficient (in the first approximation it may be accepted that $k_p = 0.5$); Q_{cp} is the conventional mean load per mill.

Using the relationships presented in [94], the approximate reductions ε''_{bi} are converted to the actual reductions ε''_i and the loads Q''_i are again calculated. On the basis of the resultant values of Q''_i it is again possible to correct the reductions using equations (7.11)–(7.13). The procedure is repeated until the required accuracy of the optimality criterion (7.6) is reached.

The number of iteration is almost completely independent of the number of the stands in the rolling mill. This is an advantage of this optimisation method in comparison with other methods. The high rate of convergence of the activation optimisation method enables the method to be used for the system of initial setting, dynamic rearrangement, automatic thickness regulation, and also carry out highly accurate and high-speed computer experiments of the optimisation of the reduction schedule. For example, the reduction schedule with the uniform distribution of the forces in the stands satisfies most efficiently the task of producing high-quality cold-rolled strips in the currently available mills fitted with a set of the means for the efficient control of thickness, profile and shape. The schedule ensures stability of the process and its minimum power requirement. This has been confirmed by practical experience in sheet rolling production [93].

7.2. Selection of the value of relative reduction in the final stand of the cold rolling mill

The currently available continuous cold rolling mills are fitted with the automatic systems of regulating thickness and tension (SARTT), the profile and shape of the strip (SARPS). Experience shows that the efficient operation of the automatic regulation systems SARTT and SARPS is ensured by the selection of the correct schedule (strategy) of rolling. For example, it was established for the conditions and range of the 6-stand 1400 mill that the stability of the rolling process is ensured only at high relative reductions (35–40%) in the final stand. In this case, the efficiency of operation of the SARTT and SARPS automatic systems and, consequently, the quality of production are greatly improved.

The application of the high reductions in the final stand makes it possible to realise the traditional procedure of regulating the tension of the strip in the final space between the stands by changing the position of the pressure devices of the 6-th stand of the rolling mill. The efficiency of operation of the SARTT system was determined by comparing the transfer ratios $\partial h_1/\partial T_0$ and $\partial T_0/\partial S$ at different reductions (h_1 is the thickness of the strip at exit from the stand; T_0 with the rear tension of the strip, S is the position of the pressure screws). The change of the rear tension $T_{k,k+1}$ of the strip is a function of the change of the speed of travel of the strip ($V_{0,k+1}$) at entry into the gap between the rolls of the k +1-st stand:

$$\frac{\partial T_{k,k+1}}{\partial t} = \frac{E_k Q_k}{L_{k,k+1}} \left(V_{0,k+1} - V_{1,k} \right), \tag{7.14}$$

Here E_k, Q_k is the modulus of elasticity and the cross-section area of the strip in the roll gap; $L_{k,k+1}$ is the distance between the stands; $V_{1,k}$ is the speed of exit of the strip from the k-th stand.

In the stationary process $V_{0,k+1} = V_{1,k}$.

The displacement of the pressure screws of the k+1-th stand results in a change of the relative reduction ε_{k+1} and this results in a change of the entry speed by the value $\partial V_{0,k+1}$:

$$V_{0,k+1} + \delta V_{0,k+1} = V_b \left[1 - \frac{h_1 + \delta h_1}{h_0} \left(1 + S + \partial S \right) \right].$$

Consequently

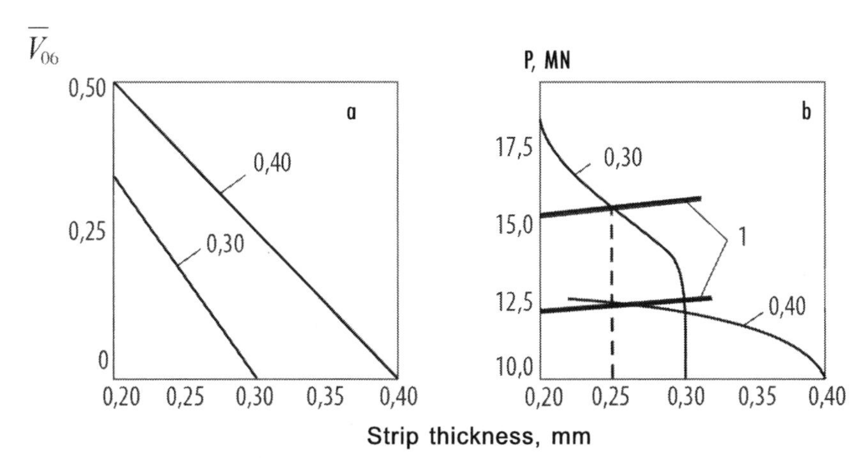

Fig. 7.1. Effect of the initial thickness of the strip and the reduction on the relative speed $\overline{V}_{06}$ of the rear end to of the strip (a) and on the rolling force P in the sixth stand of the 1400 mill (b). Notations: 1 – the line of elastic deformation of the stand; the digits at the curves – the thickness of the strip at entry into the sixth stand of the mill, mm.

$$\partial T_{k,k+1} = \frac{E_k Q_k}{L_{k,k+1}} \delta V_{0,k+1} \cdot \partial t. \tag{7.15}$$

To determine the transfer ratio $\frac{\partial T_{k,k+1}}{\partial S}$, it is necessary to investigate the value $\frac{E_k Q_k}{L_{k,k+1}} \cdot \frac{\delta V_{0,k+1}}{\partial S}$, and in order to carry out the comparative analysis of the efficiency of the effect of specific tension on the thickness of the strip at different reductions it is sufficient to examine the ratio $\frac{\delta V_{0,k+1}}{\partial S}$, since $\frac{E_k Q_k}{L_{k,k+1}} = \text{const.}$.

Investigations were carried out into the process of rolling strips with a thickness of 0.25 mm in the 6-th stand of the 1400 mill at two reduction levels: 37.5% and 16.7%. The thickness of the strip in front of the 6-th stand was equal to 0.4 and 0.3 mm, respectively. The derivatives $\frac{\partial h_{06}}{\partial S_{06}}$ and $\frac{\partial V_6}{\partial S_6}$ were estimated numerically by linear approximation of the functions P = P(ε), S = (P) in the vicinity of $h_6 = 0.25 \pm 0.02$ mm (Fig. 7.1).

According to the results of the calculations for a strip with thickness h_{16}, the transfer ratio $\frac{\partial V_6}{\partial S_6}$ at a reduction of 37.5% is more than three times greater than at a reduction of 16.7%, i.e., regulation of the tension in the last gap between the rolls by displacement of the

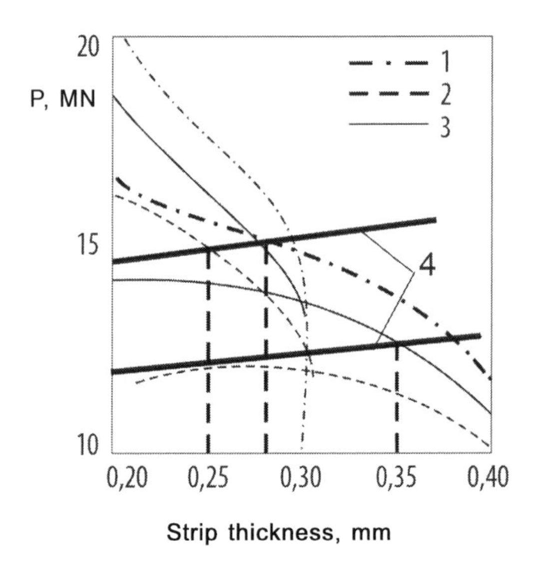

Fig. 7.2. Effect of the initial thickness, reduction and tension of the strip on the rolling force in the stand 6 of the 1400 continuous cold rolling mill. The rolling conditions are the same as in Fig. 7.1. Notations in the figure: 1 – σ_{06} = 100 N/mm^2; 2 – 150 N/mm^2; 3 – σ_{06} = 200 N/mm^2; 4 – the lines of elastic deformation of the stand.

pressure devices of the sixth stand is far more efficient. In addition to this, the increase of the ratio $\frac{\partial \bar{V}_{06}}{\partial S_6}$ at a small reduction results in a large increase of the rolling force. For example, at ε = 37.5% $\frac{\partial P_6}{\partial h_{16}} = 5$, and at ε = 16.7% $\frac{\partial P_6}{\partial h_{16}} = 5$; the value $\frac{\partial P_6}{\partial S_6}$ is equal to 0.50 and 0.13, respectively.

The principle of precise regulation of the strip thickness in the continuous cold rolling mill is based on the dependence of the rolling force on the rear tension of the strip and explained by the mismatch of the speeds of the adjacent stands. The effect of the rear tension of the strip on h_{16} at different reductions will be investigated. Figure 7.2 shows the curves of the rolling force at ε = 37.5% and ε = 16.7% for three levels of the rear specific tension; σ_{06} = 200, 150 and 100 N/mm^2. The effect of tension σ_{06} on the strip thickness h_6 was determined by the graphical procedure. The basic tension level was σ_{06} = 200 N/mm^2.

The numerical analysis results show that the decrease of the rear tension of the strip from 200 to 150 N/mm^2 at h_5 = 0.40 mm increases the rolling force, the elastic deformation of the stand and, consequently, increases the strip thickness at exit from the sixth stand by approximately 0.10 mm, and at $h_5 = 0.30$ mm – by 0.03 mm, i.e., the transfer ratio $\frac{\partial h_6}{\partial \sigma_{06}}$ at high reductions is considerably greater than at small reductions. In a specific case the ratio $\frac{\partial h_6}{\partial \sigma_{06}}$ was equal to respectively 0.0024 mm/N/mm^2 and 0.008 mm/N/mm^2, i.e., at a reduction of 37.5% the transfer ratio is 3.3 times greater than at $\varepsilon = 16.7\%$.

The experiments carried out in the 1400 rolling mill confirm these relationships. For example, at a reduction of $\varepsilon = 7.0\%$, the effect of the displacement of the pressure screws of the sixth stand on the rear tension of the strip could not be examined because of the overloading of the stand by the rolling force. It is fully justified to assume that the mechanism of the effect of reduction in the last stand of the mill on the efficiency of operation of the SARPS is close to that discussed previously. In fact, the profile of the strip is regulated by the non-uniform reduction of the strip along its width irrespective of the effect leading to the redistribution: operation of the system for hydraulic bending of the rolls (SAHBR) or the system of local cooling of the rolls (SALCR) . Analysis of the operation of SAHBR shows that in the conditions of small reductions in the sixth stand the transfer ratio 'the shape of the strip/bending force' is 10 times greater than at small reductions.

According to the results of computational experiments, the increase of the thickness of the rolled strips to 1.0 mm results in a large reduction of the efficiency of regulation of the thickness as a result of the change of the rear tension (by up to 10 times in comparison with the rolling the strips with a thickness of 0.25 mm). The efficiency of regulation of the thickness of relatively thick strips ($\geq$0.7 mm) also depends strongly on the reduction in the stand. For example, for $\varepsilon = 20.6\%$ it is four times greater than at $\varepsilon = 9.1\%$.

Thus, for the efficient functioning of the local systems of automatic regulation of the thickness, tension and profile of the strip in cold rolling the reduction in the final stand of the rolling mill should be sufficiently high and the values of the transfer coefficient in the automatic systems for regulation must be determined taking into account the level of the reduction in the stand and the thickness of the rolled strip.

7.3. Rolling in rough rolls

In the generally accepted technology, the surface of the working rolls of the final stand of the continuous cold rolling mill (with the exception of the stands used for rolling sheets) is treated to a surface roughness of 3.0–4.0 μm *Ra*. Therefore, in the initial campaign period, the rolling force is often greater than the permissible level and this results in the loss of flatness or it is not possible to produce the required thickness of the completed strips. With increase in the degree of wear of the surface of the rolls the rolling force decreases and approaches its nominal value, determined by the given reduction and tension conditions.

One of the methods of solving the problem of stabilisation of the rolling process could be the application of relatively smooth (ground) rolls in the final stand of the mill. However, this increases the weldability of the coils during annealing and also increases the probability of 'sticking' of the strip to the rolls which may lead to the formation of surface defects in the cold rolled strips.

The simultaneous use in the final stand of one rough and one ground working roll is not used widely in production practice.

To determine the possibilities of stabilizing the rolling force, thickness and shape of cold rolled strips in the process of wear of the initial roughness of the rolls in the 4-stand 1700 rolling mill investigations were carried out into the effect of the surface roughness of the working rolls on the reduction in force in the final (fourth) stand in rolling 0.51–0.54 mm low-carbon steel strips. Attempts were made to stabilise the rolling force and the final strip thickness by the parametric (programmed) change of the reduction and tension conditions.

The conditions of dynamic rearrangement of the reductions during wear of the rough rolls were calculated by the mathematical model of the continuous rolling mill. The changes of the surface roughness of the rolls were simulated by changing the coefficient K_b, included as the multiplier in the equation for the friction coefficient (6.15).

The algorithm of the calculations for calculating the reduction schedule was organised in such a manner that as the degree of wear of the surface of the working rolls of the force stand increased and, consequently, the friction coefficient changed, the values of the relative reduction ε_4 changed to ensure the constant rolling force P_4. The reduction in the first stand of the mill was kept constant,

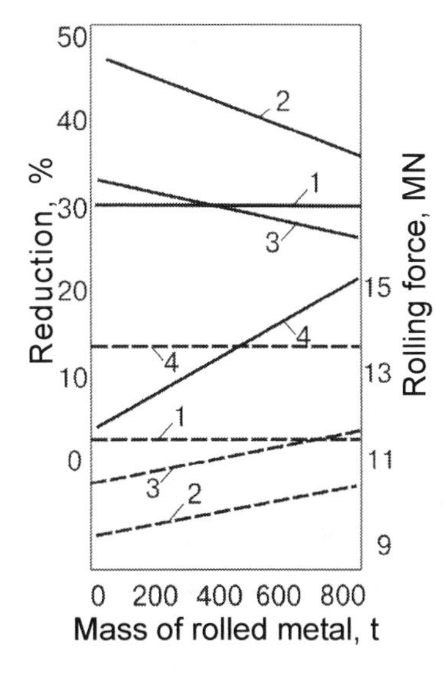

Fig. 7.3. Redistribution of reduction (solid lines) and rolling forces (dashed lines) in the stands of the four-stand 1700 mill in dependence on the mass of the rolled strips with the size of 2.5/0.51×1000 mm in the rough rolls of the final stand. The numbers at the lines denote the number of the stand.

equal to 30%, and the remaining part of the total reduction was redistributed between the second and third stand.

The results of the calculations show (Fig. 7.3) that to ensure the constant rolling force of 13.5 MN in the fourth stand of the mill rolling in the initial period of the roll campaign should be carried out with the minimum reduction (2–4%). Regardless of the large change of the relative reduction in the second and third stand, the rolling force in them changes only slightly (by 1 MN). It is characteristic that the force values increase with a decrease of the reduction in these stands. This is explained by the fact that at constant total tensions of the strip, the specific tension in the second and third gaps between the stands decrease.

Thus, it is important to maintain constant force and strip thickness in rolling in the rough rolls and this can be ensured by the dynamic rearrangement of the reduction schedule in the individual stands of the mill with increase of the degree of wear of the rough rolls.

7.4. Rolling in 'cold' rolls

The working rolls are installed in the rolling mill at the temperature usually equal to the room temperature. During continuous cold rolling the rolls are heated to approximately 70°C in the centre of the length of the barrel. The temperature difference between the central part and the edges of the roll barrel is approximately 10–30°C. This results in the formation of thermal convexity of the rolls of up to 0.1–0.2 mm. Usually, heating of the rolls lasts from 0.5 to 1.0 hour. The increase of the thermal convexity of the rolls takes place gradually during their service. This complicates the production of strips with a very flat surface and reduces the accuracy of rolling at the start of the campaign. Therefore, in practice it is often necessary to use rolls with increased initial convexity in order to obtain flat strips at the start of rolling. However, in heating of the rolls the formation of the thermal convexity is accompanied by warping of the strips causing the concentration of specific tension at the edges of the strip and increases the reliability of fracture of the strips between the stands. Breaking of the strips, caused by the instability of the profile of the rolls, causes failure of the rolls, shortens the service life of the rolls, reduces the productivity of the stand and impairs the quality of completed products. Therefore, it is necessary to find efficient solutions which would compensate the changes of the thermal convexity of the working rolls during their heating in the stand to working temperature.

The profile of the active generating line of the working rolls, which determines the shape of the gap between the rolls and the shape of the rolled strip, forms as a result of the superposition of the initial (grinding) profile, the elastic bending of the rolls and their thermal profile. Therefore, the increase of the thermal profile of the rolls can be compensated by increasing the elastic bending of the rolls. The elastic bending of the rolls depends directly on the rolling force and the latter is a function of the reduction and tension of the strip. For example, a decrease of the tension of the strip between the stands increases the rolling force and, consequently, the extent of bending of the rolls.

Calculation show that the specific tension of the strip during rolling should be reduced with increase of the mass of the rolled metal in accordance with the following dependence:

$$\sigma = \sigma_0\left(1 + ae^{-KG}\right), \tag{7.21}$$

here σ and σ_0 is the actual and initial value of the specific tension of the strip, N/mm^2; a and k are the coefficients which depend on the range of rolled metal $0.008 < k < 0.035$; $0.5 < a < 3.0$; G is the mass of the metal rolled in the rolls from the moment of installation of the rolls in the stand, t.

After installing a new set of the working rolls in the stand the specific tension between the stands in rolling the first strip should be set such that its value is higher $(1+a)$ times than the basic value σ_0. After rolling the first coil with the mass G_1 it is necessary to determine and establish a new value of the tension between the stands at $G = G_1$. The same procedure is used for converting the tension after rolling the second, third, etc coil. Since the value of σ asymptotically approaches the basic value σ_0 with increase of the mass of the rolled metal, the decrease of the tension should be interrupted when the relative difference between the values of σ and σ_0 becomes lower than 10%.

The application of this technical solution in the conditions of the 4-stand 1700 cold rolling mill reduces the number of breaks in the strips by more than 10 times. The extensive industrial application of the solution is possible both in manual control and using the means of automation, fitted with systems for determining the actual mass of the rolled metal.

7.5. Special features of the technology of rolling strips with welded joints

The passage of welded joints in strips through the deformation zone in rolling in the continuous mill is accompanied by disruption of the process stability. The perturbing factors are the variations of the mechanical properties of the metal of the strips on the welded joint, and the difference in the thickness of the ends of the contacting strips and the weld zone (Fig. 7.4)

As shown in chapter 6, when the welded joint is 'captured' by the working rolls the process parameters change – rolling force, the

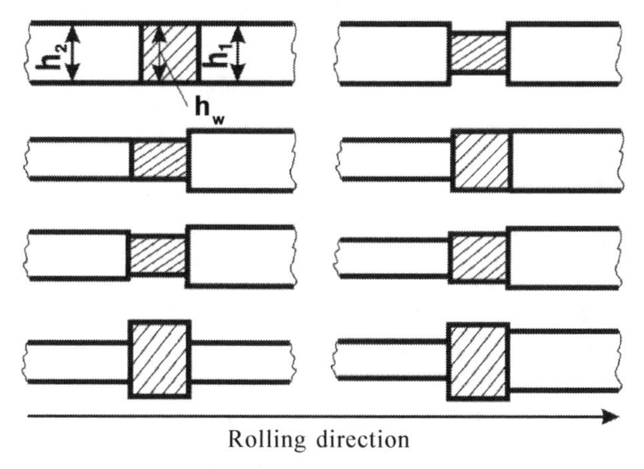

Fig. 7.4. Configurations of welded joints in strips.

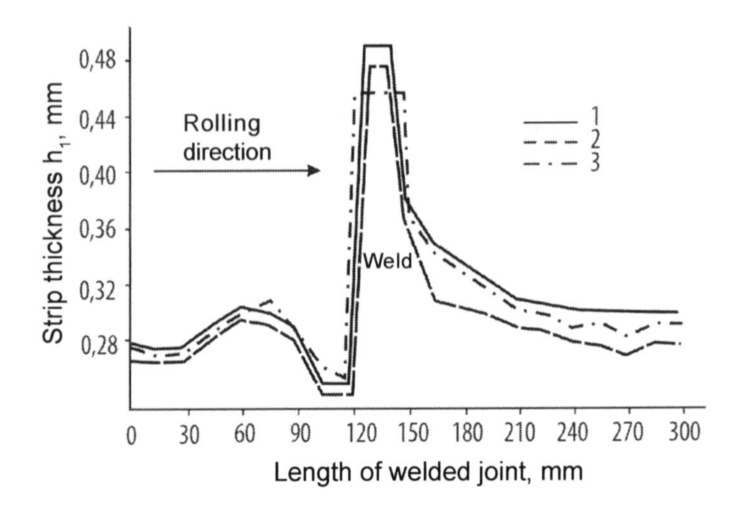

Fig. 7.5. Change of the thickness of the strip in the zone of the welded joint after rolling in the 1200 five-stand of the Magnitogorsk Metallurgical Plant; 1 – the centre of the strip; 2 – the right edge; 3 – the left edge.

forward slip and the torque at the rolls. The results of investigations of the thickness of the strip in the zone of the welded joint, rolled in the 1200 continuous cold rolling mill (Fig. 7.5), confirm the presence of vertical elastic oscillations of the rolls after rolling the welded joints.

The results of investigations of the geometry of the welded joint after cold rolling show that the thickness difference in the contacting strips in the zone of the welded joint changes continuously. The results of measurements of the longitudinal profile of the welded joints prior to and after cold rolling in the five-stand 1200 mill are presented in Fig. 7.6.

It should be noted that the initial relative thickness difference of the welded ends of the strips after cold rolling usually increases 2–3 times.

To investigate the process of rolling welded joints using the mathematical model it is necessary to define in advance the distribution of the thickness and the yield strength of the metal in the area of the welded joint and the law of strain hardening of the metal of the strip and the welded joint. In the welded joints, the rear end of the strip in the direction of cold rolling (h_1) is usually thicker than the front end (h_2). The difference in the thickness of the welded joints in the strip plate for different mills may reach 0.20

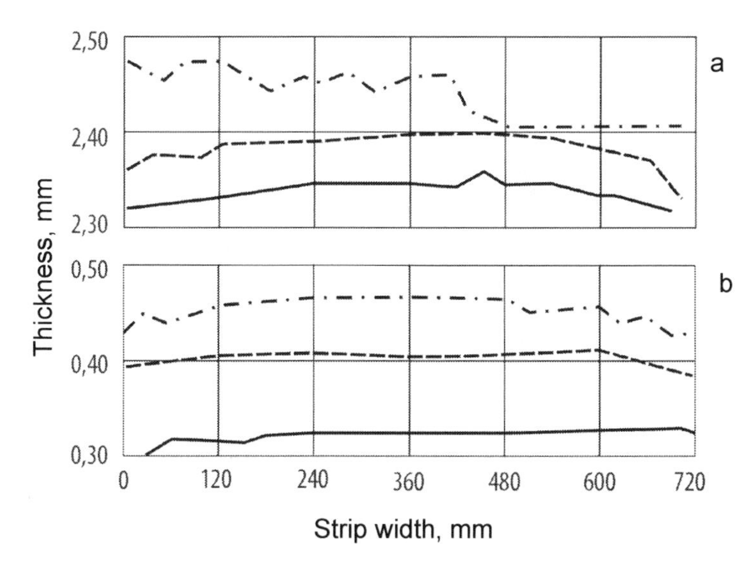

Fig. 7.6. Profiles of the cross-section of the strips and welded joints in the direction normal to the rolling axis in the initial condition (a) and after rolling (b): solid lines – the front ends of the strip; dashed lines – the rear ends; dot-and-dash lines – welded joint.

mm or more, and the proportion of the joints with this thickness difference 20%.

When welding the strips the temperature of the metal in the weld zone varies from the melting point to the initial temperature of the metal. As regards the microstructure, this region is divided to the zone of the intrinsic welded joint (melting zone) and the heat-affected zone. The zone of the intrinsic welded joint is on average 5–6 mm long, and the heat-affected zone is up to 15–16 mm long. In accordance with the difference in the microstructure, the properties of the metal of the welded joint differ from those of the metal of the contacting strips. The hardness and the yield strength of the metal (calculated value) in the region of the welded joint increase monotonically from the initial value to the maximum values in the centre of the welded joint, and then gradually decrease to the initial properties of the second contacting plane. At a hardness of the strips of 60–70 HRB units the hardness and the yield strength of the weld metal are 25–30% higher.

The data on the distribution of the mechanical properties of the metal, in particular, its hardness in the zone of the welded joint prior to and after cold rolling are presented in Table 7.1.

Table 7.1. Hardness HRB of 08kp steel in the zone of the welded joint prior to and after cold deformation

Sample No.	Prior to rolling			Sample No.	After to rolling		
	strip	weld	strip		strip	weld	strip
1	63	71	64	6	94	96	93
2	70	77	66	7	94	95	94
3	69	79	60	8	94	93	93
4	64	75	67	9	92	95	95
5	62	72	65	10	94	92	95

The experimental data indicate that when the total reduction of 60–70% is reached, the mechanical properties of the metal of the strip and the welded joint are almost completely equal. This conclusion was used as a basis for the determination of the constants a and b included in equation (6.2). The values of these constants for the welded joints in 08kp steel were equal to $a = 17$; $b = 0.7$.

Analysis of the geometry and mechanical properties of the metal in the zone of the welded joint shows the presence of the perturbing factors – large thickness difference of the strip in the longitudinal direction which (its relative value) increases during rolling, and the differences in the mechanical properties which decrease during deformation of the strip. Attention should also be given to the nature of variation of the thickness in the zone of the welded joint in the rolling direction. In many cases, the thick section of the metal of the front strip is followed by an even thicker welded joint and then by a thin area of the next strip.

7.5.1. Effect of the rolling process parameters on strip tension

To reduce the number of breaks in the welded joints, the rolling condition should be selected to ensure the lowest possible dynamic loads in the rolled strip. The effect of different factors on the inter-stand tension was estimated using the dynamic coefficient K_d, equal to the ratio of the maximum tension σ_{max} to its initial value σ_{in}:

$$K_d = \frac{\sigma_{max}}{\sigma_{in}}. \tag{7.22}$$

The critical value of the tension σ_{cr} at which the strip breaks in the final interstand gaps was determined by experiments in rolling 0.5×1015 mm strips in the 1700 continuous cold rolling mill. It was equal to 0.4 σ_T. The condition of stability of the process of rolling the welded joints can be written the following form:

$$\sigma_{max} = K_d \sigma_{in} \leq \sigma_{lim} \tag{7.23}$$

The dependence of coefficient K_a on the thickness difference of the welded joint, on the reduction in the stand, on the stress level and the rolling speed was investigated using the mathematical model described in chapter 6 and in a previous study by the authors of this book [92].

When analysing the effect of the thickness difference of the welded joint the thickness of the thicker end of the strip h_1 in the welded joint was assumed to be constant, and the thickness of the thin end h_2 was varied in the range 20%. The results of the studied presented in Fig. 7.7 were obtained for the following rolling conditions: h_1 = 0.9 mm; the thickness of the welded joint h_w = 1 mm; the length of the welded joint l_w = 30 mm; the rear tension in the steady rolling process T_{0st} = 145.2 kN; the front tension T_{1st} = 78.5 kN, rolling speed 5.0 m/s.

The simulation of the process of rolling of the strips with a welded joint shows that an increase of the thickness difference of the welded joint increases the amplitude of the oscillations of the front tension of the strip with increasing gradient (Fig. 7.7).

The nature of this dependence is explained by the effect of the magnitude of reduction in the initial thickness of the rolled strip

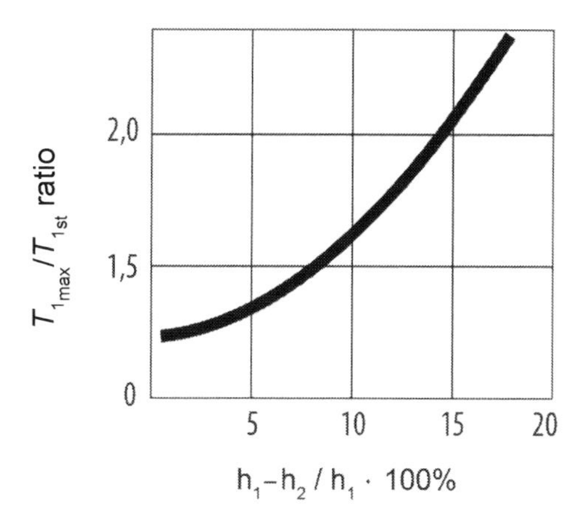

Fig. 7.7. Dependence of the 'jerk' of the front tension in rolling a welded joint on the thickness difference of the welded joint.

on the forward slip of the metal. The forward slip of the strip decreases with increasing thickness difference of the welded joint and, consequently, the front tension increases. For example, if the dynamicity coefficient in the rolling of the welded joint produced by welding the strips of the same thickness ($h_1 = h_2 = 0.9$ mm) is equal to 1.25, in rolling the welded joint with the 15% thickness difference the dynamicity coefficient K_d is equal to 2.05. When the thickness difference of the welded joint is greater than 17–18%, the dynamic loading in the strip is greater than the maximum permissible value σ_{lim} and this may cause breaking of the strip through the welded joint.

The experimental investigations of the process of rolling the welded joints in the 1700 continuous rolling mill show that when the thickness difference of the contacting strips does not exceed 10% (0.1 mm), the welded joints do not fracture. Rolling was carried out without breaking of the strips. However, if the thickness difference of the strips in the zone of the welded joints was 18.5% (0.25 mm) or more, the rolled strips fractured through the welded joint.

In the distribution of the toal reduction in the stands of the continuous rolling mill it is necessary to take into account the effect of the magnitude of partial reduction in the stand on the change of the interstand tension in rolling strips with welded joints. Figure 7.8 shows the dependence of the dynamicity coefficient of front tension on the reduction in rolling a welded joint with different thickness which confirms that the 'jerk' of strip tension rapidly increases with a decrease of reduction in the stand.

This dependence is explained by the fact that increasing reduction in the stand decreases the strength of the effect of the local increase

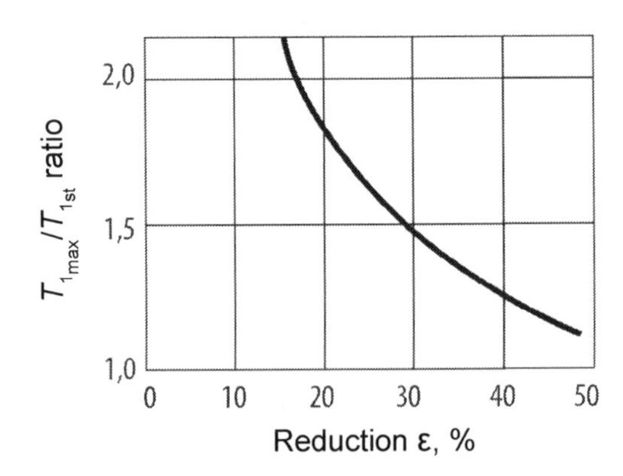

Fig. 7.8. Effect of relative reduction in the stand on the jerk of the front tension in rolling a welded joint. The conditions are the same as in Fig. 7.7.

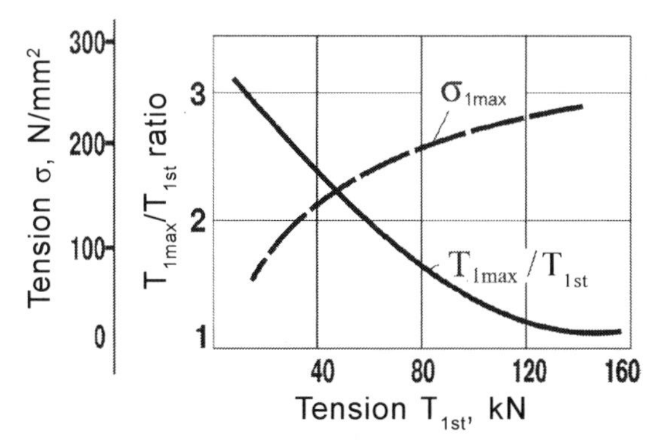

Fig. 7.9. Effect of the absolute level of the interstand tension of the strip in the rolling of welded joints on the dynamicity coefficient of total tension T_{1max}/T_{1st} and the maximum specific tension σ_{1max}. The rolling conditions are the same as in Fig. 7.7.

of the thickness of the welded joint on rolling force. Therefore, the changes of elastic deformation of the stand, reduction, forward slip and strip tension also become smaller. At small reductions the increase of pressure on the rolls from the welded joint is comparable with the rolling force of the rear end of the strip. Therefore, the relative values of these parameters change more markedly and this increases the dynamic loads in the strip. Consequently, sufficiently high (20–25%) reductions should be used in the final stands of the mill in rolling thin strips with welded joints.

The tension of the strip in the interstand gaps has a strong effect on the stability of the process of rolling the butt-welded strips. Calculations show (Fig. 7.9) that the increase of the level of the interstand tension decreases the amplitude of the relative oscillations of the tensile forces in the strip. This means that the process of rolling the welded joints at higher interstand tensions is more stable in the dynamic respect.

However, when the level of the interstand tension is increased, the maximum value of the tension approaches the limiting value σ_{lim}~300 N/mm² (Fig. 7.9). This circumstance indicates that cold rolling of the welded strips in the continuous mills can be carried at efficiently with the minimum possible interstand tension.

Figure 7.10 *a* shows the dependence of the dynamicity coefficient of the front tension of the rolling speed of the strip with a stepped decrease of the thickness in the zone of the welded joint. According to the calculated results, increase of the speed increases the front tension. At a rolling speed of 1 m/s, the coefficient K_d is equal to

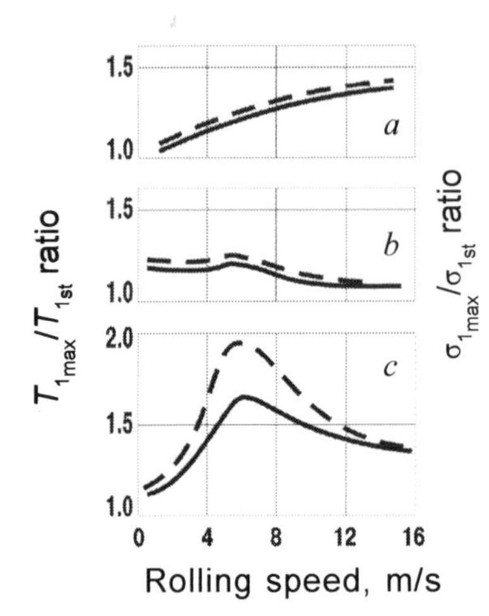

Fig. 7.10. Effect of the rolling speed on the total (solid lines) and specific (broken lines) tension of the strip in rolling welded joints of different configuration: a) h_1 = 0.9 mm; h_w = 0.9 mm; h_2 = 0.8 mm; b) h_1 = 0.9 mm; h_w = 1.0 mm; h_2 = 0.8 mm; c) h_1 = 0.9 mm; h_w = 1.0 mm; h_2 = 0.8 mm.

1.05, at 10 m/s it increases to 1.32, i.e., when the speed is increased 10 times, the maximum front tension increases by 27%.

An important factor which determines the effect of the rolling speed on the parameters of the transitional process is the duration of the effect on the rolls of the welded joint. The calculation show that the maximum deviation of the tension of the strip, caused by the welded joint, is observed at a rolling speed of 6.0 m/s and equals 20% (Fig. 7.10 *b*). With a further increase of the rolling speed the dynamicity coefficient of tension decreases. For example, at a speed of 15.0 m/s the coefficient is equal to 1.08. This means that the increase of the front tension is 8%. This dependence can be explained qualitatively by the fact that with increase of the rolling speed the duration of the effect of the welded joint on the rolls becomes smaller than the half period of the intrinsic longitudinal oscillations of the stand. Therefore, the rolls do not manage to react to the perturbation and the gap between the rolls, the forward slip, the speed of the strip and the tension change only slightly. Thus, the welded joint does not cause any significant changes in the strip tension. With increase of the rolling speed the value of the dynamicity coefficient of tension decreases. However, analysis of the transitional process in the drive–stand–strip system in rolling the actual welded joint with a thicker area and the thickness difference of the welded strip shows that these two factors together cause large variations of the front tension of the strip (Fig. 7.10 *c*). For example, if in rolling at a speed of 6 m/s a joint in strips of different thickness K_d = 1.22 (Fig. 7.10 *a*), and in rolling the welded joint K_d = 1.20

(Fig. 7.10 *b*), then in rolling the welded joint with different thickness K_d = 1.65 (Fig. 7.10 *c*).

The nature of the dependence of the front tension of the strip on the rolling speed (Fig. 7.10 *c*) can be explained by combined analysis of the curves in Fig. 7.10 *a, b*. At a rolling speed lower than 6.0 m/s the effects of the thickness difference of the strips and local increase of the thickness of the welded joint are the same. Therefore, the combined effect of these factors results in a large increase of the tension. At a rolling speed greater than 6.0 m/s, these dependences have a positive gradient and this results in a decrease of the tension.

Attention should be given to the fact that in rolling the welded joint with a local thickness increase and thickness difference of the strips the longitudinal oscillations of the role section have a large amplitude. This causes the formation of the thickness difference of the strip in the area of the welded joint and weakens the strip cross-section (Fig. 7.5).

Figure 7.11 shows the dependence of the minimum strip thickness h_{1min} at exit from the stand on the rolling speed of the welded joint which shows that in the speed range 1÷5 m/s the value h_{1min} is lowest. The rolling conditions were the same as in Fig. 7.10.

The reduction of the thickness of the metal behind the welded joint resulted in an additional increase of the specific tensile stresses in this section and increases the probability of breaking of the strip. The dependences of the specific tensions on the rolling speed of the welded joints are shown by the broken curves in Fig. 7.10. The thickness and specific tension of the strip change most appreciably in rolling the welded joints with thickness difference and with a local increase of the thickness (Figs. 7.10 and 7.11). In this case, the increase of the specific tension at a rolling speed of 6.0 m/s reaches 100% and approaches a critical value.

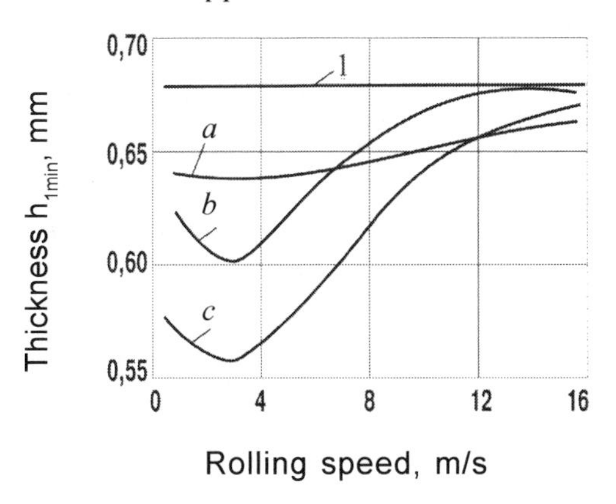

Fig. 7.11. The effect of the rolling speed and the minimum thickness of the strip leaving the stand in rolling welded joints of different configurations: the letters at the curves indicate the shape and dimensions of the welded joints as in Fig. 7.10: 1 – the nominal thickness of the strip.

It should be noted that the results quoted previously were obtained for the specific rolling conditions in which the reduction of the strip prior to the start of the deformation of the welded joint was 25%, the strip tension T_0 = 143.2 kN and T_1 = 78.5 kN. Evidently, with the change of the reduction in the stand and the level of the interstand tension the effect of the rolling speed on the dynamic loads in the strip changes (see Figs. 7.8 and 7.9). For example, when the reduction is increased to 40%, the dynamic component of the strip tension is more than halved, i.e., rolling of the welded joint with the dimensions h_1 = 0.9 mm; h_w = 1.0 mm; h_2 = 0.8 mm, l_w = 30 mm can conducted at the working speed. However, if the thickness difference of the welded joint exceeds 17–18%, the dynamic loads in the strip again approached the limiting values (Fig. 7.7).

The limiting loads in the strip in turn depend on the quality of the welded joint and the non-uniformity of deformation in the width of the rolled strip which determines the curve of the tensile stresses. The values σ_{lim} should be corrected taking into account the specific conditions of the given stand. However, the relationships described above do not change qualitatively

Thus, the computing experiments show that the welded joint results in high dynamic loading in the strip which in certain conditions in rolling may reach the limiting values and cause fracture of the strip. The main sources of perturbations are the thickness difference of the welded joint and the local increase of the thickness of the welded joint. The effect of each factor separately on the dynamic loads in the strip is small whereas their combined effect results in considerable jerks in tension.

The surface condition of the welded joint strongly influences the rolling dynamics of the welded joints, especially in the first and second stand of the continuous rolling mills. The nature of changes of the force, the rolling momentum and tension during the passage of the welded joints with different surface roughness through the deformation zone is shown in Figs. 7.12 and 7.13. The graphs show the rolling of the strips made of 08kp steel with the size of 2.5/0.5×1015 mm in the fourth stand of the five-stand 1700 mill of the Karaganda Metallurgical Plant. The dimensions of the welded joints prior to rolling: the thickness of the thick strip in the joint 0.9 mm, the thickness of the thin strip in the joint 0.8 mm, the thickness of the welded joint 1 mm; the weld length 30 mm. The relative reduction (at the thick end) 25%. The rolling speed 5 m/s. The total tension in the steady rolling process: rear tension 146 kN,

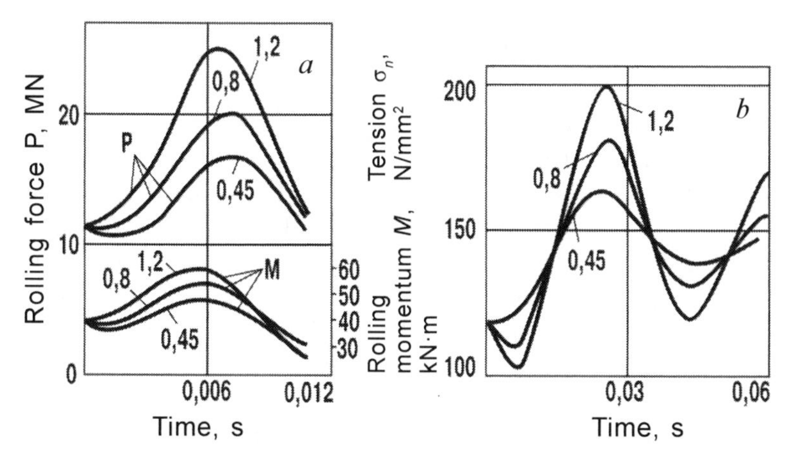

Fig. 7.12. Variation of the force P, momentum M in rolling (a) and the front specific tension σ_n (b) when welded joints with different surface roughness pass into the deformation zone. The numbers at the curves give the surface roughness of the welded joints Ra, μm.

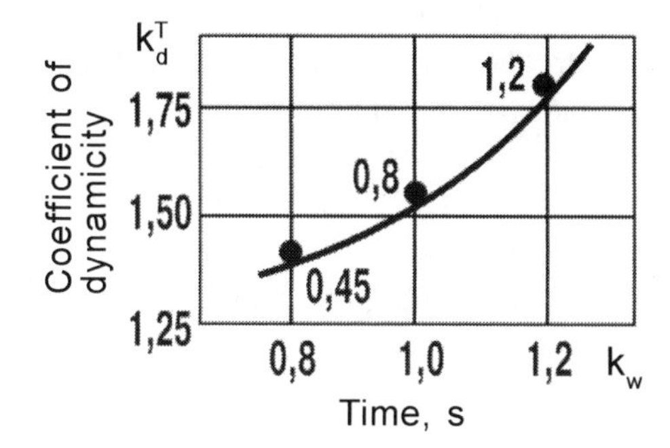

Fig. 7.13. Dependence of the dynamicity coefficient of the front tension of the strip $k_d^T = T_{s\,max}/T_{s\,st}$ in rolling welded joints on the surface roughness Ra, μm (the numbers of the points on the curves) and the coefficient of the effect of the initial surface roughness of the rolled metal k_w on the friction coefficient (the welding conditions at the same as in the example in Fig. 7.12).

the front tension 80 kN. The welded joint enters the deformation zone with the thick end. The effect of surface roughness was taken into account by the coefficient k_w of the effect of the surface roughness of the strip on the friction coefficient. In the investigated examples (Figs. 7.12 and 7.13) the surface roughness of the strips was the same and equalled 0.8 μm (for the sections of the joint $k_w = 1$), and the surface roughness of the welded joint varied in the range Ra = 0.45–1.20 μm ($k_w \approx 0.8...1.2$).

According to the results (Fig. 7.12 *a*), even a relatively small change of the surface roughness of the welded joint (by a maximum of 0.75 μm) causes a large (by a factor of 1.5–2.15) increase of the force and rolling momentum. The jerks of the strip tension also increase appreciably (Fig. 7.12 *b*). If at the surface roughness of the welded joint of 0.45 μm the coefficient of dynamicity of the total tension k_d^T was equal to 1.38, then at a weld surface roughness of 1.2 μm the coefficient equalled 1.77 (Fig. 7.13). The probability of fracture of the strip increases in this case. The dynamicity coefficient k_d^T is represented here by the ratio of the maximum value of the total tension during the rolling of the welded joint T_{max} to the nominal value of tension T_{st}, i.e. $k_d^T = T_{max}/T_{st}$.

The experimental results have been used to develop a number of new methods and introduce them into practice to increase the efficiency of the process of rolling the strips in the non-stationary conditions, in particular the rolling of welded joints.

Thus, to increase the reliability of the welded joint by decreasing the number of breaks in cold rolling after removing the flash, the welded joint should be treated to have the surface roughness different from the surface roughness of the strips. If the deformation resistance of the metal of the welded joint is greater than the deformation resistance of the ends of the strips in the joint, the surface of the welded joint should be smoother than the surface of the strips. Otherwise, the surface of the welded joint should be rougher. The preparation of the welded joint with a surface roughness different from the surface roughness of the strip makes it possible to avoid a large (almost instantaneous) increase of the rolling force at the moment of passage of the welded joint through the deformation zone, and also avoid sudden changes in the tension of the strip.

Usually, the mechanical properties of the welded joints differ along the length. The highest values of the yield and tensile stress, and the hardness of the weld metal are usually observed in the middle of the strip, the lowest values – at the edges (the reversed situation may also be encountered). The differences in the mechanical properties along the length of the welded joint may reach 20%. Because of the non-uniformity of the mechanical properties along the length of the welded joint, the pressure of the metal on the rolls in the middle and at the edges of the strip during the passage of the welded joint to the deformation zone differ. This results in the additional non-uniformity of the distribution of tension along the width of the strip and increases the probability of fracture through the joint. In order to weaken this negative effect and increase the reliability of the welded joints with the properties non-uniformly distributed along the length, the surface roughness of the welded joint should decrease uniformly from the longitudinal edges of the strip to its central part. As a result of the change of the surface roughness the friction conditions in different sections along the length of the welded joint change in such a manner that in rolling the section of the welded joint with a lower yield strength in the friction coefficient would be higher than in rolling that section of the welded joint with a higher yield strength. As a result, the distribution of the contact pressure along the length of the roll barrel becomes more uniform and the non-uniformity of the distribution of tension along the width of the strip and the magnitude of the tensile stresses at the edges of the strip also decrease. All these factors have a positive effect on the rolling capacity of the welded joints and decreased the risk of fractures of the strip through the joint. The thin sheet steel is rolled in the industrial rolling mills using lubrication. Since the amount of the lubricant present in the deformation zone is influenced by the direction of the microrelief of the surface of the rolls and the rolled metal, this effect can also be used for decreasing the strength of perturbations, formed during rolling when the welded joints pass through the deformation zone. In particular, the surface of the welded joint should have a microrelief oriented in one direction, for example, in the form of rectilinear grooves. When the mechanical properties of the material of the welded joint are higher than the mechanical properties of the metal of the events of the contacting strips, the grooves should be made along the welded joint, and when the mechanical properties of the material of the welded joint are are lower than the mechanical properties of the strips, the grooves should be made across the welded joint (along the rolling axis). When the surface of the welded joint has a one-directional microrelief, in particular with the rectilinear grooves, the friction coefficient in rolling the section of the welded joint changes which in turn changes the pressure on the rolls, the reduction and, at the same time, compensates the difference in the deformation resistances of the material of the welded joint and the strip.

The problems of the theory and technology of rolling the metal with the welded joints were discussed in the previous work by the authors of this book [92].

7.5.2. 'The effect of the speed' in acceleration and deceleration of the rolling mill

In deceleration and acceleration of the rolling mill, for example, when the welded joint passes through the rolls, the thickness of the strip, interstand tension and other parameters of the rolling process greatly change. The reasons for these phenomena usually include the effect of the rolling speed on the value of the friction coefficient and the thickness of the oil layer in the liquid friction bearings (LFB) of the support rolls of the stands of the mill. For example, in [95] the results of laboratory experiments confirm that in rolling the strips without lubrication the increase of the rolling speed increased the thickness of the strip and the rolling force whereas in rolling with lubrication the thickness of the strip and the rolling force decreased. However, in the available studies, carried out to simulate the process of continuous rolling there are no estimates of the strength of the effect of each speed factor separately on the variation of the thickness of the strip in the stands of the rolling mill. This is an important moment in the development of the system for automatic regulation of the gap between the rolls at a variable rolling speed.

A sufficient condition for the stability of the rolling process in the continuous mill both at the constant and variable rolling speed is the fulfilment of the following relationship in the individual stands of the rolling mill:

$$\frac{V_i}{V_{i-1}} = \lambda_i = \frac{h_{1i-1}}{h_{1i}}, \tag{7.24}$$

where λ_i is the reduction in the i-th stand (constant value); V_i is the rolling speed in the i-th stand; h_{1i} is the thickness of the strip at exit from the i-th stand.

The fulfilment of the left hand part of the equation (7.24) is a function of controlling the main drive of the stand. To fulfil the right-hand equality of the expression (7.24) it is necessary to correct the installation of the pressure screws in each stand of the rolling mill.

The thickness of the strip at exit from the stand is determined by the combined solution of the rolling equation and the equation of the thickness difference of the rolled strips:

$$P = P\left(h_0, h_1, \sigma_T, R, f, T_0, T_1, V,\right), \tag{7.25}$$

$$h_1 = S + \frac{P}{C_k} - 2S_{\mathrm{LFB}}, \tag{7.26}$$

where S is the gap between the rolls without the metal (installation of the pressure screws); S_{LFB} is the thickness of the oil film in the LFB.

The solution of the system of the equations (7.25)–(7.26) is associated with certain difficulties caused by the fact that the simple iteration process diverges for the rolling conditions in the industrial cold rolling mills.

Therefore, it is recommended to substitute the value of the output thickness calculated from the following equation [89]:

$$h_1^{(2)} = h_1^{(0)} + \gamma\left[h_1^{(0)} - h_1^{(1)}\right], \tag{7.27}$$

in the second iteration step into equation (7.25). Here $\gamma = C_K/C_S$; C_K, C_S are the moduli of rigidity of the stand and the strip; $h_1^{(0)}$ is the initial approximation of thickness; $h_1^{(1)}$ is the value of the thickness of the strip obtained in the first step of the iteration process.

The vertical component of displacement of the neck of the support roll S_{LFB} in the LFB was calculated using the empirical dependences proposed in [96]:

$$S_{\text{LFB}} = a_k \sqrt{V} / P, \tag{7.28}$$

Here a_k is an empirical coefficient equal to 75 kN · mm · $s^{1/2}/m^{1/2}$.

The attractive feature of the equation (7.28) is that, firstly, it is used to approximate the data obtained in the range of the forces and speeds corresponding to the actual rolling conditions of the continuous mill; secondly the equation was derived for the LFB which is used in the majority of cold rolling mills; thirdly, the equation can be usually used in the mathematical model.

The rolling process was simulated in a continuous 5-stand 1200 rolling mill (tin plate mill). The rolling of 0.25 mm thick strips of steel 08 kp produced from 2.2 mm thick strips plate was investigated. It was assumed that the position of the pressure screws remains constant when the rolling speed changes, i.e., the size of the roll gap was not corrected (S = const).

To determine the separate effect of the friction coefficient and 'floating-up' of the necks of the support of rollers in the LFB on the thickness of the rolled strip, the rolling process was simulated with and without lubrication, taking into account the change of the thickness of the oil film in the LFB and without this change (SLFB = 0), which corresponds to the application of rolling bearings for the support rolls in the stand. The experimental results show (Fig.

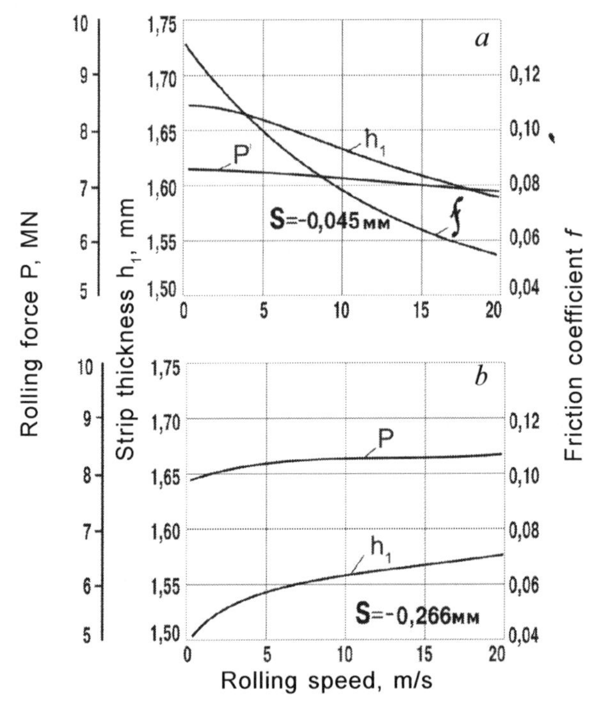

Fig. 7.14. Calculated dependences of the strip thickness h_1, force P, the thickness of the oil layer in the LFB on the rolling speed of annealed steel (h_0 = 2.2 mm) with the lubrication (a) and without lubrication (b) at the constant position of the pressure screws.

7.14 *a*) that in rolling with lubrication, the increase of the rolling speed from 0.5 to 20 m/s results in a decrease of the strip thickness at exit from the first stand from 1.67 to 1.59 mm, and the rolling force decreases from 7.3 to 6.0 MN.

The reversed situation was observed in rolling without lubrication: the rolling force increased from 7.9 to 8.3 MN, and the strip thickness from 1.51 to 1.59 mm. This is explained by the fact that in rolling the strip without lubrication the effect of the dynamic component of the yield strength becomes evident; in rolling with lubrication, this effect is compensated by the change of the friction coefficient. The nature of the calculated dependences of the strip thickness and the pressure of metal on the rolls on the rolling speed is in good agreement with the experimental data of, for example, Billigmann and Pomp [95].

The results of simulation of the rolling force and strip thickness in the first stand of the mill taking into account the change of the

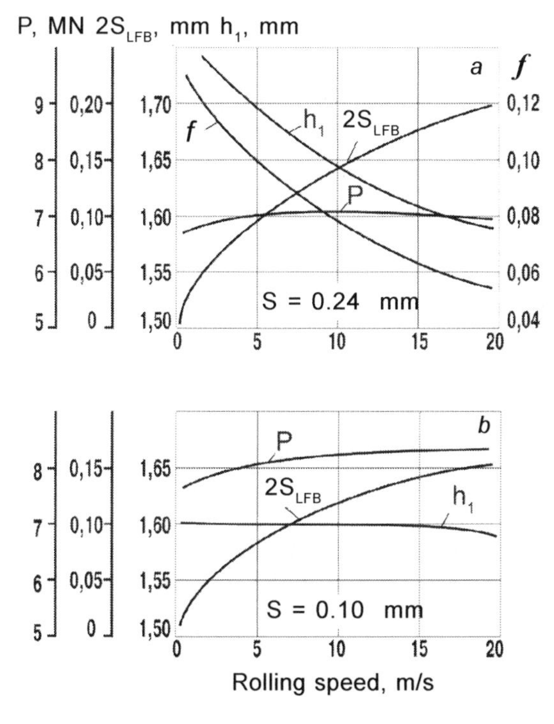

Fig. 7.15. Calculated values of the strip thickness h_1 and force P in relation to the rolling speed of annealed steel (h_0 = 2.2 mm) with lubrication (a) and without it (b) at a constant position of the pressure screws (S) and without taking into account the change of the thickness of the layer in the LFB, where S is the size of the gap between the rolls (roll gap) without the metal. The minus sign indicates that the rolls are pressed together.

thickness of the oil film in the LFB of the support rolls are shown in Fig. 7.15. Comparison of these data with the results of the simulation experiments, carried out disregarding the effect of the LFB, shows that the hydrodynamic effect in the LFB has a controlling effect on the thickness of the rolled strip and the nature of variation of the rolling force. For example, in rolling with lubrication in the speed range from 0.5 to 20 m/s the change of the output thickness by 0.08 mm is greater than in rolling without the LFB. In rolling without lubrication, as shown by the calculations, the rolling speed has only a slight effect on the change of the strip thickness. This is explained by the fact that the increase of elastic deformation of the stand as a result of increasing the deformation resistance is compensated by the increase of the thickness of the oil film in the LFB. In particular, when the rolling speed is increased from 0 to 20 m/s the strip

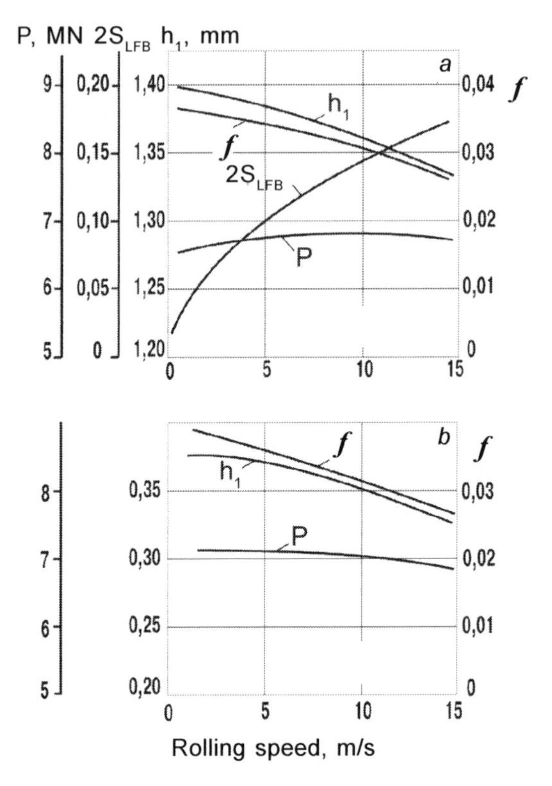

Fig. 7.16. Calculated dependences of the thickness of the strip h_1 and of force P on the rolling speed of the cold-worked strip (preliminary deformation 82%, $h_0 = 0.40$ mm) with lubrication at a constant position of the pressure screws with the change of the thickness of the oil layer in the LFB taken into account (a) and disregarded (b).

thickness decreases from 1.60 to 1.59 mm. The rolling force in this case increases from 7.60 to 8.30 MN.

When analysing the continuous rolling process using these results it is important to take into account the circumstance that the rolling speed in the first stand of the mill rarely reaches 5 m/s, and that the experiments described above were carried out in the range from 0 to 20 m/s.

The calculated dependences of the rolling force and the output thickness of the strip on the rolling speed in the fifth stand of the mill are presented in Fig. 7.16. The rolling process with lubrication was simulated with the effect of LFB taken into account and disregarded. According to these experimental results, when the rolling speed is increased from 0 to 15 m/s, the output strip thickness decreases from

0.39 to 0.32 mm in rolling with the LFB, and from 0.38 to 0.33 mm in rolling in the stand with the rolling bearings of the support rolls.

Evidently, the differences in the nature and magnitude of the variation of the strip thickness in both cases in rolling are almost non-existent. This circumstance is explained by the fact that the change of the thickness of the oil film in the LFB of the support rolls is compensated to a large degree by the elastic deformation of the stand because the rigidity of the strip in the final stands of the mill is considerably greater than the rigidity of the stand.

Thus, the variation of the thickness of the oil film in the LFB of the final stands of the mill has almost no effect on the thickness of rolled strips. The thickness of the strip changes mostly as a result of the effect of the rolling speed on the friction coefficient.

The position of the pressure screws of the stand to ensure the constant strip thickness under variable rolling conditions is determined from the equation (7.29)

$$S = h_1 - \frac{P}{C_k} + 2S_{\mathrm{LFB}}. \tag{7.29}$$

Force P was calculated from equation (7.25). The problem of determination of P and S using equations (7.25) and (7.29) is solved mathematically by a simple procedure because both unknown parameters are expressed explicitly.

The results of the calculations of the size of the gap between the rolls (installation of the pressure screws) and other process parameters of rolling tin plates in the first and fifth stand of the 1200 Magnitogorsk Metallurgical Plant mill, for example, are presented in Fig. 7.17 *a*.

Since the actual rolling speed in the first stand of the mill is not greater than 4 m/s, then according to the experimental results, the change of the size of the gap between the rolls in the stand is negligible. The rolling force does not change when the size of the gap is regulated. In the fifth stand of the mill where the cold worked thin strip is rolled, it is important to change greatly the size of the gap between the rolls in order to ensure the constant strip thickness. For example, when increasing the rolling speed from 0.5 to 20 m/s, the pressure screws should be displaced from the position –1.70 to –0.96 mm, i.e. by 0.70 mm. The negative values of the position of the pressure screws S indicate that the rolls without the metal are in pressed together.

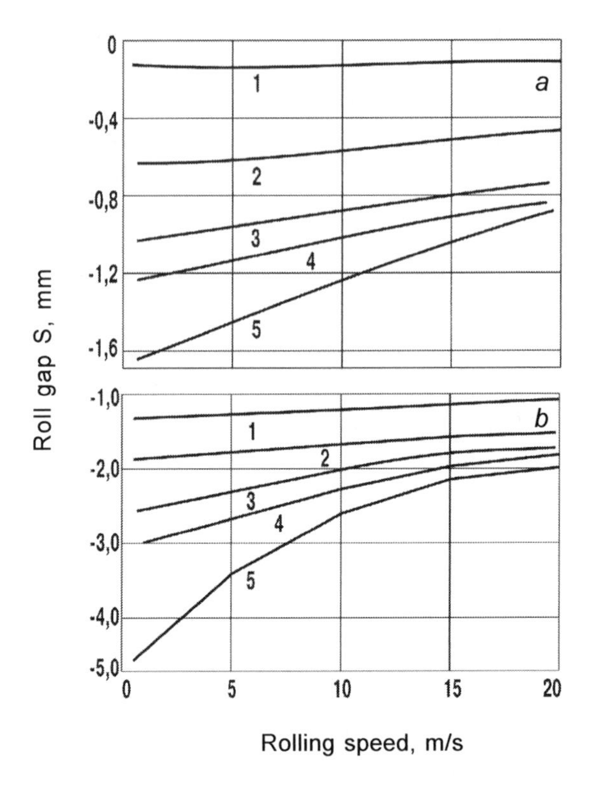

Fig. 7.17. Calculated values of the correction of the size of the gap between the rolls in dependence on rolling speed: a) 1200 mill (rolling of strips 2.2/0.25×730 mm); b) 1700 mill (rolling of strips 2.5/0.5×1015 mm). The numbers at the curves are the numbers of the stands.

Regardless of the fact that the absolute value of the friction coefficient in the fifth stand changes only slightly (from 0.0 27 at V = 0.5 m/s to 0.018 at V = 20 m/s), the rolling force greatly decreases, from 8.70 to 6.50 MN. This phenomenon is characteristic of the final stand of the continuous cold rolling mills and evidently this complicates the regulation of the size of the gap between the rolls using a program in rolling thin strips [97].

8
Stability of the technology of cold rolling of strips

8.1. Indicators of the instability of the cold rolling process

In the current conditions the quality requirements for cold rolled metals have greatly increased, especially as regards the accuracy and flatness of the surface. An indispensable condition for high quality sheet metal is the strict control of deformation modes of rolling, the implementation of the process in strictly defined ranges of the process variables. This presents certain difficulties due to the instability characteristics of strip plate and the rolling conditions. As noted above, the non-constancy and inhomogeneity of chemical composition and structure of the steel determines the magnitude of fluctuations of the initial strength yield. Often there is a scatter of the strip thickness at the entry to the stand, unstable friction conditions during rolling due to changes in the surface roughness of the rolls, rolling speed, emulsion temperature and instability of emulsion supply. Due to the simultaneous adjusting of the thickness of the strip and tension the speed, reduction and accordingly the power parameters of the rolling process in the stands are not constant, and the indicators of the accuracy and flatness of the rolled strips also vary. All of this suggests that the rolling process obeys probabilistic laws. In most cases, without specific studies it is very difficult to determine the range of variation of process parameters and to predict the reliability of ensuring the defined limits of their variation. Therefore, taking this into account, the increase in the stability of the rolling process should be considered as one of the most urgent problems of sheet rolling production.

In the industrial mills conditions the physical constants describing the properties of the rolled metal (yield strength, hardening parameters) and external conditions friction in the deformation zone (friction coefficient), as well as temperature, energy-power and kinematic parameters of the rolling process are random variables whose distribution is determined by the statistical characteristics (expectation, standard deviation, etc.). Because the dependences of the energy–power and kinematic parameters of the rolling process on the initial yield strength of the metal, strip thickness and the friction coefficient, are non-linear, then the average values of these functions are determined not only by the mean values of the mentioned arguments but also by their variances.

To solve the problems associated with the investigation of possible ranges of variation of the values of the parameters of the rolling process, the prediction of reliability of fulfilling the specified limits of their variation, it is promising to use probabilistic approaches. Such analysis of the process of hot rolling of strips is described in chapter 4 of this book. Below we consider the results of this analysis applied to the process of cold rolling of thin strips.

As noted above, the choice of modes of reduction and tension, energy–power and kinematic parameters of the process of cold rolling strip is largely dependent on the nature of hardening of deformed steel. The shortcoming of many well-known formulas for the calculation of the yield strength of cold deformed steel is that they do not take into account the impact of the instability of the chemical composition and grain size on the work hardening in cold rolling of steel. In [98] the change in the yield strength steel σ_T as a function of chemical composition, grain size of the ferrite in its structure and the degree of deformation ε is described by the relationship

$$\sigma_T(\varepsilon) = \sigma_{T_o} + 33.5\varepsilon^{0.6+0.005d^{-0.5}}, \quad (8.1)$$

where σ_{T_o} =81+53[C]93[Mn]+16.2$d^{-0,5}$, N/mm^2; [C] and [Mn] are carbon and manganese contents in steel, %; d is the diameter of the ferrite grains in the steel structure, mm.

This relationship allows in the presence of frequency distributions [C], [Mn] and d to calculate the distributions σ_{T_o} and $\sigma_T(\varepsilon)$. The values of the distribution parameters σ_{T_o} and d in the structure of hot-rolled strip with a thickness of 2.0–2.5 mm in steel 08kp used for the production of cold-rolled sheet and tin plate, are in the following table (left of the slash – grain size, mm, right – yield strength, N/mm^2, sample sizes – 62 melts of strips 2.0–2.2 mm thick and 60 melts 2.4–2.5 mm thickness):

The values of the distribution parameters	2.0–2.2 мм	2.4–2.5 мм
Minimum	0.015/250	0.015/250
Maximum	0.036/325	0.031/300
Average	0.0204/276.6	0.02043/276.9
Standard deviation	0.00614/18.1	0.00523/13.1
Coefficient of variation	0.30/0.64	0.256/0.47

Note that the range of variation and the variation coefficients of these variables are significant. At a strip plate thickness of 2.0–2.2 mm the instability of the structure and properties is greater than for the strip plate with a thickness of 2.4–2.5 mm.

Calculations of the distribution of the yield strength of cold rolled steel, depending on the total degree of reduction performed by the Monte Carlo method using equation (8.1) and the actual values of the initial yield strength and the ferrite grain size in the hot rolled steel show that increasing the degree of deformation ε increases the scatter in values of $\sigma_T(\varepsilon)$ and equals 110 N/mm^2 after a total deformation $\varepsilon = 70\%$ and 120 N/mm^2 after rolling with a reduction of $\varepsilon = 90$ %. This is a significant effect and must be taken into account in technological and strength calculations.

Histograms of the yield strength distribution after cold rolling with a reduction of 70 and 90% are shown in Fig. 8.1.

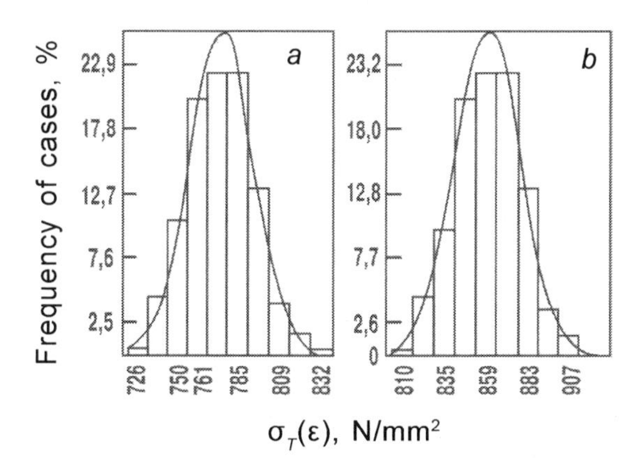

Fig. 8.1. Histograms of the distribution of yield strength $\sigma_Y(\varepsilon)$ of the cold-rolled 08kp steel after reduction of 70 (*a*) and 90% (*b*).

The characteristics of distributions of initial yield strength and the ferrite grain size in the steel structure correspond to the strip plate with a thickness of 2.0–2.2 mm. Distributions $\sigma_T(\varepsilon)$ at $\varepsilon = 70\%$ and $\varepsilon = 90\%$ are close to the normal distribution (Fig, 8.1) despite the fact that the second term in the equation describing the dependence of $\sigma_T(\varepsilon)$ on σ_{T0} and d is a non-linear function. After rolling with a reduction of 70% the average value of $\sigma_T(\varepsilon)$ is equal to 774.5 N/mm^2 and standard deviation $\sigma_T(\varepsilon)$ is equal to 18.45 N/mm^2. After 90% total deformation the characteristics of these values were 860.6 and 18.6 N/mm^2.

As already mentioned in Chapter 6 of one of the most suitable equations for describing the dependence of the friction coefficient in the deformation zone during cold rolling sheet of steel sheets in multistand sheet- and tin plate-rolling mills is the formula:

$$f = 0.1(1 + Ra_n)[1 + 0.25(Ra_r - 0.85)][0.8 + 0.2\exp(-0.65C_e)]\times \\ \times\varepsilon^{0.393-99.8h/R+0.0007\sigma_T+0.0193V} \times\exp(-3.5\varepsilon), \tag{8.2}$$

where Ra_r and Ra_n are the surface roughness of the rolls and the strip, μm; C_e is the concentration of the emulsion, %, h is the thickness of the rolled strip, mm; R is the radius of the rolls, mm; V is the rolling speed, m/s.

In this formula the variables ε, σ_T, V, C_e, Ra_r, Ra_n and h values are random, because, first, the structure and properties of steel are not constant even in the length of the single strip, and, secondly, the modern mills are equipped with multifunctional automatic control systems of thickness, tension, strip flatness and the rolling process constantly controlled. Hence, the value of the friction coefficient in the deformation zone of each stand is a random quantity. Having the information on the distribution of these variables, one can use the formula (8.2) to find the distribution of the values of the friction coefficients in each mill stand.

Histograms of the distributions of the friction coefficient in the stands of the five-stand 1700 cold rolling mill, calculated by the Monte-Carlo method from the experimentally found distributions of the parameters ε, σ_T, V, C_e, Ra_r, Ra_n and h are shown in Fig. 8.2. Table 8.1 shows the experimentally recorded (in the 1700 mill) distributions of variable parameters of the rolling process. In addition to the values listed in the Table, it should be noted that the distribution C_e was characterized by a mean value $\bar{X} = 4\%$ and the average standard deviation $S = 0.3\%$, the carbon content in steel

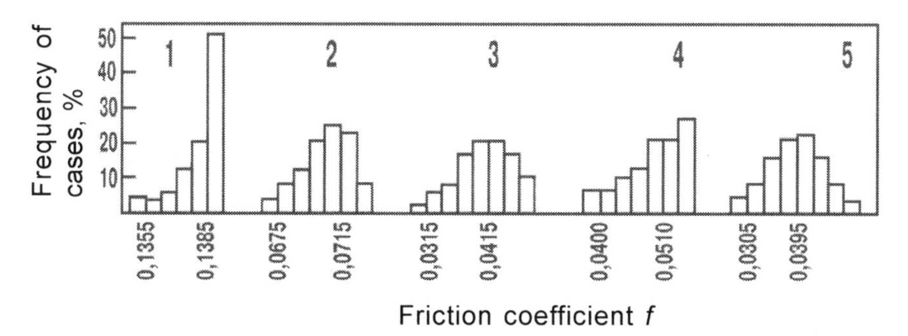

Fig. 8.2. Histograms of the distribution of the friction coefficient f in the stands 1–5 of the 1700 mill in cold rolling of strips of 2.5/0.5×1015 mm steel 08kp. Numbers 1, 2, 3, 4, 5 – stand numbers.

Table 8.1 Parameters of the distribution of process variables in the cold rolling process in the five-stand 1700 mill of strips with sizes 2.5→0.5 × 1015 (the numerator) and 4.5 → 2.0×1415 mm (the denominator) of steel 08kp

Parameters	Stands		
	1	2	3
Ra_r, μm	2.5/0.3	0.8/0.1	0.8/0.1
Ra_n, μm	1.5/0.15	1.5/0.15	1/0.1
ε, fractions of units	0.325/0.025	0.317/0.035	0.268/0.074
	0.168/0.028	0.119/0.019	0.102/0.027
V, m/s	3.64/0.325	5.34/0.455	7.3/0.965
	6.41/0.35	7.3/0.24	8.15/0.34

Table 8.1 continued

Parameters	4	5
Ra_v, μm	0.8/0.1	1/0.15
Ra_n, μm	1/0.05	1/0.05
ε, fractions of units	0.263/0.09	0.24/0.06
	0.162/0.05	0.229/0.05
V, m/s	9.9/1.17	13/3.4
	9.7/0.63	12.6/0.98

$\bar{X}$ = 0.08%, S = 0.01%; manganese $\bar{X}$ = 0.35%, S = 0.05%, the ferrite grain size of ferrite $\bar{X}$ = 0.026 mm, S = 0.005 mm. The radius of the working rolls is 300 mm.

According to the data (Fig. 8.2), the magnitude of the variation of the friction coefficient increases from the first to the last stand. In these conditions, the magnitude of the friction coefficient of the fifth stand can vary 1.5 time. This is because the instability of the parameters of the strip rolling process also increases from the first to the last stand. Such a pattern is observed both in the rolling of thin (0.5 mm) and relatively thick (2.0 mm) strips (Table 8.1). For example, when rolling strips of thickness 0.5 mm, the average standard deviation of the rolling speed is increased 10 times in the transition from the first to the fifth stand. Increasing reduction increases the yield strength variation of rolled metal (Fig. 8.1).

In the last mill stand, where the variables are undergoing the largest change in rolling, the distribution of the friction coefficient is close to normal. In general, the distribution of f may differ significantly from the normal distribution, since the dependence (8.2) of the friction coefficient on the process parameters of strip cold rolling is essentially non-linear. In the fifth mill stand the coefficient of variation of f is 9.6%. The coefficient of variation of f increases as the average value of this indicator decreases (Fig. 8.2).

The obtained statistical data on the distribution of the friction coefficient in different stands of the cold rolling mill significantly improve the accuracy of the mathematical models of the process of cold rolling of strips, result in a 20% increase in the accuracy of calculation of its power parameters, improve the quality of stand setup, and the selection and optimization of technological regimes.

8.2. Calculation of instability parameters of the process

The instability of the parameters of the rolled strips and of the cold rolling process, as the hot rolling process on in the wide-strip hot-rolling mills (WSHRM), was simulated by the Monte Carlo method, which allows one to set randomly the scatter of these parameters with a given dispersion in accordance with the normal distribution law. The cold rolling process parameters were calculated using the mathematical model described above, which provides an opportunity for optimization of the technological modes using a predetermined criterion, in particular, the ratio of the rolling forces when the strip thickness is fixed at entry to the stand and at exit from it. The optimization criterion (see section 7.1) was the condition of equality of these forces in the stands, which, as already noted, summarizes the requirements for ensuring the stability of the process and the optimum flatness of the strips. The rolling speed in the last mill stand was considered constant. In the other stands the strip thickness was controlled so as to ensure equal rolling forces. Occasional arguments considered in the calculation were the thickness of the strip plate before the first stand, the original yield strength of rolled metal $\sigma_{T\,in}$ and the friction coefficient in the deformation zone of each stand. The distribution of these variables was taken on the basis of experimental data obtained in the 1700 industrial mills.

The methodology for calculating the parameters of instability of the cold rolling process, taking into account the impact of random factors, was implemented as follows. At the first step in the initial approximation the nominal variables (initial and final thickness of the strip, the original yield strength of the metal, the friction coefficient in mill stands) equal to their mathematical expectations were defined, and the reduction mode was optimized by the criterion of equality of the forces. As a result, the basic reduction mode in the strip mill stand was determined. Further, in each experiment using the values of these random variables with variances of the defined distribution, the process parameters were calculated for the specific input data. The basic relationship of the reduction between the stands was kept constant. In this way, the work of the stand in the conditions of functioning of the automatic control system of strip thickness was simulated.

In calculations the working roll diameter of the five-stand 1700 mill was taken to be 600 mm, in the four-stand mill – 500 mm. The rolling speed at the final stand was assumed to be 15 m/s. For the 1700 mills we simulated the process rolling of strips 1000 mm wide with a thickness of 0.5–2.0 mm thickness using the 2.0–5.0 mm strip plate. At the mathematical expectation of the rolled strip thickness $\bar{h}$ = 0.5 mm the accepted standard deviation was S_h = 0.0035 mm; at $\bar{h}$ = 2.0 mm S_h = 0.035 mm, which corresponds to the field of tolerances in accordance with GOST 19904 for the indicated standard strip dimensions. The distribution of the initial thickness of strip plates was characterized by the averages $\bar{H}$, equal to 2 and 5 mm, and the standard deviation S_H – 0.06 and 0.12 mm. The initial yield strength distribution of the strip plate was characterised by $\sigma_{T\,in}$ = 230 N/mm^2 and $S_{\sigma_{T\,in}}$ = 20 N/mm^2. In the initial optimization of the reduction (using the nominal values of the parameters) at the entry to and exit from of the stand the tension was taken to be 50 N/mm^2, interstand tension 150 N/mm^2. In the process of examining different combinations of $\bar{\sigma}_{Tin}$, h_{in} and f_i the interstand tension was assumed to be constant in absolute values. The parameters of the distribution of the friction coefficient in the stands 1–5 of the mill were the following:

Stands	1	2	3	4	5
$\bar{f}_i$	0.08	0.06	0.05	0.05	0.07
S_f	0.0058	0.0044	0.0037	0.0037	0.0051

In the analysis of the rolling process on 4-stand mill the distribution of the friction coefficient in the first three stands was assumed to be the same received, and in the fourth – the same as in the fifth stand of the 5-stand mill.

Table 8.2. Characteristics of the rolling of strips in the five-stand 1700 mill (in the numerator and the denominator – in rolling strips with thicknesses 2 → 0.5 mm and 5 → 2 mm, respectively)

Parameters	Stands				
	1	2	3	4	5
$\bar{\varepsilon}$, %	27.8	28.9	29.1	25.7	7.6
	18.3	17.7	17.2	17.4	13.2
S_ε, %	0.46	0.45	0.45	0.47	0.59
	0.49	0.50	0.50	0.50	0.52
$\bar{h}$, mm	1.45	1.03	0.73	0.54	0.50
	4.09	3.37	2.79	2.30	2.00
S_h, mm	0.036	0.019	0.096	0.0046	0.0035
	0.034	0.057	0.041	0.035	0.035
$\bar{P}$, MN	8.85	8.9	8.91	8.92	8.98
	7.90	7.92	7.92	7.92	7.94
S_P, MN	0.74	0.57	0.53	0.69	1.42
	0.66	0.43	0.39	0.41	0.42
$\bar{N}$, kW	1131	2539	2859	2765	1666
	406	4378	4610	4963	7270
S_N, kW	143	111	119	147	96
	251	171	178	208	186
$\bar{M}$, kN·m	65.9	103.6	82.8	58.6	30.7
	16.4	140.0	121.9	108.5	134.6
S_M, kN·m	7.23	4.90	3.78	3.27	1.80
	9.99	6.55	5.56	4.97	3.59

Table 8.3. Strip rolling process parameters in a 4-stand 1700 mill

Parameters	Stands			
	1	2	3	4
$\bar{\varepsilon}$,%	31.3	33.5	34.5	16.5
	21.3	21.2	21.8	17.5
S_{ε}, %	0.544	0.526	0.519	0.662
	0.59	0.50	0.59	0.62
$\bar{h}$,mm	1.38	0.91	0.60	0.5
	3.94	3.10	2.4	2.0
S_h, mm	0.032	0.015	0.0057	0.0035
	0.077	0.048	0.036	0.035
$\bar{P}$, MN	8.68	8.69	8.67	8.7
	8.03	8.07	8.07	8.08
S_p, MN	0.68	0.51	0.47	0.98
	0.60	0.41	0.37	0.43
$\bar{N}$, kW	1504	3181	3711	1663
	1235	5631	6314	8768
S_N, kW	157	131	146	140
	283	217	243	253
$\bar{M}$, kN·m	70.2	98.3	75.6	41.9
	40.8	140.1	123.1	136.4
S_M, kN·m	6.5	4.6	3.2	2.3
	9.0	6.6	5.3	4.3

The estimated distribution parameters of the rolling process bands are given in Tables 8.2 and 8.3. According to the data, the degree of deformation and the rolling force vary most markedly in the last mill stand. Significant fluctuations ranges of the force in the last stands of both mills indicate that these stands in particular must be equipped with means of operational impact on the strip flatness. It should be noted, however, that the distribution of the force in the last stand of the mill during rolling of thin strips (0.5 mm) has a pronounced asymmetry. The probability of relatively large deviations from the most frequently occurring values of the force should be considered when selecting modes of deformation in terms of the required margin of the strength of equipment in the last stands of the mill.

Stable production of flat strips (even within the same melt) is difficult without effective high-speed means of regulation of their flatness, especially in the last mill stand. According to the calculated results the regulation of the rolling process on the basis of the criterion of equality of the forces in the mill stands along with reduced rolling force scatter leads to the increase in the scatter of the total moments and rolling power (~1.5 times). At the same time the algorithm, which ensures the equality of powers, leads to the opposite result.

Thus, to obtain flat thin strips (thickness 0.5–1 mm) it is recommended to use the SART algorithms realizing the equality of rolling forces. For thicker strips (2–3 mm thick) where the loss of flatness is unlikely, the process can be regulated on the basis of the condition of maintaining the equality of rolling forces between the rolling mill stands apart from the first stands.

The instability of the rolling power and torque in the first rolling stands is caused by significant fluctuations in the initial yield strength and thickness of the strip plate. Since in the calculations the rolling speed in the final stand was firmly set, the regulation of the process corresponds to a change in the speed of the previous strip mill stand. It reflects on the rolling power, especially in the first stand. In connection with this the drives of the first roll stands work in an essentially unstable operating mode, i.e., during rolling of even one roll the system can be operated in the drive mode and in the generator mode.

Comparison of the distributions of process parameters of the rolling of thin (0.5 mm) strips from 2.0 mm thick strip plate at the five- and four-stand mills allows one to conclude that the general patterns are the same for them, but in the last stand of the 4-stand mill the coefficient of variation of force is lower. The 4-stand mill is characterised by higher stability of the parameters of the rolling process. The level of rolling forces in both rolling mills about the same despite the fact that in the 4-stand mill the reductions are greater. This is due to the fact that the work rolls of the 4-stand mill have a smaller diameter (500 mm as compared with 600 mm in the 5-stand mill) and this reduces the stresses in the metal in the deformation zone due to the shorter length of the arc of contact. Summary moments and rolling power in the stands 4-stand mill are slightly greater (on average 26–27%), but the total rolling power remains practically the same (the difference does not exceed 0.2%).

In connection with the modern equipment of cold rolling mills incorporating effective means for controlling the profile and shape of rolled metal and cooling of the rolls unequivocal view that it is preferred have working rolls of the larger diameter since they have higher thermal stability and provide a better self-regulation of the reduction in the width of the strip seems obsolete [99].

Thus a comparative analysis showed that in terms of the stability of the strip rolling process in 4- and 5-stand mills in the working conditions of the system for the automatic regulation of thickness and tension (SARTa-T) the 4-stand mills working rolls with a diameter of 500 mm are preferred to the 5-stand with the rolls with a diameter of 600 mm. The possible scatter of the parameters of rolling strips of equal size is about 20% lower in the conditions of the 4-stand mill. This is due to the use of smaller working rolls diameters. More details on this issue are discussed in [100, 101].

8.3. Dynamic loads in drive lines and vibrations in the stands of continuous cold rolling mills

Speaking about the stability of the process of cold rolling of strip, it is necessary to address the subject of resonant oscillations in the working stands and overloading in the main lines of the continuous mills. Due to the dynamic loads, vibrations of individual components and the rolling mill on the whole, there is intensive wear of its equipment, primarily mechanical joints (gears, clutches, frame doorways, roll pillows, etc.), the precision of rolled strips is degraded and there are emergency situations. As shown above, the rolling of thin strips at a speed exceeding 15 m/s increases the instability of the basic process parameters – rolling forces and interstand tensions. This creates prerequisites for moving the work rolls with cushions within the gaps in the stand windows, vibrations. To exclude this phenomenon is necessary to create conditions under which the cushion of the rolls are constantly pressed to the front or rear vertical planes of the stand window. The horizontal forces acting on the chocks of the working rolls should be constant in magnitude and direction. The authors of [102] on the basis of studies carried out on the 1700 cold rolling mills recommend to eliminate resonant vibrations in the stands by the following measures: increase as much as possible the degree of deformation of the metal in the stand; reduce the rear and increase the front interstand tension; limit the instability of forces and tensions using automatic control systems in the mill; during acceleration or deceleration of the stand pass through the resonance zones of speed with maximum acceleration (deceleration) .

The above recommendations are certainly legitimate, but they are very general. Giving concreteness to these recommendations is possible only applied for each individual rolling mill stand, given its design features, the range of rolled steel and the technological regimes used.

The conditions under which the vibrations occur were examined in more detail on the example of the five-stand 2030 mill. Vibrations occur in this mill usually by rolling strips of 0.8 mm thickness at a speed less than 15 m/s. If no action is taken, the vibrations of the stand increase rapidly, the stand begins to 'buzz'. Due to fluctuations in the rolling forces and interstand tensions the accuracy of rolled strips is disrupted – longitudinal thickness difference can reach significant values (0.2 mm or more). Alternating light and dark bands appear on the surface that are perpendicular to the rolling axis in increments of 150–200 mm. All this leads to fractures of the strips with all the negative consequences.

Analyzing this phenomenon frequently encountered in the rolling mill, the authors of [103] note that with the vibrations are prevented by different methods – changes of the dynamic properties of the cold rolling mill stands by the use of damping elements, which will be discussed below, by optimization of reductions and tensions, and by the selection of appropriate properties and quantity of the supplied lubricating and cooling liquid (LCL) (coolant). Entering of the vibrations in the resonance stage is usually prevented by reducing greatly the rolling speed to a safe level.

Many researchers believe that one of the main causes of vibrations is that in rolling at a high speed and, as a consequence, high (180–200°C) temperatures of the rolls, the strip and the temperature in the deformation zone cause thermal decomposition of the lubricant components. This is evidenced by increased contamination of the coolant with soot products. Furthermore, increasing temperature decreases the amount of lubricant entering the roll gap and also the thickness of the lubricating film in the contact surfaces of the rolls and rolled metal. As a result, the friction coefficient in the deformation zone increases, the amount of generated heat increases, and this leads to a further increase in temperature. The results of research carried out at the same 2000 mill by the experts[1] of the

[1]I.Yu. Prikhod'ko, V.V. Akishin, P.P. Chernov and E.A. Parsenyuk

Ferrous Metallurgy Institute of the National Academy of Sciences of Ukraine also showed a significant effect of the quality of the lubricant used in the rolling process on the appearance of vibrations and the possibility through the use of effective coolants to increase the rolling speed of strips without the appearance of resonant oscillations. Thus, the hypothesis that the growing vibrations (elf-oscillations) are excited due to the establishment of critical (200°C and higher) temperatures and deteriorating friction conditions in the deformation zone is valid.

In the experiments [103] on an industrial 2030 mill in 2030 the basic technological process parameters were measured during strip rolling in the vibrations mode together with the mechanical vibrations in the middle part of the frame of the rolling stands. It was found that the emergence and development of vibrations when rolling strips of various sizes have the same character.

For example, rolling with a speed of 15.6 m/s of 2.3/0.5 × 1450 mm steel 08Yu strips was stable stably, without vibrations. The process parameters were as follows:

Stands	The thickness of the strip after the stands, mm	Interstand tension, N/mm^2	Rolling force, MN
-	2.381	39	-
No. 1	1.824	153	10.41
No. 2	1.134	153	9.52
No. 3	0.751	150	9.01
No. 4	0.533	171	10.56
No. 5	0.528	27	10.97

Vertical oscillation with increasing amplitude were encountered in the stand 4 when increasing the rolling speed to 18.9 m/s.

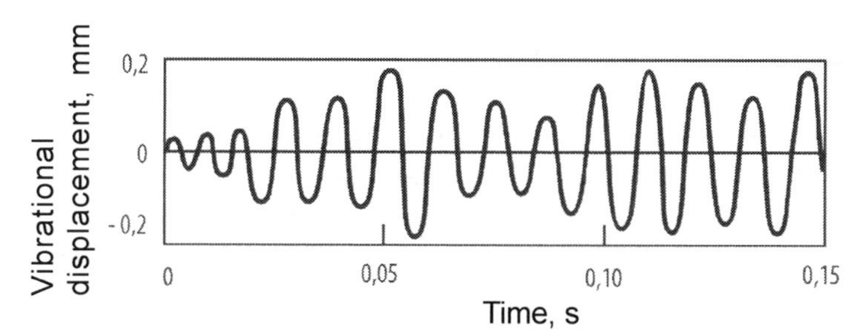

Fig. 8.3. Vibrational displacement of the frame of stand 4 at the stage of the formation of vibrations [103].

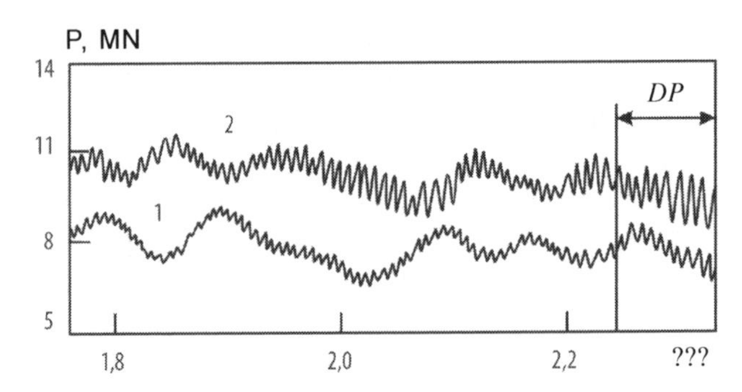

Fig. 8.4. Rolling force P in the stand 3 (curve 1) and 4 (curve 2) in the stage of development of vibrations (DP – dangerous phase) [103].

Vibrations then spread to the stands 3 and 5. Figures 8.3 and 8.4 show oscillograms of the start and development of the oscillations. The oscillograms recorded sensor signals of vibrational displacements and pressure in the hydraulic cylinders of the pressure devices of the stands.

According to the results the 2030 continuous 5-stand cold rolling mill in 2030 as a dynamic system has relatively strong damping properties. The damping coefficient is within the range 90–95 MN/m/s. The oscillation amplitude of the rolling force in the stand – the source of vibrations – could reach values of 0.3–0.4 MN within 0.2–0.4 s. Through the front and rear interstand strip tension vibrating stand spread vibrations to the neighbouring stands and then to the whole mill. However, the impact on the previous stand was stronger than on the next. Rolled strip experienced transverse vibrations in the gap between the stands. The pre-failure situation occurred in the conditions when the amplitude of synchronous vibrations in two neighboring stands exceeded 0.2–0.4 MN. The risk of breakage of the strips was considerably lower at vibrations of the same stand, even if the amplitude of rolling force oscillations was significantly higher (up to 1.2 MN).

Spectral analysis of the composition of the signals of vibrational displacement of the stand frame and rolling force (pressure in the hydraulic cylinder of the pressure device) during the vibrations of stand 4 of the 2030 mill showed that the main vibration frequencies are in the range from 0 to 20 Hz and from 90 to 150 Hz. Low-frequency components are apparently caused by torsional vibrations of the main drive lines of the stands with a frequency of 8–12 Hz and transverse vibrations of the strip in the gap between the stands with frequency intervals of 12–20 Hz. Vibrational processes in the main drive lines of the stands of the cold rolling mills are discussed in more detail below. High-frequency oscillations at the developed vibrations of the stands equalled 30–35 % of the total variance of the rolling force. The maximum high-frequency oscillations of the stands in this mill were within 115–120 Hz. The main share of high-frequency oscillations in the spectrum is occupied by the vertical oscillations at the natural frequencies stands. In other words, self-oscillations occur at intrinsic frequencies of the vertical oscillations of the mill stands.

Theoretical and experimental studies of the Ferrous Metallurgy Institute of the National Academy of Sciences of Ukraine showed that the onset of resonant vibrations in the continuous rolling mill stands is closely related to the kinematic conditions in the rolling gap, namely, the forward slip during rolling. The magnitude of the forward slip is largely determined by the front and rear strip tension. Therefore, the level and the ratio of the tensions have a decisive influence on the emergence and development of vibrations of the stands. Reducing the tension in the second interstand interval (between stands 2 and 3), and the third and fourth intervals from 130–150 to 90–100 N/mm^2 when rolling strips with a thicknesses of 0.5–0.6 mm in the 2030 mill of the Novolipetsk Metallurgical Concern allowed the threshold speed to be increased by 0.83–1.66 m/s without causing resonance vibrations. However, to avoid the loss of stability of the strips due to their vibrations in the lateral direction the interstand tension between the fourth and fifth stands, where the variation of tension due to the operations of the automatic control system of strip thickness has a maximum, should not be less than 110 N/mm^2. In the general case, the value of the interstand tension should be between 0.20 and 0.15 of the values of yield strength of the metal in this interstand gap.

Reduced the forward slip by increasing the back tension leads to the excitation of vibrations ('buzzing') of the mill at a lower speed than in the case of rolling with increased front tension. Moreover, in the conditions where the rolling process takes place with a large backward creep and in the deformation zone there is no forward creep slipping of the working rolls on the surface of deformed metal and in relation to the supporting rolls may take place. This phenomenon contributes to the resonant vibrations including damage due to the surface of the supporting rolls. In relation to the five-stand 2030 mills it is recommended to realise the rolling process in the stands 2–4 with a forward slip of at least 0.5% and at a minimum level of the interstand tensions. Similar conclusions were also reached by the authors of [102] based on the study of vibrations on the cold rolling mills

According to the recommendations of [103] the most effective method of dealing with the vibrations is to equip the strip cold rolling mills with systems of early diagnosis of the origin of vibrational processes related to the control systems of the rolling speed reacting rapidly to their signal, as well as other means to prevent possible emergencies. Such an efficient system that takes into account the balance of horizontal forces acting on the rolling stand, which allows to diagnose resonant vibrations at an early stage of their occurrence and thereby prevent 'buzzing' of the mill was considered in [104]. Its application allowed to significantly reduce the amount of regulatory effects on the rolling speed in the investigated 2030 mill.

Experimental investigations of dynamic loads in the drive lines of the mills associated with the need for a complex system of measurement and registration, require longer stopping of the mill for bonding on the drive shafts of the sensors, etc. In general, it is time-consuming and costly research but provides results very valuable for science and practice. Therefore, the experimental data obtained by such studies in industrial mills will be analysed.

Measurements of the torque directly on the spindles of the working rolls of the 4-stand 2500 and 5-stand 1700 cold rolling mills showed that when the rollers grip the strip peak loads form in the drive lines 1.8–4.5 times greater than the load in the steady state rolling mode [105]. The highest peaks occur when rolling thin strips. In particular, in the 2500 rolling mill producing strips with a thickness of 0.7 mm rolled from 2.5 mm strip plate the peak moments in the spindles of the stands No. 1 and 2 reached 600 and 400 kN·m. In the steady rolling process these moments were respectively 150 and 120 kN·m. In the 1700 mill when rolling strips of 0.5 mm thickness from 2.5 mm thick strip plate the peak moments in the spindles of the stands 3 and 4 reached 400 kN·m, whereas the steady moments did not exceed 120 and 85 kN·m.

Typical oscillograms of the torques formed in the spindles of the working rolls of both continuous mills are shown in Fig. 8.5. These oscillograms clearly illustrate that the moment at engagement is two or more times greater than the moment in the steady state rolling M_{steady}. The influence of elasticity cause damped oscillations drive lines arise in them damped oscillations. The frequency of the arising

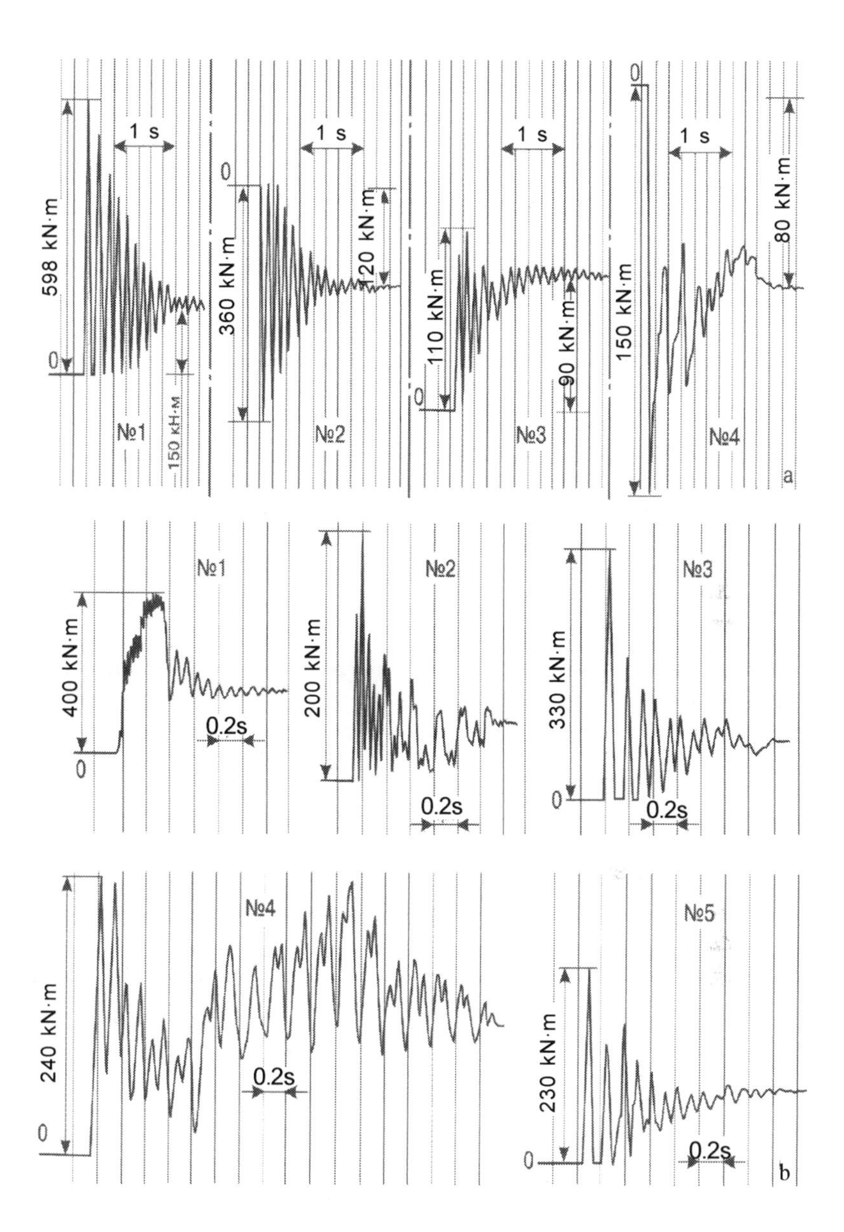

Fig. 8.5. Typical oscillograms of the torque at engagement in the upper spindles of the working roll of the stands 1–5 of the 2500 continuous cold rolling mill (a – steel 08kp, 2.5 → 0.8×1320 mm) and 1700 mill (b – steel 10kp, 4 → 1.5×1315 mm).

elastic vibrations of the drive lines for continuous cold rolling mills in 2500 and 1700 are, respectively, 8 and 5 Hz.

The systems for protection of the equipment of the main lines of the mills for maximum current in engagement of metal by the rolls

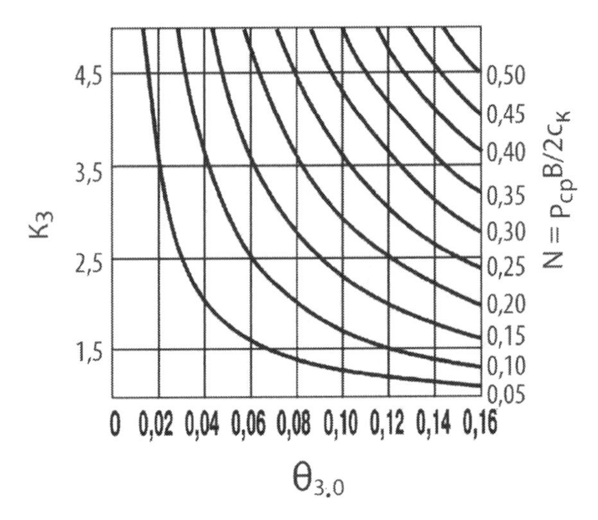

Fig. 8.6. Nomogram for determining the coefficient K_3 of overloading of equipment at capture; for notations see the text.

are ineffective. The duration of their response is significantly longer than the duration of increase of elastic moments in the drive lines in engagement (maximum peak moment in the spindles of the 2500 and 1700 mills is 0.03 and 0.05 s). Therefore, this protection does not have enough time to protect the mechanical equipment of the drive lines from dynamic overloads.

The rolling moment at engagement is greater than in the steady-state process, owing to the greater reduction of the leading edge of the strip and the effect of the elastic properties of the stand. To determine the maximum rolling moment at engagement in the sheet cold rolling mills a nomogram (Fig. 8.6) was constructed, rather universal, thanks to the construction in the dimensionless parameters. The values calculated for this stand from the specified structural and regime parameters $N = p_{av}B/2c_k$, $\theta_{3.0} = \sqrt{(\Delta h + x_0)/R}$ and M_{steady} using the nomogram, can be used can to find the coefficient K_3, and then determine the maximum deformation at the moment of engagement: $M_{\text{PRmax}} = K_3\ M_{\text{steady}}$. Here: R – radius of the roll; B – width of the strip; Δh – reduction in the stand; p_{av} – average specific pressure in the deformation zone; x_0 – the largest deformation of the stand $x_0 = P/c_k$; P – rolling force (pressure of metal on the rolls) with the filled deformation zone; c_k – stand stiffness.

The nomogram can also solve the inverse problem if it is required to optimally distribute the reductions on the stands taking the permissible loads into account.

Experimental studies and theoretical calculations have shown that the maximum rolling moments at engagement essentially depend on the reduction mode in the stands of the mill. Thus, if the relative reduction in stand No. 1 in rolling 0.5–1.2 mm strips in the 1700 mills was less than 20%, in the spindles of the stands No. 4 and 5,the maximum values of the rolling moments at engagement reached 300 kN·m. The steady rolling moments did not exceed 90 kN·m. With increasing relative reduction in stand No. 1 and a simultaneous decrease in the stand No. 5 maximum torque values in rolling stands decreased. It was found that the maximum rolling moments in the stands will be the lowest if the relative reductions in all stands except the last will be about the same, and in the last stand somewhat lower.

When rolling thin strips (0.7 mm in the 2500 mill and 0.5 mm in the 1700 mill), the maximum torque values at engagement were the lowest at the following distribution of relative reductions (%) in the stands:

Stands / Mills	No. 1	No. 2	No. 3	No. 4	No. 5
2500	25 – 26	27 – 28	27 – 28	15 – 18	–
1700	30 – 32	32 – 34	27 – 28	25 – 27	15 – 16

In these reduction modes the uneven distribution of peak and steady torques on the drives of the stands was the smallest and loading of mechanical equipment – the most rational. In particular, the 1700 mill at a specified reduction mode in rolling steel 08kp strips 0.5 mm thick from the 2.5 × 1015 mm strip plate the maximum torques in spindles of the stands 1 and 5 were, kN·m:

№ 1	№ 2	№ 3	№ 4	№ 5
114	127	150	142	138

Such moments are lower than the values acceptable for the equipment of the drive lines and their distribution in the stands is fairly even.

The rolling moment acting on the drive line of the mill causes it elastic oscillations (Fig. 8.5). Analysis of the experimental oscillograms showed that the torque in the spindle at engagement consists of the rolling moment and the dynamic components due to the elastic vibrations imposed on it. Statistical processing of the torque oscillograms obtained in rolling 120 strip rolls of various sizes in the 2500 mill and 150 strip rolls in the 1700 mill shows found that the maximum peak loads in the drive lines of the continuous cold rolling mills can be calculated with sufficient accuracy for practical purposes (8–10%) by the formula $M_{max} = M_{PRmax} K_d$, where M_{max} is the maximum peak torque in the main line; K_d is the dynamic factor, depending on the strip end engaging speed, engagement angle and the natural frequency of the drive line.

The values of the dynamicity coefficients for the drive lines of obtained from the results of experimental data for loads were as follows(No. 1–5 μ stands):

Mill \ Stand	No. 1	No. 2	No. 3	No.	No. 5
2500	1.1 – 1.3	1.2 – 1.4	1.2 – 1.5	1.3 – 1.7	–
1700	1.1 – 1.2	1.2 – 1.5	1.6 – 1.9	1.1 – 1.3	1.2 – 1.4

The dynamicity coefficient K_d for other continuous cold rolling mills can be determined from the given constructional and regime parameters according to the formula:

$$K_d = 1 + \frac{1}{\sqrt{1 + \frac{\sqrt{R(\Delta h + x_0)}}{V_{eng} T_i}}},$$

where V_{eng} is the strip end engaging speed; T_i is the period of intrinsic oscillation of the drive line.

Operating experience of the continuous 4- and 5-stand 1700 cold rolling mills showed that in the drive lines of the pressure devices associated with the automatic control systems of thickness there were frequent breakdowns of the safety units, which led to the forced outages mills and reduced the speed of the strip thickness control. To identify the causes of this phenomenon and develop measures to address it, the 1700 mill was used to perform special experimental and theoretical studies [106] using drive lines of the pressure screws of the final stands operating in the most adverse conditions (largest number of breakdowns of the safety units).

When rolling strips of the basic range, oscillograms were produced of the current and the speed of rotation of the left and right adjusting screws, the pressure under the pressure screws and the moments in the drive shaft in the area between the engine and the gearbox. Analysis of the oscillograms shows that the loads in the drive line during acceleration and braking have the form of damped oscillations. The elastic torsional vibrations occurring in the drive shaft do not affect the electromagnetic processes in the engine and the armature current does not respond to them due to the inertia of the electromechanical system. Therefore, the drive current cannot be used to estimate the maximum loads in the line.

Maximum dynamic moments in the shaft line in the transient conditions significantly (sometimes 5–7 times) surpassed the static moments of the drive and triggered safety devices. The high-frequency nature of dynamic loads adversely affects the fatigue strength of the equipment. When braking, gaps open in the kinematic pairs of drive lines with their subsequent impact closure, resulting in dramatically increased loads. In acceleration the maximum dynamic loads significantly depend on the nature of changes in the electromagnetic torque of the drive (current diagram). Introduction of the preoperational stage (half of the maximum armature current is kept constant for 0.1 s) significantly reduces the dynamic loads during acceleration, but at the same time reduces the speed of the drive of the pressure screws and degrades the quality of regulation of thickness and the profile of rolled strips.

To estimate the maximum dynamic loads and develop measures to reduce them, the dynamic processes occurring in the elastic electromechanical drive system of the pressure screws were studied theoretically and experimentally. Based on these studies, it was suggested [106] to introduced an elastically deformed device with rubber–metallic elements to substantially reduce the rigidity of the drive line of the pressure screws of the 1700 cold rolling mill and increase its damping properties. A drawing of the device is shown in [106]. Application of the developed elastic-damping coupling using four rubber–metallic bushings instead of the toothed coupling made it possible to reduce the stiffness of the 'motor–reducer' section from $0.335 \cdot 10^6$ N·m/rad to $0.0367 \cdot 10^6$ N·m/rad, and the gap by 1.5–2°.

To assess the efficiency of the developed elastic-damping system the power parameters of the drive line of the pressure screws of the 1700 cold rolling mill were oscillographically recorded. After installing the elastic-damping system dynamic loads arisen earlier during acceleration and braking, disappeared. Torque changes in the transient modes were smooth and elastic vibrations and overloading absent. The torque in the line in the transient mode and the motor current changed synchronously so the overload protection system of equipment worked efficiently. This gives the opportunity to avoid the application in some mills shear elements in the pressure screws couplings, and to protect the equipment by set the limits on the maximum torque of the motor. In addition, the installation in the drive lines of the pressure screws of the elastic-damping system significantly improves the working conditions of the equipment and can improve the performance of the thickness controller and, consequently, the accuracy of the rolled strips.

9

Features of the rolling method of production of sheet steel

Currently, the main method of producing sheet steel is rolling. The strength calculation of the coilers of rolling mills and finishing units are based on the dependence of the pressure of the coil of the coiler drum on the parameters of the coiling process of the strip. The stresses produced in cold rolled coils after removal from the coiler substantially affect the quality of sheet products, as they can cause the loss of stability of turns and formation of the 'bird' defects lead to welding of the contacting turns of the strip during the subsequent heat treatment and the formation of 'kinking' or 'welding' defects. Increasing the mass of the coils and reducing the thickness of strips in modern cold rolling mills increase the probability of formation of these defects.

The stress–strain state of rolls, a tendency to welding of turns in annealing of coils and the formation of these defects depend in a complicated way on the coiling conditions and surface roughness of the strips, deformability of the coiler drum and other factors. All this greatly complicates the selection of the best modes of tensioning during coiling. Lessons learned in one mill are often insufficient to make the best decision for another stand. Selecting the preferred modes of coiling strips should be based strictly on a theoretical analysis of the stress–strain state of the rolls, which takes into account all the nuances of technology.

9.1. A mathematical model of the stress–strain state of coils of cold rolled strips

In real conditions the stress–strain state of the coils is affected not only by the physical and mechanical properties of the material of coiled strips but also it surface condition and the degree of flatness (warpage, waviness). Of great importance are the magnitude of the surface roughness of the contacting turns and its change under loading and the presence of lubricant at the surface. This determines the tightness of contact of adjacent turns in coils. To increase the accuracy of calculations of the stress state of the coils one must considered the contact phenomena at the boundaries of adjacent turns of rough strips and often avoid using the artificial procedure – replacing the real multilayer body of the coil by a continuous one.

It should be noted that the choice of the most favourable modes of tensioning the coiled strips on the drum should be based on solving the inverse problem – determine the mode of tensioning the strips in coiling according to a given law for the stress distribution in the coil on the drum or after removal from the drum of the coiler.

Formulation and general solution of the forward and inverse problems[1] allow taking into account the effects marked by attracting additional information on the nature convergence of rough surfaces adjacent turns in rolls at random mode changes strip tension during winding.

It is assumed that both in coiling and after removing the coil from the coiler drum the strip turns are in an elastic state. There is no slipping of the turns relative to each other. The strip turn in the coil are treated as concentric rings. Stresses within one turn are considered permanent in the circumferential direction, but varying from turn to turn. Thus, the problem is reduced to axisymmetric. Due to the fact that in industrial conditions the cold rolled coils consist of a large number of turns (from a few hundred to two or three thousand), the error introduced in the calculation by the mentioned assumptions is negligible.

It is believed that the turn of the strip in the coil, as well as the coiler drums have a cylindrical anisotropy. The thickness and elastic properties of each turn can be identical or different. The problem is solved in the linear–elastic formulation. The stress–strain state of the strip element in the coil or drum is planar.

In the ideal (no gaps) fitting in the contacting surfaces of the turns of the strip in the roll the movement of the outer surface of the i-th turn u_i^{out} is equal to the movement of the inner surface of the (i +1)-th coil u_{i+1}^{in}, i.e. $u_i^{out} = u_{i+1}^{in}$. Here u^{out} and u^{in} are the displacements of the outer and inner surfaces, respectively. However, in actual

[1]The problem was formulated and solved by V.I. Timoshenko.

contact of adjacent turns the contact of the adjacent turns in the coil is discrete. The degree of gapping of the strip contacting surfaces is characterized by δ – the radial distance between the cylindrical surfaces of adjacent turns (Fig. 9.1).

Since, in general, the surface of the strip is rough, then the quantity δ represents the distance between the troughs of the roughness profile lines of the contacting surfaces, characterizes the average value of the gap between the turns and depends on the size of microirregularties. The value δ may be defined as the inelastic layer thickness between the turns (e.g., the lubricant layer, the cushioning material, etc.). Increasing the contact pressure crushes the asperities causing them to come closer. Consequently, the conjugation condition of the contacting surfaces have the form (Fig. 9.1)

$$u_i^{\text{out}} = u_{i+1}^{\text{in}} + \Delta\delta_i, \tag{9.1}$$

where $\Delta\delta_i = \delta_i - \check{\delta}_i$ is the value of convergence of the surfaces; $u_{i+1}^{\text{in}} = r_{i+1}^{\text{in}} - \check{r}_{i+1}^{\text{in}}$;
$u_i^{\text{out}} = r_i^{\text{out}} - \check{r}_i^{\text{out}}$; $\check{r}_{i+1}^{\text{in}}, \check{r}_i^{\text{out}}, \check{\delta}_i$ – the geometrical parameters of turns after deformation.

Note that, if between the i-th and $(i+1)$-th turns of an elastic material there is an inelastic layer, for example oil, the interaction between the turns can be modeled well by equation (9.1) by setting the appropriate dependence of $\Delta\delta_i$ on the interturn pressure. The function will characterize $\Delta\delta_i$ either the change of poor contact of the surfaces of adjacent turns or the deformation

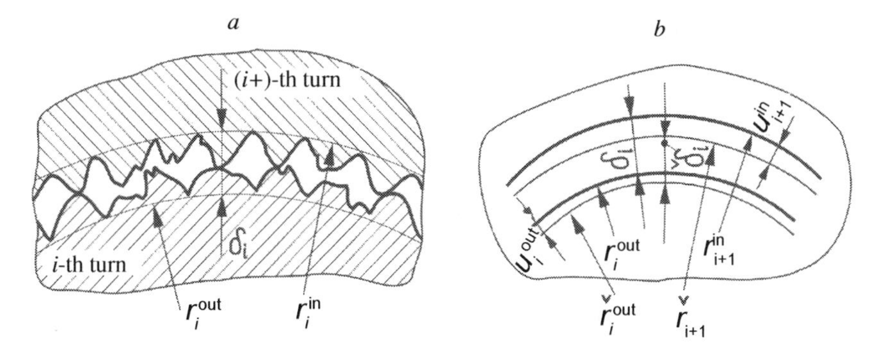

Fig. 9.1. Scheme of contact of deformed turns in the strip coil: *a* – discrete contact of rough surfaces of adjacent turns; *b* – the convergence of the contacting surfaces under load.

law of the inelastic interlayer occurring between the i-th and (i+1)-th elastic turns.

The amount of displacement of the i-th turn, considered as the anisotropic body, under the action of internal and external pressures is determined by the following equation:

$$u = \frac{k_i - \mu_i}{E_{t_i}} \cdot \frac{q_{i-1} r_{i-1}^{k_{i+1}} - q_i r_i^{k_{i+1}}}{r_i^{2k_i} - r_{i-1}^{2k_i}} r^{k_i} + \frac{k_i + \mu_i}{E_{t_i}} \times$$
$$\times \frac{q_{i-1} r_i^{k_{i-1}} - q_i r_{i-1}^{k_{i-1}}}{r_i^{2k_i} - r_{i-1}^{2k_i}} \cdot \frac{r_{i-1}^{k_{i+1}} \cdot r_i^{k_{i+1}}}{r^{k_i}}, \tag{9.2}$$

where $k_i = \sqrt{\frac{E_{t_i}}{E_{r_i}}}; E_{t_i}, E_{r_i}, \mu_i$ – moduli of elasticity in the tangential and radial directions and Poisson's ratio of the material i-th coil; r_{i-1}, r_i, r – internal, external, and the current radius of i-th turn; q_{i-1} and q_i – internal and external pressure acting on the i-th turn.

As noted above, in general it is assumed that the thickness of the i-th strip of the turn h_i is not equal to the thickness of the strip of the (i +1)-th coil h_{i+1}, i.e. $(r_i - r_{i-1}) \neq (r_{i+1} - r_i)$. Furthermore, in the general case $E_{t_i} \neq E_{t_{i+1}}, E_{r_i} \neq E_{r_{i+1}}, \mu_i \neq \mu_{i+1}$ Thus, the solution is performed for strips with longitudinal thickness difference and different properties along its length.

Substituting in equation (9.1) values u_i^{out}, u_{i+1}^{in} their expressions obtained using relations (9.2) give an equation that relates the pressure q on the contact surfaces of the three neighboring turns q_{i-1}, q_i, q_{i+1} (Fig. 9.2):

$$A_i q_{i-1} - D_i q_i + B_i q_{i+1} = -F_i. \tag{9.3}$$

The coefficients A_i, B_i, D_i in equation (9.3) are defined by the geometric dimensions r_{i-1}, r_i, r_{i+1}, and the physical constants E_t, E_r, μ of the i-th and (i+1)-th turns. The expression A_i, B_i, D_i for the general case are easy to find from the dependence (9.2). Free term F_i of equation (9.3) is determined by the convergence of the i-th and (i+1)-th turns $\Delta\delta_i$ under load, wherein $F_i \to 0$ when $\Delta\delta_i \to 0$.

In a subsequent solution the boundary conditions are used, allowing in equation (9.3) written for inner (first) or outer coil turn to eliminate one of the unknowns [107]. As a result, the following relationship is obtained:

$$q_i = L_{i+1} q_{i+1} + K_{i+1}, \tag{9.4}$$

where $L_{i+1} = B_i(D_i - A_i L_i)$; $K_{i+1} = (A_i K_i + F_i)/(D_i - A_i L_i)$.

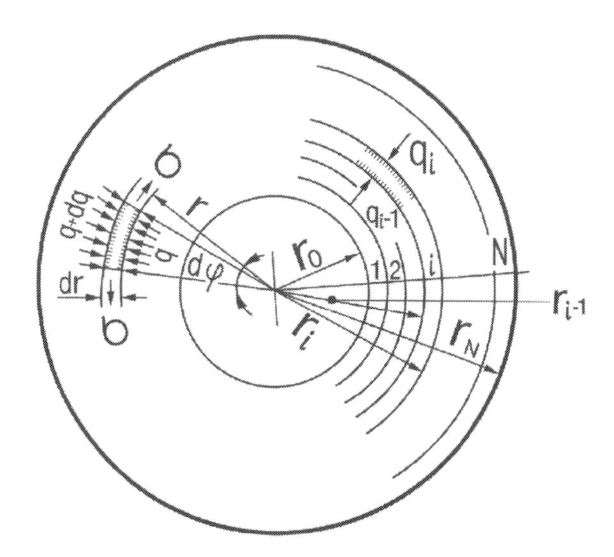

Fig. 9.2. Design scheme: N – number of turns of strip in a coil. The remaining notations are explained in the text.

Equation (9.4) is solved by the 'trial' method [107]. Initially, all the values of the trial coefficients L_i and K_i are found and then, using arrays L_i and K_i also the condition to determine q of the outer turn – stress values q_i. Tangential (circumferential) stresses σ in the turns are determined from the equilibrium condition of the turns

$$\sigma_i = \frac{q_i r_i - q_{i+1} r_{i+1}}{r_{i+1} - r_i}.$$

Calculation of the stress–strain state of coiled strip during coiling on the drum reel

We find the stresses in the coil in winding on the drum a strip of constant thickness with isotropic properties. We assume that the variation in the thickness of the metal and changes of the mechanical properties are negligible. The Young modulus and Poisson's ratio of cold-rolled steel can only change as a result of a radical change in its chemical composition.

We define the increment of stresses Δq and displacements Δu of the i-th turn in the coil consisting of j turns under the effect of the j-th turn. When naming variables, the first index will indicate the number of the considered turn, the second – the number of the turn, the impact of which is determined. The pressure between the i-th and (i+1)-th turns of the coil consisting of j turns is denoted by $q_{i,j}$

and $\Delta q_{i,j} = q_{i,j} - q_{i,j-1}$. Further, writing equation (9.3) for $q_{i,j-1}$, and $q_{i,j}$ and subtracting the first from the second term we receive

$$A_i \Delta q_{i-1,j} - D_i \Delta q_{i,j} + B_i \Delta q_{i+1,j} = -F_{i,j}. \tag{9.5}$$

As the pressure (radial stress) changes from $q_{i,j-1}$ to $q_{i,j}$ the gap between the turns is changed by the $\Delta\delta_{i,j}$ value equal to the difference of the size of the gap δ_i at pressures $q_{i,j-1}$, and $q_{i,j}$ (see Fig. 9.1). Since in the coiling of each subsequent turn the increment Δq of inter-turn pressure is small, then

$$\Delta\delta_{i,j} = \delta_i(q_{i,j-1}) - \delta_i(q_{i,j}) = \frac{d\delta}{dq}\Big|_{q_i,j-1} \Delta q_{i,j}.$$

For an isotropic material of the strip $(E_{t_i} = E_{r_i} = E = \text{const}, \mu_i = \mu = \text{const})$ and at the same thickness of the turns the expressions for the coefficients in the above equation have the form:

$$A_i = \frac{2r_{i-1}^2}{r_i^2 - r_{i-1}^2}; \qquad B_i = \frac{2r_{i+1}^2}{r_{i+1}^2 - r_i^2};$$

$$D_i = \frac{r_{i-1}^2 + r_i^2}{r_i^2 - r_{i-1}^2} + \frac{r_i^2 + r_{i+1}^2}{r_{i-1}^2 - r_i^2} - \frac{E}{r_i}\frac{d\delta}{dq}\Big|q_{i,j-1}; \; F_{i,j} = 0.$$

When calculating the stresses in the roll during its winding (in the position 'on the drum'), the first boundary condition is that the increment of radial deformation of the drum $\Delta u_{b,j}$ from the action of j-th turn is equal to the increment of the radial deformation of the inner surface of the first turn $\Delta u_{1,j}$ minus the change in the gap size between the first turn and drum $\Delta\delta_0$, i.e. $\Delta u_{b,j} = \Delta u_{1,j}^{in} - \Delta\delta_0$.

If the drum is represented in a thick-walled cylinder with an equivalent ratio $\lambda = a_{in}/a$, where a_{in} and a is the inner and outer radii of the cylinder, the variation of the radial deformation of the drum $\Delta u_{b,j}$ due to changes in the pressure acting on it by the amount $\Delta q_{0,j}$ will be determined formula [107, 108]

$$\Delta u_{b,j} = B\Delta q_{0,j}, \tag{9.6}$$

Where $B = \frac{a}{E_b}\left(\frac{1+\lambda^2}{1-\lambda^2} - \mu_b\right)$; E_b *and* μ_b – modulus and Poisson coefficient of the drum material.

In general, the coiler drum can be regarded as an anisotropic body, and its deformation described by a more complex model [107].

Given that $\Delta\delta_{0,j} = \frac{d\delta}{dq}\Delta q_{0,j}$, we can write

$$\Delta u_{1,j}^{in} = \alpha_0 \Delta q_{0,j} + \beta_0 \Delta q_{1,j}, \tag{9.7}$$

where

$$\alpha_0 = \frac{r_0}{E(r_1^2 - r_0^2)}\left[\left(r_0^2 + r_1^2\right) - \mu\left(r_0^2 - r_1^2\right)\right] - \frac{d\delta}{dq}\bigg|_{q_{0,j}};$$

$$\beta_0 = -\frac{2r_0 r_1^2}{E(r_1^2 - r_0^2)}.$$

Equating the expressions (9.6) and (9.7), we find the dependence

$$\Delta q_{0,j} = L_1 \Delta q_{1,j},$$

where $L_1 = \dfrac{\beta_0}{-\dfrac{a}{E_b}\left(\dfrac{1+\lambda^2}{1-\lambda^2} - \mu_b\right) - \alpha_0}$.

Substituting $\Delta q_{0,j}$ in equation (9.5), written for the first turn (i = 1), we obtain

$$A_i \Delta q_{0,j} - D_1 \Delta q_{1,j} + B_1 \Delta q_{2,j} = 0; \qquad \Delta q_{1,j} = L_2 \Delta q_{2,j},$$

where $L_2 = -\dfrac{B_1}{A_1 L_1 - D_1}$.

Further, in accordance with the general case (9.4) we have

$$\Delta q_{i,j} = L_{i+1} \Delta q_{i+1,j}, \quad i = 1, 2, 3, \ldots\ldots, j-1.$$

As a second boundary condition we use the equilibrium equation of the upper (peripheral) turn. Assuming that the compressive interturn pressure is positive, we write

$$\Delta q_{j-1,j} = q_{j-1,j} = \sigma_{0,j} h / r_{j-1},$$

where $\sigma_{0,j}$ is the strip tension during winding j-th turn (tension given by the coiler at the moment of formation of the j-th turn); h is the thickness of the strip.

Complete (total) pressure on the i-th turn (at $r = r_i$) of the coil from the effect of all upper turns will be the sum $q_{i,N} = \sum_{j=i+1}^{j=N} \Delta q_{i,j}$, where N is the number of turns of the strip in the roll. When calculating the stress state in the turns of the strip roll the inner cycle of the program in the variable i from 0 to j–1, and the outer cycle to the variable j – from 1 to N. The result will be found all values

$q_{i,j}$. The pressure of the strip roll on the coiler drum q_b is determined by the action of all the turns of the strip roll $q_b = q_{0,N} = \sum_{j=1}^{j=N} \Delta q_{0,j}$.

Stress analysis in a strip a roll after its removal from the coiler drum winder

When removing from the drum the strip roll is unloaded – reducing stress due to the radial displacement of the turns. The index 'c' denotes the state of the strip roll after removal from the coiler drum. Then, writing equation (9.3) for pressures on the strip roll on the drum q_i and after removal from the drum q_i^c given δ_i or δ_i^c, respectively, and subtracting one from the other, we get the same equation for Δq_i^c – change of the the interturn pressure when removing the strip roll from the drum:

$$A_i^c \Delta q_{i-1}^c - D_i^c \Delta q_i^c + B_i^c \Delta q_{i+1}^c = -F_i^c .$$

In the last equation, the coefficients of A_i^c, B_i^c are expressed in the same way as when calculating the coefficients of the same name for the strip roll on the coiler drum. In the coefficient D_i^c there is no third term. The free term in the right-hand side has the form $F_i^c = -\Delta\delta_i^c E / r_i$, where $\Delta\delta_i^c = \delta_i^c(q_i^c) - \delta_i(q_i)$. Functions gap changes during loading δ_i and unloading δ_i^c in general may be different. The linear dependence of the form $\Delta\delta_i^c = \frac{d\delta^c}{dq}\Big|q_i \Delta q_i^c$ here can not be accepted, since when unloading the strip roll the interturn pressure may be reduced by a large amount (by an order and more).

The boundary conditions for the strip roll removed from the drum: on the inner surface of the first turn the pressure is reduced to zero, i.e. $\Delta q_0^c = +q_b$ at $r = r_0$ on the outer surface of the upper turn the pressure remains equal to zero, i.e. $\Delta q_N^c = 0$.

In accordance with equation (9.4)

$$\Delta q_i^c = L_{i+1}^c \Delta q_{i+1}^c + K_{i+1}^c , \tag{9.8}$$

where $L_{i+1}^c = \frac{B_i^c}{D_i^c - A_i^c L_i^c}$; $K_{i+1}^c = \frac{A_i^c K_i^c + F_i^c}{D_i^c - A_i^c L_i^c}$. Thus, the first value of the trial coefficients L^c and K^c are expressed as follows

$$L_2^c = \frac{B_1^c}{D_1^c}; \quad K_2^c = \frac{-A_1^c q_b + F_i}{D_1^c}.$$

The algorithm does not change the decision. Initially, all the calculated values of the coefficients L_i^c and K_i^c. Then, knowing Δq_N^c, from equation (9.8) we have Δq_{N-1}^c, etc. The interturn residual pressure in the strip roll q_i^c removed from the coiler drum is defined as

$$q_i^c = q_i - \Delta q_i^c. \qquad (9.9)$$

The described algorithm is easily implemented for any changes in the strip tension $\sigma_0(r)$ during its winding. The gap size $\delta_i^c(q_i^c)$ is determined more accurately by iterations usign the determined values of q_i^c.

It should be noted that the circumfenetial stresses in the outer and inner turns of the strip roll after removing from the coiled are equal to zero. When the rear end of the strip leaves the last mill stand the tension disappears and the stresses in the upper layers roll fall. A similar phenomenon due to the slippage of the front (inner) end of the strip when removing the roll of the coiler drum winder is observed in the inner layers. However, a detailed analysis revealed that the slippage is substantial only in the three to four outer and inner turns. Practically it does not affect the magnitude of the inter-turn pressure in the strip roll.

Asymptotic solutions

In a first approximation, solutions to the given problem for particular cases can be searched in a closed form.

In the case of $h_1 = h = \text{const}$, $h_i/r_i \ll 1$ for all i the difference equation (9.3) written for the strip roll, all the turns of which have the same isotropic properties, can be represented as a differential operator. We assume that the equation (9.3) is obtained by converting the following differential equation:

$$\alpha\frac{d^2q}{dr^2} + b\frac{dq}{dr} + cq = -F.$$

The differentials are replaced by finite differences. The error of this approximation is reduced in proportion to h^2:

$$\frac{d^2q}{dr^2} = \frac{q_{i+1} - 2q_i + q_{i-1}}{h^2}; \quad \frac{dq}{dr} = \frac{q_{i+1} - q_{i-1}}{2h};$$

$$\alpha_i\frac{q_{i+1} - 2q_i + q_{i-1}}{h^2} + b_i\frac{q_{i+1} - q_{i-1}}{2h} + c_iq_i = -F_i.$$

Citing similar in the left side of the last equation and comparing it with equation (9.3), we see that the coefficients of the variables q_{i-1}, q_i, q_{i+1} in one and the other equations are connected as follows:

$$A_i = \frac{1}{h^2}\left(a_i - \frac{hb_i}{2}\right); \quad B_i = \frac{1}{h^2}\left(a_i + \frac{hb_i}{2}\right); \quad D_i = \frac{1}{h^2}\left(2a_i - h^2 c_i\right).$$

Solving these expressions for the coefficients a_i, b_i, c_i, we find

$$\alpha_i = (A_i + B_i)h^2/2; \quad b_i = (B_i - A_i)h; \quad c_i = A_i + B_i - D_i.$$

In accordance with the results of [107] the expressions for the coefficients A_i, B_i, D_i may be represented in a compact form:

$$A_i = (r_i^2 - 2r_i h + h^2)(2r_i + h); \quad B_i = (r_i^2 + 2r_i h + h^2)(2r_i + h); \quad D_i = 4r_i^3.$$

then $\alpha_i = 2r_i^3 h^2$; $b_i = 2(3r_i^2 - h^2)h^2$; $c_i = 0$. Taking into account the expressions of these coefficients differential equation takes the form

$$r^3 \frac{d^2 q}{dr^2} + (3r^2 - h^2)\frac{dq}{dr} - \frac{E(4r^2 - h^2)}{4rh}\Delta\delta(r,q) = 0.$$

since $h^2 \ll r^2$, for convenience of the subsequent decision, we write

$$\frac{d}{dr}\left(r^3 \frac{dq}{dr}\right) - E\frac{r}{h}\Delta\delta(rq) = 0. \tag{9.10}$$

In general, due to presence of the term $\Delta\delta(r,q)$, equation (9.10) is non-linear and in its numerical solution we use the algorithm described above. However, for this particular set of functions $\Delta\delta(r,q)$, in some cases it is possible to find an analytic solution. In particular, for ideal fitting of the turns (without gaps) $\Delta\delta(r,q) = 0$ and the third term in equation (9.10) vanishes. Then we obtain a simple equation for q, the solution of which is expression $q = \frac{C_i}{r^2} + C_2$. Arbitrary constants C_1 and C_2 are determined using boundary conditions for the cases of the strip roll on the drum of the strip roll removed with coiler drum.

In particular, for a strip roll on the drum the boundary conditions of equation (9.10) written for Δq_j, are $\Delta q_j = \frac{\sigma_o(r_j)h}{r_j - h}$ at $r = r_j - h$ and $\Delta u_{in} = \Delta u_b$ at $r = r_{in}$, where r_{in} and r_j are the radii of the inner and the j-th turns of the strip roll during winding of the strip. As a result, we find:

$$\Delta q_i(r) = \frac{\sigma_0(r_j)h}{r_j} \cdot \left\{ 1 - \frac{1/r^2 - 1/r_j^2}{\left(\frac{1}{r_{\text{in}}^2} - \frac{1}{r_j^2} \right) + \frac{2}{r_{\text{in}} \left[r_{\text{in}} (\mu - 1) - BE \right]}} \right\}. \tag{9.11}$$

Using the expression $\Delta q_j(r)$, according to formula (9.11) we can find function $q_j(r)$. For this it is necessary to sum the increments $\Delta q(r)$ from the action of all turns overlying the considered turn with the radius r. That is, it is necessary to gradually assume $r_j = r_i + h$, $r_i + 2h$, $r_i + (j-i)h$. Turning to the integration, we write

$$q(r) = \int_r^{r_{\text{out}}} \Delta q\left(r_{i,j}\right) dr_j \tag{9.12}$$

or

$$q(r) = \frac{1 - r_{\text{in}}^2 / r^2 + C}{1 + C} \int_r^{r_{\text{out}}} \frac{\sigma_0(r_j) r_j}{r_j^2 - A^2} dr_j,$$

where $A^2 = r_{\text{in}}^2 / (1 + C)$; $C = 2 / \left[(\mu - 1) + \frac{E}{E_b} \left(\frac{1 + \lambda^2}{1 - \lambda^2} - \mu_b \right) \right]$;

r_{out} is the outer radius roll $r_{\text{out}} = r_N$.

Putting in the last expression $r = r_{\text{in}}$, we find the pressure of the strip roll on the coiler drum $q_b = q(r_{\text{in}})$:

$$q_b = \frac{C}{1 + C} \int_{r_{\text{in}}}^{r_{\text{out}}} \frac{\sigma_0(r_j) r_j}{r_j^2 - A^2} dr_j. \tag{9.13}$$

In the simplest case when $E = E_b$, $\mu = \mu_b$ and winding a strip in a roll with a constant tension $\sigma_0(r_j) = \text{const}$, expression (9.13) is converted into a well-known Sims formula

$$q_b = \frac{\sigma_0}{2} \left(1 - \lambda^2 \right) \ln \frac{1}{1 - \lambda^2} \left(\frac{r_{\text{out}}^2}{r_{\text{in}}^2} - \lambda^2 \right).$$

Equation (9.10) recorded relative to Δq^c, with the boundary conditions for the strip roll removed from the coiler drum leads to the Lamé formula the radial stresses in the cylinder loaded with internal pressure, at a value equal to q_b but with opposite sign:

$$\Delta q^c(r) = -q_b \frac{r_{\text{in}}^2}{r_{\text{out}}^2 - r_{\text{in}}^2} \left(1 - \frac{r_{\text{out}}^2}{r^2} \right).$$

In accordance with the expression (9.9) the stress–strain state of the strip roll after removing it from the coiler $q^c(r)$ can be determined by the superposition of the stress field in a strip roll on the drum $q(r)$ according to the formula (9.12) with the determined stresses $\Delta q^c(r)$. We obtain

$$q^c(r) = \frac{1 - \frac{r_{\text{in}}^2}{r^2} + C}{1+C} \int_r^{r_{\text{out}}} \frac{\sigma_o(r_j) r_j}{r_j^2 - A^2} dr_j + \frac{r_{\text{in}}^2}{r_{\text{out}}^2 - r_{\text{in}}^2} \times$$

$$\times \left(1 - \frac{r_{\text{out}}^2}{r^2}\right) \frac{C}{1+C} \int_{r_{\text{in}}}^{r_{\text{out}}} \frac{\sigma_0(r_j) r_j}{r_j^2 - A^2} dr_j. \tag{9.14}$$

Solution of the inverse problem (synthesis problem)
In industrial practice, the inverse problem is of special interest – determining the tension mode of the strip during winding the strip roll, providing the required stress state of the the strip roll on the drum or after removing it from the coiler, i.e. finding the function $\sigma_0(r)$ by determining the law of variation $q(r)$ or $q^c(r)$. In the first case, the solution is unique and is obtained by transforming and differentiating equation (9.12):

$$\sigma_0(r) = \frac{2A^2}{r^2 - A^2} q(r) - r\frac{dq(r)}{dr}. \tag{9.15}$$

In the inverse problem for the strip roll removed from the winder roll we denote

$$C_3 = \frac{C}{1+C} \frac{r_{\text{in}}^2}{r_{\text{out}}^2 - r_{\text{in}}^2}; \quad C_4 = \int_{r_{\text{in}}}^{r_{\text{out}}} \frac{\sigma_0(r) r}{r^2 - A^2} dr.$$

Then, solving the equation (9.14) with respect to $\sigma_0(r)$, we obtain

$$\sigma_0(r) = -r\left(1 - \frac{A^2}{r^2}\right) \frac{d}{dr}\left(\frac{q^c(r)}{1 - A^2/r^2}\right) + 2C_3 C_4 A^2 \frac{r_{\text{out}}^2 - A^2}{r^2 - A^2}. \tag{9.16}$$

When substituting the last equation to expression C_4 we obtain the identity. Consequently, the solution of equation (9.14) with respect to $\sigma_0(r)$ is ambiguous, i.e. the given diagram of changes in the radial stresses $q^c(r)$ in a strip roll after removal from the coiler can be obtained under different conditions of tensioning the strips during winding $\sigma_0(r)$, defined by different choices of constant C_4 in the formula (9.16). This conclusion is consistent with the findings obtained by other means in [109].

In practice, from a variety of modes of changing the tension $\sigma_0(r)$, satisfying equation (9.16) one must choose such modes that satisfy technological constraints imposed on the function $\sigma_0(r)$. In particular, the function $\sigma_0(r)$ can not take negative values and values higher than $K_3\sigma_B$ (σ_B – ultimate tensile strength of the material of the coiled strip; K_3 – the coefficient, $0 < K_3 \leq 1$), etc. In determining the most favourable tensioning modes constant C_4 in equation (9.16) can be used to obtain a suitable expression for the function $\sigma_0(r)$.

It is assumed that it is required that after the removal of the strip roll from the coiler the radial stresses in it are equal to zero, i.e. $q^c(r) = 0$ throughout the range of variation of r. In this case, the first term in equation (9.16) vanishes and

$$\sigma_0(r) = \frac{C_5}{r^2 - A^2} = \frac{C_5}{r^2 - r_{\text{in}}^2 / (1+C)},$$

where $C_5 = 2C_3C_4A^2(r_{\text{out}}^2 - A^2)$.

We believe that during the coiling of the first (inner) turn of the strip roll the tension in the strip is equal to unity: $\sigma_0(r_{\text{ext}}) = 1$. From this condition we find the constant C_5. Then $C_5 = \frac{C}{1+C} r_{\text{in}}^2$ and $\sigma_0(r) = \frac{C \cdot r_{\text{in}}^2}{(1+C)r^2 - r_{\text{in}}^2}$. Thus, in the examined case the strip tension during coiling should decrease proportionally to the square of the radius of the wound coil.

More flexible and easier to implement than the selection method is an approximate solution of the considered direct and inverse problems.

The method of approximate analysis of the stress state of strip rolls.

The main idea of this method consists in that the radial q and hoop σ stresses in the strip roll are regarded as the sum of stresses q_0 and σ_0, arising from the tension of the unwound strip assuming the absence of elastic strains of the strip roll and additional stresses $\bar{q}$ and $\bar{\sigma}$, due to the elastic displacement of the turns in the radial direction:

$$q = q_0 - \bar{q}; \qquad \sigma = \sigma_0 + \bar{\sigma}.$$

As in the usual notation of Hooke's law the stresses q and σ are considered tensile, and in the diagram in Fig. 9.2 they are presented as compressive, in the first of these equations the minus sign is placed in front of $\bar{q}$.

Using the equations of equilibrium of the turn element, the equations expressing the relationship of stresses and strains, and the boundary conditions, allows to obtain the following dependence of q_{i-1}, q_i, q_{i+1} on $\sigma_0(r)$:

$$A_i q_{i-1} - D_i q_i + B_i q_{i+1} = A_i' \sum_{i-1} - D_i' \sum_i + B_i' \sum_{i+1} = -F_i, \tag{9.17}$$

$$\text{where } \sum_{i-1} = \int_{r_{i-1}}^{r_{out}} \sigma_0 dr; \qquad \sum_i = \int_{r_i}^{r_{out}} \sigma_0 dr; \qquad \sum_{i+1} = \int_{r_{i+1}}^{r_{out}} \sigma dr.$$

The expressions for the coefficients A_i, B_i, D_i, A_i', B_i', D_i' are presented in our paper [107].

In solving the problem in this formulation, but with the gap on the right side of equation (9.17) taken into account, one should add a term containing as a factor the convergence of the surfaces of the turns of the strip roll.

Equation (9.17) can be regarded as a system of equations with respect to q_i, if Σ_i is known, i.e. function $\sigma_0(r)$. We can solve the inverse problem – for a given change in the variation $q(r)$ to find $\sigma_0(r)$. Generally it is necessary to solve a system of equations of the form

$$A_i^0 f_{i-1} - D_i^0 f_i + B_i^o f_{i+1} = -F_i, \tag{9.18}$$

where f_i should be taken as equal to q_i in the direct problem and equal to Σ_i in the inverse problem, and the coefficients A_i^0, D_i^0, B_i^0 – equal to respectively A_i, D_i, B_i or A_i', D_i', B_i'.

Equation (9.18) is solved numerically by the above procedure using the 'trial' method and the boundary conditions for cases of 'strip roll on the drum' or 'the strip roll after removal from the coiler drum' [107].

When cooling or heating the strip rolls of cold-rolled metal their stress–strain state changes. Due to the significant non-linearity of the dependence of the gaps between the turns on contact pressure and stress in the strip rolls these dependences will vary even in uniform heating of all the turns. Especially when cooling strip steel rolls after cold rolling and also in heating and cooling the strip rolls during subsequent annealing the temperature of the turns varies differently in the thickness of coiled strip rolls. However, even under such conditions it is not difficult to consider the influence of the temperature factor in the developed models. In solving this thermoelastic problem in the model one must enter the dependence of the thermal expansion of the material of the strip on temperature and in the conditions of contact of the inner and upper surfaces of the lower turn in the strip roll to take into account the change in the gap due to thermal changes in the thickness of the turns. In another approach, one can find the thermal radial stresses in the strip roll, considering it as a solid cylinder, and summing them with the contact pressure between the turns of the strip roll after removal from the coiler, to determine the stress state of the strip roll during heating during annealing. The solution of the problem in this formulation is given in [110]. This problem was solved in greater detail and more comprehensively later in [111–115]. Therefore, when considering the lower temperature and the stress–strain state of cold rolled strips we use the solution published in [111–113].

The main feature that should be considered in the calculation of the stress–strain state of the rolls when changing their temperature is that the thermal resistance on adjacent surfaces of the turns are determined to a large extent by the interturn pressures and the temperature change in the strip roll depends on its elastic stress state. The interaction between the temperature and stress–strain state of the strip roll is very strong. In addition, the level and distribution of stresses in the strip rolls depend on the flatness (warpage, waviness) of the contacting turns of the strip, the surface roughness, as well as the presence of layers of materials in which the relationship between stresses and strains is not subject to Hooke's law and which under the influence of high temperature undergo phase changes, restructuring, physico-chemical transformations, etc. This may lead to the convergence of both contacting surfaces, in particular due to the compression of surface microirregularities due to compression of the turns and the appearance of gaps in the temperature distribution and heat flux. These features taken into account by using the concept of generalized non-ideal contact, which allows on the one hand to take into account in terms of compatibility of strains the convergence adjacent contacting surfaces as a function of the interturn pressure and, on the other side, breaks in the temperature and heat flux on adjacent turn surfaces [112].

Following publications [112, 113], the previously adopted notation is supplemented as follows: T – temperature of the strip in the strip roll; T_g – gas temperature in the furnace for annealing cold rolled steel rolls; τ – time; ρ – density; λ – the coefficient of thermal conductivity; R – coefficient of thermal resistance; r – radial coordinate; z – coordinate along the length of the strip coil (in the direction of width B of the strip); α – coefficient of heat transfer; α_t – coefficient of linear expansion; indices: i – the number of layers; '+' or '–' is assigned to the 'top' and 'bottom' from the contact boundary of the turns; 'in' – represents a parameter on the inner surface; 'out' – on the outer surface.

The temperature field in the finite width of the strip roll with thickness of the turns h_i, where $i = 1, N$, is described by the system of equations:

$$(\rho c)_i \frac{\partial T}{\partial \tau} = \frac{\partial}{\partial r}\lambda \frac{\partial T}{\partial r} + \frac{\lambda}{r}\frac{\partial T}{\partial r} + \frac{\partial}{\partial z}\lambda \frac{\partial T}{\partial z}, \tag{9.19}$$

$$\text{at } r_i < r < r_{i+1}, i = 1, N, -B/2 < z < B/2.$$

The boundary conditions on the heat exchange surfaces of the butts of the strip roll (axis z):

$$\lambda \frac{\partial T}{\partial z} = -\alpha_-(T_{g-} - T), \text{ at } z = -B/2;$$

$$\lambda \frac{\partial T}{\partial z} = -\alpha_+(T_{g+} - T), \text{ at } z = B/2. \tag{9.20}$$

Integrating equation (9.19) with respect to z within the range of $-B/2$ to $B/2$, we introduce the average temperature in the roll width:

$$T_{av}(r) = \frac{1}{B}\int_{-B/2}^{B/2} T(r,z)dz \tag{9.21}$$

Using in equation (9.19) the value $T_{av}(r)$ we get (index 'av' is omitted):

$$(\rho c)_i \frac{\partial T}{\partial \tau} = \frac{\partial}{\partial r}\lambda \frac{\partial T}{\partial r} + \frac{\lambda}{r}\frac{\partial T}{\partial r} + \lambda \left.\frac{\partial T}{\partial z}\right|_{z=B/2} - \lambda \left.\frac{\partial T}{\partial z}\right|_{z=-B/2},$$

$$r_i < r < r_{i+1}, i = 1, N. \tag{9.22}$$

Taking into account the boundary conditions (9.20), equation (9.22) takes the form:

$$(\rho c)_i \frac{\partial T}{\partial \tau} = \frac{\partial}{\partial r}\lambda \frac{\partial T}{\partial r} + \frac{\lambda}{r}\frac{\partial T}{\partial r} - (\alpha_+ + \alpha_-)/B \times T + (\alpha_+ T_{g+} + \alpha_{+-} T_{g-})/N,$$

$$r < r_{i+1}, i = 1, N. \tag{9.23}$$

With further solving the problem equation (9.23) is written for each turn.

On adjacent surfaces of the turns the conditions of non-ideal thermal contact have the form

$$T_j^+ - T_j^- = R_j^+ \left(\lambda \frac{\partial T}{\partial r}\right)_j^+; \left(\lambda \frac{\partial T}{\partial r}\right)_j^+ = \left(\lambda \frac{\partial T}{\partial r}\right)_j^-, \tag{9.24}$$

at $r_{j+1} = r_j + \delta_j^+$, $r = r_{j-1}$, $j=1, N-1$.

Heat transfer boundary conditions on the inner and outer surfaces of the roll:

$$\lambda \frac{\partial T}{\partial r} = -\alpha(T_g - T), \text{ at } r = r_{\text{in}}; \quad \lambda \frac{\partial T}{\partial r} = \alpha(T_g - T), \text{ at } r = r_{\text{out}}. \quad (9.25)$$

The authors of [113] reduced the non-ideality of thermal contact to the usual thermal resistance. The coefficient of thermal resistance R_t generally depends on the roughness and contamination of contact surfaces and their compression force. R_t is usually expressed by the semiempirical formulas or dependences constructed from experimental data. The work [113] presents an empirical formula [116]:

$$R_t = \frac{1}{\dfrac{2\lambda A\left(\dfrac{P_k}{3\sigma_k}\right)^n}{\pi r_m} + \dfrac{\lambda_c R}{\delta_e}}, \quad (9.26)$$

where r_m – radius of the contact patch (r_m = 30...40 μm); λ – the coefficient of thermal conductivity of the strip material; A – coefficient taking into account the real microrelief of the contacting surfaces; P_k – unit load on the contact surfaces, N/mm²; σ_k – tensile strength of the softer material at the temperature of the contact area, N/mm²; n = 1.1...1.6 – exponent at P_k = 0...20 N/mm²; λ_c – thermal conductivity of the medium at the contact zone $\frac{Bm}{m \cdot K}$; $K = 1.4 - \frac{1.5}{P_k^{0.25} \cdot 10^5}$ – experimentally determined coefficient; δ_e – equivalent thickness of the gap (for approximate calculations δ_e shall be equal to the average height of asperities of the contacting surfaces).

Using (9.26) we can find the coefficient of thermal resistance for any pair of contacting surfaces in the temperature range to 1000 K and a pressure P_k = 0.5...25 N/mm². Because of uncertainty in the choice of empirical terms of equation (9.26), the authors of [113] used in the calculations the experimental data for steel 45–steel 45 pairs at T = 376 K taken from the literature [117] :

For ease of use in the calculation these data were approximated [113] by a power function.

P_k, N/mm²	0	5	10	15	20	25	30
$R_t \cdot 10^4$ m²·deg/W	5	3	2.25	1.8	1.6	1.4	1.25

When calculating the stresses in the strip roll arising under tension of coiling the strip and thermal deformations, displacement u_i of the surface of the i-th turn is defined as the sum of the displacements under the effect of radial stresses $u_{i,q}$ and the influence of thermal stresses $u_{i,T}$:

$$u_i = u_{i.q} + u_{i,T} \tag{9.27}$$

The amount of displacement $u_{i,q}$, as shown above, is determined by expression (9.27), and the quantity $u_{i,T}$ can be found by solving the problem of the stress–strain state of the cylinder at a variable temperature field [118]:

$$u_{i,T}^{\text{in}} = \frac{2\alpha_T r_{i-1}\rho_{i-1}}{1-\rho_{i-1}^2}\int_{\rho_{i-1}}^{1}(T-T_0)\rho d\rho,$$

$$u_{i,T}^{\text{out}} = \frac{2\alpha_T r_i}{1-\rho_{i-1}^2}\int_{\rho_{i-1}}^{1}(T-T_0)\rho d\rho.$$

Here $\rho_{i-1} = r_{i-1}/r_i$, T_0 – initial level of temperature change in the i-th turn.

Taking into account the expressions (9.27) and the last formulas of the conjugation condition of contacting surfaces of the turn of the strip in the roll takes the form

$$u_{i,q}^{\text{out}} = u_{i+1,q}^{\text{in}} + \Delta\delta_i(q_i) + \alpha_T r_i (T_{i+1} - T_i). \tag{9.28}$$

Substituting into (9.28) the expressions $u_{i,q}^{\text{out}}$ and $u_{i+1,q}^{\text{in}}$ recorded on the basis of the relationship (9.2), after transformations we obtain the equation which differs from equation (9.3) is only by the free member F_i which, with the temperature taken into account, is more complicated

$$F_i = E\frac{\Delta\delta_i}{r_i} - \alpha_T E(T_{i-1} - T_i). \tag{9.29}$$

i.e. the constant term F_i is now determined not only by the $\Delta\delta_i$ value of convergence of the i-th and (i +1)-th turns under load q_i, and by the temperature difference between the i-th and (i +1)-th turns. Obviously, in the case of a uniform temperature field in the strip roll when $T_{i-1} = T_i = T_{\text{st.roll}}$ the free term expression F_i takes the form as in equation (9.3).

Subsequent calculations in the calculation of the stress–strain state roll in the states ‘on a coiler drum’ and ‘after removal from the drum', taking into account the non-constant temperature in the thickness of the turn do not undergo any changes compared to the above decision. The only difference is that for the case of removal of the strip roll from the coiler the free term is $F_i = \alpha_T E(T_{i-1} - T_i)$.

Thus, the stress–strain state of the strip roll when its temperature changes during cooling (heating) is described by the non-isothermal equations in the form of equation (9.5), taking into account the above expressions for F_i. Subtracting from the recorded non-isothermal equation a similar equation (9.5) for the isothermal state of the strip roll we obtain an equation for the determination of the radial stresses increment Δq_i^T in which free term has the form:

$$F_i = \frac{\Delta\delta_i^T}{r_i} - \alpha_T \left(T_{i-1} - T_i\right), \tag{9.30}$$

where $\Delta\delta(q_i^T) = \delta_i(q_i^T) - \delta_i(q_i)$; $q_i^T = q_i + \Delta q_i^T$; q_i^T, q_i are the radial stresses in the non-isothermal and isothermal strip roll.

As a result, the problem of determining the thermal and stress-strain state roll reduces to a system of equations (9.19)–(9.29) to determine the effect of the temperature field and the system of equations of the type (9.5) to find the radial stresses in the strip roll. These two systems of equations are interconnected via the dependence of the interturn thermal resistances on pressure and the free term of equation (9.5) on the temperature distribution over the thickness of the strip roll. Thermal resistance coefficient R_{T_i} and the dependences of the convergence of the contacting turns of the strip on the interturn pressure are determined on the basis of experimental data, including those presented in this book.

An algorithm for solving the problem in question, as set out in particular in [113], is as follows.

Initially, the initial temperature distributions and the parameters of the stress–strain state of the strip roll are set. Then, step by step, the problem is solved using the method of splitting by physical processes. Namely, in the first half-step in time there are interturn thermal resistances from the values of temperature and contact pressure given from the previous moment. Then, the thermal state of the roll is calculated. In the second half-step, the stress–strain state of the roll is calculated for a given temperature field, taking into account the change in conditions for contacting surfaces of adjacent turns. That is, the problem is solved in the quasi-stationary approximation.

The authors of [113] pay special attention to the fact that after removing the strip roll from the drum, when the roll cools down or when the initial temperature field is uneven, there can be stratification zones in the roll, in which the surfaces of the turns lag behind each other. Between them there are gaps. There is no mechanical and thermal contact between the turns. In this case, when calculating the above algorithm, zones are identified for the thickness of the winding of the roll, in which the interturn pressures take negative values. In these zones there is no force contact of the turns. Therefore, further, the strip roll is conditionally divided into parts in which a force contact is kept between the turns, and calculations are made separately for each such part by the same algorithm.

Thus, the proposed mathematical models allow us, both in strict and in approximate formulations, to calculate the stress–strain state of rolls of rough strips.

9.2. Numerical evaluation of the conditions for contact of strip turns in a roll

To perform numerical calculations, it is necessary to determine the dependence of the convergence of turns in a strip roll on the state of the surface of the strip and the load. Extensive experimental and theoretical studies of the contact of rough surfaces were carried out by N.B. Demkin, I.V. Kragelsky and other scientists. However, in the known works, contact problems are solved mainly with respect to mechanically treated surfaces with regular roughness and regular geometric shape. Several characteristics of the surface roughness (contour area, coefficients of the reference curve, coefficient characterizing the elastic deformation of microirregularities, etc.) are introduced into the calculation models, the determination of which is time-consuming and the accuracy of the results is not high. The surface of the-rolled sheet has an irregular microrelief, which is distinguished by a considerable heterogeneity in the width and length of the sheets (strips). The inhomogeneity of the roughness is combined with the comparatively high non-flatness (warpage, waviness) of the sheet metal (the magnitude of non-flatness often exceeds the thickness of the sheets). In addition, in production conditions, during the rolling and finishing of strips, a process lubricant is used, which also affects the characteristics of the contact. In the known studies of the convergence of thin cold-rolled sheets in a package, when they were compressed, the surface condition of sheet steel was assessed only qualitatively 'the surface was 'dry' or 'oiled'. The roughness effects, which play a decisive role here, were not considered at all.

In connection with the foregoing, the possibility of using the known methods of the calculation of the approach of the contacting surfaces of the sheets is limited and it becomes necessary to carry out the corresponding studies pertaining directly to the sheet metal. We note that in coils of pickled hot-rolled strips, the average gap is 3–5% of the thickness of the coiled strip.

The value of the interturn gaps in the outer layers of the roll can reach 16–17% [119]. Naturally, the presence of gaps and the nature of their variation in the coil thickness must be taken into account when calculating the stress–strain state of the rolls.

The authors of this book investigated the influence of the roughness on the contact compliance of the sheet surface by examining the compression of packets of samples of cold-rolled steel and copper with thicknesses of 0.6–2.3 mm. In some cases the surface of the samples was degreased, in others – smeared with a 5% emulsion of OM emulsol. The roughness of the samples varied within the range Ra = 1.0–6.5 μm.

According to the obtained data (Fig. 9.3), the absolute value of the approach of the contacting surfaces is directly proportional to the compressive force of the packet, the thickness of the sheets, the surface roughness and inversely proportional to the yield strength of the metal. The dependence of the approach on the force is essentially non-linear, and the dependence on the roughness is close to linear. This allows the influence of the roughness value on the convergence of the sheets under load to be expressed by the coefficient K_{Ra}. For the basic roughness value, at which $K_{Ra} = 1$, it is convenient to take $Ra = 1$ μm. Sheet steel with such a roughness has become most widespread. The convergence value $\Delta\delta_{Ra}$ of sheets having a roughness of any value is found by multiplying the value $\Delta\delta$ of the sheet approach, the surface roughness of which is equal to the base value $Ra = 1$ μm, by the coefficient K_{Ra}, i.e. $\Delta\delta_{Ra} = K_{Ra}\,\Delta\delta$. Based on the results of the studies, the K_{Ra} coefficient was found to depend on the roughness value Ra, μm: $K_{Ra} = 0.91 + 0.09\,Ra$. The advantage of this dependence is that in it the influence of the surface roughness of the sheets on the approach is reflected taking into account their actual non-flatness (warpage, waviness), taking into account the alignment of the sheets under loading.

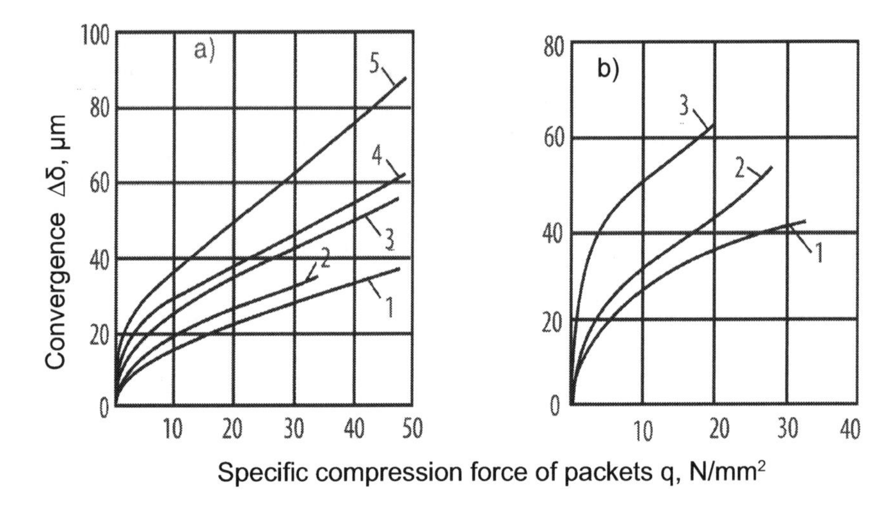

Fig. 9.3. Dependence of convergence of the contacting surfaces of sheets on compression force of packets, plate thickness h and surface roughness Ra: a – steel, $\sigma_T = 700$ N/mm^2; 1 – $h = 0.7$ mm, $Ra = 1.4$ μm; 2 – $h = 1$ mm, $Ra = 1$ μm, lubrication with a 5% OM emulsion; 3 – $h = 1$ mm, Ra – 1 μm; 4 – $h = 1.15$ mm, $Ra = 2.4$ μm; 5 – $h = 2.3$ mm, $Ra = 2.4$ μm; b – copper, $\sigma_T = 99$ N/mm^2; 1 – $h = 1$ mm, $Ra = 1$ μm; 2 – $h = 2$ mm; $Ra = 1$ μm; 3 – $h = 2$ mm, $Ra = 6.5$ μm.

The total deformation of the package of sheets under compression is determined by two components: elastic deformation of sheet metal and a decrease in the gap between the sheets Δδ. Our experiments confirmed the well-known view that the relative magnitude of the reduction of gaps between the sheets (ratio $\Delta\delta/h$) decreases with increasing thickness of the sheets in the package (Fig. 9.4). The following approximate expressions were obtained for dependences $\Delta\delta/h = \varphi(h, q)$, shown in Fig. 9.4:

$$1)\Delta\delta / h = -0.997 + 0.759h^{-0.5} + 0.706q^{0.2} \qquad (R = 0.96);$$

$$2)\Delta\delta / h = 0.0955 - 0.0397q^{0.2} + 0.0303h^{-0.5} \qquad (R = 0.91);$$

$$3)\Delta\delta / h = (0.000955 + 0.00025q^{0.2} - 0.000482h^{-0.5})q \quad (R = 0.94);$$

where R – coefficient of multiple correlation of 0.96 with $q \geq 6$ N/mm^2 and all values of h, mm; 0.91 and 0.94 – for all values of q and $h \geq 1.15$ and $h \leq 0.6$ mm.

In the course of the experiments it was established that in the presence of the emulsion on the surface of the metal the strip rolls are more dense than the strip rolls from the strips without lubrication. The gaps between the turns and convergence decreased by about 10–15% (see Fig. 9.3). The reason for this is that coating the strip with a thin emulsion layer improves the adhesion of the turns in the

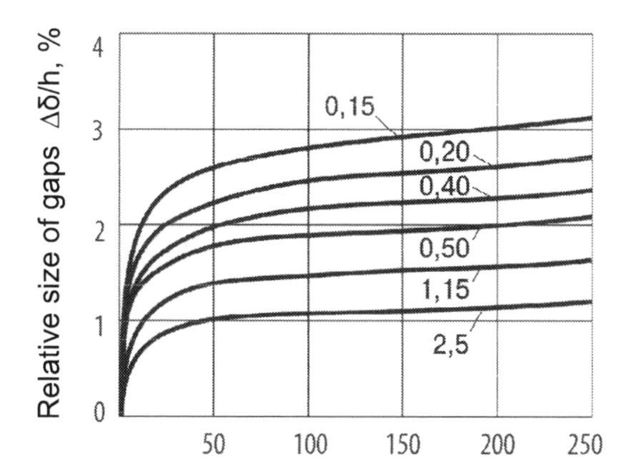

Fig. 9.4. Change in the relative size of the gaps $\Delta\delta/h$ between the sheets under compression of packets [120]. The numbers on the curves – sheet thickness, mm.

strip roll. If the strip is applied a thick layer of viscous oil, then this lubrication will affect the stress-strain state of the strip roll already as the interturn layer with specific properties.

9.3. Welding of strip loops in rolls during metal annealing

One of the main reasons for the appearance of 'fracture' defects in production of cold-rolled sheet steel is interturn sticking–welding strips in the rolls during their annealing.

The thermal resistance in the contact area of the turns, and, consequently, the speed and uniformity of the heating of the rolls during annealing, depend on the roughness of the surface of the strip. With a rough surface of the strip, the contact of adjacent turns in a strip roll (sheets in a bundle) is discrete. Reducing the touch area should facilitate the separation of coalescing surfaces. However, increasing the pressure at the contact points due to a smaller number of these points contributes to the welding process. As the roughness increases, the amount of technological lubrication remaining on the strip increases, which also affects the welding of the contacting surfaces of the sheet metal. The actual contact area of the adjacent turns and their welding during annealing depend on the roughness of the strip surface, its temperature and the contact interturn pressure, which is determined by the stress–strain state of the roll. Moreover, the temperature of the turns and the rate of their heating depend on the distribution of the interturn pressure in the rolls, which in turn is a function of the temperature and the conditions for their heating in the bell furnaces.

Studies carried out at the Cherepovets Metallurgical Combine showed that when processing working rolls of the fourth stand of a cold rolling mill with the shot with a size of 0.35–2.0 mm, an increase in the number of passes from three to four reduced the amount of rejects due to fracture and 'seizure'. The strip rolled in rolls shot blasted with the shot 0.5–1.0 mm in diameter, was welded less than in shot blasting with the shot 1.1–2.0 mm in size. As the quality of the surface roughness of the sheet steel worsened during the increase in the number of rolled products on one pair of the shot blasted rolls of the last stand of the continuous mill from 200 to 1000 tons, the defective metal portion increased from 1 to 1.5% due to 'seizure'. With an increase in the volume of rolled metal up to 1500 tons, the number of sheets affected by these defects reached 3.9%.

Similar results were obtained at the Magnitogorsk Iron and Steel Works. When shot blasting the rolls of the fourth stand of the 2500 continuous cold rolling mill with the shot with the size of 0.8–1.0 mm and 1.5–2.0 mm for a different number of passes, coiling of the strips into strip rolls with a tension of 50 N/mm^2 and annealing the cold-rolled metal at temperatures of 690±10°C, the metal sorting according to the 'fracture' defect, depending on the roughness, was: 0.72; 0.68; 0.23% with a surface roughness Ra = 0.6; Ra = 1.0; Ra = 1.7 μm, respectively (shot blasting with the 0.8–1.0 mm shot) and 0.75; 0.52; 0.34 for roughness Ra = 1.1; Ra = 1.3 and Ra = 1.7 μm (shot blasting with 1.5–2.0 mm shot). Obviously, the reason for the increase in the welding of coils in rolls and the appearance of defects by 'fracture and 'seizure', there was not so much a change in the roughness Ra as a decrease in the density of microirregularities of the surface of the strip in cases when larger shot was used for shot blasting the rolls and wear of the surface of the rolls during operation.

When examining the effect of the surface roughness of sheets after cold rolling on welding, samples of sheet steel with various surface roughness were annealed in muffles under an argon atmosphere at different contact pressures between sheets, temperatures and holding time. The degree of welding was evaluated by the shearing force (breaking of the contact) of the sheets 'welded' during annealing. The breaking force was directed along the contact plane of the samples. At a pressure of 0.1 N/mm^2 annealing was carried out at temperatures of 600, 640, 680, 720, 735, 750°C. At an annealing temperature of 720°C, the specific pressures were 0.05; 0.1; 0,15; 0.2 N/mm^2.

The dependence of the shearing force of the welded samples on the annealing temperature *T* and the surface roughness begins to appear at $t \geq 700$°C (Fig. 9.5). As the contact pressure increases, the effect of the sheet roughness on welding is more pronounced (Fig. 9.6).

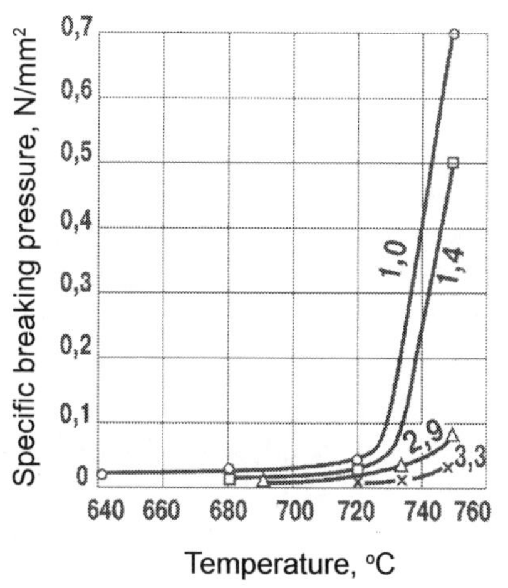

Fig. 9.5. Dependence of the specific breaking force of welded sheets on their surface roughness and annealing temperature (contact pressure 0.1 N/mm^2, lubricant – emulsion of polymerized cottonseed oil). Figures in curves – the roughness of the surface of samples *Ra*, μm.

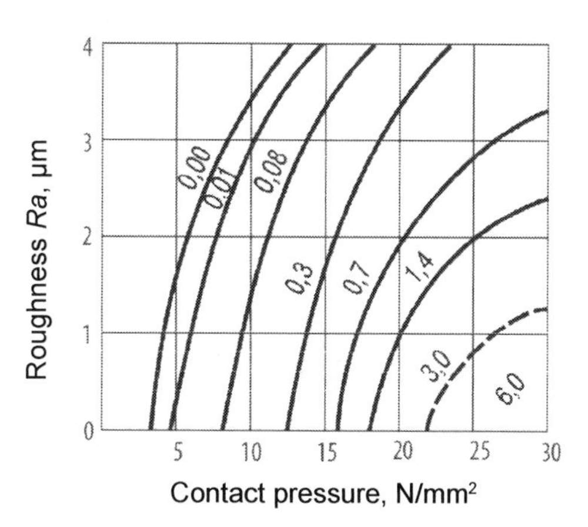

Fig. 9.6. Dependence of the breaking force of the sheets welded during annealing on the roughness of their surface and the contact pressure (annealing temperature 720°C. Figures at the curves – magnitude of breaking force, N/mm^2). Dotted line – projected results.

In the process of manufacturing research at the 'Zaporozhstal' plant unwinding a roll of the annealed strips with surface roughness *Ra* greater than 1 μm, rolled with a tension of 30 and 50 N/mm², did not show any significant clumping of the turns. The metal, rolled with a tension of 70 N/mm², showed weak coalescence of the turns; it was reinforced in rolls wound at a tension of 100 N/mm². In all cases, the sticking was greater in the inner layers of the rolls, where the contact pressures are higher. It follows that when the coils are tensioned to 50 N/mm² and the annealing temperatures are reduced to 700°C, a decrease in the roughness of the strip surface to *Ra* = 1 μm does not lead to welding of the coils in the strip rolls. At the same time, we note that the experiments at Zaporozhstal' were carried out on small strip rolls (up to 15 tons).

At the same time, special studies and experience of metallurgical enterprises have shown [43] that with an increase in the roughness of the surface of working rolls of the last stand of the cold rolling mill and rolled metal, the contamination of its surface increases. As an example, Fig. 9.7 shows the dependences of the metal damage detected by the Magnitogorsk Metallurgical Combine by the defects caused by the welding of the coils during their annealing and the contamination from the roughness of the surface of the cold-rolled black tin.

Despite the fact that the problem of the appearance of defects on the surface of the bands arising from the welding of the coils in rolls during annealing has been known for a long time, it acquired

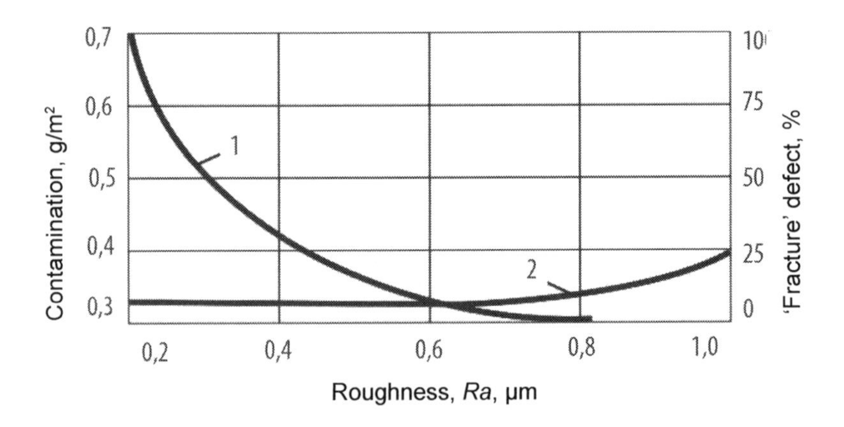

Fig. 9.7. The effect of the surface roughness of the work rolls of the fifth stand of the 1200 continuous rolling mill on the formation of kinking defects (1) and the impurity (2) of the surface of cold rolled strip 0.25 mm thick [121]

particular acuity in connection with the general transition to the technology of production of sheet steel from continuously cast slabs, in which the cold-rolled metal is annealed at elevated temperatures. In the process of heating and cooling rolls of cold-rolled strips during annealing in bell furnaces, interturn welding depends, first, on the stress–strain state of the rolls after winding and removing the coil from the drum. Secondly, from the redistribution of these stresses during the storage and transportation of coils of cold-rolled steel to the thermal department. Thirdly, changes in the temperature field in rolls during their heating and cooling during annealing and the associated variation in the interturn pressure due to the superposition of the fields of the initial stress–strain state and temperature stresses. Consequently, the modes of winding the rolls, their subsequent heating and cooling should be such that during the period of high (maximum) temperatures during annealing, the interturn pressures are minimally possible, at which welding of surfaces of adjacent turns of a strip in a roll does not take place.

The results of experimental studies of the influence of temperature and duration of annealing of pressed steel sheet specimens, compression pressure of samples, metal roughness and other factors on the degree of adhesion (welding) of contacting surfaces, which were evaluated by the force of their disengagement, are presented in [122]. Samples were annealed in an atmosphere of a nitrogen-hydrogen mixture. Figures 9.8 and 9.9 show the dependence of the separation force of samples of sheet steel for deep drawing on the pressing pressure of the contacting specimens and annealing temperature and time. The dots in the figures, enclosed in

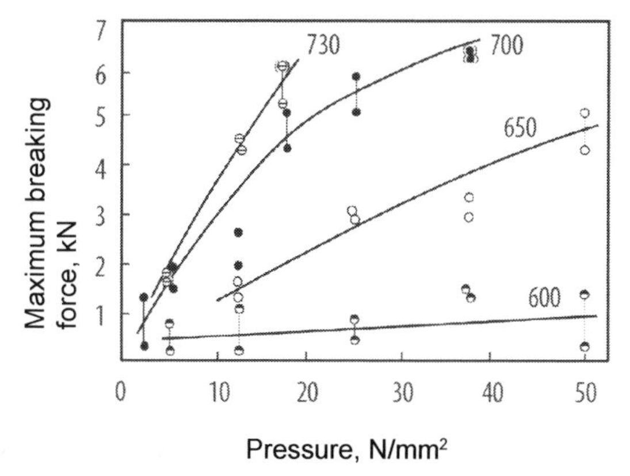

Fig. 9.8. Effect of pressure and annealing temperature (figures at the curves) on welded samples. Annealing time 90 min [122].

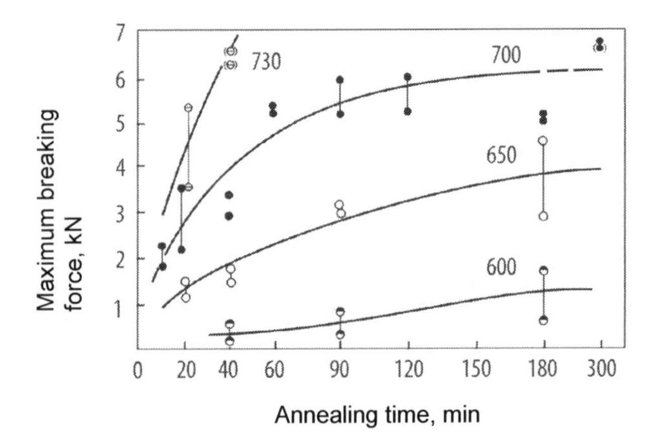

Fig. 9.9. Effect of annealing time and temperature (figures at the curves) of the weld samples. Compression force in annealing the samples 25 N/mm^2 [122].

parentheses, designate the samples that did not separate on the contacting surfaces, but the rupture of the base metal occurred. The conclusions that follow from these experimental data are as follows.

Influence of pressure on welding of contacting metal surfaces increases with increasing annealing temperature. Temperatures below 600°C do not significantly affect the degree of adhesion of the contacting surfaces, even at relatively high compression pressures. The level of radial stresses in the strip roll at high annealing temperatures affects the interturn welding more than at low temperatures. The relationship between the contact pressure and the degree of welding of the contacting surfaces is close to linear. In the opinion of the authors of Ref. [122], the process of interturn welding of strips in rolls during annealing occurs in the major part during the cooling of the metal. The most dangerous for the welding of coil strips in rolls is the combination of high interturn pressure and long duration of annealing of metal at maximum temperatures.

The experimental results under consideration confirm the conclusions on the decisive influence on the welding of coils of a strip in rolls of the roughness of the surface of metal. The dependence of the degree of welding of the contacting samples of sheet steel with different surface roughness are shown in Fig. 9.10. Samples rolled in polished rolls had a surface roughness varying along and across the rolling direction (Ra = 0.15/0.43 μm). The remaining samples had an isotropic surface roughness. Before annealing, the surfaces of all samples were cleaned with acetone from lubrication and dirt residues. As shown in Fig. 9.10, this increase in the roughness of cold-rolled

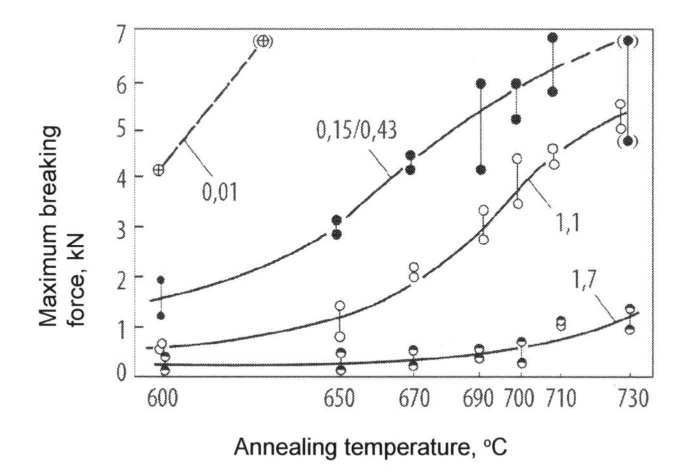

Fig. 9.10. Influence of surface roughness of the metal on welded samples annealed for 90 min. Compression pressure of samples 25 N/mm². The numbers at the curves – the value of the roughness *Ra*, μm [122]

steel significantly reduces the risk of welding of the contacting surfaces even in the range of critical temperatures (650–700°C). The authors of Ref. [122] believe that in the production practice, the roughness effect is even more pronounced than in laboratory experiments. On the whole, it can be concluded that the detailed materials of Ref. [122] confirmed the main conclusions (see Figs. 9.5, 9.6), made in our earlier [42, etc.] publications on this topic.

The examples given above are taken from the experience of specific metallurgical plants and, naturally, in a general sense, should be regarded as an illustration, an experimental confirmation of the result, conclusions, and regularities examined. For rolling mills on which strip rolls of a larger or smaller mass are processed, where the flatness of the strips is different, where other values of the tension on the coiler are used, the quantitative parameters of the effects considered will differ from those indicated in the examples. However, the nature of regularities will not change. We also note that some technological aspects related to the effect of temperature on the stress–strain state of strip rolls of cold-rolled strips, the occurrence of interturn welding of coils during annealing will be dealt with in the following sections of the book.

9.4. Experimental studies of stresses in strip rolls

Influence on radial stresses in a strip roll of the surface microgeometry and the thickness of the strip, the friction conditions between the turns, the amount of tension at coiling of the roll and other parameters of this process is of scientific and practical interest for a wide range of specialists in sheet rolling production. The results of determining this influence with the help of the mathematical model proposed above for coiling rolls of strip steel are discussed in the previous section of the book. Here we consider the experimental data obtained, which apart from the independent significance can also indicate the correspondence or inconsistency of the actual situation in the industry with the results of calculations based on the proposed models.

Materials concerning the stress–strain state of rolls in the state 'on the drum of the coiler' and after removal from the drum, will be analyzed mainly on the basis of studies performed at the Karaganda Metallurgical Combine and published in [123, 124].

For the experimental study of the stressed state of coils by the specialists of the Karaganda Metallurgical Combine, a special method was developed for measuring the interturn pressures at an arbitrary point of a strip roll at any stage of winding–unwinding operations, including after unloading the coiler drum [124]. The pressure sensor used was a resilient element made from spring steel of special configuration with resistance strain gauges attached to it. To minimize the distortion in the pressure measurement caused by the presence of a sensor between the turns, the height of its elastic element was taken as small as possible under the manufacturing conditions (2.2 mm); the design of the sensor provided maximum 'stiffness' of the load characteristic (spring deflection at maximum load did not exceed 0.15 mm); The elastic element was loaded according to the beam scheme on free supports with concentrated load application in the centre. At the point of measurement, a so-called false turn with a window for mounting the sensor was rolled into a roll. The thickness of the h_f (Fig. 9.11) was less than the height of the unloaded sensor by 0.20 mm.

Calibration of the sensor on the press at a maximum force of 24.5 kN showed the linearity of the load characteristic. The sensor connection for the purpose of equilibration was carried out using a bridge-based non-amplifying circuit. The signal from the sensor was taken with a string collector. The results of the measurements were recorded with an automatic recording instrument.

To obtain a pressure averaged over the width of the strip, four strain gauges, glued to the elastic element at regular intervals and connected in series, were used; The length of the elastic element was greater than the width of the strip (125 mm). The sensor was placed in the window of the 'false turn' and together with it rolled

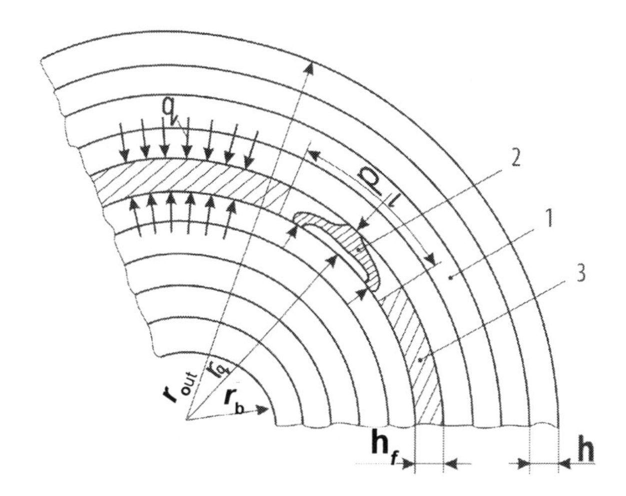

Fig. 9.11. Measuring interturn pressure in the strip roll: 1 – strip roll thickness h; 2 – sensor; 3 – 'false turn'.

up into a roll. The measurement of the interturn pressure in a strip roll of of thickness h, the inner radius r_b, and the outer radius r_{out} is shown in Fig. 9.11. When testing the technique after installing the sensor into a roll and winding without tension 5–10 turns of the strip the sensor was re-calibrated by a portable press. To exclude the 'shielding' effect of the 'false turn', the length of the window l was selected experimentally in such a way that a minimal discrepancy of the measured values is provided for initial and repeated calibration

Interturn pressure is calculated by the formula:

$$q = P / l \cdot B,$$

where P – measured load; B – width of the strip.

The measuring instrument scale was graduated in units of N/mm^2.

The pressure of the first turn on the coiler drum q_b can be calculate virtually with no error by the formula [119]:

$$q_b = \sigma_0 h / r_b,$$

where σ_0 – coiling tension; r_b – radius of the coiler drum.

Comparison of the pressure values calculated with the help of the sensor and calculated by this formula for different thicknesses and specific tensions of the strips showed that the difference between them does not exceed ±10%. Experiments indicated a high reliability of the sensor and good reproducibility of the results.

During the experiments, the effect on the level of interturn pressures in strip rolls of coiling tension, thickness, roughness of

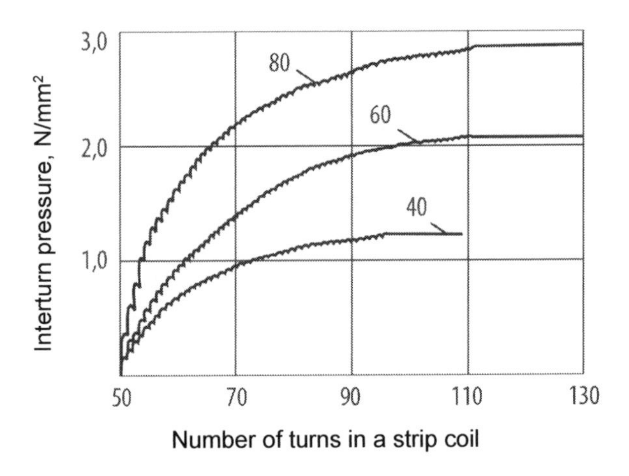

Fig. 9.12. Changing the pressure in the interturn coil depending on the number of wound turns (material – steel 08kp size 0.5 × 120 mm). The numbers on the curves – the tension coiling, N/mm^2.

the surface and non-flatness of the bands was investigated. The experiments were carried out on a 200 laboratory mill equipped with coilers and an automatic coiling tension control system. The strip plate was a roll of cold-rolled ribbon with a thickness of 0.15–1.0 mm and a width of 100–120 mm.

The diagram of the change in pressure on the 50th turn of the roll, depending on the number of coils wound over it for cases of looking through the strip with different tension (Figure 9.12), shows that the interturn pressure increases in a stepwise fashion after winding each next turn. The maximum increase in load occurs after winding the first turn; then the increment gradually decreases and after winding a certain number of turns, the pressure practically no longer increases. The nature of the variation in the interturn pressure is similar to the change in the pressure of the coil on the coiler drum during the winding of the strip [125]. The maximum value of the interturn pressure and the intensity of its change during winding increase with increasing tension (Fig. 9.12). In experiments with a relatively small value of tension (40–100 N/mm^2) and a comparatively small number of turns, the interturn pressure as a function coiling tension increased almost linearly.

As is known, in industrial mills the strip coiling is tested by tensioning by no more than 70–80 N/mm^2. Usually the tension is 2–2.5 times less. Therefore, for practical calculations, it is possible to use a linear approximation of the dependence of pressure on the coiling tension.

To quantify the effect of these parameters, the pressure was measured between turns in a strip roll located in the immediate vicinity of the coiler drum, when winding strips of the same thickness rolled on polished and shot blasted rolls. Coiling was carried out with feed and without feed of a lubricant on the strip (30% emulsion based on emulsol OM).

According to the data (Fig. 9.13) the surface roughness significantly affects the force between the turns. The effect of the presence of the lubricant on the strip is markedly manifested only for a comparatively small strip surface roughness ($Ra < 0.5$ μm). Experimental results demonstrate again the need to consider the roughness when selecting coiling modes of the rolls of cold rolled strips. The rolls in the last stands of continuous wide-strip cold rolling mills are usually shot blasted. Roughness decreases during rolling. Therefore, to ensure the uniform stress state in the strip rolls one should reduce the level of coiling tension with decreasing roughness rolled strips or use other ways to reduce the pressure to

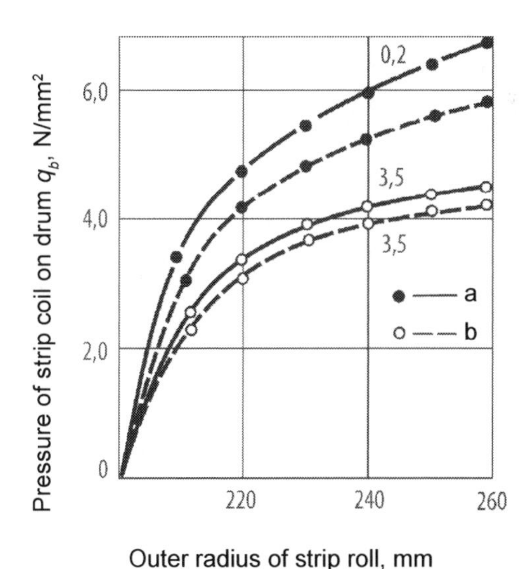

Fig. 9.13. Dependence of interturn pressure on the radius of the strip roll when coiling 0.65 mm thick strips with different surface roughness *Ra*, μm (figures at the curves): a – winding with strip lubrication; b – without lubrication (coiling tension – 100 N/mm^2)

exclude interturn welding of the strip in coils during their subsequent annealing. With the same winding thickness the interturn force and pressure of the strip roll on the drum of the coiler increases with increasing thickness of the strip. These conclusions made for the material confirm the findings drawn on the basis of calculations using a theoretical model of the process of winding strip rolls.

The greatest difficulties in practice are experienced when ensure the optimal stress state of the strip rolls in coiling thin strips. Therefore, in [123] the authors experimentally investigated the effect of the thickness on the contact pressure in coiling mostly thin strips (0.15–0.8 mm). In the studied range the dependence of the interturn pressure on the strip thickness is close to linear (Fig. 9.14).

In a production environment on most mills set the value of specific tension unchanged for the entire assortment of rolled strips, whereby the rolls of thin strips have a greater tendency to axial shift turns unwinding. At the temper mill in 1700 the Karaganda Metallurgical Plant during the studies [123] incidence education telescoping rolls value greater than 20 mm depending the thickness of the strips was as follows:

The thickness of the strips, mm	0.5–0.7	0.8–1.0	1.1–1.5	1.6–5.0
Number of cases,%	65	61	37	19

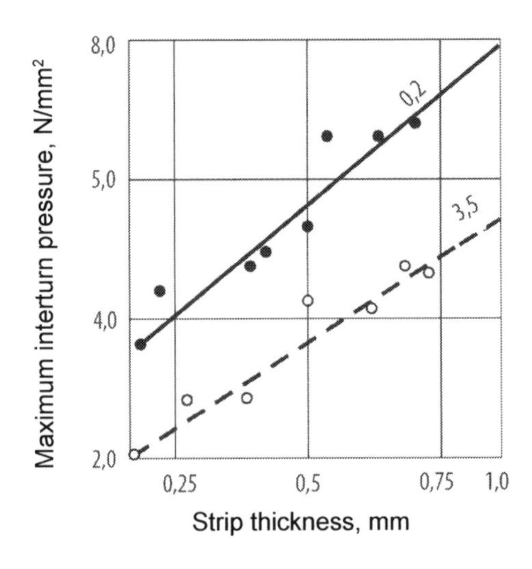

Fig. 9.14. Dependence of the maximum interturn pressure in strip roll on the thickness of coiled strips: coiling tension – 100 N/mm^2; numbers on the curves – strip roughness *Ra*, μm.

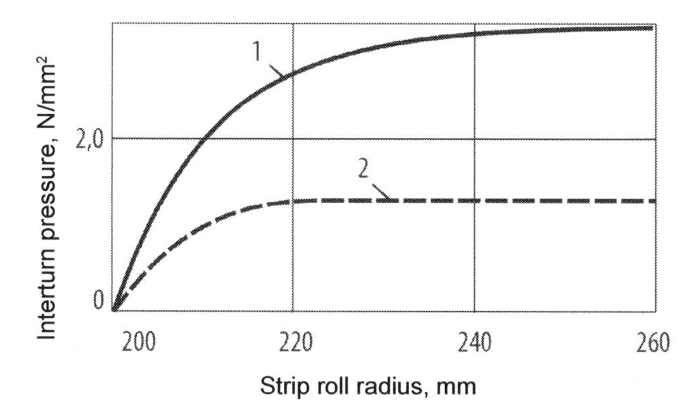

Fig. 9.15. Change interturn pressures during coiling strip 0.3 mm thick with tension 100 N/mm^2 depending on the radius of the roll: **1** – flat strip; **2** – wavy

Experiments have shown that the profile and shape of the strip exert a decisive influence on the level of interturn pressures. Large gaps form between the turns of a non-planar strip in the roll, the reduction of which consumes considerable work when winding subsequent turns. Therefore, when winding the strip even with a slight distortion of the shape, the interturn pressure increases significantly more slowly. The overall level and the maximum value of the interturn pressures in this case are less than when winding under the same conditions with respect to a flat strip (Fig. 9.15). However, in the case of a non-planar strip in rolls, as a rule, in individual sections, because of the corrugation and buckling of the metal, the contact pressure is several times greater than the average stress level. These areas are the zones of welding the turns during annealing.

The uneven distribution of the interturn pressures across the width of the strip can be associated with local disturbances in its profile and shape, for example, due to the distortion of the cross-sectional profile in the edge zones of the rolls. In these areas, the formation of kinks is possible due to local coalescence of the turns during annealing. When winding strips with a wedge-shaped cross-section, the uneven distribution of pressure along the width of the strip can reach such a degree that it will cause the appearance of a 'bird' on the side of the thickened edge. This question, due to its importance, will be considered below in further detail.

The pattern of distribution of interturn pressures along the section of the strip roll before and after unloading the coiler drum, and also after its repeated loading (decompression), obtained by direct

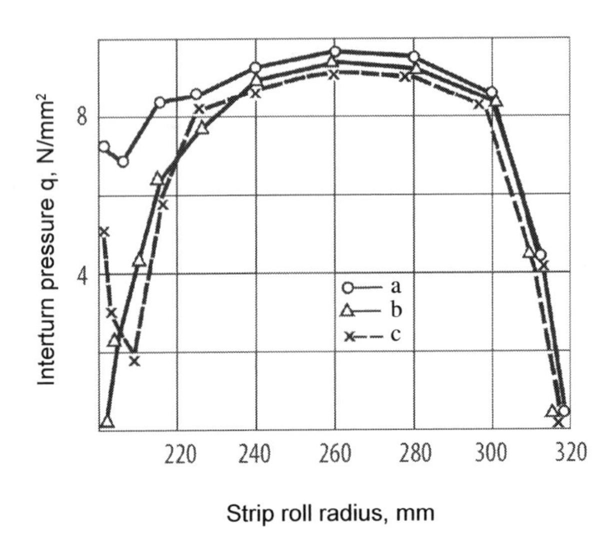

Fig. 9.16. Interturn pressure distribution over the cross section at the coiling roll band 0.38 mm with a tension of 200 N/mm². The surface roughness $Ra = 0.8$ μm: **a** – after winding; **b** – after compression of the drum; **c** – after repeated it uncompressing

measurement, is shown in Fig. 9.16. In the experiments carried out, the maximum interturn pressure was recorded in the average thickness of the coil winding. The pressure between the coils located in the immediate vicinity of the coiler drum (practically the pressure of the coil on the coiler drum) turned out to be less than the maximum by 10–30%. This character of the distribution of pressures in a strip roll is apparently due to a change in the influence of the load on the interturn clearances with repeated loading and increased compliance of the mechanisms of the coiler drum of the 200 mill.

The maximum pressure in the roll after unloading the drum slightly changed. Reducing interturn pressure was observed only in a small part of the inner turns (20–30% of the total thickness of winding). If the drum with the strip roll located on it is decompressed the original stress state produced in the winding process, is completely restored. In certain parts of the inner turns the contact pressures are reduced even more. This is obviously due to the fact that the zones of the inner turns, especially in large rolls, have low axial resistance with subsequent unwinding, in particular for skin pass rolling. It was established that in the case of winding rough strips (Ra = 1.5–3.0 μm) the initial stress state after re-loading the drum recovering better.

This circumstance can be used in practice to prevent the formation of 'telescopic' areas in the strip rolls during unwinding.

9.5. Effect of process parameters on the winder stress–strain state of rolls

To determine the winding conditions of strips that are optimal from the standpoint of the requirements of sheet steel production and to ensure the high quality of this product, it is first of all necessary to reveal the main regularities of the influence of various factors on the stress–strain state of the strip rolls. To this end, we consider the results of a parametric study of stresses in rolls.

In Fig. 9.17 the values of the roll pressure on the coiler drum, calculated using the developed mathematical model, are compared with the experimental results of [120, 125]. According to the graphs in Fig. 9.17, when calculating the stresses in strip rolls without taking into account the change in the size of the gaps, the pressure

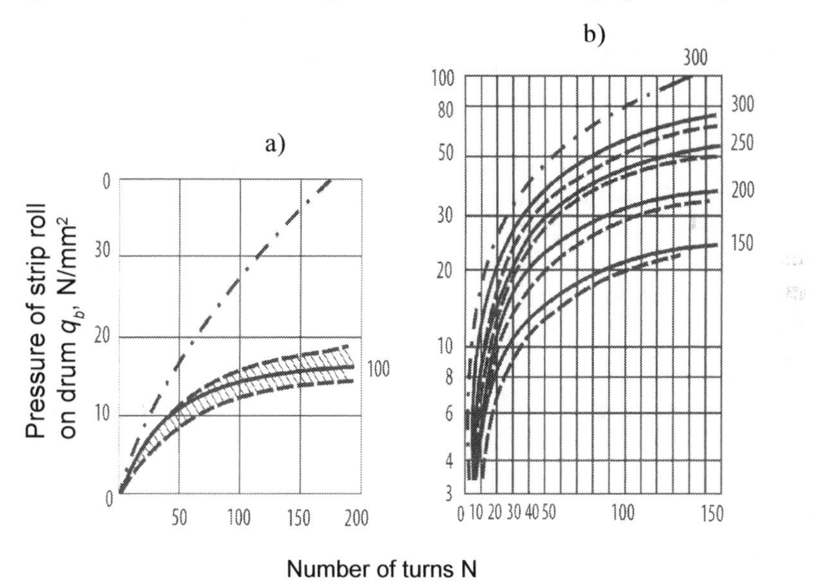

Fig. 9.17. Dependence on the pressure of strip rolls q_b on the coiler drum on the number of turns N at a constant winding tension. The numbers on the curves – tension σ_0, N/mm^2: a – according to [120] (the radius of the drum 250 mm, λ = 0.7 strip material – steel 08 kp, strip thickness 1 mm, width 145 mm, shaded zone – experimental data for $\sigma_0 \approx$ 100...110 N/mm^2; b – the data of [125] (drum radius 77 mm, λ = 0.74, strip material – transformer steel, strip thickness 0.33 mm, width 22.4 mm. Dashed lines – experimental dependence; solid lines – calculated to reflect changes in the gaps; dot-dash – calculated without considering this factor.

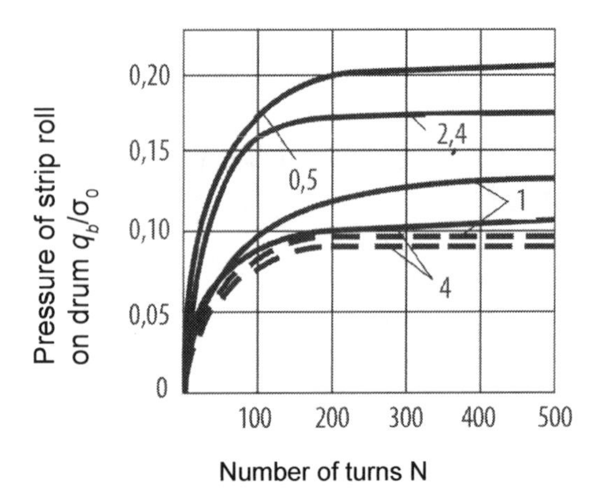

Fig. 9.18. Dependence on the pressure of strip roll on the coiler winder drum q_b/σ_0 (drum radius 380 mm; $\lambda = 0.7$) on the number of turns N when winding 2 mm thick strips (top two curves) and a thickness of 0.75 mm (four lower curves) with constant tension $\sigma_0 = 100$ N/mm^2 (solid lines) and $\sigma_0 = 50$ N/mm^2 (dashed). The numbers on the curves are the roughness value *Ra*, μm.

on the drum with increasing number of wound turns continuously increases. The results are overestimated, the more the number of turns of the strip on the roll. Calculations performed with allowance for the real conditions of the contact interaction of the strip surfaces in a strip roll (with allowance for gaps) show that the pressure value rises to a certain (critical) value, after which it practically does not change (Fig. 9.18). This is due to the fact that the force from the action of subsequent (after a critical number) turns is almost completely spent on changing the size of the gaps between the turns. The calculated pressure dependences on the coil drum on the number of wound strip layers are in full agreement with the experimental data of other works, for example [74].

After removing the strip roll from the coiler drum, the interturn pressure decreases by more than an order of magnitude compared to the radial pressure in the strip roll on the drum. With an increase in the number of turns, the level of the interturn pressure in the roll removed from the coiler continuously increases.

The distribution of radial and tangential stresses in a roll during winding of strips with constant tension (σ_0 = const) under conditions corresponding to rolling on a 1680 continuous four-stand mill of the Zaporozhstal' Combine is shown in Fig. 9.19. A characteristic feature of the stressed state of rolls in the 'on the drum' position is that, due to a loose fitting of the turn, tangential (circumferential)

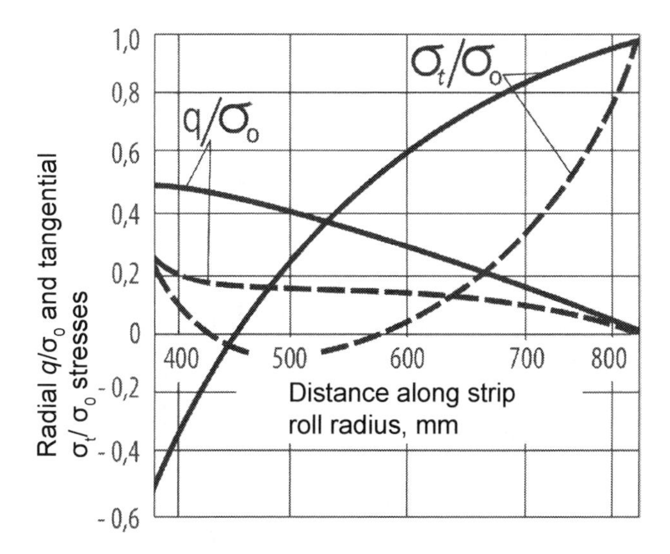

Fig. 9.19. Distribution of radial q/σ_0 and tangential σ_t/σ_0 stresses in the strip roll on the coiler drum with ideal fitting of the turns (solid) and the gaps between the turns (dashed lines) (σ_0 = 100 N/mm^2, λ = 0.7, *Ra* = 1 μm).

compressive stresses $\sigma_t/\sigma_0 < 0$ are experienced in the middle of the winding thickness, while the turns in the inner and outer layers are in the stretched state $\sigma_t/\sigma_0 > 0$. This is due to the fact that the turns in the middle of the winding thickness are able to move radially by an amount greater than the deformation of the drum due to the reduction of the gaps. With an ideal contact of the turns (without gaps), the compressive tangential stresses appear in the layers adjacent to the coiler drum.

Influence of surface roughness. With an increase in the roughness of the surface of cold-rolled strips under constant winding conditions, both the pressure of the strip roll on the drum and the level of the interturn pressure in the rolls after removal from the coiler are reduced (Figs. 9.18, 9.20). According to the graphs in Fig. 9.20, with an increase in the roughness of the surface of the strips, the maximum interturn pressure in the strip rolls is shifted towards the outer radius.

With increasing tension during winding, the effect of the roughness of the surface of the coiled strips on the pressure of the strip rolls on the coil drum and the level of the interturn pressure in the coil increases after removal from the coiler (Fig. 9.21). In the example shown in Fig. 9.21 *a*, a change in the tension value from 50 to 100 N/mm^2 enhances the roughness effect by about 5%. Consequently,

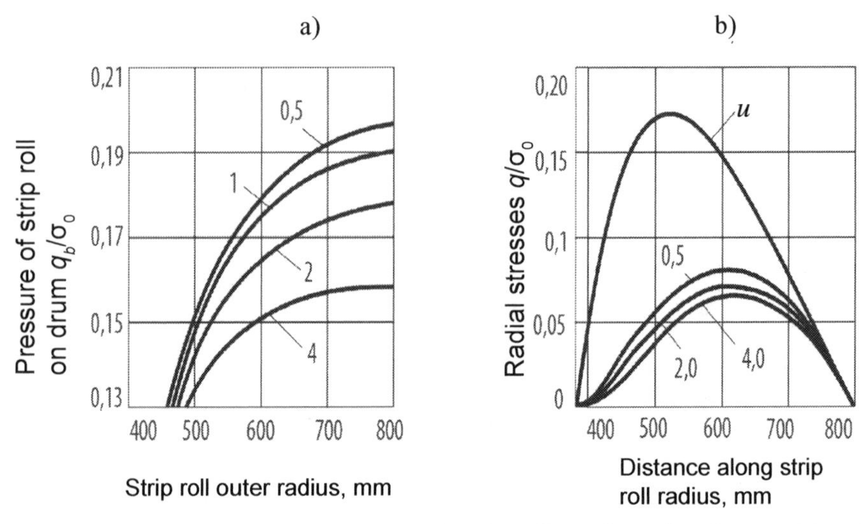

Fig. 9.20. Dependence on the pressure of strip roll on the coiler drum q_b (a) and interturn pressure after removal from the coiler drum winder q^c/σ_0 (b) on the magnitude of surface roughness of 2 mm thick strips (σ_0 = 100 N/mm^2, λ = 0.7, the curve u – in perfect contact of turns). The numbers on the curves – the value of the roughness Ra, μm.

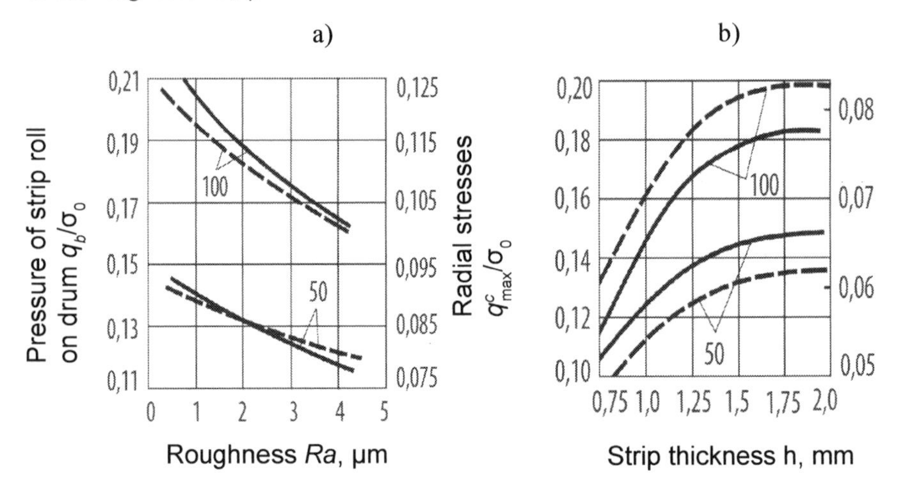

Fig. 9.21. Depending on the pressure of strip rolls on the coiler drum q_b/σ_0 (dashed line) and the interturn maximum pressure in the strip roll after removal from the coiler q^c_{max}/σ_0 (solid) on surface roughness (a) and the thickness (b) of strips for the same inner r_b and outer r_{out} radii of the strip roll. The numbers on the curves – the tension during winding σ_0, N/mm^2: a – rolling conditions as in the 1700 five-stand mill of the KarMK (h = 2 mm; r_b = 300 mm, r_{out} = 1350 mm; λ = 0.7); turns in contact without gaps q_b/σ_0 = 0.935; q^c_{max} = 0.585; b – the conditions of the 1680 four-stand mill at the Zaporozhstal' plant (r_b = 380 mm; r_{out} = 800 mm; λ = 0.7 mm Ra = 1 μm); turns in contact without gaps q_b/σ_0 = 0.522; q^c_{max} = 0.179.

the issue of choosing the necessary roughness of the strip surface is particularly acute in those plants where rolls are wound at relatively high strip tension.

The role of roughness also increases with increasing mass of strip rolls. In this case, the mutual influence of the roughness and the weight of the strip rolls is more pronounced on the interturn pressure in the strip rolls after removal from the coiler than at the pressure of the coils on the coiler drum. For example, an increase in the roughness of the strip surface from $Ra = 0.5$ to $Ra = 4.0$ μm reduces the maximum value of the interturn pressure in small strip rolls by approximately 14.2% (Fig. 9.20) and by 19.8% in large rolls (Fig. 9.21 *a*). The same roughness variation reduces the pressure on the drum by approximately the same amount – by 18.6 and 19.6%, respectively. It also confirms that if for strip rolls of mass ~15 tons the presence of gaps and roughness leads to a decrease in the interturn pressure by approximately 2.3 times as compared with the case of ideal contact of the strip surfaces, for the strip rolls weighing 45 tons – 7 times.

lubricating the strips with an emulsion increases the pressure of the strip rolls on the drum and the interturn pressure in the strip rolls after removal from the coiler, both at a relatively smooth ($Ra = 0.5$ μm) and rough surface ($Ra = 4.0$ μm). For conditions, for example, at the 1700 mill in the Karaganda Metallurgical Combine (Fig. 9.21 *a*) at $\sigma_0 = 50$ N/mm^2, the increase in q_b and $q^c_{\max}$ due to the oiling of the strips by the emulsion is 5.5–6.5%. Larger values refer to the roughness $Ra = 0.5$ μm.

Figure 9.22 shows the calculations of the stress–strain state of rolls of cold-rolled strips, applied to the conditions of the 2030 cold rolling mill of the Novolipetsk Metallurgical Combine (inner and outer roll diameters of 650 and 1950 mm, drum thickness coefficient $\lambda = 0.7$, strip thickness $h = 0.5$ mm, the elastic moduli of the material of the strip and the coiler drum are the same and equal to $E = 210000$ N/mm^2, the tension of the strip during winding of the strip roll does not change and is equal to $\sigma_0 = 25$ N/mm^2) [113]. According to the presented results, with an absolutely smooth surface of the strip and tight contact with the coils, the pressure on the coiler drum and the radial stresses on the coil after removal from the coiler are almost ten times higher than those for the winding of the strip with a rough $Ra = 1.5$ μm surface. The maximum value of the interturn pressures, provided that the turns are perfectly in contact with the coil after

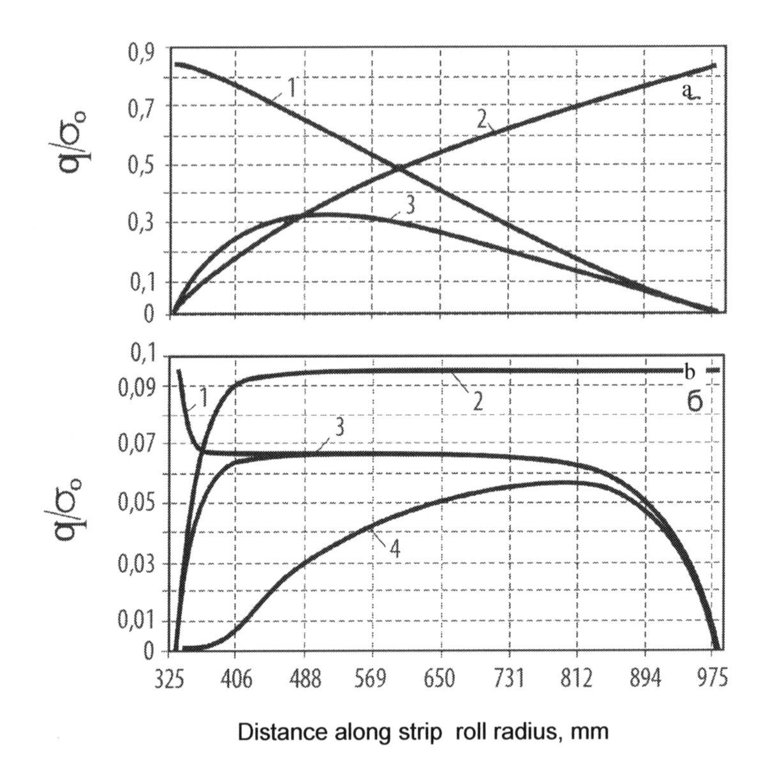

Fig. 9.22. Distribution of interturn pressure in the roll diameter for a 0.5 mm thick strip with a completely smooth (a) and rough $-Ra = 1.5$ μm (b) surface. The numbers on the curves: 1 – strip roll on the drum; 2 – change in pressure on the drum when winding strip roll; 3 – strip roll removed from the drum to reflect changes of the interturn gaps when removed from the drum; 4 – strip roll removed from the drum without changes in the interturn gaps when removed from the drum [113].

being removed from the coiler, is approximately three times less than the pressure on the coiler drum.

These results are obtained for the case when in calculations the dependence of the value of the interturn gaps on the compressive force of the coils when removing the roll from the coiler drum remains the same (unchanged) as when winding the strip roll. That is, it is believed that the surface microrelief remains unchanged after the first loading and retains its original elastic properties. Under actual conditions, during the first loading, a partial crushing of the microirregularities of the contacting surfaces can occur, and under repeated loading or unloading the dependence of the interstitial gaps on the contact pressure will change. This effect was taken into account by the authors of Ref. [113]. Curve 4 in Fig. 9.22 *b* shows the distribution of interturn pressures obtained on the assumption that when winding the strip on the drum plastic irreversible crushing of the microirregularities occurs and when the roll is removed from the drum, unloading of the strip roll occurs without restoring the properties of the initial microrelief of the contacting surfaces of the turns. In this case, the interturn pressures in the strip roll after being removed from the coiler drum are much less than in the case when the elastic properties of the contacting surfaces are completely restored. The authors of Ref. [113] assumed in the first approximation that when the loading is repeated, the gaps decrease twofold in comparison with the primary loading.

The above results confirm the conclusion that the roughness of the surface of cold-rolled strips leads to a significant decrease in the magnitude and uneven distribution of the interturn pressures in a strip roll. The higher the roughness of the strip, the greater the possible elastoplastic deformation of the microirregularities and, consequently, the lower the level of interturn pressures in the strip roll. This effect is especially pronounced when the thickness of the bands is reduced and the number of turns in the roll is increased, i.e. when the area of contacting the rough surfaces of the strip increases. In general, the effect of roughness manifests itself the stronger, the greater the absolute value of the interturn pressure in rolls.

We note that the roughness of the surface of the strip substantially affects the stability of the rolls to the loss of their shape. With an increase in the roughness of the metal surface, the coefficient of friction between the turns of the strip on the roll increases. This prevents slippage, the shift of one turn relative to the other, which occurs during the sagging of the roll under the action of its own mass. The higher the level of interturn pressures in a strip roll, the closer the contact of surfaces of adjacent turns and the greater the coefficient of friction. As a result, the danger of the loss of stability of strip rolls due to the relative slippage of the loops is weakened. This circumstance should also be taken into account when it is necessary to prevent slippage of the turns under the influence of dynamic loads caused by intensive acceleration or slowing down of rotation of the strip rolls during winding or unwinding. In more detail, the causes, the mechanism for the appearance of various types of loss of stability of rolls of cold-rolled strips and methods for preventing such defects are considered in the publications [123, 126].

Effect of the thickness of coiled strips. With an increase in the thickness of the coiled strips, the pressure of the strip roll on the coiler drum increases (Fig. 9.21 *b*). The interturn pressure in the strip

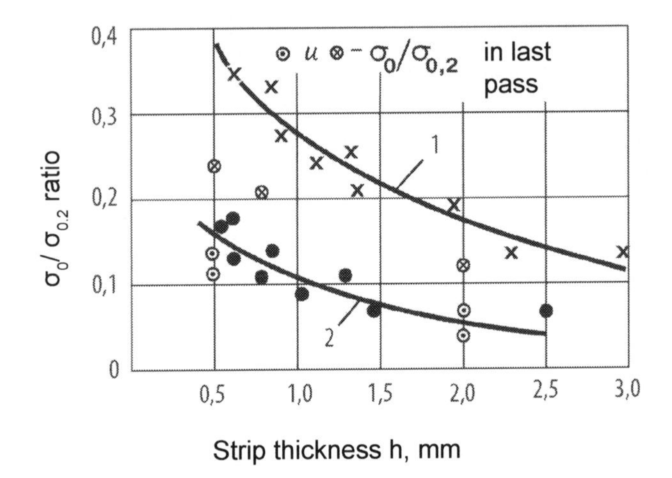

Fig. 9.23. Dependence of tension during winding of rolls on strip thickness rolled on reversing mills 1680 (1) and 1200 (2) plant Zaporozhstal (strip width of 1000 mm, Ra = 0.5–1.5 micron)

rolls after removal from the drum also increases with increasing h, if the gap between turns remains unchanged.

In the production practice, the tension when winding cold-rolled strips, as a rule, is reduced with increasing thickness of the metal. For example, in reversible cold rolling mills, when winding the strips after intermediate passages, in most cases the tension allowed by the motor is applied. In the last pass, a lower tension is set (Fig. 9.23), but its value should ensure a stable work of the system of maintaining the constancy of tension with increasing diameter of the strip roll.

The role of roughness is enhanced by decreasing the thickness of the strips transformed to coils. In particular (see Fig. 9.18), with a strip thickness of 2 mm, an increase in their roughness from Ra = 1 to Ra = 4 μm leads to a decrease in pressure on the drum by about 15.8%, and at a thickness of 0.75 mm, the same growth of the roughness reduces the pressure on the drum by as much as 24.6%. The maximum value of the pressure between turns of the strip roll after removal from the coiler is reduced by 11 (see Fig. 9.20) and 16.5%, respectively.

The reason for the noted regularities is that the decrease in the thickness of the strips reduces the density of winding the strip rolls, since the value of the contact approach of the turns under the load depends on their thickness (see Figs. 9.3 and 9.4).

Effect of the size and deformability of the coiler drum. At a constant value of the tension of the strips, the pressure of the strip roll on the drum increases with decreasing drum diameter [108]. The pressure of the roll on the drum is greater the lower the compliance of the drum. Since the compliance of the drum is determined, on the one hand, by its design and, on the other hand, by the elastic properties of the material from which it is made, the stresses in the strip rolls depend on the ratio of the elastic moduli of the strip and drum materials. With a decrease in this ratio, the pressure of the strip roll on the drum increases. Consequently, coiler drums made of cast iron will experience less pressure than steel drums. Therefore, when winding aluminum or copper strips with the same tension, the pressure of the strip roll on the drum will be greater than when winding steel strips [108].

According to the graphs in Figs. 9.17, 9.18, 9.21, the pressure of the strip roll on the coiler drum is non-linearly dependent on the value of the strip tension at the time of winding. The main reason, first, is that the convergence of the turns in the strip roll is essentially non-linearly dependent on the load (see Fig. 9.3, 9.4). Secondly, the deformation of the coiler drum is non-linearly dependent on the pressure of the wound strip [107, 108]. The deformability of the coiler drum is characterized by the coefficient of its thickness λ. According to the results of studies of the power parameters of coiling devices made at VNIImetmash (Scientific Research Institute of Metallurgical Engineering), the value of the coefficient of thickness λ of both slotted and gapless drums should be chosen depending on the pressure of the strip roll on the drum. At pressures q_b = 5–10 N/mm^2, λ = 0.85–0.9, with pressure q_b increasing up to 20 N/mm^2 λ decreases to 0.55–0.7, and then remains unchanged. In our calculations, these experimental data were used in the form of the following approximating relationships: $\lambda = 0.85$ for $q_b \leq 10$ N/mm^2; $\lambda = 1-0.015\ q_b$ at $10 \leq q_b \leq 20$ N/mm^2; $\lambda = 0.7$ for $q_b \geq$ 20 N/mm^2. The proposed algorithm for calculating stresses in strip rolls allows us to refine the values of λ at each step (after winding each turn). Since the intensity of pressure growth of the strip roll on the drum depends on the tension of the strip σ_0 and, hence, the chosen value of λ depends on the σ_0 value the stresses in the strip roll are non-linearly related to the tension σ_0. According to the graphs shown in Fig. 9.24, at relatively high tension (σ_0 = 200 N/mm^2), and as a result, at the sharp increase in the pressure on the drum, the coefficient λ quickly reaches its limit value of 0.7, so that the curve

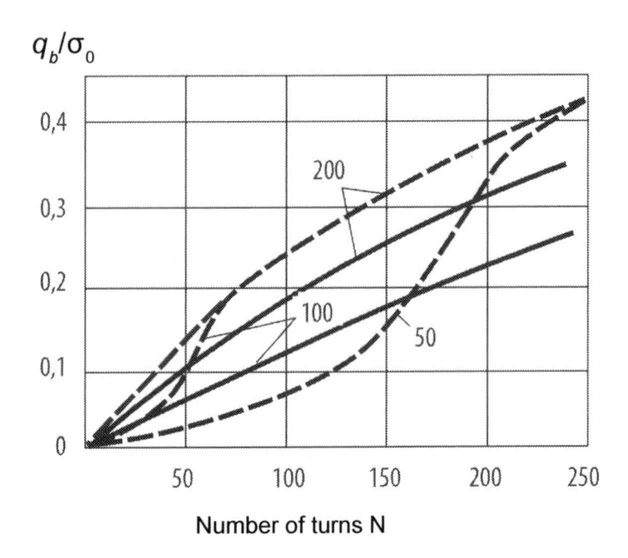

Fig. 9.24. Dependence of the pressure of the strip roll on the drum q_b/σ_0 with a radius of 250 mm on tension during winding a 1 mm strip and with different number of turns N in the strip roll. The numbers on the curves – the tension σ_0, N/mm². Solid lines – the value of sliding of segments of the drum u_{b_0} = 0.077 mm [120] and λ = 0.7, dashed curve – value λ was specified at each step of the iterative cycle according to the relation $\lambda = 1–0.015\ q_b$.

$q_b = \varphi(N)$ always has curvature (convexity) of the same sign. If the pressure increases slowly, as at σ_0 = 50 N/mm², then the function $q_b = \varphi(N)$ changes its sign of curvature as N increases.

The effect of λ on the coil pressure on the coiler drum is shown in Fig. 9.25. In this example, the value of λ for each case was assumed to be different, but unchanged in the process of winding one strip roll. According to the graphs, with increasing λ, the pressure of the strip roll on the drum decreases. Moreover, when winding relatively smooth strips (Ra = 0.5 μm), the dependence $q_b = \varphi(\lambda)$ is manifested stronger.

The stressed state of the rolls after being removed from the winder drum also depends on the compliance of the drum. If during the winding of the roll there is a radial deformation of the drum, even when the roll is on the coiler, the inner coils move in the radial direction. As a result, the possibility of free radial displacement of the inner turns after removal from the drum will be largely exhausted. The pressure of the outer turns on the inner ones remains unchanged, since regardless of the radial deformation of the drum and the inner turns of the roll in the upper turn, at any time of winding the circumferential tension will be equal to the tension of the wound strip. As a result, with the unchanged tension of the strips, the level of the interturn pressure in the rolls after removal from the coiler will be the greater the greater the deformation the drum received during the winding process. Thus, for conditions corresponding to rolling in the 1700 KarMK mill (Fig. 9.25), the maximum value of the interturn pressure in the roll after removal from the coiler q^c_{max}/σ_0 is 0.091/ 0.0782; 0.092/0.0786; 0.093/0.0793, respectively, for λ equal to 0.5, 0.7 and 0.85 (to the left of the slash – the roughness value of the strips Ra = 0.5 μm, to the right Ra = 4 μm). The change in the values of q^c_{max} as a function of λ is small, but qualitatively it manifests itself clearly.

In known works, for example [74, 120], the deformation of the drum is considered as a function of the specific pressure of the strip roll on the drum. In a finer analysis, it is also necessary to take into account that in the initial period of winding the amount of slipping of the segments and consequently the radial deformation of the drum is determined by the total pressure on the drum. And since the total pressure on the drum is directly proportional to the width of the strip, the interturn pressure in strip rolls wound at the same specific tension of the strips will be the greater the larger the width of the strips. This conclusion is of fundamental importance.

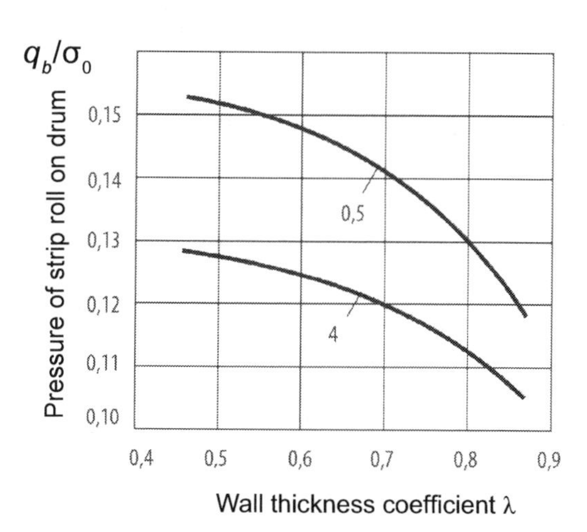

Fig. 9.25. Dependence of the pressure of strip rolls on the drum q_b/σ_0 on the wall thickness coefficient of the drum λ and the surface roughness of wound strips: the radius of the drum 300 mm, h = 2 mm; N = 525, σ_0 = 50 N/mm². The numbers on the curves – the value of the roughness Ra, μm.

The interturn pressure in a strip roll removed from the drum is related to the diameter of the drum by an inversely proportional dependence. Therefore, strip rolls of the same mass and rolled with the same tension but having a smaller inner diameter are more prone to welding of the turns during annealing the formation of 'bird' and 'fracture'. For more details see [108, 110]. Since the interturn pressure increases with decreasing inner radius of the strip roll, the roughness of the strip surface is more pronounced with smaller dimensions of the coiler drum (strip roll).

Influence of the coiling temperature of strip rolls during cold rolling. Theoretical analysis showed that the change in the temperature of the strips during their cold rolling and coiling significantly affects the stress–strain state of the rolls. It is known that the temperature of the rolled strips depends on many parameters of the rolling process – the total degree of deformation and the distribution of the reduction in the stands of the continuous mill, the level of tension, the friction conditions, the rolling speed, the quantity and properties of the lubricating fluid, etc. The parameters most influencing the temperature of rolled and coiled rolls are the rolling speed and the amount of lubricating–cooling fluid (LCF) supplied to the rolls and strip during rolling. These process parameters are relatively easy to control over relatively wide ranges, which makes it possible to consider the possibility and expediency of using them to influence the temperature of coiled strips in strip rolls in order to optimize their stress–strain state. However, this issue has not yet been investigated. Very little experimental data is available. Therefore, special attention should be paid to the results of investigations carried out in the industrial conditions of the 2030 continuous wire-strip mill of the Novolipetsk Metallurgical Combine [114, 115]. We will analyze them from the cognitive and practical points of view.

When carrying out of these experiments on the 2030 mill the rolling speed bands of steel 08Yu with a thicknesses of 0.7–0.9 mm from 3.0–3.5 mm strip plate ranged from 5.0 to 19.2 m/s. The total volume of LCF supplied to the mill was varied from 680 to 1159 m^3/h. In the 1st mill the coolant flow rate was 32, in the 2nd from 157 to 290, in the 3rd 193–330, 4th 66–312 ad the 5th mill 132–195 m^3. The strip roll temperature was measured at their ends with an infrared camera (thermovision).

The results showed that when rolling 0.7/3.0×1245 mm strips, increasing the rolling speed from 5.0 to 19.2 m/s increased the temperature of the strip rolls from 66 to 112°C. Reduction of the LCF flow from 1159 to 680 m^3/h for rolling the 1.5/3.5×1255 mm strips increases the temperature by 40°C at a rolling speed of 17.2 m/s and by 20°C at a rate of 10 m/s. The change of the volume of the coolant supplied to the mill stand by a factor of 2–3 exerts a stronger effect on the temperature of the rolled strip at a higher speed. Changes in the rolling speed and the quantity of coolant supplied in these ranges approximately equally affected the strip temperature (40–50°C by each factor). The impact of a 50°C strip temperature increment on the stress–strain state of the strip rolls of the cold-rolled steel (at a thermal expansion coefficient of $\alpha = 1.2 \cdot 10^{-5}$°C^{-1} and the modulus of elasticity $E = 210000$ N/mm^2) after removing the rolls from the coiler drum and averaging the temperature in the thickness of the wound strip was equivalent to the effect of increasing the coiling tension by 126 N/mm^2.

Practically, according to the technology adopted at all the rolling mills, the rolling speed is periodically changed. For example, the speed of rolling sections of the strip with welds is reduced. After passing of such section through the mill stand, the rolling speed is intensively increased to the previous values. As a result, the temperature changes along the length of the coiled strip and, as a consequence, as the temperature of the turns of the strip roll is equalized, the radial and tangential stresses are redistributed after removal from the coiler. How strongly this factor influences the danger of sticking–welding of the turns during subsequent annealing, as well as the danger of shearing of the turns in the axial direction (the appearance of telescopic sections) is shown in [114] using specific examples. Temperature stresses arising from the inconstancy of the rolling speed and temperature variations along the length of the strip can lead subsequently to the formation of local zones along the thickness of the winding of coils where the level of the interturn pressures exceeds the maximum permissible values or, conversely, the contact of the surfaces of the turns weakens and they can shift relative to each other. These phenomena are extremely undesirable, since they lead to the appearance of defects that worsen the quality of the finished sheet rolling products.

The effect of temperature changes during annealing on the stress–strain state of strip rolls of cold-rolled strips. It should immediately be noted that this essentially important question has not been thoroughly studied and not covered in the technical literature. Meanwhile, judging from the above theoretical calculations, the influence of the state of the surface of the strips (the magnitude and elastic–plastic properties of the rough layer, the presence of contaminants on the surface, the emulsion residues, etc.) and the initial interturn pressures in the roll on the formation of the field of thermal stresses during heating and cooling of a strip roll during annealing of cold-rolled steel in hood-type annealing furnaces is of considerable interest in this case.

The results of calculations of the temperature field and the interturn pressures in rolls are shown in [113] during their cooling after annealing from the holding temperature of 700°C to 400°C. In the first variant, it was assumed that the surface of the strip was absolutely smooth, which ensures perfect mechanical and thermal contacts of adjacent turns. In the second, the strip is rough Ra = 1.5 μm. In addition, it was assumed that at a holding temperature of 700°C, the temperature of all windings along the winding thickness is the same, i.e. at the beginning of cooling, the strip roll is in an isothermal state. The change in the temperature of the shielding gas during the cooling of strip rolls in the bell furnace was taken according to the experimental data in accordance with the graphs in Fig. 9.26.

The temperature and interturn pressure change with cooling of the strip rolls as shown in Figs. 9.27 and 9.28. According to the results of calculations, in the process of cooling the strip rolls wound from strips with an ideally smooth surface, radial thermal stresses

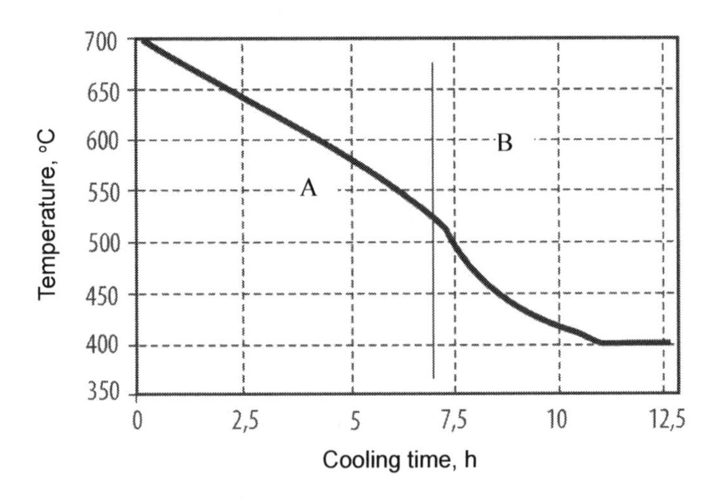

Fig. 9.26. Change of the temperature of the shielding gas in cooling strip rolls under a bell (region A) and after removing the bell under a muffle (region B) in a bell furnace [113].

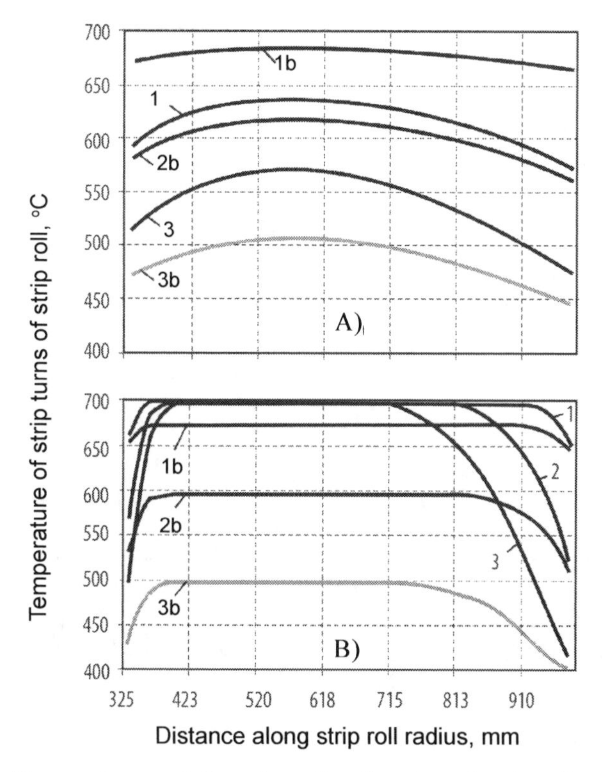

Fig. 9.27. Temperature variation along the diameter of strip rolls of 'infinite' (A) and finite (1000 mm) width (B) in the course of cooling over time of, respectively, 2.5 (curve 1); 7.5 (2) and 12.5 h (3): A – calculations were carried out without taking into account the thermal resistance in contact of the adjacent turns, the strip surface is assumed to be perfectly smooth; B – thermal resistance are taken into account, rough surface, Ra = 1.5 μm [113].

may appear in them and their magnitude is commensurable with the pressure of the strip roll on the coiler drum (Fig. 9.28-A). In the case of loosely adhering to turns caused by the roughness of the strip surface and a significant decrease as a result of the interturn pressures in the strip rolls after removal from the coiler drum, as noted above, a strong temperature unevenness in the thickness of the winding (Fig. 9.27-B) and in the radial stresses near the inner and the outer radii of the strip roll (Fig. 9.28-B) becomes evident.

Loss of stability of strip rolls when they are removed from the coiler. In the production of sheet steel, there are two types of the loss of stability of strip rolls. In the first case, the internal turns become unstable (buckled) and a 'bird' type defect is formed; in the second case, the strip rolls under the influence of their own weight 'sag' and take the form of an oval. The tension of the strips

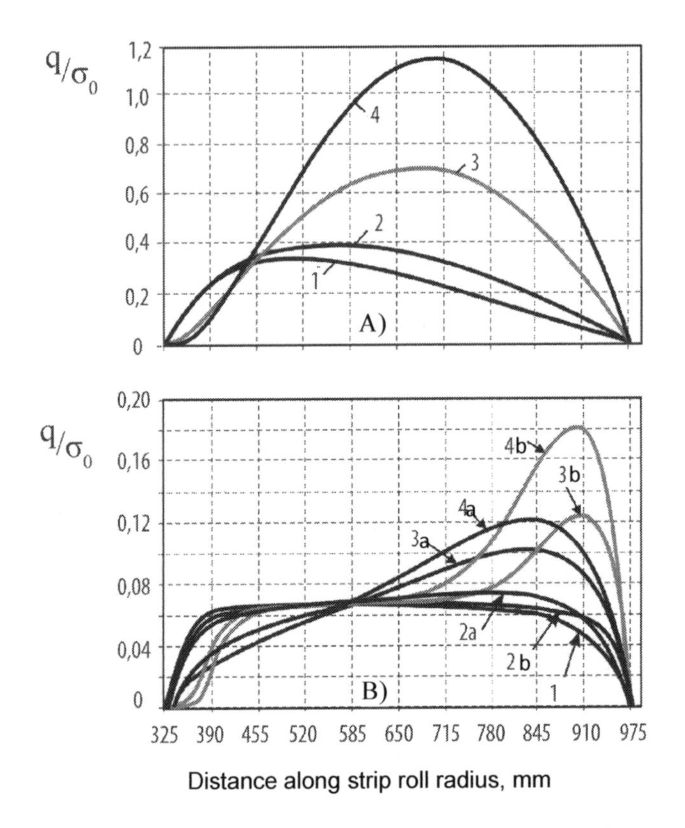

Fig. 9.28. Interturn distribution of stresses along the diameter of the strip roll of finite (1000 mm) width a) and 'infinite' width b). The numbers on the curves – cooling time, respectively, 0 (roll isothermal curve (1) 2.5 (2) 7.5 (3) and 12.5 hours (4): **A** – perfectly smooth surface of the strip; **B** – rough, *Ra* = 1.5 μm. [113]

during winding should also exclude the formation of 'birds' (upper limit), and the possibility of 'sagging' (lower limit). The resistance of the strip roll to the loss of stability is directly proportional to the magnitude of the moment of inertia of the cross section of the strip and inversely proportional to the radius of the turns. Since the moment of inertia of the cross-section of the strip is proportional to the cube of its thickness, in order to prevent loss of stability of the rolls, especially 'sagging', it is necessary to ensure the greatest possible cohesion of the turns by increasing the coefficient of friction between the contacting surfaces. In this regard, it is recommended that the surfaces of the rolled strips have a higher roughness. Experiments carried out on the 1680 cold rolling mill of Zaporozhstal' and the 1700 mills of the Karaganda Metallurgical Combine showed that when the surface roughness of the steel is

Ra = 1–2 μm, there is no relative slippage of the adjacent turns of the strip. With such a roughness of the metal surface, the 'sagging' of the strip rolls is observed only in cases when strips are coiled with a tension of less than 30 N/mm^2.

With an increase in the roughness of the surface of the strips, the magnitude of compressive tangential stresses in the inner turns of the strip roll decreases. For example, with respect to the conditions of the 1700 KarMK mill (see Fig. 9.21), with an increase in the roughness of the strips from 0.5 to 4 μm *Ra*, the value in the inner coil turns decreases by approximately 13%. Therefore, with an increase in the roughness of the surface of the strips the risk of the formation of a 'bird' defect (bulging of inner turns) also decreases.

Summarizing the results of their own research and the data given in the publications, the authors of [123, 124, etc.] note that the value of the critical stress exceeding which leads to the formation of a 'bird' (the first kind of loss of stability) is directly proportional to the modulus of elasticity of the strip material and approximately to the square ratio of the thickness of the strip to the radius of the coiler drum (inner radius of the roll).

To increase the stability of the roll to sagging (the second kind of loss of stability), it is necessary to ensure the maximum possible adhesion of the turns among themselves, which is achieved by increasing the coiling tension and increasing the roughness of the strip.

Telescoping (the third kind of loss of stability) is most often formed when unwinding large-sized strip rolls, in particular during skin pass rolling, and depends on the mass of the strip roll, its dimensions and linear acceleration of unwinding the roll.

The dependence of the telescopicity of the strip rolls with different outer radii on the acceleration of unwinding, obtained from the results of measurements at the 1700 skin pass rolling mill of the Karaganda Metallurgical Combine, is shown in Fig. 9.29. The experimental data presented here support the conclusion that the value of the acceleration must be chosen taking into account the dimensions of the rolls. For large-sized rolls (r_{out} > 1.1–1.2 m), the acceleration value should not exceed 0.8 m/s^2 [123].

The stability of the strip rolls of cold-rolled strips is affected by processing lubrication. It leads to an increase in the interturn pressures in the strip roll and to a decrease in the coefficient of friction. The action of the second factor prevails. Therefore, if there is a large amount of lubrication (emulsion) between the turns, the risk

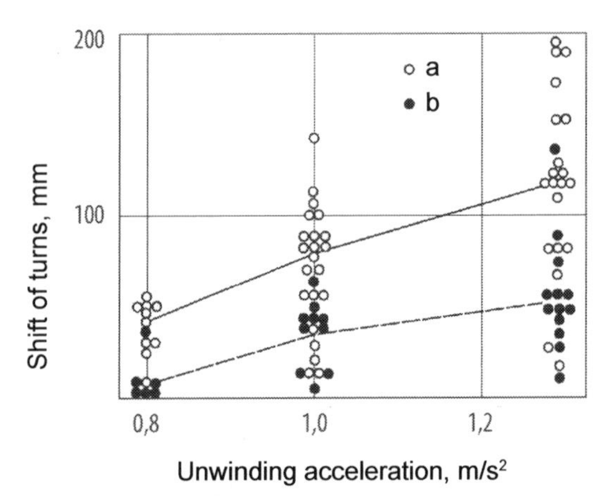

Fig. 9.29. Dependence of the axial shift of turns in strip rolls on acceleration of unwinding in the 1700 skin pass rolling mill: a and b – strip rolls of radius larger than 1.2 m and less than 1.0 m.

of axial shift of the turns and the formation of telescopic areas in the strip rolls increases. In this connection, the rolling technology must provide for measures to prevent the emulsion from getting between the turns of the coiled strip (emulsion blowing off the surface of the strip, etc.).

In addition, some aspects of the stability loss of the strip rolls of hot-rolled and cold-rolled strips during their cooling, transportation and annealing will be considered below.

The mechanism of formation of scratches during unwinding with tension of strip rolls of hot-rolled strips. One of the reasons for the appearance of scratches on the surface of sheet steel is the mutual displacement of the turns of the hot-rolled strip in a strip roll on the uncoiler of the cold rolling mill - the 'coiling' of the strip roll. This phenomenon is most pronounced in the plants whose picking units are equipped with coiling machines. The relative displacement of the turns of the strip roll was investigated on the uncoiler of the 2500 cold rolling mill of the Magnitogorsk Iron and Steel Works. Two mutually perpendicular lines were drawn with red paint at the butts of the strip roll and then a change in the arrangement of these lines was recorded visually and with the help of filming with the strip was directed to the mill and during the cold rolling of the strips. It was found that the greatest mutual displacements of the turns occur in the outer part of the unwinding part of the strip roll due to the jerks of the strip observed at the time of its feed in the mill. In the loosely wound rolls, the displacement of the coils covers

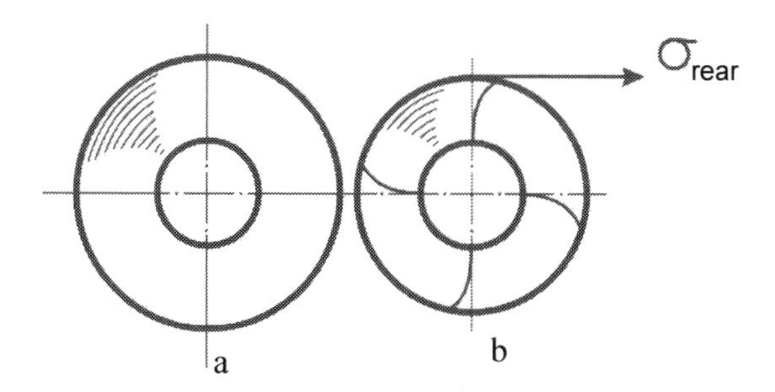

Fig. 9.30. Displacement of the lines marked on the end of the strip roll during its unwinding: a – starting position; b – in the process of unwinding.

approximately 2/3 of the winding thickness from the outer surface of the strip rolls. The inner parts of the rolls (~1/3 of the thickness) are compacted due to the separation action of the drum of the decoiler of the cold rolling mill, so when the strip roll is unwound, the displacement of the turns is much smaller. Figure 9.30 shows the displacement of the applied lines during the unwinding of the roll.

The process of 'dragging' the roll on the drum of the decoiler is continuous. As the upper turns are loosened, the subsequent inner turns of the roll are shifted and tightened. This is evidenced by the gradually increasing forward slope of the lines applied to the butts of the strip roll. As a result, almost all turns are subject to mutual displacement.

The mechanism of the formation of scratches on the surface of the strip when the coil is dragged is discussed in detail in [43].

9.6. Selecting tension modes when winding rolls of cold-rolled strips

Winding strip rolls with varying strip tension. The distribution of stresses in rolls wound with a change in tension during the winding according to linear increasing and decreasing laws was considered in an approximate formulation in [108]. Dependences of the pressure of the strip rolls on the coiler drum and the interturn pressure in the strip rolls after removal from the coiler on the number of turns of the strip under different tension regimes, calculated taking into account the roughness of the metal surface, are shown in Figs. 9.25, 9.31. The change in tension during the winding process is a linear function of the length of the strip in the roll:

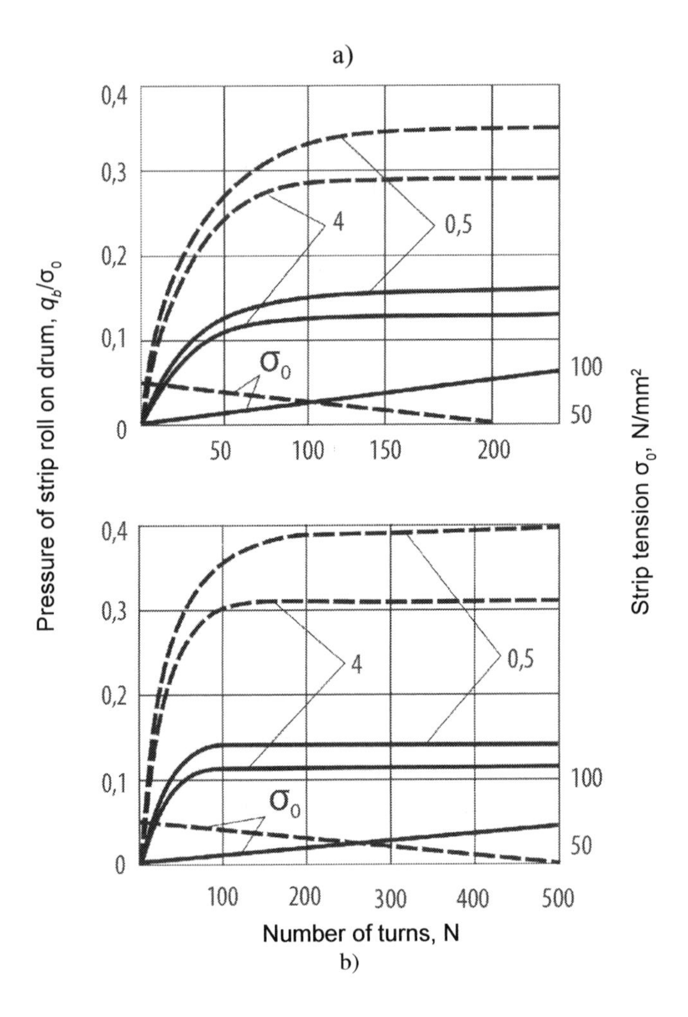

Fig. 9.31. Dependence on the pressure of strip rolls on the coiler drum q_b/σ_0 on the number of turns N of strip 2 mm thick with increasing (solid lines) and decreasing (dashed lines) tension modes σ_0: a – rolling conditions as in the 1680 mill of the Zaporozhstal' plant (the radius of the drum 380 mm, $\lambda = 0.7$); b – rolling conditions as in the 1700 mill of the KarMK (drum radius 300 mm, $\lambda = 0.7$). The numbers on the curves – roughness value *Ra*, μm.

$$\sigma_0(r) = \sigma_0\left[1 + (z-1)\frac{r^2 - r_{in}^2}{r_{out}^2 - r_{in}^2}\right],$$

where σ_0 – the tension of the strip at the beginning of coiling; z – index of tension mode change (at $z > 1$ tension increases, with $z < 1$ decreases; at $z = 1$ $\sigma_0 = \text{const}$).

In this type of function $\sigma_0(r)$ the potential energy for the same strip rolls and the average values of the strip tension is constant for increasing and decreasing modes.

According to the graphs in Fig. 9.31, the strip roll pressure on the coiler drum is less when the tension is increased during the winding process than for the decreasing tension regime. The value of q_b is significantly affected by the gradient of the change in tension $d\sigma_0(r)/dr$. Since in the winding of relatively small strip rolls (Fig 9.31 *a*), the tension increases faster than when winding large-sized rolls (Fig. 9.31 *b*), the value of q_b in the first case is greater even in spite of the increased radius of the coiler drum. The tension used for winding the inner turns of the strip roll adjacent to the drum is important.

After removal of the small srip rolls from the coiler would with an increase in tension according to the linear law $d\sigma_0(r)/dr > 0$, the value of the interturn pressure is much larger than in the cases σ_0 = const and at the decreasing tension mode $d\sigma_0(r)/dr < 0$ (see Figs. 9.21 *a* and 9.32 *a*). For the strip rolls of large mass (Fig. 9.32 *b*), the difference in pressure is insignificant. Consequently, the value of the interturn pressure in the strip rolls after removal from the coiler is determined not so much by the tension drop σ_0 at the beginning and at the end of winding, as by the gradient of the change in tension $d\sigma_0(r)/dr$. At high negative values of the tension change gradient, a state is achieved when the absolute radial deformation of the inner

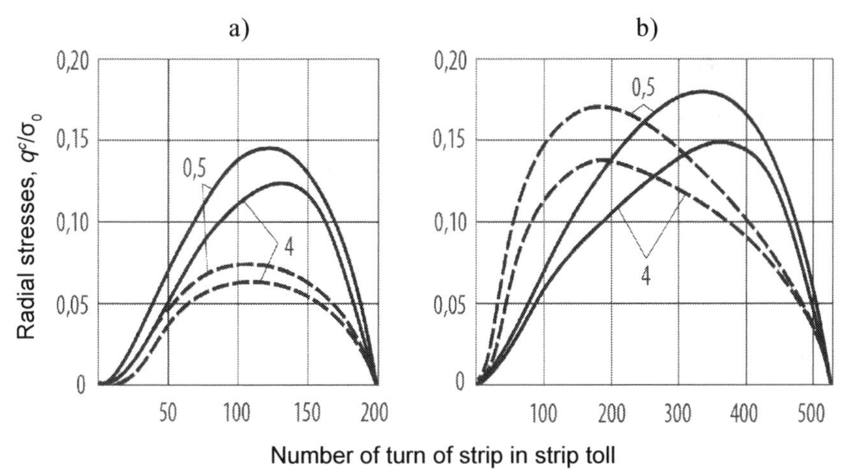

Fig. 9.32. Interturn pressure in the thickness of strip rolls wound with various tension modes, after removal from the colier winder. Designations are the same as in Fig. 9.31.

turns wound with increased tension becomes larger than that of the outer turns and, as a result, after separation from the drum the strip roll stratification will occur.

The possibilities of reducing the interturn pressure in large-sized strip rolls due to a monotonic decrease in tension during the winding process are limited (Fig. 9.32 *b*). Effective modes are those in which the tension varies periodically. As an example, consider the sinusoidal law of tension variation. The function of changing the tension, depending on the length of the strip, is given in the form

$$\sigma_0(r) = \sigma_{0_n}\left[1 + A\sin\left(2\pi n\frac{r^2 - r_{\text{in}}^2}{r_{\text{out}}^2 - r_{\text{in}}^2} + \varphi_0\right)\right],$$

where σ_{0_n} – nominall tension; A – amplitude of change of tension; φ_0 – initial phase; n – number of periods of oscillation of tension during winding the entire roll (frequency).

For the sinusoidal change of tension as a function of the number of wound turns this dependence has the form

$$\sigma_0(r) = \sigma_{0_n}\left[1 + A\sin\left(2\pi n\frac{r - r_{\text{in}}}{r_{\text{out}} - r_{\text{in}}} + \varphi_0\right)\right].$$

In the numerical examples these parameters were used to set the following values: $A = 0.05$; 0.2; $n = 1$; 2; 10; $\varphi_0 = 0$; $\pi/2$; π. The obtained graphs of the change in the tension in the strip roll in the winding process are shown in Fig. 9.33.

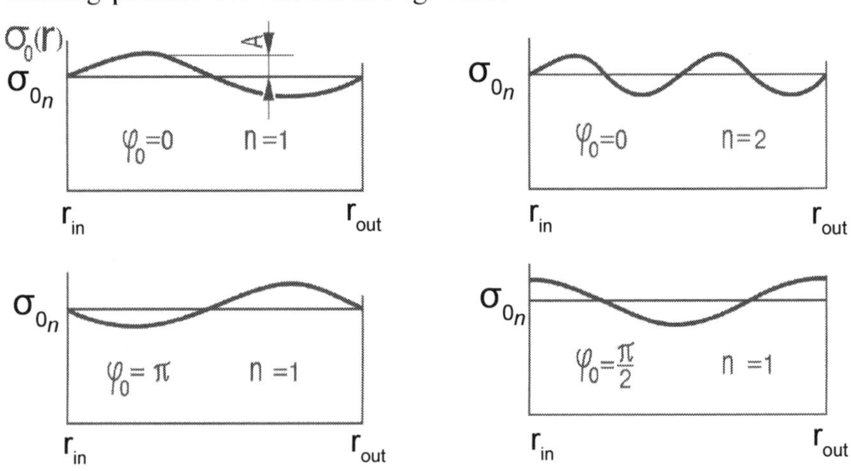

Fig. 9.33. Examples of the periodic variation in strip tension during the winding of strip rolls ($A = 0.2$).

Table 9.1. Dependencres of the roll pressure on the drum q_b and the maximum value of interturn pressure in the strip roll after removal from the coiler q^c_{max} on the parameters of the sinusoidal variation of the strip tension during winding (I – for ideal contact of turns, R – for roughness of the surface of the strip Ra = 1 μm)

Parameters of strip tension mode			q_b/σ_{0n}		q^c_{max}/σ_{0n}	
A	n	φ_0	I	R	I	R
The radius of the drum 380 mm; λ = 0.7; h = 2 mm; N = 210; σ_{0_n} = 50 N/mm^2						
0.05	1	0	0.530	0.140	0.176	0.0630
		$\pi/2$	0528	0.144	0.171	0.0645
		π	0.517	0.134	0.183	0.0705
	2	0	0.528	0.141	0.178	0.0680
		$\pi/2$	0.525	0.141	0.178	0.0663
		π	0.519	0.133	0.186	0.0653
	10	0	0.525	0.139	0.178	0.0663
		$\pi/2$	0.524	0.138	0.179	0.0668
		π	0.522	0.136	0.181	0.0674
0.2	1	0	0.549	0.151	0.179	0.0568
		$\pi/2$	0.541	0.164	0.146	0.0629
		π	0.497	0.125	0.197	0.0821
	2	0	0.541	0.154	0.154	0.0719
		$\pi/2$	0.532	0.155	0.180	0.0691
		π	0.506	0.122	0.205	0.0668
	10	0	0.528	0.144	0.177	0.0662
		$\pi/2$	0.524	0.138	0.176	0.0677
		π	0.519	0.131	0.185	0.0695
$\sigma_0(r)$ = const = 50 N/mm^2			0.523	0.137	0.179	0.0667
The radius of the drum 300 mm; λ = 0.7; h = 2 mm; N = 525; σ_{0_n} = 50 N/mm^2						
0.2	1	0	0.988	0.141	0.625	0.1002
		$\pi/2$	1.008	0.175	0.558	0.0816
		π	0.883	0.135	0.546	0.1019
	2	0	0.982	0.144	0.594	0.0986
		$\pi/2$	0.983	0.173	0.538	0.0996
		π	0.889	0.132	0.585	0.1016
	10	0	0.958	0.155	0.587	0.0912
		$\pi/2$	0.945	0.157	0.587	0.0906
		π	0.913	0.123	0.585	0.0932
$\sigma_0(r)$ = const = 50 N/mm^2			0.935	0.138	0.585	0.0894

The results of calculations of stresses in the strip rolls on the drum and after removal from the coiler drum are presented in Table 9.1. The characteristic diagrams of the distribution of the radial and tangential stresses in strip rolls of various sizes and weights are

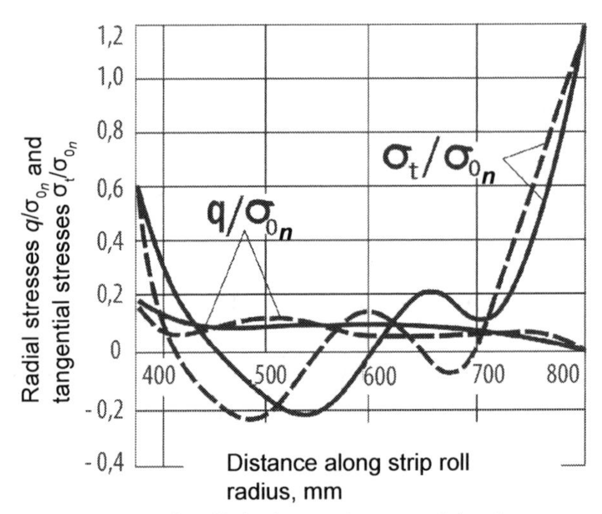

Fig. 9.34. Distribution of radial q/σ_{0n} and tangential σ_t/σ_{0n} stresses in strip rolls on the coiler drum wound with sinusoidal changes of strip tension: σ_0 = 50 N/mm^2; A = 0.2; n = 2; $\varphi_0 = \pi/2$ (solid lines); $\varphi_0 = \pi$ (dashed); λ = 0.7; strip surface roughness Ra = 1 μm.

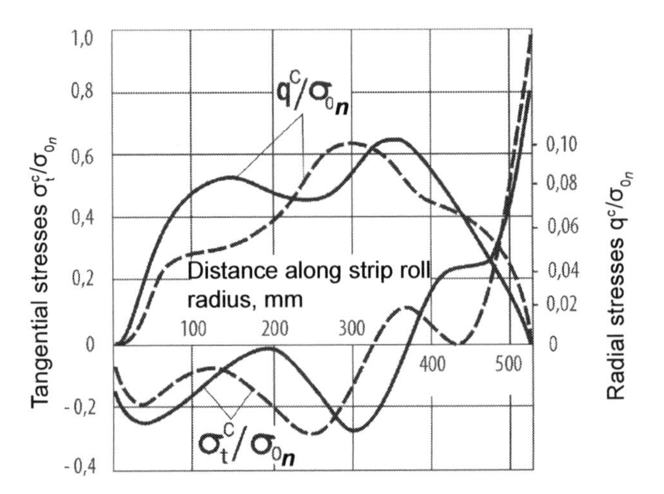

Fig. 9.35. Distribution of radial q^c/σ_{0n} and tangential stresses σ_t^c/σ_{0n} in strip rolls, wound with sinusoidal changes strip tension, after removal from the coiler: inner radius of strip roll 300 mm, outer – 1350 mm; λ = 0.7; σ_0 = 50 N/mm^2; surface roughness of the strips Ra = 1 μm; A = 0.2; n = 2; φ_0 = 0 – solid line; φ_0 = $\pi/2$ – dashed curves.

shown in Figs. 9.34 and 9.35. According to these figures, changing the tension of the strips at their coiling can significantly affect both the pressure of the strip roll onto the drum and the interturn pressure in the strip roll after removal from the coiler. For example, in Table 9.1 the difference between the values of q_b and q^c_{max} in winding strip rolls with the sinusoidal change $\sigma_0(r)$ and the same values obtained by $\sigma_0(r)$ = const reaches ~20%.

The curve of the tangential stresses in the strip rolls, wound with the sinusoidal variation of $\sigma_0(r)$, has the form of a deformed sinusoid with alternating zones of positive and negative values. The interturn pressure in the strip rolls after removing from the coiler is distributed in the thickness of the coiled strip more uniformly than in the case with $\sigma_0(r)$ = const. The curve of the radial stresses may have several maxima.

Analysis of the data of Table 9.1 shows that with increasing amplitude the effect of the sinusoidal dependence $\sigma_0(r)$ increases. Higher values of the pressure of the strip rolls on the drum are observed in cases when the angle of the initial phase φ_0 is equal to 0 or $\pi/2$. This is due to the fact that the value of q_b is mainly determined by the tension at the winding of the inner turns adjacent to the drum. Accordingly, for $\varphi_0 = \pi$, q_b assumes the smallest values.

The effect of the sinusoidal mode weakens as the frequency of the change in the tension increases. Thus, for $n = 10$ (Table 9.1), the values of q_b and q^c_{max} already differ little from the values of these quantities for $\sigma_0(r)$ = const.

The dependence of the interturn pressure in the strip rolls after removal from the coiler on φ_0 has the reverse character: q_c is minimal for $\varphi_0 = 0$ and $\varphi_0 = \pi/2$.

Reducing the interturn pressure in the strip rolls after removal from the coiler in the case of winding them according to the sinusoidal regime of the change in the strip tension occurs because the turns wound with a lower tension receive less radial movement and over the entire thickness of the strip roll they receive part of the load from the action of the turns that are stretched more during winding.

The analysis of the winding regimes of strips on various continuous cold rolling mills revealed the main regularity, which is that the amount of tension during winding is set inversely proportional to the mass of the rolls. Thus, in experiments in the 1680 four-stand mill of the Zaporozhstal' plant (weight of rolls up to 15 tons), the tension value σ_0 of the winding of the strips was assumed to be unchanged for the entire range of rolled metal and set equal to 50 N/mm^2, at the 2500 mill of the Magnitogorsk Iron and Steel Works (weight of rolls up to 30 tons) σ_0 = 40–45 N/mm^2, and in the 1700 mill of the Karaganda Metallurgical Concern (weight of rolls up to 45 t) σ_0 = 40–43 N/mm^2. In many other mills even lower tension is used when winding strip rolls of cold-rolled strips. For example, in the 2030 mill of the Novolipetsk Metallurgical Combine in some cases the tension between the last stand and the coiler is set at 21–25 N/mm^2.

Thus, the conducted studies have shown that the traditional method of winding coils with a constant strip tension is effective only in cases where the production conditions do not threaten the loss of stability of the roll and the welding of the turns during subsequent annealing (for example, after intermediate passes on the reversing mills). When winding rough strips with constant tension, the tension value should be set in an inverse proportion to the weight of the rolls. The minimum permissible value of tension is determined by the rolling technology (speed, total reduction, temperature of rolled metal, flatness, variability of thickness, etc.) at a particular mill.

Advantages in comparison with the known are winding modes, providing there is a periodic change in the tension value [123]. Such modes provide relatively low interturn pressure in strip rolls after removal from the coiler and reduce the risk of the loss of stability of the inner turns. As a result of the performed studies, a method for winding the strip was developed, according to which the tension value is changed during the whole winding time by a sinusoidal dependence with an amplitude in the range 0.1–0.3 of the nominal tension value. In this case, the frequency of the change in the value of the tension should not be equal to or a multiple of the frequency of the natural oscillations of the drive line of the coiler drum. A strip tension control device was developed for coil winding for implementation of the method for the 1700 cold rolling mill at the Karaganda Metallurgical Concern [127].

In order to prevent the loss of stability of the inner turns of cold-rolled tinplate (in the production of tinplate the risk of the formation of a ‘bird’ is particularly great), it is advisable to wind the strip onto the coiler drum in such a way that the first 5–20 turns are formed from a strip with a thickness exceeding 1.3–3 times the nominal thickness, and the subsequent 20–50 turns are formed from a strip with a thickness gradually decreasing to a nominal thickness. This increase in the thickness of the most stressed (the greatest value of the circumferential compressive stresses) inner turns of the strip increases the stability of the strip rolls [50, 126].

The rationale for this solution is that the critical radial pressure q_{cr}, at which the loss of stability of the turns occurs, strongly depends on the thickness of the strip. The value of the critical pressure q_{cr} is a function of the thickness of the strip h and the radius of the turn r. Approximately one can assume that

$$q_{cr} = 0.65E\left(\frac{h}{r}\right)^{\frac{11}{5}},$$

where E is Young's modulus of the strip material.

In this expression, the exponent with a ratio of h/r varies from 2 to 3. However, this does not matter. In any case, it can be seen that by increasing the thickness of the most stressed (the most compressed) inner turns, it is possible to substantially increase the stability of cold-rolled strips.

In the production, for example, of tinplate, the inner layer of turns in the strip roll, as a rule, 30-70 mm thick, loses stability. With a strip thickness of 0.2 mm, it consists of 150–350 turns. The stability of this inner layer of windings can be increased by more than 10 times if this layer consists of turns of a strip three times as thick as 0.63 mm. Hence, the thickened metal layer will have about 50-110 turns. In practice, it is sufficient to increase the stability of the inner layer 1.5–2.0 times. It turns out that the layer of thickened strip should consist of 20 turns. That is why it was proposed to form 5–20 inner turns of a roll from a strip that exceeds the nominal thickness.

In order to increase the uniformity of the density of strip rolls wound from strips of different widths, it is advisable to set the inverse width of the strip in the transition from winding the strips of one width to the other. This solution follows from the above conclusions regarding the influence of the width of the strips on the compliance of the drum and, respectively, on the stress–strain state of the strip rolls after removal from the coiler.

One of the modern trends in sheet rolling production is the creation of equipment for processing large-sized rolls. Mills for cold rolling strips of 60–80 tons and even 200–250 tons are being designed. In the process of unwinding enlarged strip rolls with starting and braking modes of the coiler–uncoiler, there is a danger of relative slippage of the turns and damage to the surface of the metal. Consequently, the method of winding large-sized strip rolls must exclude interturn slip when they are unwound. As the speed of the coiler drum increases with the coiled roll operating in the decoiler mode, slippage does not occur if the inequality $M_{fr} > M_{dyn}$ is satisfied, where M_{fr} is the moment of frictional forces on the surface of an arbitrary turn with radius r; M_{dyn} is the dynamic moment of the mass of a strip roll located above the considered turn with radius r. The calculations carried out using the inverse problem model showed that in order to fulfill this condition, the amount of tension in the winding of the roll must be varied in inverse proportion to the radius of the wound turn.

Incidentally, we note that the tension when winding at the strip plate in the pickling line should be higher than when unwinding the strip rolls in the cold rolling mill. In this case, the risk of relative slippage of the turns and damage to the surface of the strips (formation of scratches, scoring and other defects) is reduced when unwinding coils in a cold rolling mill [119].

To reduce slippage of the roll relative to the coiler–uncoiler drum, the surface of the drum shell in contact with the strip must be made rough (for example, by shot blasting or spark erosion).

It was shown above that the roughness of the surface of the strip rolls strongly affects the stress–strain state of the rolls. On the basis of this property, a method has been developed for winding strips onto a coiler drum, characterized in that the tension value is changed during winding in direct proportion to the roughness of the strip surface and is set equal to

$$\sigma_{0R} = \sigma_{0R_1}(Ra/Ra_1)^n,$$

where σ_{0R} is the tension value when winding a strip with a surface roughness of Ra; σ_{0R_1} is the tension value when winding a strip with a surface roughness of Ra_1; $0.02 \le n \le 15$ is the exponent.

This winding method ensures the uniformity of the interturn pressure in rolls of strips with different surface roughness and, as a result, improvement of the quality of the finished metal. In addition, an increase in tension between the last stand of the cold rolling mill and the coiler during the rolling of strips with a high roughness makes it possible to reduce the rolling force in the last stand, which has a favourable effect on the stability of the process and the accuracy of rolled metal.

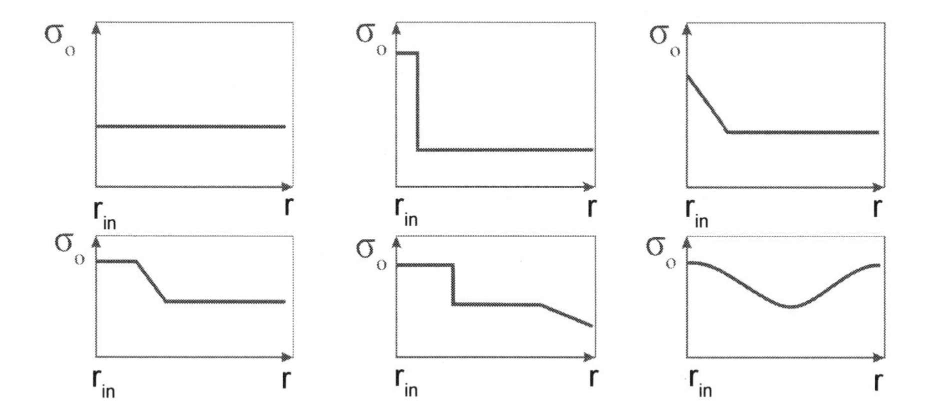

Fig. 9.36. Possible modes of changing the tension σ_0 of rolled and coiled strips: r_{in} and r – inner radius and the actual roll radius.

In addition to the above described method of winding the strips, which involves periodically changing the tension to prevent the loss of stability of the inner turns of the strip roll after removal from the coiler drum, a winding method can be used in which the first five to ten windings are wound with increased tension, for example, for carbon steels, five times higher than the processing tension, followed by a gradual decrease in the tension to the processing level after 50–100 rotations of the coiler (Fig. 9.36). However, along with advantages, this method also has drawbacks. In particular, an increase in the tension in the winding of the first turns leads to an increase in the pressure of the strip roll on the coiler drum, which is undesirable [119].

Technical solutions aimed at reducing flaw defects caused by an uneven distribution of the roll temperature in the radial direction are worthy of note, by compensating for the tension of the strips when winding the rolls.

The speed of cold rolling of strips in the modern sheet and tinplate rolling mills, is not constant. Typically, the front and back butts of the strips and the welded joints are rolled and wound into rolls at a speed reduced to 5–7 m/s. There may be other reasons for the reduction in rolling speed, for example, the presence of metal defects. At the same time, the working (nominal) rolling speed can be 25–30 m/s or more.

The amount of heat released in the deformation area, the thermal power of rolling is proportional to the rolling speed. Consequently, at a constant power of the cooling system of the rolling mill, the temperature of the rolled and coiled strips varies with the rolling speed.

The above fluctuations in the temperature of the strips must be taken into account when adjusting the winding process of strips to strip rolls, since the temperature deformation of the turns in the strip rolls is commensurable with the deformation caused by the tension of the strip between the last stand and the coiler.

The relative elongation of each turn of the strip roll in winding on the coiler drum can be determined from the relation

$$\varepsilon = \frac{\sigma_0}{E} + \alpha \cdot t_0$$

Here ε – elongation of the turn of strip roll in the circumferential direction at the time of winding the strip roll; σ_0 – specific tension of coiled strip, N/mm^2; t_0 – the nominal excess of the strip temperature above ambient temperature, °C; α – the coefficient of thermal expansion of steel, 1/deg; E – Youngs' modulus of steel, N/mm^2.

During winding the strip on the coiler drum strip temperature change. Let increment values of t_0 and σ_0 be respectively Δt and $\Delta\sigma_0$. Substitute them in the first expression and define the value $\Delta\sigma_0$ from the condition that the increment of the elongation of the turn is equal to zero

$$\varepsilon + \Delta\varepsilon = \frac{\sigma_0 \pm \Delta\sigma_0}{E} + \alpha(t_0 \pm \Delta t).$$

Subtracting the first expression from the last one gives

$$\Delta\varepsilon = \frac{\pm\Delta\sigma_0}{E} \pm \alpha \cdot \Delta t.$$

Equating the left-hand side of this expression is zero, we find

$$\Delta\sigma_0 = -\text{sign}\,\Delta t \cdot \mathrm{E} \cdot \alpha \cdot \Delta t,$$

where $\text{sign}\,\Delta t = \frac{\Delta t}{|\Delta t|}$ is the sign of the change in temperature of the strip.

Consequently, the found dependence of $\Delta\sigma_0$ on Δt can be used as a basis for adjusting the strip tension when winding a roll. When implementing such a regulation, it is necessary to measure the temperature (increment of temperature) of the strip at the exit from the rolling mill and adjust the value of the tension of the coiled strip into a function of the deviation of its temperature. At the same time, the increase in the heating temperature of the strip during rolling should correspond to a decrease in the tension of the strip, and a decrease in temperature to an increase in the tension.

In cases where there is no possibility of reliable temperature control (in the range 0-200°C) of the strip moving with a high speed, the regulation of the winding tension of the roll can be carried out by an indirect parameter – the rolling speed, which is fixed quite accurately. When the rolling speed changes from a certain nominal value V_0 to a value of V, the temperature of the coiled strip changes from t_0 to t. The temperature increment $\Delta t = \beta \cdot \Delta V$, where the coefficient β is the ratio $\beta = dt/dV$ or approximately $\beta = \frac{\Delta t}{\Delta V}$. The coefficient β takes different values depending on the rolling speed of the strips. In the range of low (charging) velocities (up to 5 m/s), the coefficient β assumes maximum values equal to approximately $\beta = 10\frac{\text{deg}}{\text{m/s}}$ and at high rolling speeds (30–40 m/s) this coefficient has minimum values $\beta = 0.5\frac{\text{deg}}{\text{m/s}}$.

Substituting now the expression for Δt in the equation determining $\Delta\sigma_0$ and considering that by increasing the rolling speed the strip temperature increases (and vice versa), we obtain, $\Delta\sigma_0 = -\text{sign}\Delta V \cdot m \cdot \beta \cdot \Delta V$, where $\Delta V = \frac{\Delta V}{|\Delta V|}$ is the sign of the incement of the rolling speed from the nominal value V_0 to V; m – coefficient of proportionality, equal to the product of Young's modulus by the coefficient of thermal expansion.

The presented calculations show that in order to avoid the negative effect of the non-uniform temperature distribution in the thickness of the winding of the strip roll due to fluctuations in the rolling speed on the stress–strain state of the rolls of cold-rolled strips, it is expedient to adjust the tension of the strips depending on the rolling speed. At the same time, an increase in the rolling speed of the strips corresponds to a decrease in the strip tension, and a decrease in the rolling speed of the strips increases the tension. This results in a reduction in the interturn contact stresses in strip rolls due to the elimination of the negative effect on the stress level of the variability in the speed and temperature of the rolled strips. As a result of this adjustment, the probability of loss of stability of the inner turns of the strip should be reduced after the roll is removed from the coiler drum.

It should be noted, however, that even at a constant rolling speed, the temperature of the strip at the mill exit is, as a rule, unstable, for example, due to a change in the degree of deformation in the last stand, vibration resistance of the rolled steel, and other factors. The most significant influence among these factors is the degree of deformation of the strip in the last stand of the cold rolling mill.

In addition, the dependence $\beta(V) = \frac{dt}{dV}$ corresponds to the set (family) of the dependences of the type $t(V) + C = \int \beta(V) dV$.

When using one tension control channel σ_0 of the strip when winding the roll by deviating the strip speed from the set value, an error in determining the temperature increment towards its overestimation or underestimation is possible. Therefore, in order to improve the accuracy of the control of σ_0, it is also possible to determine the temperature of the strip based on information about the roll temperature and the value of the relative reduction in the last mill stand.

Note that the proportionality coefficient between $\Delta\sigma_0$ and Δt is the product of the empirical constants ($m = E \cdot \alpha$), that is, the Young modulus of the $E = 2.1 \cdot 10^5$ N/mm^2 and the coefficient of thermal expansion of steel $\alpha = 11.9 \cdot 10^{-6}$ 1/deg. Since the method in question is intended primarily for use in ferrous metallurgy in the production of steel strips, the characteristics of steel must be used to determine this coefficient. For a particular sheet rolling mill, the value of the coefficient should be selected from the specified range, taking into account local conditions, the adopted technology (the range of rolled strips, the nominal values of the rolling speed and the tension of the strips between the last stand and the coiler of the mill, the cooling system of the rolls and rolled metal, etc.).

Once again, we emphasize that the temperature of the coiled strip, changes in the temperature of the strip during the winding process, including because of the regulation of the rolling speed, are a very potent influence on the stress–strain state of cold-rolled strips and the propensity to lose stability of its inner windings after removal from the coiler drum – the formation of defects such as 'bird'. Neglect of this factor, especially on high-speed multistand mills, can negate the effectiveness of any other channel for adjusting the tension of rolled and coiled rolls of thin strips.

In the process of production of sheet steel, strip rolls of cold-rolled strips, after removal from the coiler drum, are turned 90° and, in a vertical position, transported to the thermal compartment where the metal is annealed in hood-type annealing furnaces. During transportation, the coils cool down, the temperature of the turns of the strip in the roll are lowered. In this case, the nature of the temperature distribution along the thickness of the winding changes, which accordingly changes the diagram of the radial and tangential stresses. From a technological point of view, it is desirable that changes in the stress–strain state of rolls are accompanied by a decrease in the interturn pressure, especially in the zones of its maximum values. Then the preconditions for sticking, welding of the contacting turns of the strip in the roll during the subsequent annealing will be relaxed or completely eliminated.

Immediately note that earlier in our work [119] it was concluded that to ensure the stability of rolls of hot-rolled strips to sagging when transported and stored in a horizontal position, it is necessary to increase the temperature of the coiled strips along their length from the front to the rear. In this case, when the strip roll cools, the thermal strains of the outer turns in the radial direction will be greater than in the inner turns. The inner turns will be compressed by the outer interturn gaps

The density of rolls and their resistance to subsidence will increase. The temperature difference between the front and back butts of the strip should be within 50–200 degrees [119] and be selected from the condition for ensuring the required mechanical properties and structure of the hot-rolled metal. The validity of this recommendation was confirmed by the results of industrial experiments on the stability of rolls of hot-rolled strips to sagging, performed at the industrial 2000 wide-strip hot rolling mill of the Novolipetsk Metallurgical Combine [128, 129]. This will be discussed in more detail in the following sections of the book.

With reference to the cold-rolled steel production process, this method of coiling of strips into strip rolls is unacceptable, since in this case the task of increasing the density of winding and resistance to sagging of rolls is not worth it. On the contrary, measures should be taken to reduce the density of winding, the interturn pressure, to exclude the possibility of loss of stability of another type – the formation of a 'bird' and the welding of strips of the strip in a strip roll during subsequent annealing. Consequently, the regulating effect on the winding process of rolls of cold-rolled strips should be different, in contrast to the above-mentioned method of winding coils of hot-rolled metal. We note that in our review [130] we considered a method for coiling hot rolled strips into rolls, consisting in the fact that when winding a roll, the temperature of the strip at its head is kept high, and in the tail section low. The difference between the temperatures of the front and back butts of the strip here should also be 50–200°C, and the temperature along the length of the strip varies linearly. But the gradient of the function $d\sigma_0/d\ell$, where ℓ is the length of the strip of another sign, is negative. This method was developed to compensate for the negative effect of the uneven cooling of the various turns of a coiled strip on the mechanical properties of hot-rolled steel.

Thus, in the production of hot-rolled steel, depending on its chemical composition, the conditions of deformation in roughing and finishing cages, the cooling conditions on the withdrawing delivery table of the wide-strip hot rolling mill, the temperature of coils wound into rolls can be both increased and reduced in length to solve specific technological tasks and ensuring the required quality of the finished product.

The conclusions drawn above about the significant influence of the temperature of both hot-rolled and cold-rolled strips on the level of interturn pressures in rolls, on the resistance of rolls to loss of stability show that the effect of temperature can be used in the form of a channel for controlling the winding of coils and cold-rolled metal, i.e., acting on the temperature of rolled and coiled coils of cold-rolled strips, to control the stress–strain state of the rolls on the coiler drum, after removing from the coiler drum and as they cool down before annealing. This idea was realized by the authors of the patent of the Russian Federation RU 2236917C1, in which it was proposed to reduce the temperature along the length of the rolled and coiled strip in the steady-state cold rolling process. Moreover, the difference between the temperature of the strip at the beginning and at the end of the winding of the roll is set in the range 5...100°C. The technical effect when using the invention consists in the elimination of defects such as 'coil break', "sticking spots–welding", 'scratches', 'risks' on the surface of finished sheets and strips.

The elimination of these defects is due to the fact that, unlike the other known methods for producing rolled strip metal, here a reduction in the strip temperature during the rolling process with the increase of the diameter of the coiled roll is increased, according to the predetermined law, is used as the processing method. After averaging the temperature of the turns of the strip in the roll during its cooling, the inner initially hotter turns, cooling down, contract in length, and the outer turns, initially colder, warming, are further elongated. Correspondingly, the values of the elastic deformation of the turns in the radial direction are changed. Due to this, the winding density is weakened. The main thing is that the temperature averaging and the reduction of the winding density occur during the period when the roll is in the vertical position after it has been turned from the horizontal position before annealing in the hood-type furnace. During this period, there is no danger of the roll sagging under the influence of its own mass, which could have happened if the density of the roll were insufficient due to a lower tension of the strip when winding it. In the latter case, a loosely rolled roll, being in a horizontal position, would sag immediately after it was removed from the coiler drum. Thus, after averaging the temperature for the thickness of the winding of the roll, the level of the interturn pressures decreases, which significantly affects the interaction of the surfaces of the strip in the roll during annealing. The smaller the value of the interturn pressures, the smaller the adhesion (sticking, welding) of surfaces under the action of high temperatures.

In general, according to the authors of the patent in question, the problem is solved of the formation during the annealing of rolls of the level of the distribution of forces that cause the adhesion of adjacent turns of the strip at which they do not exceed the critical values corresponding to the welding (sticking) of the contacting surfaces.

Reducing the temperature of the strip along the winding of the roll by less than 5°C does not allow achieving the desired effect. The resulting reduction in the interturn pressures does not ensure a reduction in the degree of cohesion during the annealing process to the safe level necessary to prevent sticking and welding of the strips in a strip roll. If the temperature of the strip from the beginning of the coiling of the roll to its end is reduced by more than 100°C, it is possible that zones with loosely adhering turns may form in the strip roll. This leads to slippage of the turns relative to each other during the unwinding of the rolls on the temper mill or in the line of the cutting unit. As a result, there are scratches and grooves on the metal surface, which impair its quality. In addition, the probability of the loss of stability of rolls in the form of their sagging under the action of their own mass increases. The law of variation of the strip temperature during the winding of the roll is chosen within the range of 5...100 ° C depending on the coiling tension, the surface roughness and the non-flatness of the rolled strip.

The considered method for producing rolled strip products was realized at the 2030 five-stand mill of the Novolipetsk Metallurgical Combine. Strips of 1000 mm width, 0.5 mm thick from a 2.5 mm thick strip plate were rolled at a speed of 20 m/s, and rolled into strip rolls weighing 25 tons with a constant tension of 30 N/mm^2. The surface roughness of the bands was Ra = 0.7 μm. As the diameter of the roll increased from 650 mm to 1950 mm, the temperature of the rolled and coiled strips was reduced linearly from 135°C to 30°C by increasing the total flow rate of the cooling fluid supplied to the mill stand from 480 to 1400 m^3/h.

Figures 9.37, 9.38 and 9.39 show various variants (presented in the RU 2236917C1 patent) of the variation of strip temoerature during winding of rolls and the resultant distributions of the interturn pressures (radial contact pressures) in the rolls after averaging the temperature in the thickness of the roll and the distribution of the specific forces of separating the turns in the roll after annealing in hood-type furnaces. It may be seen that when winding the rolls at a constant temperature (135°C) of the strip the interturn pressure reaches the maximum in the radius zone of the strip of 475 mm and decreases in the outer turns at a radius greater than 750 mm. After heating for 50 h and holding the rolls at a high temperature during annealing the specific force of separation of the adjacent turns is greater than the safe level which is 3 N/mm^2. This leads to seizig of

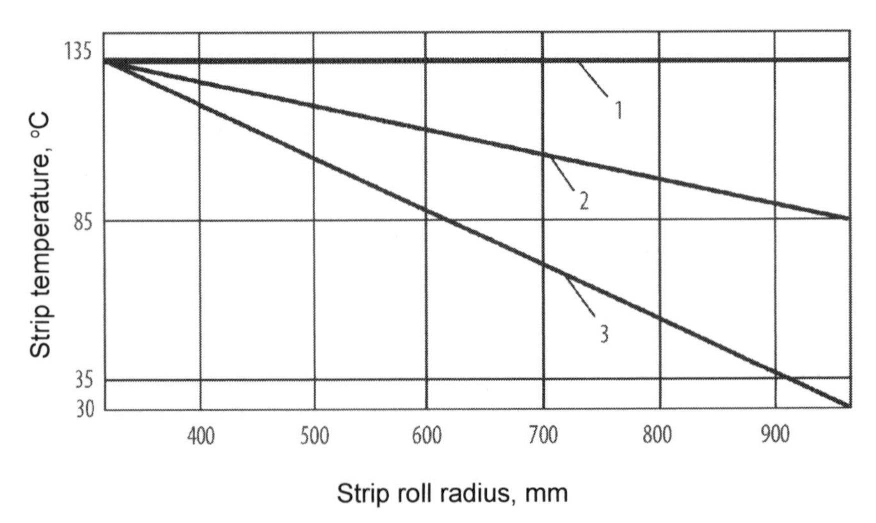

Fig. 9.37. Examples of changes in strip temperature during strip roll winding. Line 1 – at a constant strip temperature along strip roll winding; 2 – while reducing the temperature by 50°C; 3 – 105°C. Rolling conditions are indicated in the text.

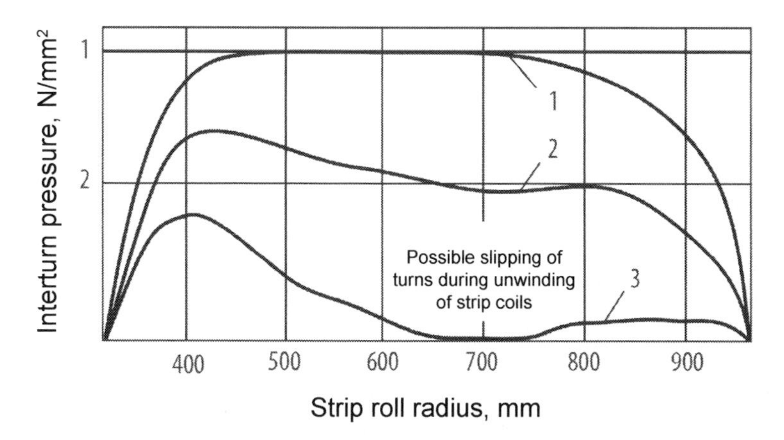

Fig. 9.38. Interturn pressure distribution along the radius of the strip rolls removed from the coiler drum after averaging temperature in them. Curve 1 – coiling of strip rolls at a constant strip temperature; curve 2 – the strip with a reduction in temperature in the course of winding at 50°C; curve 3 – with a 105°C reduction. Rolling conditions are indicated in the text.

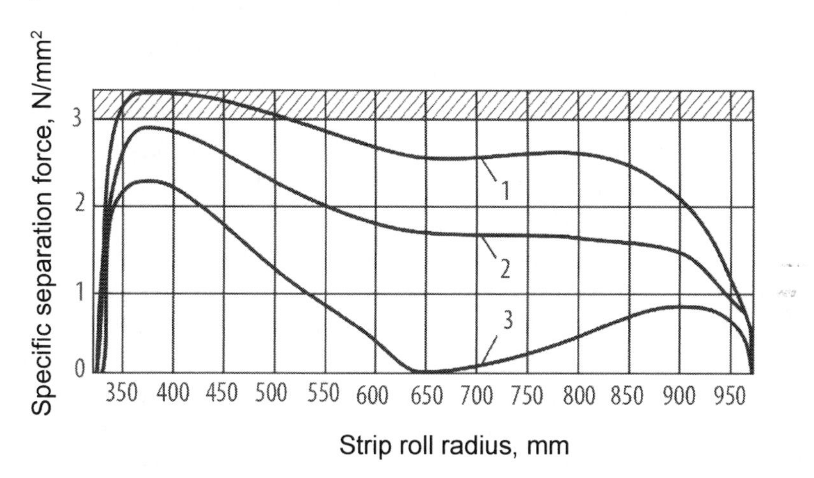

Fig. 9.39. Distribution of specific separation force of turns in strip rolls after annealing in a bell furnace. Curve 1 – in the case of winding strip rolls at constant temperature; curve 2 – with a reduction of the strip temperature during winding strip rolls by 50°C; curve 3 – with a decrease of 105°C. The shaded region is the region of danger of seizure (welding, sticking) of surfaces of the adjacent turns. Rolling conditions are indicated in the text.

the adjacent turns and skin pass rolling was then accompanied by the formation of kinking defects and 'sticking spots-welding'

Judging from the graphs in Figs. 9.38 and 9.39, the decrease in strip temperature during the winding of the strip roll by more than 100°C is excessive. There is a danger of slippage of the turns during the unwinding of the strip roll during skin pass rolling and the formation of defects 'scratches', 'grooves'. Elimination of defects such as 'coil breaks', 'sticking spots – welding', 'scratches', 'risks' was achieved even with a change in the temperature of the coiled strip from 135°C to 85°C, i.e. by 50°C (curves 2 in Figs. 9.38 and 9.39).

It should be noted that, regardless of the strip tension regimes when winding coils in cold rolling or skin pass rolling processes during cooling and storage of the strip rolls in horizontal or vertical positions, the winding temperature along the winding thickness is redistributed and levelled, and the overall level of radial and tangential stresses is reduced. There is a relaxation of stresses and in cases of a long time of storage of the strip rolls of sheet steel in a horizontal position they can separate and sag. Naturally, during the heating of coils of cold-rolled steel during annealing the relaxation of stresses is accelerated and the level of interturn pressures decreases significantly. When annealing in hood-type furnaces, the strip rolls are placed in an upright position, where there are no prerequisites for their sagging, but there is a danger of damage to the butts of the strip roll.

Talking about the relaxation of stresses in the strip rolls of cold-rolled strips after removing them from the coiler drum of a continuous or reversible cold rolling mill, it must be borne in mind that due to softening the elastic strains become residual. The magnitude of the residual strains of the strip rolls depends on the temperature and duration of the process of softening, and most importantly, it is different in the width of the strip and the thickness of the winding of the roll, since the level of stresses in different turns is different. The residual strains of the metal in the process of stress relaxation change the initial profile and shape of the strip. During the stay of a cold-rolled strip in a strip roll, its flatness can improve. That is why, after annealing in rolls, the flatness of the strips is often better than that of a metal subjected to continuous annealing.

Analysis of the above methods of winding rolls of cold rolled strips allows to draw a conclusion about the possibility and expediency of combining various technical solutions, which will make it possible to use the advantages of each of them more effectively. For example, analyzing the effect of the roughness of the strip surface on the stress–strain state of the rolls discussed above, the experts of the Institute of Ferrous Metallurgy of the National Academy of Sciences of Ukraine proposed to take into account the roughness of their surface when implementing the step-by-step tension regime of turns wound in the strip rolls. Namely, the number of turns of the strip in the roll, first wound with increased tension (Fig. 9.36), should be increased with decreasing roughness of the metal surface. It is also proposed that when winding inner turns of the strip rolls with increased tension (Fig. 9.36), the temperature of rolled and coiled strips should be increased in direct proportion to the amount of tension excess.

The opinion was expressed at some companies that the tendency to weld the turns of the strip rolls during the subsequent annealing of cold-rolled steel, which can be numerically estimated from the tearing force of welded surfaces of adjacent turns, is directly dependent on the value of the total reduction of the strip during cold rolling. According to the above experimental data and the results of calculations, the degree of welding of the contacting metal surfaces directly depends on the contact pressure, i.e., on the interturn pressure in the roll after removing it from the coiler drum, which is a function of the strip tension when winding the coil. The greater the tension of the strip during winding, the higher the value of the interturn pressure in the roll, and the greater the force required to tear the strips welded during annealing in the roll. It turns out that in order to reduce the degree of welding of the coils in a roll during subsequent annealing, it is necessary at a higher value of the total reduction of the strip during the rolling process to establish a lower tension of the strip between the last stand and the coiler. Accordingly, when the total deformation of the strip decreases, the tension of the strip in coiling into a roll should be increased.

The opposite point of view on the choice of the tension of cold-rolled strips in strip rolls is stated in the article [126]. The authors of this work statistically processed the results of rupture tests at a temperature of 710°C of samples of 08Yu sheet steel rolled with a total reduction of 50–70% and showed that the value of the yield strength of steel at this temperature is directly proportional to the yield strength of the strip plate and the total degree of deformation in cold rolling. Further, assuming that the welding of coils of cold-rolled strips in strip rolls during annealing is determined by the plastic deformation of the microirregularities of the contacting surfaces under the effect of temperature and the interturn pressure, it was concluded that if the yield point of steel at the annealing temperature is higher, the microirregularities will crumple less and the danger of welding the turns will decrease. That is, the higher the yield strength of steel at the annealing temperature, the smaller is the danger of welding the contacting surfaces and, consequently, the tension of the strips when coiling the rolls in a cold rolling mill can be large. This logic leads to the conclusion that the tension of strips wound into strip rolls can be increased in cases of a stronger initial strip plate and with a greater total deformation of the strips during the cold rolling process. As a result, it was proposed [126] that the tension between the last stand and the coiler of a cold rolling mill (in particular, the 2500 four-stand mill of the Magnitogorsk Metallurgical Combine) be set equal to 0.8–0.9 yield strength of cold-rolled steel at a temperature of 710°C.

The authors of [126] also confirmed the previously expressed opinion that the coiling of cold-rolled strips into strip rolls at constant tension is inefficient and has drawbacks. Namely, it is difficult for the coiler to grip the front end of the strip. In addition, the outer turns are loosened when the strip roll is removed from the coiler drum. Then, when tightening the turns during the unwinding of the roll, scratches form on the surface of the strip. To exclude these negative phenomena, it was suggested that the first five to ten turns be wound with a tension 1.5–1.7 times greater than the nominal one, and the outer eight to ten turns – with a tension equal to 1.2–1.3 times the nominal tension at which the main part of the strip roll is produced. This tension mode is similar to that shown in Fig. 9.36 (in the middle in the upper row). The only difference is that when winding the last outer turns, the tension of the strip is sharply increased. It is emphasized that in general, the tension level of the coiled strip should be minimal, but it is important to exclude the displacement of the strips in the strip roll when they are unwound in the skin pass rolling mill.

Following the logic of the authors of this paper [126], it can be concluded that the tension conditions of the cold-rolled strips when winding into strip rolls should take into account the properties of steel (the steel grade), the temperature–deformation conditions of hot rolling and many other factors that affect the plastic deformation resistance of the surface microrelief of strips in rolls when they are heated and held during annealing in hood-type furnaces, on the tendency to weld the surfaces of the strip contacting under pressure at high temperatures. This approach is most suitable when choosing the tension of cold-rolled strips when winding them in strip rolls. The only question is how important is the impact of each of the above factors on the stress–strain state of rolls during annealing and interturn welding of the strip. And whether due to weakly influencing factors, it is necessary to complicate the technology of cold rolling and the rolling of rolled strips into rolls.

We note that the approach to the consideration of the phenomenon of welding the strip turns in the strip rolls during the annealing of cold-rolled steel, proposed in a paper worthy of note [126], does not take into account the effect of restoring the elastically deformed microirregularities of the metal surfaces and the changes due to this stress state of the roll when removed from the coiler drum and then heated during annealing.

The known fact of reducing the interturn welding in the strip rolls during annealing with increasing contamination of the surface of cold-rolled steel should, of course, not be associated with the tension regimes of the strips when they are being coiled into strip rolls in a cold rolling mill. The amount of the grease component in the emulsion, the degree of its purification, measures to remove the remnants of the emulsion from the surface of cold-rolled strip wound into the strip rolls should ensure high surface finish of the finished sheet rolling product, its marketable appearance and its suitability for subsequent processing.

During rolling, the actual relative reduction deviates from the nominal due to the thickness of the strip plate. Taking into account the fact that initial thickness differences in the cold rolling of strips in modern mills are practically eliminated, the deviation of the relative reduction from the nominal one can be defined as the ratio of the input thickness difference to the nominal thickness of the strip plate.

The actual total relative reduction of each cross section of the rolled strip in the method of correction of the winding tension of the strip roll is considered as the algebraic sum of the nominal relative reduction and the indicated deviation, which must be introduced into the system of the actuator synchronously with the output from the last stand of the corresponding section of the strip. In this case, the obtained actual relative reduction is compared with the value of the relative reduction to which the initial tension setting corresponds. If there is a deviation, a signal is given to the actuator to correct the amount of winding tension.

Thus, the application of the method considered allows, when changing the range as well as when the thickness of the strip plate deviates from the nominal value, this correction of the tension on the coiler thanks to which high-quality dense winding is achieved, and when the strip rolls are subsequently annealed, the extent of welding of the strip turns decreases.

The mechanism of the effect of the total reduction of the strip in the cold rolling process on the extent of welding of the turns in strip rolls during the subsequent annealing of cold-rolled steel is primarily due to an increase in the temperature of the rolled strip with an increase in the degree of its deformation. A temperature increase affects the level of the interturn pressures in the roll in the same way as the increase of the winding tension.

In production conditions, it is convenient to apply the winding method according to which rolls of small diameter, for example single-stranded, are wound at increased tension. If the strip roll consists of two or three strips, then after winding about ⅓ of the strip roll, the tension of the strip is reduced by 1.2–2 times. Under such winding regimes, the interturn pressures are approximately the same in both small and large strip rolls. As already noted, many rolling mills do this. Examples of the applied tension regimes in winding cold-rolled strips are shown in Fig. 9.36.

In parallel with the use of effective methods for winding strip rolls of cold-rolled strips, it is advisable to reduce metal welding during annealing by applying weld inhibitors to the surface of the strips. Thus, an aerosil (finely dispersed silica) can be introduced in an amount of 0.001–0.005% in the lubricating–cooling liquid during rolling as an inhibitor of welding. In order to increase the protective properties of the metal from welding and improve the stability of the lubricating fluid, it has been proposed to additionally introduce polyacrylamide in its composition, and as a welding inhibitor use silica powder with the following component ratio,% by weight: polyacrylamide 0.01–0.05; silica (converted to SiO_2) 0.05–0.2; lubricating–cooling liquid – the rest.

During the annealing of cold-rolled strips at temperatures above 680°C aerosil or silica sol reacts with oxides of iron, aluminum, manganese and forms a film on the steel surface, which protects the turns of rolls from welding. Polyacrylamide is introduced to stabilize the silicic acid sol (silica) in lubricating–cooling liquids. Studies have shown that, for example, additions of silica sol in an amount of 0.2–0.3% and polyacrylamide in an amount of 0.05% to a lubricating–cooling liquid based on emulsol OM during cold rolling of 08kp steel provide a reduction in the peel strength of welded sheets during subsequent annealing from 2.9 to 0.19–0.2 N/mm^2. Such additives to lubricating–cooling liquids used for rolling

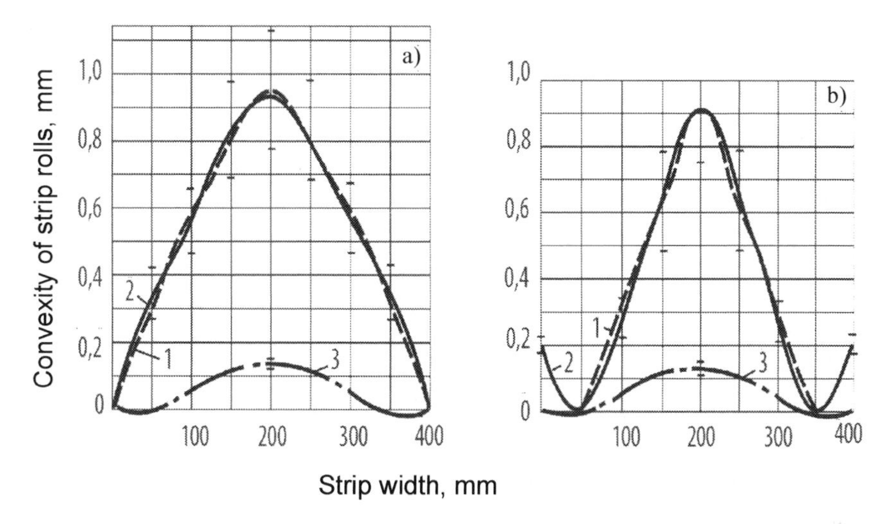

Fig. 9.40. Profiles of the generatrices of the strip rolls (1 – approximated; 2 – measured) and coiler drum (3 – measurements) of the 540 mill while rolling flat strips (a) and strips with undulations (b) [132].

have been developed by the companies of Germany, USA, Japan. The results of laboratory experiments to evaluate the effect of means based on liquid potassium glass on the welding of strip turns in strip rolls during annealing are presented in [122]. At concentrations of this substance in water of ~0.5%, the separation force of the contacting surfaces of the metal decreased by 2–3 times. However, it should be borne in mind that such additives can contaminate the equipment of the rolling mill and worsen the cleanliness of the surface of the finished sheet rolling product.

When winding the strips with irregularities in flatness and transverse thickness differences in the places of distortion of the profile and shape, the contact pressures between the turns substantially increase. This creates the prerequisites for local (in certain sections of the strips along their width) adhesion–welding of surfaces of adjacent turns. When unrolling strip rolls bands form in these areas – slip lines, less frequently fractures. In the case of a strong local distortion of the profile and the shape of the strips after rolling, the plastic deformation of the strip occurs with the formation of a 'trough' and the formation of slip lines on its surface during the unwinding of the strip rolls. When passing through the tensioner and the rolling rolls, a fold may form on the trough site.

The distortions of the shape of the cold rolled strip caused by the unsatisfactory profile of the cross-section of the strip plate lead to stronger sticking–welding of the turns in the coils and coarse bands – slip lines compared with distortions of the shape due to violation of the cold rolling technology, for example, local roll overheating. The most disadvantageous from the point of view of local sticking-welding of the strip turns in the strip rolls is the concave cross-section profile of the strip plate. In this case, the defects in question are formed at the edges, which significantly increases the risk of breaks of strips during the subsequent skin pass rolling. The best for the production of thin cold-rolled strips and tinplate is the hot-rolled strip plate which has a convex profile of the cross-section with a transverse thickness difference of 0.02–0.05 mm. Local distortions of the profile of the cross-section of the rolled strip plate should not exceed 0.02 mm, and the thickness difference of the edges should not be more than 0.03 mm. More details about this were mentioned in the previous chapters of the book. Here we consider only the effect of the convexity of the profile of the cross-section of cold-rolled strips, the transverse thickness difference and local thickenings on the stress–strain state of the strip rolls and the consequences resulting from these factors.

Strong thickening on local sections of hot-rolled metal (strip plate) at cold rolling with high degrees of deformation (up to 90%), as a rule, is transformed into local undulation of the cold-rolled strips with all accompanying negative effects. This is particularly evident in the production of very thin strips and tinplate [50, 126]. But even small local thickenings along the width of the strip plate during cold rolling are not smoothed out but are transformed into similar local thickenings of the cross-section profile of the cold-rolled strips. The magnitude of the decrease in the height of local thickenings, as well as the values of the total convexity of the profile of the cross section of the strip plate, is proportional to the coefficient of the total stretching of the metal during cold rolling, i.e., in the case of keeping the flatness of the metal unchanged, the initial relative transverse thickness difference of the strip plate is inherited by the profile of the cross-section of the cold-rolled strips. When coiling into the strip rolls, the adjacent turns of the strip naturally contact primarily in the areas of thickening of the metal, where stress concentration occurs, and the interturn pressures increase.

The production practice showed that on the strips of tinplate after cold rolling almost always have minor local thickenings hereditarily passing from the strip plate. Local thickenings on the width of the profile less than 0.3–0.5% of the thickness of the strip, as a rule, do not lead to distortion of the shape during rolling and skin pass rolling. However, when winding a thin (0.25 mm or less) tinplate with sufficiently high tension as a result of superposition of these thickenings, local ring-shaped bulges form on the roll, which can lead to distortion of the shape of the strips. After the unwinding of the strip rolls, the distortions in the shape of the bands have the form of local undulations. To prevent such a defect, at the Karaganda Metallurgical Combine the tinplate strips are coiled into strip rolls at a specific tension of no more than 35–40 N/mm^2.

When rolling extremely thin strips and tapes on the reversible mills, due to the uneven distribution of tangential and radial stresses in rolls across the width of the strips caused by their transverse thickness difference, folds and the relative displacement of turns of strip rolls may form in the transverse direction, i.e. there may be the loss of stability of the strip rolls. In the opinion of the authors of [132, 133], this phenomenon is associated with the emergence of the convexity of the generatrix of the strip roll due to the convex profile of the cross section of the strip. As the strip roll is wound, the convexity of its generatrix increases, which is accompanied by an increase in longitudinal stresses in the middle part of the coiled strip. These elevated longitudinal stresses cause folding on the strip and on the strip roll. If the strip roll generatrix has a concave shape due to uneven wear of the coiler drum, the liner installed on it or the concave cross-sectional profile of the strip, the folds on the strip are formed under the influence of stresses already directed perpendicular to the rolling axis to the centre of the strip. The wedge shape of the strip cross-section leads to a displacement of the turns in the strip roll, sliding them to the left or to the right with respect to the rolling axis.

The results of measurements of the profile of the generatrix of the strip rolls during the winding of steel strips with thicknesses of 0.10–0.15 mm and waviness along the edges are presented in [132, 133] (wave amplitude 2.38 ± 0.1 mm, wavelength 148.8 ± 45.3 mm) and a transverse thickness difference of 2.0–2.5% of the thickness. The profiles of the generatrix of the strip rolls were measured on a four-roll mill 135/480 × 450 and a 20-roll mill 540. The profile of the coiler drums had a convexity of 0.13 ± 0.26 mm. The profile of the generatrix of the strip roll during the winding of the practically flat strip had a convexity of 0.94 ± 0.17 mm (Fig. 9.40 *a*). When winding the strips with waviness along the edges, the strip rolls

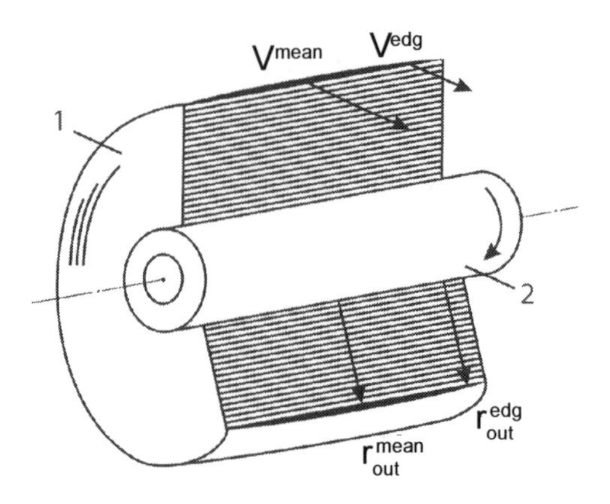

Fig. 9.41. Mechanism of winding cold-rolled strips with a convex cross-sectional profile: 1 – strip turns in a strip roll; 2 – coiler drum. The remaining notation is given in the text.

had a convexity in the middle part along the width of the strips of 0.91 ± 0.16 mm (Fig. 9.40 *b*).

When experiments were carried out on a reversible mill 135/480 × 450, sleeves with a concave profile were mounted on the drums of the right and left coilers. The diameter of the sleeve along the rolling axis was 1.4 mm less than at its edges. The transverse thickness difference of the 0.1 mm thick strip was approximately 2 μm. The tension of the strip when winding the strip rolls was 198 N/mm^2. Under these conditions, the difference in the diameter of the roll with 9087 turns in the middle of the strip and at a distance of 5 mm from the edge was 1.65 mm. The profile of the strip roll in the middle of the strip was convex. Due to the transverse thickness variability of the strip, the turns in the roll were in contact in the middle part of the strip, and at the edges gaps were formed, close in size to the magnitude of the thickness differences of the strips.

Studies have shown that when winding a strip roll, the first turns of the strip fit tightly to each other along the entire width. As the strip roll is wound on the drum due to the convexity of the profile of the cross section of the strip the tension near its edges weakens and becomes equal to zero, which leads to the formation of interturn gaps along the edges of the roll. After winding the third part of the strip roll, the zone of tight contact of the turns is located on a section equal to 0.7 of the strip width. In such cases, the winding tension is distributed unevenly across the width of the strip. The magnitude of longitudinal stresses in the middle of the strip increases 3.5 times, which can lead to its breakage. Radial and tangential stresses are concentrated in the middle part along the width of the strip roll. There is a growing danger of welding the turns during annealing or the loss of stability of the strip rolls and the formation of 'bird' defects.

The results of the authors of [132] are in agreement with the conclusions of other researchers. Figure 9.41 shows the mechanism of strip roll formation as an illustration to the above. As can be seen, because of the convexity of the cross-section of the strip, the outer radius of the upper turn in the middle of the width of the strip r_{out}^{mean} is larger than the radius near the edge of the band r_{out}^{edg}, i.e., $r_{out}^{edg} > r_{out}^{mean}$. As a result, the speed of the strip surface during the formation of the upper turn in the middle part will be greater than at the edges of the strip roll: $V^{mean} > V^{edg}$. Accordingly, with the constant value of the total tension created by the coiler, the specific tension along the width of the strip will not be the same. In the middle part along the width of the strip, the winding tension will be greater than at the edges. For example, when winding a strip roll from strips with a convexity of 2% of strip thickness with an average tension of 40 N/mm^2 in the middle part of the width, the tension level reaches 90 N/mm^2. After removal of the strip roll from the coiler coiled in such conditions, its stress–strain state varies in different ways in different sections along the width of the strip, and this affects the flatness of the metal.

Obviously, if a coiled strip with a convex cross-section profile has a certain warpage in the width in the middle part, then the distribution of tensile stresses along the width of the strip can be uniform even if there is a convexity of the generatrix of the strip roll. In this case, the stress–strain state of the strip roll will be the same in all sections along its width. When the strip roll cools down after removing from the coiler drum and then heating it during annealing, the level of radial and tangential stresses will decrease. Changes in the temperature field will be accompanied by a redistribution of stresses in the radial and axial directions of the strip roll, which, of course, will be influenced by the initial warpage of the cold-rolled metal.

It was shown above that the magnitude of the interturn gaps depends substantially on the contact pressure of the adjacent turns of the strip and varies most strongly at the initial moment of loading, when the radial forces in the strip roll are still relatively small (Fig. 9.4). Since the maximum interturn pressures in coils of strips with a convex cross-section profile are concentrated in the middle part along the width of the strip roll, it is at these sites that the surfaces of the contacting coils come closest to each other. The magnitude of this approach, i.e. the amount by which the interturn clearances decrease under the action of the load is commensurable with the magnitude of the convexity of the profile of the cross section of the strip. For example, if the convexity of the profile of a roll of hot-rolled strips 2.5 mm thick is 0.05 mm, then the convexity of cold-rolled strips 0.5 mm thick will be approximately 0.01 mm, which is 2% of their thickness. According to the experimental data shown in Fig. 9.4, the gaps between the turns of the strip are reduced by a close amount under the action of the interturn pressure, which occurs when winding the rolled strips into strip rolls. Therefore, the strip rolls consisting of two or three thousand turns of cold-rolled strips with a convex cross-section profile do not have a significant convexity.

It should also be noted that the resistance to bending in the transverse direction of the tension-wound strip increases with increasing thickness h, Young's E of the material, and decreases with increasing width B of the strip. Drawing an analogy with the resistance to bending under a load of a beam of rectangular cross section, the bending probability of the strip can be assumed to be proportional to B^2/Eh^2. In accordance with this assumption, when winding comparatively thick and narrow strips into strip rolls, the effect of the convexity of their cross-sectional profile on the convexity of the generatrix of the strip roll is weakened. The non-uniformity of the distribution of the specific tension along the width of the strips and the concentration of the interturn pressure in the middle part of the strip rolls decrease with all the ensuing consequences.

The considered mechanism of formation of strip coils in the winding of strips with a convex profile of the cross section and the resulting problems of the possible loss of stability of the rolls are most relevant for the production of tinplate, with an increase in the width of the rolled strips and a decrease in their thickness. From all that has been said there follows the obvious conclusion about the need to measure and adjust the tension along the width of the rolled and coiled strips. Such a technical solution has been proposed in the method for winding cold rolled strips [134].

The authors of [135, 136] calculated the stress–strain state of strip rolls of cold-rolled strips in the positions 'on the coiler drum' and after removal from the drum using the dependences (9.12) published in our book [127]. Next, comparison was made of the zones of maximum stresses in strip rolls after removal from the coiler and when they were heated during annealing with the places of the most probable appearance of 'coil break 'defects during the skin pass rolling of the strips. As expected, the welding of turns in the strip rolls during annealing and the formation of 'coil break' defects are observed approximately in the middle part of the winding thickness, where the radial stresses reach the highest values. The above conclusions on the need to regulate the tension in the width of coiled cold-rolled strips, depending on the profile of their cross-section in these publications, are supported by experimental data obtained at the 2030 five-stand mill of the Novolipetsk Metallurgical Combine. To create a relatively uniform stress field along the width and thickness of the winding of coils of various thickness strips, it is necessary to reduce the specific tensions in the thickened zones along the width of the strips and increase in areas where the thickness of the strips is smaller. The tension along the width of the bands is proposed to be changed by adjusting their flatness. The greatest danger of the appearance of 'coil break' defects is when the strips with an asymmetric (wedge-shaped) cross-section profile are wound into strip rolls with a uniform distribution of the specific tension along their width. Strips with a convex cross-section profile must be wound up into strip rolls, creating maximum tension along the edges of the strips, and minimal – in the middle part along their width. Recommendations [135, 136] on the prevention of breaks in the unwinding of the strip rolls in the tempering mill are similar to those given in [42, 43]. It is proposed to install an additional roller at the entrance to the tempering mill, which must be bent by a strip and move in a horizontal plane in the direction of the stand as the strip roll is unwound. The use of such a scheme of unwinding the roll in the tempering mill makes it possible to reduce the tearing stresses of the upper turns of the strip when they stick together and reduce the stresses from the bending of the unwound strip. Due to the movement of the roller in the horizontal plane, the angle of separation of the turn of the strip from the strip roll remains constant. Sharp bbutts of the strip when separating the turns from the strip roll in the case of their welding during annealing, leading to 'coil break' with thinning, are excluded.

As shown in the paper [132], in industrial practice, in the production of particularly thin metal, in order to eliminate or mitigate the negative effects caused by the convexity of the cross-section of coiled strips, the drum of the coiler or the sleeve on which the coil is wound is specially profiled. For example, the sleeve has a concave profile. As the roll is wound, as shown above, the pressure on the drum increases. As a result, its surface bends. Due to the sliding of the segments, the rigidity of the winding drum changes. Proposals have been made for the regulation, change of the convexity of the coiler drum during the winding of the strip roll with a special hydraulic device located inside the drum. However, so far there is no information on the implementation of such decisions. Apparently they are difficult to perform, and their effectiveness is not as great as it would be desirable.

The defects like 'coil break' and "bands–slip lines" appear mostly when unwinding the strips from strip rolls (at the moment of stripping the strip from the body of the strip roll). It is recommended to avoid bending of the strips with a radius of curvature less than the critical value, which depends on the mechanical properties, the thickness, the specific tension of the strip and is quantified for the prevention of kinks (for a metal with a yield strength of 240 N/mm^2 with a strip tension of 70 N/mm^2) by the ratio $R_{curv} \geq 235h$, where R_{curv} is the radius of curvature of the strip at bending, mm; h is the thickness of the strip, mm [42, 43].

The allowable curvature radius decreases with increasing yield and a decrease in specific strip tension. The scheme of strip roll unwinding in the temper mill should provide a minimum level of total bending deformation.

In the annealing process and during transportation of the strip rolls in the thermal department of the cold rolling plants it is important to minimize the possibility of damage to the surface of the rolls. Areas of mechanical damage of the generatrix and the butt surfaces of the strip rolls are additional zones of sticking–welding of the turns.

The coilers and uncoilers of the high-speed temper and rolling mills in the cold rolling shops work, as is well known, in intensive modes with significant dynamic overloads arising primarily because of the non-concentricity of the strip rolls and drums, which in turn causes fluctuations in the strip tension. These oscillations lead to shock loads on the electrical and mechanical parts of the equipment, the appearance of longitudinal thickness differences and distort the given stress in the strip roll. Wobbling of the strip rolls during unwinding at the temper mill worsens their axial stability and causes the appearance of slip lines due to the uneven distribution of tension in the strip width [123].

Studies of the NIITyazhmash (Scientific Research Institute of Heavy Engineering, Sverdlovsk) showed that the amplitude of strain vibration substantially depends on the speed of rolling (temper rolling). The maximum value of the amplitude is observed in resonance areas at the first natural frequency of the coiler–strip–stand system. High-frequency (reverse) vibrations of tension on the coiler and the uncoiler are not eliminated by the existing tension regulators, since the current of the drive responds to instant changes of tension with a large delay. Lowering the oscillation amplitude of the tension can only be achieved with decreasing beating of the strip roll and the coiler drum (decoiler), which in turn is ensured by fulfilling a number of requirements to the coilers: equal stiffness when receiving radial loads to avoid uneven deformation of the drum during coiling; maximum concentricity relative to the axis of rotation of the drum surface; maximum rigidity of the machine supports.

When designing the winding structures and determining the speed regimes of rolling or temper rolling, it is necessary to take into account the dynamic loads that arise as a result of the oscillatory process in the coiler–strip–stand system. The main technological measures that help to reduce tension fluctuations during winding-unwinding operations include: minimizing insufficient coiling in strip rolls; periodical grinding of the shafts of coiler drums in assembly with sectors to exclude their non-concentricity appearing during operation; exclusion from working intervals of rolling or temper rolling of strips of the speeds corresponding to resonance zones.

Beating of rolls on the unwinding machine and coiler of the temper mill adversely affects the properties of the metal. Due to the beating of the rolls, there is a periodic change in the tension of the treated strip, which causes inconsistent strip deformation and temper forces. A consequence of this is the unevenness of the mechanical properties and surface roughness along the length of the strip.

All of the foregoing allows us to conclude that the choice of modes for winding operations should be carried out on the basis of an integrated approach (calculation of the stress state of rolls, verification of the regimes by their stability criteria, selection of the magnitude and law of change in the tension of the winding taking into account specific technological parameters – dimensions and mass of strip rolls, the surface of coiled strips). When selecting the winding modes, it is necessary to take into account the following patterns of stress formation of strip rolls:

– the dependence of the interturn pressure on the winding tension by a value of up to 100 N/mm^2 is close to linear;

– the value of the interturn pressures is directly proportional to the thickness of the strip (with the same thickness of winding); in the thickness range from 0.15 to 0.7 mm this dependence is also close to linear;

– distortions in the profile and shape of strips wound in strip rolls cause a substantial redistribution of the interturn pressures in the width and thickness of the strip roll;

– with an increase in the roughness of the strip surface, the interturn pressure decreases, the presence of lubricant on the strip increases the interturn pressure only for a small roughness of the metal surface (Ra < 0.5 μm); after removing the strip roll from the coiler drum, a significant reduction in the interturn pressures is observed in the inner turns (20–30% of the total thickness of the winding) of the strip roll;

– the factor strongly influencing the stress–strain state of coils is the temperature distribution along the length of the coiled strip.

Increasing the stability of winding–unwinding operations and improving the quality of the strips is facilitated by a decrease in the tension oscillations arising from the non-concentricity of the rolls and the drum of the winding machine. To increase the stability of large-sized strip rolls during temper rolling, it is advisable to limit the acceleration at the initial moment of unwinding (not more than 0.8 m/s^2).

9.7. Stress-strain and temperature condition of hot-rolled strips

Technological modes of production of hot-rolled strips usually involve the end of rolling in the last finishing stand at temperatures of 840–930°C, cooling of the strips on the outgoing roller table and winding it strip rolls in the coilers at 550–700°C, depending on the steel grade. This has already been mentioned in the previous chapters of the book. The rolls of hot-rolled strips are then fed to the conveyor and transferred to the warehouse for cooling before further processing. The conveyor also serves as a refrigerator. On high-performance mills, it is not always possible to reduce the speed of the conveyor to increase the cooling coil length. Therefore, when designing mills, the total length of conveyors is often increased, which entails quite understandable losses and complexities.

The strip rolls received in the hot-rolled products warehouse should be cooled to 30-50°C. In the 2000 wide-stirp hot rolling mills (WSHRM) of the Novolipetsk and Cherepovets metallurgical plants, for example, the maximum mass of rolls is 36 tons. The duration of natural cooling of such rolls is 3–5 days. Depending on the size of the strip rolls, the time of natural cooling can be equal to 6–8 days. Such long cooling requires large areas and lengthens the production cycle. The process of obtaining commercial rolled products can be greatly shortened and the area of storage facilities can be reduced by the use of accelerated cooling of the strip rolls.

The development of WSHRM goes along the path of increasing the mass of rolls of rolled strips. In the future, the mass of WSHRM rolls can be 60–70 tons. It is possible to predict cooling conditions for such strip rolls only on the basis of a theoretical model adequate to the actual process of cooling them in industrial conditions. In our studies [137–139], various methods of accelerated cooling of coils weighing up to 36 tons under the conditions of one of the existing WSHRM were investigated, a method for calculating cooling conditions was developed, the adequacy of the theoretical model and the reliability of the results of experimental studies were evaluated, and based on these materials the conditions of the accelerated cooling of the strip coils weighing up to 70 tons analysed.

In industrial practice, the rolls of hot-rolled and cold-rolled strips are usually transported in an upright position. Such a method of transportation often leads to their damage, and also requires an increase in the mass of the load-lifting devices of bridge cranes. When transporting the strip rolls in a horizontal position (the roll axis is horizontal), these drawbacks can be avoided. However, in this case, there is a danger of the loss of stability of the rolls, the danger of their sagging. When choosing the method for transporting rolls of hot-rolled strips in the projected sheet rolling shops, the problem arises of determining the stress–strain state and the stability of the rolls to sagging. This problem was solved [137, 139] analytically with the use of the finite element method and by experimental studies in the production conditions of the Novolipetsk Metallurgical Combine.

Influence of cooling conditions of rolls. Experimental studies of various methods of forced accelerated cooling of horizontally located rolls of hot-rolled strips were carried out in the industrial conditions of the operating WSHRM of the above combine. For this purpose, special devices have been made (Fig. 9.42) for air-cooled coil cooling

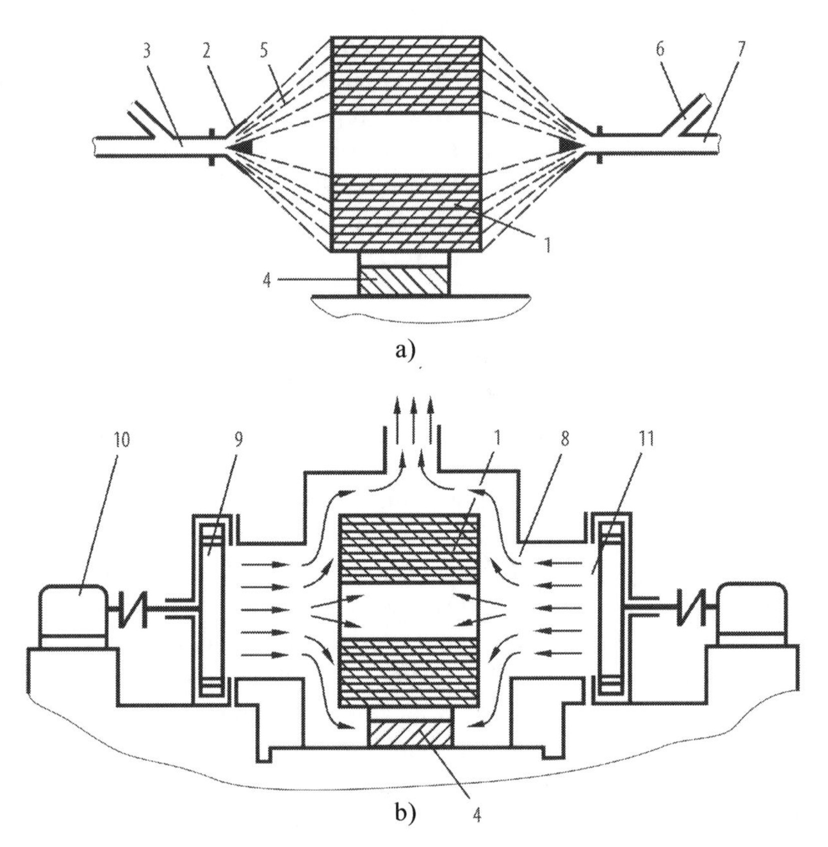

Fig. 9.42. The experimental apparatus for an air–water (a) and accelerated forced air cooling the strip rolls (b); 1 – roll; 2 – nozzle; 3 – mixer; 4 – cradle; 5 – air-water mixture jet; 6 – water supply; 7 – air supply; 8 – stand; 9 – fan; 10 – electric motor; 11 – airflow.

when the coolant is fed to its end surfaces and for forced blowing of a horizontally placed roll by air from the end surfaces. During the experiments, the temperature in the middle of the strip roll was recorded with a thermocouple installed between its turns, and the total duration of cooling of the strip roll (at the most 'lagging' point) was determined. Rolling and winding of strips into strip rolls was carried out according to the technology adopted in this mill.

The results of studies of cooling strip rolls of hot-rolled strips in different ways are shown in Fig. 9.43. The analysis of the obtained results showed that practically all strip rolls of a mass of 29–31 tons, rolled from strips of different thicknesses from carbonaceous and low-alloyed steels, cooled down for 10-12 hours with water-cooled cooling (at the maximum flow of water and air to create a mixture). With a decrease in water consumption and a constant air flow, the

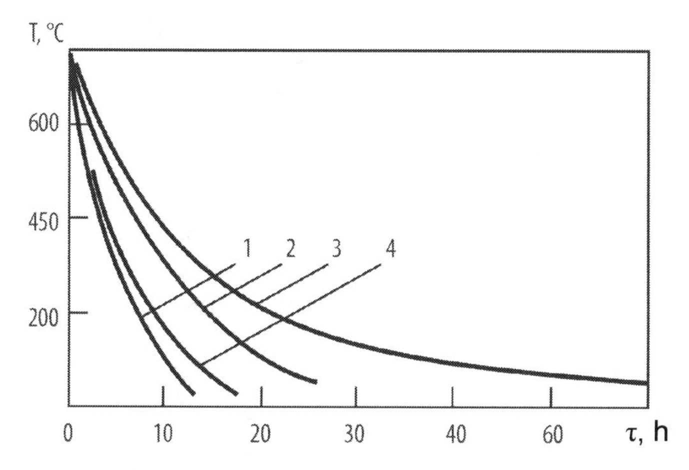

Fig. 9.43. Change of the strip roll temperature (*T*, °C) at the installation site of the thermocouple (in the middle of winding the roll and the middle part of coiled strip width) depending on the duration (τ, h) and cooling methods: 1 – water–air; 2 – forced air; 3 – natural; 4 – forced air cooling of strip rolls weighing 15 tons; for 1, 2 and 3 – strip rolls weighing 29–31 t; strip rolls wound from the 4–6×1450–1550 mm strips.

cooling time augmented by an average of 30% and amounted to 12-13 hours. When it was blown from two sides of the end surfaces of the strip roll with fan air, it was 23–24 h for 29–31 tons of strip rolls and 13–14 h when their weight was reduced to 15 tons.

Thus, in the process of industrial tests, it was found that the water–air cooling of horizontally located strip rolls is more effective than forced air cooling and accelerates cooling of the strip rolls by a factor of 2. In comparison with the natural cooling ([128], Fig. 9.43, curve 3), forced air cooling of the strip rolls is more effective 2.0–2.5 times, and water–air 4.5–5.0 times.

In the theoretical calculation of cooling, the strip roll was taken as a multilayer cylinder characterized by an average (in the temperature range) heat capacity, density, and coefficient of thermal conductivity in the axial direction.

The coefficient of thermal conductivity in the radial direction depends on the density of the strip roll and the thickness of the strip. The process of cooling hot coils was described by a two-dimensional equation of thermal conductivity, which has the following form in the cylindrical coordinates:

$$cQ\frac{dT}{d\tau} = \lambda_r\left(\frac{d^2T}{dr^2} + \frac{1}{r}\cdot\frac{dT}{dr}\right) + \lambda_z\frac{d^2T}{dz^2};$$

with the initial conditions $T(r, z)_{r=0} = T_0$; the boundary conditions on both cooling surfaces – convective heat transfer:

$$\lambda_i \frac{dT^k}{dx_i} = -\alpha^k \left(T^k - T^k_{\text{med}}\right),$$

where i – index of directions (radial or axial); k – the index of the heat transfer surface; α^k – coefficient of heat transfer from the surface k, W/(m^2·K); λ_i – coefficient of thermal conductivity in the direction i, W/(m·K); c – specific heat, J/(kg·K); Q – density, kg/m^3; T^k – cooling surface temperature, °C; T^k_{med} – ambient temperature from the respective surfaces, °C.

This equation with the indicated initial and boundary conditions was solved by numerical methods, representing it in finite-difference form using the variable direction method [140]. Calculations of cooling of the strip rolls by natural, forced air, water–air and water methods were carried out. The coefficients of heat transfer by convection were determined specifically for each cooling method, starting from the initial temperature of the coolant (water and air) equal to 20°C, the speed of their movement and the surface temperature of the rolls of 500–600°C. For natural and forced cooling by air, the heat transfer coefficient values α_1 = 17.5 W/(m^2 · K) and α_2 = 46.5 W/(m^2 · K), respectively, the heat transfer coefficient by convection with water-cooled cooling α_3 = 2300 W/ (m^2 · K). When the strip roll of the hot-rolled strip is immersed in water, α_4 = 20–30 W/(m^2 · K), depending on the surface temperature. A larger value of α_4 was taken for a roll with a surface temperature of 500–550°C. The coefficient of thermal conductivity in the axial direction for low-carbon steel in the interval 50–700°C was assumed to be 46.4 W/(m · K). The coefficient of equivalent thermal conductivity of the roll in the radial direction was determined from the dependence:

$$\lambda_e = (1 - a/100)[\lambda_{\text{air}}/(1-\eta) + \alpha h/\eta] + \alpha\lambda_{\text{m}}/100,$$

where a – the degree of contact, % (a = 3% for the strip roll at a winding density η = 0.9–0.98); α – coefficient of heat transfer by radiation through the gas interlayers, α_{m} = 96 W/(m^2·K); λ_{m} – thermal conductivity of metal, $\lambda^{\text{mean}}_{\text{m}}$ = 46.4 W/(m·K); λ_{a} – thermal conductivity of air in the interlayer, $\lambda_{\text{a}} = 4.8 \cdot 10^{-3}$ W/(m·K); h – thickness of the strip roll, mm. Its values used in the calculations are as follows:

Strip thickness mm	2	6	10	15	30
λ_e, W/(m·K)	1.74	2.17	2.64	3.21	4.64

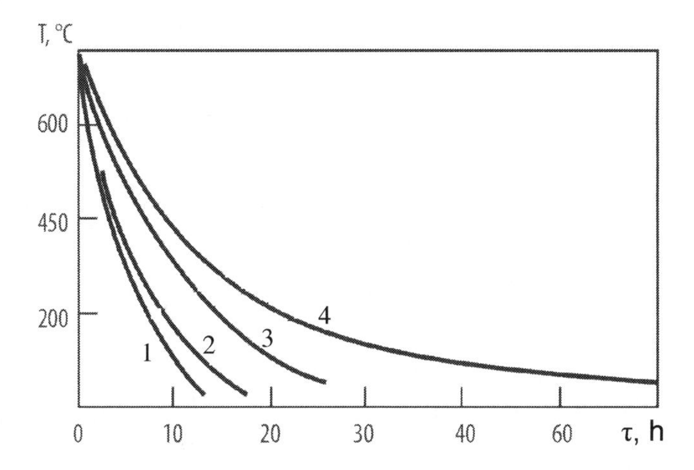

Fig. 9.44. Calculated values of the temperature change inside the strip roll at different ways of cooling: 1 – water; 2 – air–water mixture; 3 – forced air; 4 – natural air

The results of calculating the temperature of the turns of the strip rolls with different methods of forced accelerated cooling are shown in Fig. 9.44. Thickness of the rolled strip h = 6 mm; inner diameter of the roll 850 mm; outer – 2080 mm; the width of the strip is 1400 mm. Experimental studies of the conditions for forced accelerated cooling of rolls and their theoretical calculation showed that the developed mathematical model corresponds to the process under consideration. For example, the productivity of the process of air–water cooling of strip rolls during experimental studies was 2.0–2.1 t/h, and according to the calculated data 1.9–2.0 t/h.

As expected, the calculated data confirmed the fact that the most prolonged is the cooling of the strip rolls in the air. The productivity of the process when cooling strip rolls weighing up to 70 t does not exceed 0.55 t/h. The most effective methods are water–air and water cooling. In comparison with cooling of strip rolls in air with water-cooled cooling, its duration is reduced by 3.5–5.0 times, with cooling in water – by 4-7 times, with forced cooling by air – by 1.5-2.5 times. The cooling rate of the outer and inner coils of rolls as compared to the cooling of strip rolls in air increases in forced air cooling by almost 2 times, with water–air cooling by 3-3.2 times, and in water cooling by at least 30–40 times. The efficiency of water cooling is 30–33% higher than that of water–air cooling. However, the cooling rate of the outer and inner turns in water is an order of magnitude higher than that of the central turns in the winding thickness, which leads to a significant difference in the mechanical properties and microstructure along the length of hot-rolled strips.

Thus, from the point of view of process productivity, it is most expedient to use air–water cooling when cooling rolls of hot-rolled strips. The productivity of the process with this cooling of the strip rolls weighing up to 70 t will be 2.4–2.6 tons per hour. Water cooling is preferable to use in the final stage, starting, for example, from 350–300°C, i.e. when the structural transformations have ended in the metal. However, in the high-performance WSHRMs, the organization of water cooling is associated with considerable difficulties, and water–air cooling of rolls is easier to implement.

To study the influence of methods of forced accelerated cooling of the strip coils on the structure and mechanical properties of sheet products, samples were taken during the course of the studies for the length of the strips from rolls subjected to forced cooling and cooled naturally in the air. The results of mechanical tests of sheet metal samples from the experimental and comparative strip rolls are the following (WA, N, FA – cooling with a water–air mixture, natural and forced by air, H, M, T – head, middle and tail parts of the strip) are given in our paper [137].

According to the data obtained, during water–air cooling of strip rolls, the strength properties of the metal in the main part of the strips are 10–40 N/mm^2 higher than under natural cooling. A smaller effect on the strength properties of the metal strips, rolled up into a strip roll, is exerted by their forced cooling by air. The yield stress σ_T and ultimate strength σ_B of the test metal, as compared with the metal of air-cooled strip rolls, increased by 5–10 N/mm^2. Studies of the microstructure of samples taken from coils subjected to water-air cooling showed that the test metal, in comparison with the metal cooled in air, has a finer ferrite (1–2 points) and larger bands of the the structure. This is probably due to the fact that when the rolls are slowly cooled in air, carbon is redistributed, especially in the middle part (in the thickness of the winding) and the removal of internal stresses takes place, which leads to a decrease in the strength characteristics of hot-rolled steel.

To estimate the amount of scale formed on the surface of the hot-rolled strip and the duration of its subsequent pickling, depending on the cooling methods of the strip rolls, laboratory studies of samples taken from the head, middle and tail sections of the strips in experimental rolls were carried out.

Samples measuring 30×5 mm with a working surface of 0.003 m^2 were investigated. The samples were pre-washed in alcohol. The samples were pickled in a 15% solution of hydrochloric acid at 75

±2°C. The amount of scale on the hot-rolled strip was determined by cathode pickling in a 10% solution of sulphuric acid with the addition of 2 g/l of the C-5U inhibitor. Samples were taken from strip rolls subjected to water–air, forced air and natural cooling in horizontal and vertical strip roll positions.

The process of scale formation on the surface of the strip is characterized by two main stages: the intensification of scale growth due to the activation of the delivery of the oxidant (air, vapour, water) to the surface of the metal; a decrease in the intensity of scale formation with an increase in the cooling rate of the strips in rolls due to the rapid passage of the temperature range at which scale growth is observed.

The most uniform scale distribution along the length and width of the strip was observed for a strip roll subjected to forced air cooling. The greatest unevenness was found for a strip that was cooled in the air in an upright position. The most uniform pickling of the strip was ensured after water–air cooling of the strip rolls. Forced air cooling of the strip rolls did not ensure uniform pickling due to the formation at the edges of the strip of areas with lots of hematite (Fe_2O_3) in the scale. At the same time, the amount of scale on the strip did not increase. The increase in the fraction of hematite in the scale is explained by the significant activation of the process of delivery of air oxygen to the metal surface at an insufficiently high cooling rate.

Detailed results of the experimental studies of the distribution of the amount of scale over the length and width of hot-rolled strips and the duration of their subsequent etching, depending on the methods for accelerated cooling of strips in rolls, are given in our paper [137]. The most favourable in terms of the amount of scale and the duration of pickling are air–water cooling and natural cooling of the rolls in the horizontal position.

Thus, summarizing all of the above, it can be concluded that the duration of the pickling of the strips and the amount of scale on the strip essentially depend on the methods of accelerated cooling of the rolls. Water–air cooling of rolls leads to a 16–18% decrease in the amount of scale on the strip and 8–13% of the duration of pickling of sheet metal, which allows increasing the productivity of continuous-pickling units and reducing the consumption of pickling acid.

The stress–strain state of strip rolls of hot-rolled strips in the process of transportation. Uneven cooling causes uneven deformations along the winding thickness, which, depending on the

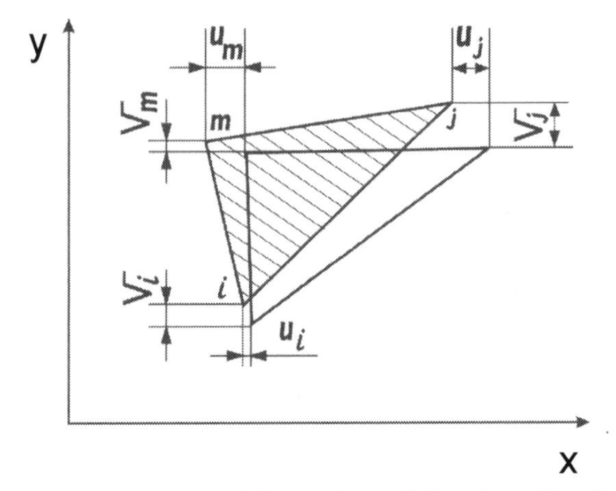

Fig. 9.45. Movement of the vertices of the triangular element *ijm*.

initial temperature distribution and cooling conditions, can lead to both stratification of the turns and to the appearance of compressive radial temperature stresses, i.e. to the compaction of the strip roll. Stratification of the hot-rolled strip in a roll is extremely undesirable, as the sagging of a roll under the action of its own mass sharply increases. Proceeding from this, it is necessary to determine the cooling conditions and the initial temperature distribution over the section of the rolls of hot-rolled strips, which ensure the occurrence in them of the radial compressive stresses.

In general, the mechanism of the effect of temperature on the stress–strain state and the propensity of strip rolls to sagging is the same as for the rolls of cold-rolled strips. However, there are some differences. The analyzed issue is extremely important in the production technology of both hot-rolled metal products and rolled stock for cold-rolling mills. Therefore, we will study it in detail. In this case, we show the possibilities and ways of solving the problem using the finite element method.

In solving this problem, the two-dimensional medium was divided into triangular elements [141]. The displacement of each of the vertices of the triangle *ijm* (Fig. 9.45) was expressed by the components u_i, v_i, u_j, v_j, u_m, v_m, forming a six-dimensional vector $\{\delta\}^l$ for some element *l*. Using the analytical dependences [141], the displacements u and v within the triangle element considered were expressed through the displacements of its nodes, and the deformations inside the element through the displacement of its vertices. Hooke's law was used for the transition from deformations of the element to stresses.

The relationship between nodal forces and displacements (the basic equation of the finite element method) has the form

$$[K]^{l}\{\delta\}^{l}=\{F\},$$

where {F} is the vector of nodal forces of the element; $[K]^{l}$ is the stiffness matrix of the element..

The equilibrium condition of the totality of finite elements is obtained by summing these relations in all respects,

$$[K]\{\delta\}=\{R\}$$

where {R} – vector of external forces applied to the system.

Solving the obtained system with an allowance for boundary conditions, the movements of nodal points were determined, and then deformations and stresses within each element were found.

To implement the finite element method, a set of special programs was developed. During calculations, rolls of hot-rolled strips were represented as an anisotropic hollow cylinder.

In many, especially in early studies, it was assumed that due to the small thickness of the strip, it is not necessary to treat each turn individually. Instead, we can introduce a hypothetical material whose behaviour corresponds to the behavior of the aggregate of a large number of turns as a whole. The properties of this material must take into account the experimentally established non-linear relationship between deformations and stresses in compression in a direction perpendicular to the location of the turns, and also their mutual slippage. That is, the roll is considered as an anisotropic body, the elastic properties of which are different in the axial and radial directions. The degree of this anisotropy depends on the density of the winding of the roll.

The foregoing approach makes it possible to replace the consideration of a roll as an aggregate of a large number of linear elastic elements – turns, in the interaction of which a constructive non-linearity arises, by investigating a solid body from a material possessing a physical non-linearity of elastic properties. At the same time, it is possible to take into account the stresses created in the strip roll when winding at tension, treating them as initial ones.

Even in this formulation of the problem, finding the exact analytical dependencies connecting the pressure between turns with technological factors of winding, is fraught with great difficulties. In our early paper [142] we give a solution convenient for practical calculations for the case of a strip with constant tension, obtained with allowance for the following assumptions: there is no relative slip between the turns in the strip roll; the turns are considered as concentric rings; the circumferential tension within a single turn remains constant, but varies from turn to turn; the strip roll is a solid body with only elastic deformation of the turns. Using these assumptions in the subsequent work [123], the problem was solved for the case of an anisotropic strip roll with an arbitrary change in the strip tension during winding. We obtained formulas for calculating the pressure of the roll on the drum q_b and the radial stresses at any point of the roll after removing the strip roll from the drum q^c, which were published in [123]. However, it should nevertheless be borne in mind that the above solution, taking into account the characteristics, the elastic properties and variation of the gap between the turns of a strip in a roll when considering its stress–strain state [107, 108, 144], allows studying and detecting effects that are not predictable in other approaches. At the same time, the proposed solution of the problem by the finite element method also has certain advantages, which will be shown below. The distribution of stresses arising in an anisotropic hollow cylinder, calculated by the finite element method under the action of its own mass, is shown in Fig. 9.46.

The stresses in strip rolls when coiling strips with tension, were calculated as initial with respect to the stresses caused by deformation of strip rolls under their own weight. If the resulting stress, obtained by superposition of the fields of the initial stresses caused by winding $(\sigma''_r, \sigma''_\theta)$ and the stresses from its own weight $(\sigma'_r, \sigma'_\theta, \sigma'_{r\theta})$ and expressed by the formulas $\sigma_r = \sigma''_r + \sigma'_r, \sigma_\theta = \sigma''_\theta + \sigma'_\theta, \tau'_{r\theta} = \tau'_{r\theta}(\tau''_{r\theta} = 0)$, satisfy the following condition at each point of the strip roll

$$\left.\begin{array}{l}\sigma_r < 0 \\ |\tau_{r\theta}| \leq f|\sigma_r|\end{array}\right\},$$

where f is the friction coefficient, then there is no separation and slippage between the turns. In this case, the strip roll is deformed as a continuous elastic anisotropic body, and the deformation of the strip roll is insignificant. Thus, by checking the fulfillment of the last condition, it is possible to establish the stress values ensuring the strip roll stability to sagging.

The initial stresses and those formed during winding of the strip rolls were calculated using the techniques given in [107, 108, 119, 123]. Zones where this condition was violated were constructed in the course of a numerical solution automatically according to the finite element grid using a graphical output program. In Fig. 9.47

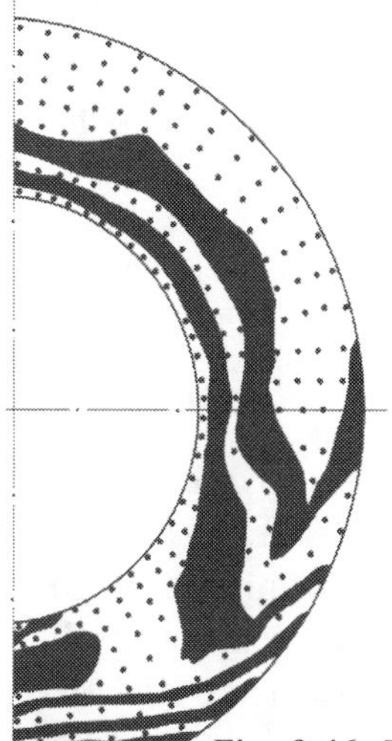

Fig. 9.46. The field of isochromes (lines of equal principal stress difference) is an anisotropic ring.

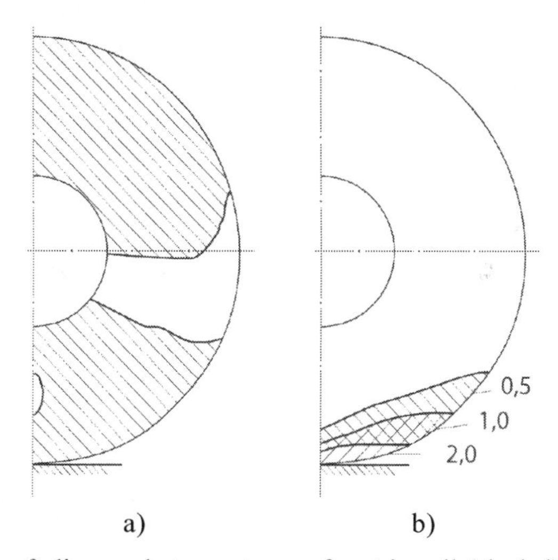

Fig. 9.47. Zones of slippage between turns of a strip roll (shaded): a – excluding the effect of initial stresses due to winding with tension; **b** – for various tensions. Numerals in the figure – the tension, N/mm^2.

a, it is seen that in the absence of initial stresses (corresponding to strip winding at low tension), slippage occurs between the turns practically throughout the entire section of the roll. As the tension of the coiled strip roll is increased, the dimensions of the slip zone decrease (Fig. 9.47 *b*) and already at a tension of 0.5–2.0 N/mm^2 (at a friction coefficient of 0.3) it is localized in the immediate vicinity of the support.

Thus, at a tension of the order of 0.5–2.0 N/mm^2, due to the cohesion of the turns with each other almost throughout the entire section, the strip roll is deformed as a continuous anisotropic elastic cylinder. In this case, the sagging of the roll is insignificant.

If the coiling machines of the hot rolling mill do not provide the necessary tension, the initial compressive radial stresses in the strip rolls are absent and there may be gaps between the turns. In this case, when studying the stress–strain state of the strip rolls, their complex structure must be taken into account.

The scheme of the finite element method for an elastic body considered above assumes a linear relationship between stresses and strains. The problem of deformation of loosely wound strip rolls is non-linear. To solve such problems, the finite element method was developed for procedures that do not require modification of the computational scheme of the method. If a solution of the linear problem is found, then it is possible to obtain a solution of the non-linear problem by means of an iterative process, at each step of which the material constants are chosen so that non-linear determining equations are satisfied.

In this paper we used the variable rigidity method (variable parameters) [141]. According to the scheme described, the deformation of strip rolls rolled without tension with a tight (without gaps) adherence of the coils was calculated. In such rolls, for most of the section, the turns slip relative to each other.

The results obtained make it possible to conclude that the considerable sagging of coils observed in practice is due to the loose fit of the turns. Therefore, of particular interest is the possibility of calculating strip rolls having gaps between turns. Introduction to the consideration of gaps does not require significant changes in the formulation of the problem. They are reduced to the assumption of small compression deformations in the radial direction of the strip roll without the occurrence of compressive stresses.

Distribution of temperature and temperature stresses during cooling coils of hot-rolled strips [139]. To calculate the temperature distribution when cooling the strip roll, we introduce a cylindrical coordinate system $r\theta z$ (Fig. 9.48 *a*). We assume the roll cools evenly that along the entire lateral surface and, consequently, the temperature field is axisymmetric. An axisymmetric nonstationary temperature field in a hollow cylinder of finite length that has an anisotropy of the thermophysical properties is described by a differential heat equation

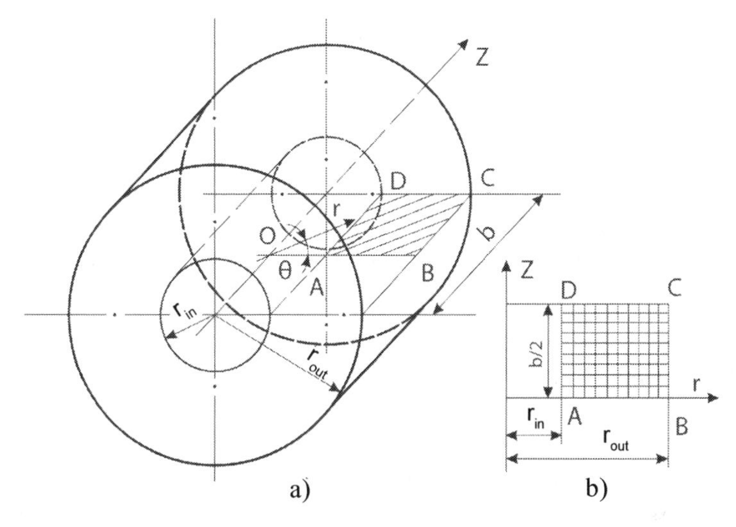

Fig. 9.48. Scheme for calculating the temperature distribution during cooling of the strip roll: a – system of cylindrical coordinates $r\theta z$ in the strip roll; b – the solutions of the heat equation.

$$\frac{dT}{dt} = a_r \left(\frac{d^2T}{dr^2} + \frac{1}{r}\frac{dT}{dr} \right) + a_z \frac{d^2T}{dz^2},$$

where a_r and a_z are the diffusivity coefficients in the radial and axial directions, respectively.

With cooling in air the boundary conditions in the form of convective heat exchange Newton's law are acceptable. The cooling conditions are assumed to be identical at both butts of the strip roll. In view of this, the temperature distribution is symmetrical relative to the plane extending perpendicularly to the axis.

The heat conductivity equation was solved in a rectangular domain *ABCD*, bounded by the inner and outer radii r_{in} and r_{out}, the axis *Or* and the straight line $z = b/2$ (Fig. 9.48 *b*). The boundary conditions on the sides of the rectangle are of the form:

$$\left.\frac{dT}{dr}\right|_{r=r_{in}} = \frac{\alpha_1}{\lambda_r}\left(T - T_m\right) - \text{on the side } AD;$$

$$-\left.\frac{dT}{dr}\right|_{r=r_{out}} = \frac{\alpha_2}{\lambda_r}\left(T - T_m\right) - \text{on the side } BC;$$

$$\left.\frac{dT}{dz}\right|_{z=0} = 0 - \text{on the side } AB;$$

$$-\left.\frac{dT}{dz}\right|_{z=\frac{b}{2}}=\frac{\alpha_3}{\lambda_z}\left(T-T_{\mathrm{m}}\right)-\text{on the side } DC.$$

Here α_1, α_2, α_3 are the heat transfer coefficients on the inner, outer and the end surfaces of the strip roll; λ_r, λ_z are the the coefficients of thermal conductivity of the roll in the radial and axial directions; T_m is the ambient temperature; b – the width of the strip coiled into a roll.

The thermal conductivity coefficient of the strip roll in the radial direction λ_r was represented by the equivalent thermal conductivity coefficient λ_e determined experimentally. According to [145], the average value of the equivalent thermal conductivity coefficient λ_e for the temperature range 20–700°C is 3.0 W/(m·°C). The thermal conductivity coefficient of the strip roll in the axial direction λ_z is approximately equal to the thermal conductivity coefficient of solid metal for the specified temperature range. The average heat transfer coefficient on the inside surface depends on the ratio of the strip width to the inner diameter of the roll and is in the range 11.1 ÷ 17.6 W/(m²·°C) for b/D_{in} = 1.18÷1.76.

The thermal conductivity equation with the above boundary conditions was solved numerically using a standard program that implements the explicit scheme of the finite difference method. The domain of solutions of the equation (rectangle *ABCD*) was divided into 10 intervals along the axis r (Fig. 9.48 *b*). Temperature was calculated in 110 nodes of the obtained finite-difference grid with a time step of 360 s. The values of temperature in the nodes were printed every 60 min of the cooling time.

Figure 9.49 shows the temperature at the nodes and isotherms constructed using the results of the calculation with the following initial data: r_{in} = 0.425 m; r_{out} = 1.295 m; b = 1.5 m; $\rho = 7.6\cdot10^3$ kg/m³; $a_r = 6.85\cdot10^{-7}$ m²/s; $a_z = 9.1\cdot10^{-6}$ m²/s; λ_r = 3.0 W/(m·°C); λ_z = 40 W/(m·°C); α_1 = 14.1 W/(m²·°C); $\alpha_2 = \alpha_3$ = 60 W/(m²·°C); T_m = 20°C; T_0 = 700°C, which corresponds to the strip roll weighing of 55 t produced from a strip 1500 mm wide, coiled at a constant temperature T_0 = 700°C. According to the graphs in Fig. 9.49 the temperature in the strip rolls during their cooling was unevenly distributed.

Since in the previous section of the book the problem of the stress–strain state of coils of hot-rolled strips under the action of the intrinsic mass was considered as planar, then in the first approximation it suffices to consider the effect of radial and tangential temperature stresses averaged over the length of the strip

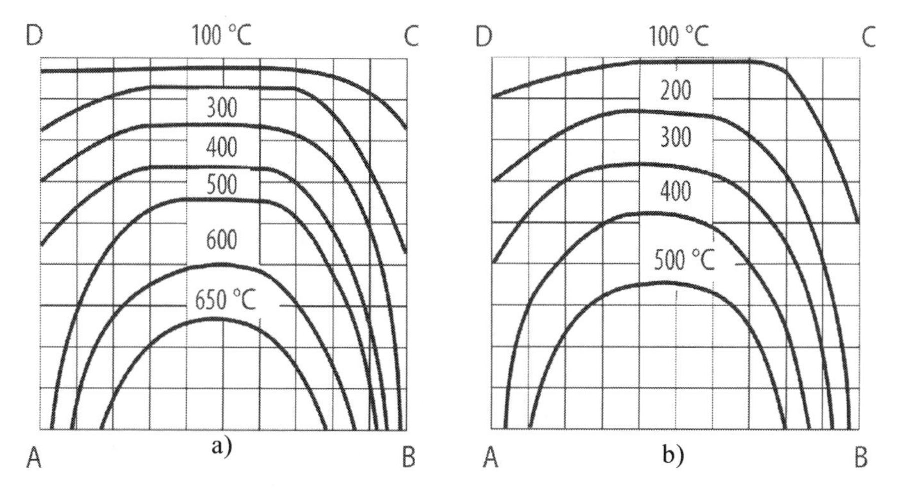

Fig. 9.49. Isotherms in the cross section of a strip roll 2 hours (a) and 4 hours (b) after the start of cooling.

roll on its stress–strain state. For this purpose, the temperature along the length of the strip roll was averaged:

$$\bar{T}(r) = \frac{2}{b}\int_0^{b/2} T(r,z)\mathrm{d}z.$$

The known distribution of the average temperature $\bar{T}(r)$ was used to determine the average thermal stresses along the length of the strip, using the formula for a hollow cylinder with free ends [129, 147]:

$$\sigma_r^{\mathrm{T}} = \frac{\alpha_{\mathrm{T}} E}{1-\nu}\frac{1}{r^2}\left(\frac{r^2 - r_{\mathrm{in}}^2}{r_{\mathrm{out}}^2 - r_{\mathrm{in}}^2}\int_{r_{\mathrm{in}}}^{r_{\mathrm{out}}} \bar{T}(\rho)\rho d\rho - \int_{r_{\mathrm{in}}}^{r} \bar{T}(\rho)\rho d\rho\right);$$

$$\sigma_\theta^{\mathrm{T}} = \frac{\alpha_{\mathrm{T}} E}{1-\nu}\frac{1}{r^2}\left(\frac{r^2 + r_{\mathrm{in}}^2}{r_{\mathrm{out}}^2 - r_{\mathrm{in}}^2}\int_{r_{\mathrm{in}}}^{r_{\mathrm{out}}} \bar{T}(\rho)\rho d\rho - \int_{r_{\mathrm{in}}}^{r} \bar{T}(\rho)\rho d\rho - \bar{T}(r)r^2\right),$$

where α_{T} is the coefficient of linear thermal expansion.

Figure 9.50 *b* and *d* schematically shows the distribution of radial σ_r^{T} and tangential $\sigma_\theta^{\mathrm{T}}$ stresses calculated by the above formulas. When cooling strip rolls rolled at a constant temperature, tensile radial stresses are observed at the inner surface of the strip rolls, which lead to stratification of the turns, which then affects the extent of the roll sagging. If, when winding the roll, the temperature of the outer turns is higher than that of the inner ones, the area where the

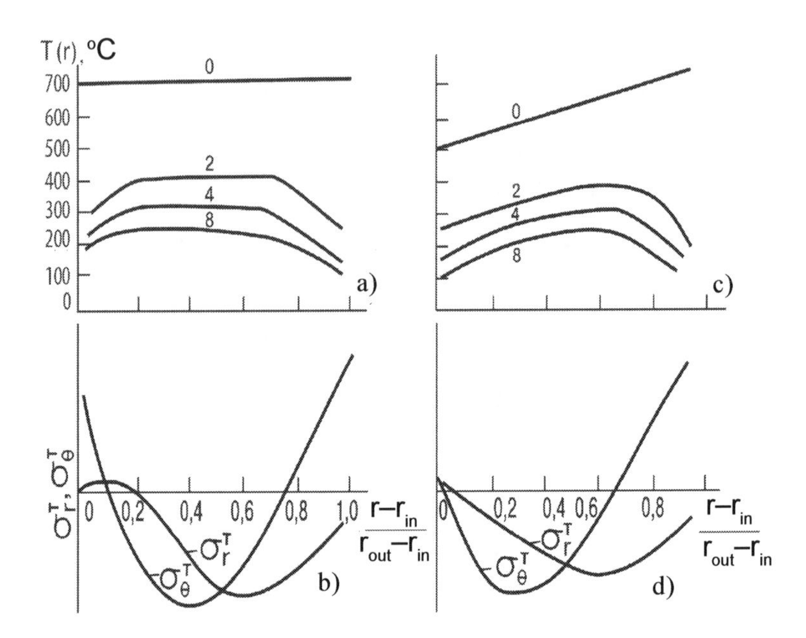

Fig. 9.50. Distribution of temperature and thermal stresses: a, b – when cooling the strip wound at a constant temperature; c, d – when cooling of the strip, the initial temperature of which from the front to the rear was distributed linearly; figures for curves – duration of strip roll cooling, hour.

inner turns are stratified becomes smaller. According to the curves in Fig. 9.50 *d*, at a temperature difference between the outer and inner surfaces of the strip roll at the initial moment of the cooling process of ~200°C, there is practically no stratification zone of the turns.

The obtained results confirm the conclusion [119] that to ensure stability of the rolls to sagging, a lower temperature of the leading end of the strip is required in comparison with the rear (by ~150-200°C).

9.8. Rational technology of cooling and storage of hot-rolled strips

The sagging of strip rolls of hot-rolled strips during transportation and storing them in the horizontal position was experimentally investigated at the 2000 hot rolling mill of the Novolipetsk Metallurgical Combine. The radial deformation of the strip rolls was recorded using a TH-1 theodolite on rolls of 3sp steel strips measuring 1.8 × 1250 mm with a mass of 20 t, 7.0 × 1780 mm with a mass of 30 t and 10.0 × 1500 mm with a mass of 30 t. The deformation of the rolls, wound from strips of another assortment was evaluated visually.

Strip rolls of hot-rolled strips after rolling and winding at a temperature of t_w = 660÷680°C were removed from the coiler with the help of a staple and laid horizontally on a flat surface. The theodolite was place at a distance of 10 m from the strip roll perpendicular to its end surface. The angles under which the strip roll to be examined was visible in the instrument's lens were measured every hour until the strip roll cooled down completely (for 20 hours). The diameter of the strip rolls in the horizontal and vertical sections was calculated from the measured angles. The radial deformation of the strip roll ('sagging') was estimated from the change in the diameter of the strip roll in the vertical section. The amount of sagging was taken as the difference between the diameter of the strip roll at the time it was removed from the winder drum and the diameter of the completely cooled roll.

The results of the experiments showed that strip rolls of hot-rolled strips, after laying them in the horizontal position on a flat surface, immediately sag under the action of their own mass (for strip rolls from strips of 1.8 × 1250 mm this sagging was 40 mm). Later, as strip rolls cool in air, the amount of their sagging changesd insignificantly. So, for 20 hours it increased by only 10 mm and with further cooling practically did not change.

The sagging of strip rolls depends on the thickness of the rolled strips. In particular, for strip rolls of 1.8 × 1250 mm strips it was 50 mm (with outer and inner strip roll diameters of 1880 and 850 mm, respectively); the saging of the rolls of the 10.0 × 1500 mm strips was smaller and amounted to 24 mm (with outer and inner strip roll diameters of 2025 and 850 mm, respectively); the sagging of the strip rolls of the 7.0 × 1780 mm strips was 30–35 mm (with the outer and inner strip rolls diameters of respectively, 1890 and 850 mm), i.e., with a decrease in the thickness of the strips, the deformation of the strip rolls increases:

The thickness of the strips, mm	2	7	10
Deformation of strip rolls%	4–5	3–4	2–3

Experimental studies have shown that to ensure the stability of rolls of hot-rolled strips to sagging, first of all, a tight (without gaps) adhesion of the strips of in a roll is necessary. The design of coilers

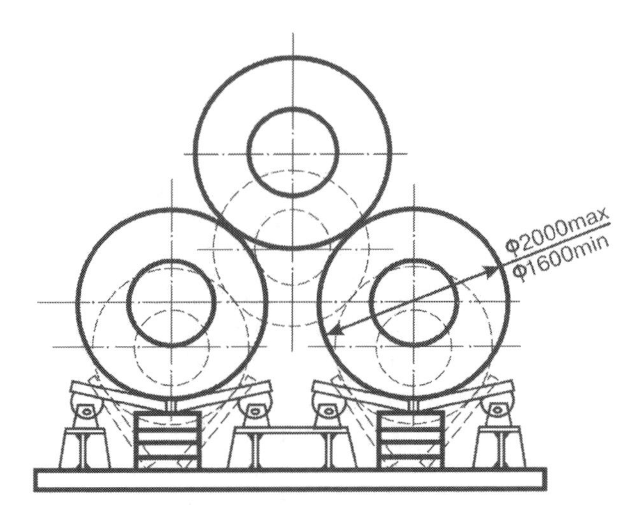

Fig. 9.51. Stand for determining the strain of horizontal hot-rolled strips.

of the 2000 WSHRM of the Novolipetsk Metallurgical Combine ensures the density of the strip rolls of 0.9–0.95. With such a factor of filling the strip roll, the radial deformation (sagging) depends on the thickness of the coiled strips: the thinner the strip rolls, the greater the roll sagging, i.e. there is an inverse relationship. At a tension of ~ 0.5-2.0 N/mm^2, a tight (without gaps) adhesion of the turns in the strip roll is ensured. The filling factor of such strip rolls is close to 1 and their deformation is no longer dependent on the thickness of the strip. Transportation and storage of the strip rolls of hot-rolled strips rolled with a tension of 0.5–2.0 N/mm^2 occurs without the loss of stability.

The results of experimental studies of the sagging of the strip rolls and the values of radial strain obtained by calculation are consistent. For strip rolls wound in a roll 2 mm thick, the calculated deformation value of the strip rolls was 46 mm, and the experimental value was 50 mm. For the strip rolls made of a strip 10 mm thick, the calculated and experimental deformation values were 21 and 24 mm, respectively. A good coincidence of the results indicates the suitability of the developed algorithm for calculating the deformation of the strip rolls of hot-rolled strips.

Two special stands were constructed (Fig. 9.51) [129], having inclined lodges for laying hot coils to study the sagging of the strip rolls of hot-rolled strips with their horizontal arrangement on special shelves (pallets) in one or three tiers. The angle of inclination of the walls of the lodgement could be changed in the range from 25 to 40°.

The studies included determination of the radial deformation of hot strip rolls when storing in one tier for cooling; measurements of the deformation of strip rolls in two–three-tier warehousing in a horizontal position; determination of the optimal angles of the slope of the walls of V-shaped lodges at different storage methods; comparison of the cooling speed of the strip rolls in the horizontal and vertical positions.

The strips rolled on the 2000 WSHRM were rolled up into rolls weighing 25–27 t (at a thickness of 2 mm) and 30–32 t (at a thickness of 6 mm or more) according to the technology adopted at the mill. Then, the rolls were removed from the winder drum using a pulling trolley and, using a staple suspended from the bridge crane, were installed horizontally on the stands of the stand with a given angle of inclination of the walls to 1–3 tiers. After the coils were completely cooled (after 2–3 days), the inner and outer diameters of the strip rolls were measured in the vertical and horizontal plane using a tape measure and a specially manufactured ruler. Deformation under various storage schemes was determined for strip rolls of thin

Tilt angle of the walls, deg	40	35	30	25
Roll sagging, mm	15–20	30–40	50–60	70–80

(2–4 mm thick) and thick (6–10 mm) strips. The amount of sagging of the strip rolls was taken as the difference between the diameters measured in the horizontal and vertical planes.

When storing the rolls in a horizontal position, the deformation of the strip rolls installed on the shelves of the stand with the angle of inclination of the walls 40°, 35°, 30° and 25° was determined in one layer (Fig. 9.52). The results of the studies showed that as the angle of inclination of the cradle walls was reduced, the sagging of 2 mm-thick strip rolls increased:

The strip rolls of thick (6 mm or more) strips, regardless of the angle of the walls, practically did not deform – the difference in the outer diameters in the vertical and horizontal planes did not exceed 5 mm. Internal diameters of strip rolls of thick strips also did not change when cooled on the stand shelves. The change in the inner diameter of the strip rolls of thin strips obeyed the same pattern as the outer ones:

Tilt angle of the walls, deg	40	35	30	25
Deformation, mm	0	10	20	25

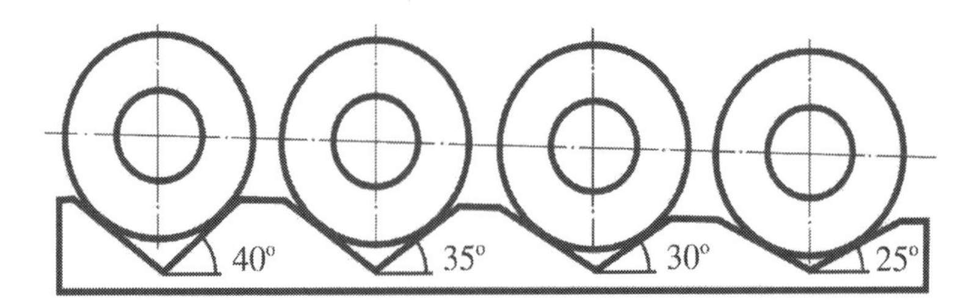

Fig. 9.52. Storage scheme for strip rolls of hot-rolled strips with different lodgment angles of the walls.

Thus, for transporting the strip rolls in a horizontal position from the WSHRM coilers to the production warehouse on the lodges installed on the conveyor, and also the storing of 2–4 mm strips into one tier of strip rolls, the optimal angle of the slope of the walls of the lodges is 40°. The decrease in the angle of inclination leads to the sagging of the rolls along the outer and, to a lesser extent, the inner diameters.

According to the results of experiments in the lower tier, the deformation along the outer diameter of the strip rolls of strips 6–10 mm thick was insignificant (within 5–15 mm) and did not depend on the angle of inclination of the walls of the cradles. On the inner diameter of the rolls are not deformed. In the second tier the rolls were not deformed at all. For strips 2-4 mm thick with a wall angle of 35 °, the deformation of coils of the lower tier by the outer diameter was only 5-10 mm without changing the inner diameter. The rolls of the second tier did not deform either along the outer or inner diameters. In the case of two-tier storage of coils of strips 2 mm thick, when the rolls of the lower row were placed not on the bedding but on a flat surface, the deformation amounted to 10% of the value of the outer diameter. In this case, there was a noticeable change in the shape of the rolls and the formation of an ovality along the outer and inner diameters. When storing coils of hot-rolled strips in three tiers on lodges with a slope angle of 35°, the results were obtained, as in the case of two-story warehousing.

Multi-tier storage of rolls for cooling before further processing in comparison with single-row allows more rational use of warehouse areas. However, in multi-tiered warehouses, there is a greater risk of the rolls sagging: the lower row is further deformed by the rolls that lie on top [128]. To assess the subsidence of coils in a two-story warehousing in a horizontal position, experiments were carried out, which included setting up five rolls so that three rolls of the lower row were located on the lodges, and the other two made up the second tier (Fig. 9.53).

The experimental values of the deformation in the radial direction of the rolls are shown in the table.

Storage scheme rolls	Thickness of coiled strip, mm	Coiling temperature, °C	Tilt angle of the walls of storage rack, deg	Measured parameters				Deformation rolls Δ, mm
				$D_{hor.out}$ mm	$D_{ver.out}$ mm	$D_{hor.in}$ mm	$D_{ver.in}$ mm	
Single tier	2.0	620	40	2000	1985	850	850	15
	2.0	620	35	1890	1860	850	840	30
	2.0	620	30	2060	2010	870	850	50
	2.0	620	25	2020	1950	870	845	70
	6.0	640	40	2010	2000	840	830	10
	6.0	650	35	1990	1990	850	850	0
	6.0	650	30	2000	2000	850	850	0
	6.0	640	25	1960	1950	850	850	10
Two-tier	2.5	640	35	2000	1990	840	840	10
	2.5	640	35	1945	1945	840	840	0
	2.5	640	35	1970	1970	840	840	0
	2.5	640	40*	1970	1970	840	840	0
	2.5	640	40*	1970	1970	840	840	0
Three-tiered	2.0	610	35	1960	1960	840	840	0
	2.0	610	35	1995	1990	840	840	5
	2,0	610	35	1990	1990	840	840	0
	2,0	610	40*	1990	1990	840	840	0
	2,0	610	40	1990	1990	840	840	0
	2,0	610	40**	1990	1990	840	840	0

Note: $D_{hor.out}$, $D_{ver.out}$ – outer diameter rolls in horizontal and vertical planes; $D_{hor.in}$, $D_{ver.in}$ – inside diameter. * The rolls located on the second tiers; **The rolls situated on the third tier.

Regulation rolls during transportation and storage significantly affects the duration of cooling from the temperature of the winding up to a temperature environment. The importance of this factor increases as the mass rolls, since the duration of the cooling coils weighing 60–70 tons can reach 8–10 days. Effect of the conditions and duration of cooling rolls the amount of strain on the horizontal arrangement has been insufficiently studied. Of particular value in

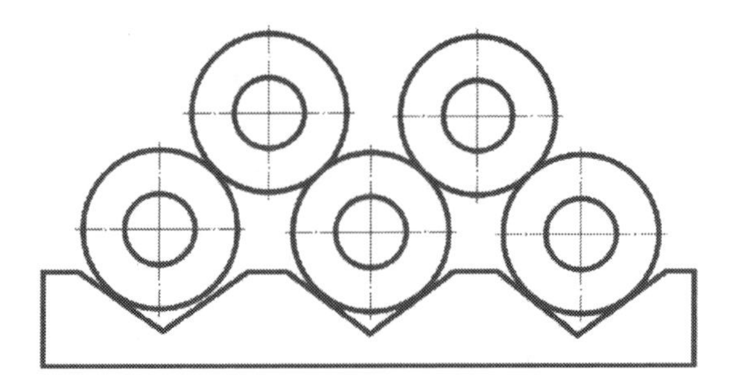

Fig. 9.53. Storage of strip rolls horizontally on racks of the stand in two tiers. Tilt angle of the walls of the shelves for strip rolls in the lower tier is 35°.

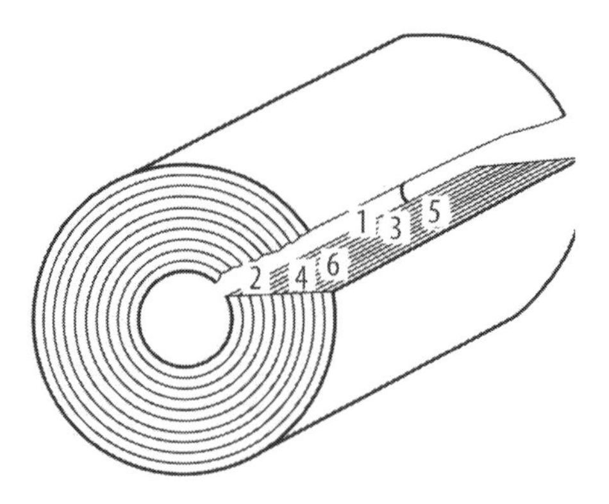

Fig. 9.54. Arrangement of thermocouples in the strip roll sections extending through the centre of the strip (1, 3, 5) and at a distance of 100 mm from the butts of the strip roll (2, 4, 6).

the study of this issue are the results experimental studies carried out in industrial environments, due to their technical complexity, performance, on the one hand, and the high cost and complexity on the other.

In the work [128] the authors compared the nature of the temperature distribution in the thickness of the winding, and the duration of cooling at constant conditions of the vertically and horizontally spaced rolls of hot-rolled strips. The temperature of the turns of the strip in the strip rolls was measured using thermocouples (Fig. 9.54) by the method [148].

When winding the strip into a roll, the coil drum of the 2000 WSHRM of the Novolipetsk Metallurgical Combine was periodically stopped and tubes of a high-conductivity material were placed between the turns. The thermocouples introduced into the holes of the tubes at different depths made it possible to measure the temperature

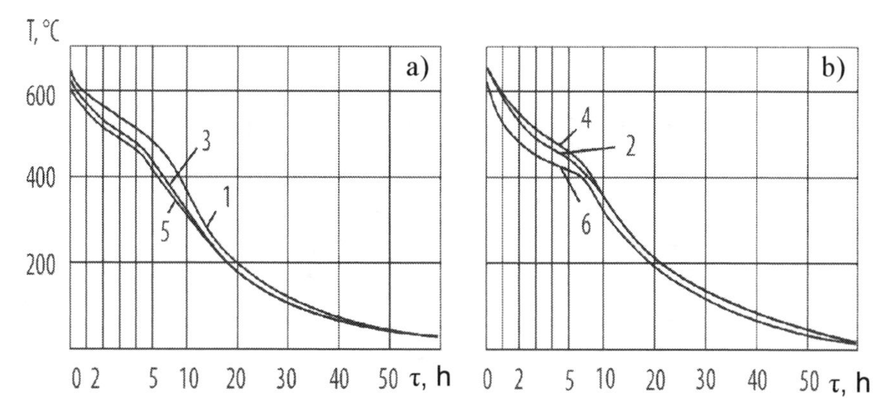

Fig. 9.55. Cooling curves of turns of the strip roll weighing 30 t during cooling in air in a horizontal position: the inner radius of strip roll 425 mm; outer radius 995 mm; coiled strip thickness 10 mm (curve numbers correspond to the location of the thermocouples in Fig. 9.54)

of the windings along the thickness and width of the winding. The thermocouples were installed after removing the strip rolls from the coiler and laying them on the shelves. The thickness of the strip (10 mm) was chosen so that,by the time of the first stopping of the coiler drum the tail portion of the strip emerged from the last stand of the finishing mill group. The short duration (1–2 s) of each stop of the winder did not significantly change the thermal state of the rolls. Temperature measurements after connecting the recording instrument were carried out every hour until the strip rolls cooled down completely.

The results of measurements (Fig. 9.55) indicate a difference in temperatures in the mean parts of the strip roll (curves 3 and 4) and zones adjacent to the outer and inner turns (curves 1, 2, 5, 6), in the initial cooling period. The temperature drop at the points located in the middle part of the strip roll and in the section at a distance of 100 mm from the ends reached 50–80°C. During cooling the temperatures were equalized. After 6 hours, the difference was 30–40°C; after 35 hours only 5–10°C. After 50 hours from the beginning of cooling the maximum temperature in the roll (point 3) did not exceed 50°C. The temperature at other points was 5–10°C less. At the beginning of the cooling period, the inner turns (curves 1 and 2) had a higher temperature (30–40°C) than the coils adjacent to the outer surface (curves 5 and 6). After 6–7 hours, the temperature of the outer and inner layers of the strip in the roll became almost the same.

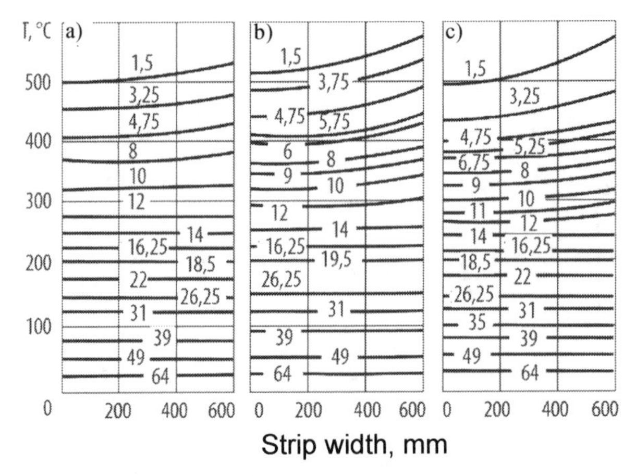

Fig. 9.56. Temperature distribution along the roll axis in the outer (a), middle in winding thickness (b) and inner (c) turns (numbers on the curves – the cooling time of the strip roll, h; experimental conditions as in Fig. 9.55).

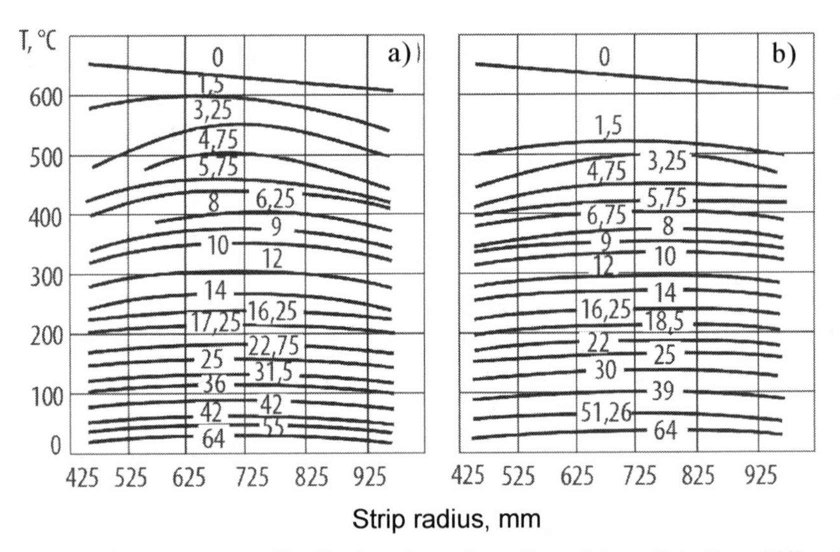

Fig. 9.57. The temperature distribution along the radius of the roll in the middle of the width of the strip (a) and at a distance of 100 mm from the butts of the strip roll (b); notation and experimental conditions as in Fig. 9.56.

The curves of the distribution of temperature over the width of the strip and along the radius of the strip roll (in the thickness of the winding) are shown in Figs. 9.56 and 9.57. The greatest difference of the temperatures in the radial and axial directions was observed in the initial period of cooling during the first 6–7 hours. After 12-14 hours, the temperature in the width was equalized; in the radial direction the temperature difference was maintained. This pattern was observed both at a distance of 100 mm from the butts of the strip roll, and in the middle part along the width of the strip. The temperature drop across the section of rolls decreased from 70-80°C after 3–4 hours of cooling to 5°C after 30–35 hours. The temperature difference of 5 ° C between the middle and end turns was maintained throughout the cooling period.

The temperature in the middle of a strip 10 mm thick in the middle turns of horizontally and vertically located rolls weighing 30 t was measured during cooling from a coiling temperature of 650–700°C:

Cooling time, h	10	20	30	40	50	60	70	80	90
Turn temperature (°C) for strip roll in position:									
Horizontal	360	200	120	90	60	40	30	25	25
Vertical	500	270	160	130	100	80	65	50	40

According to the results of experiments, the position of a roll of hot-rolled strip significantly affects the duration of its cooling. So, the rolls with a mass of 30 t in a horizontal position, cooled to a temperature of 40°C for 60 hours. With the vertical position of the rolls, the duration of their cooling increased to 90 hours, i.e. 1.5 times.

Increasing the cooling speed of horizontally arranged rolls is achieved as a result of the fact that the main amount of heat is diverted from the strip roll through the end faces in both directions. With the vertical position of the strip rolls, heat transfer only passes through one end face. Therefore, the cooling time of the upper and lower strip rolls in the stack is significantly different. Warehousing in a horizontal position in several tiers contributes to a more even cooling of all the strip rolls in the stack.

Based on the foregoing, we can conclude that the optimal procedure when transporting rolls on lodgements from the WSHRM coilers to the warehouse of products and storing them in 1–3 layers in a horizontal position is the one with the angle of inclination of the walls of the lodgements 35–40°. At these angles the strip rolls on the outer and inner diameters are practically not deformed, while retaining a cylindrical shape. A decrease in the angle of inclination of the walls of the lodgments from 40 to 25° leads to the sagging of

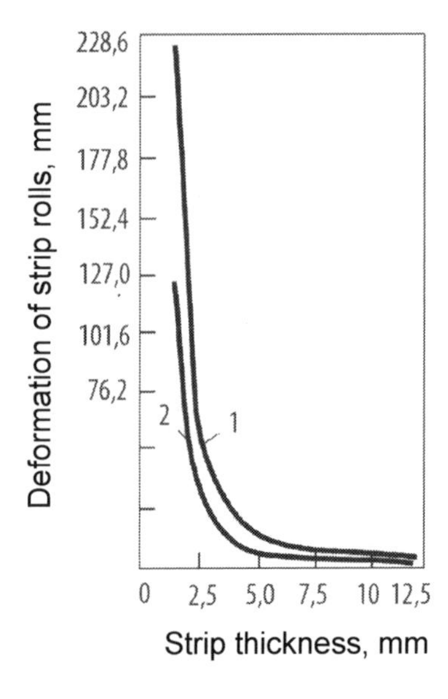

Fig. 9.58. Deformation of loosely wound strip rolls of varying thickness, with an average temperature of 540°C: 1 – specific mass of rolls 213.6 kg/cm; 2 – the same, 106.8 kg/cm.

the strip rolls wound from thin (less than 5–6 mm) strips under the action of their own mass. The deformation of strip rolls of strips 6 mm thick or more does not depend on the angle of inclination of the walls of the lodgements. The storage of the rolls of strips with a thickness of less than 10 mm horizontally in one or more tiers on a flat surface is undesirable, since it distorts the shape of the rolls in the inner and outer diameters.

Different ways of transportation, storage and forced cooling of the strip rolls of hot-rolled strips, technological methods used to preserve the shape of the strip rolls, and proposals for the conservation and utilization of heat of hot metal rolls are detailed in the review [130]. Despite a considerable period of time that passed after the publication of the survey, no new revolutionary solutions in this field of sheet rolling production have emerged. The above materials of theoretical and experimental research will be supplemented only with separate, most interesting results, generalized in the review [130].

The influence of the thickness of the strips on the sagging of loosely rolled coils of hot-rolled steel is shown (according to foreign literature sources) in Fig. 9.58 [130]. As can be seen, with an increase in the thickness of the bands, the amount of sagging decreases: for strips thicker than 3 mm, it does not exceed 50 mm (with outer and inner roll diameters of 2050 and 750 mm, respectively). The deformation of tight-rolled strip rolls of hot-rolled strips 2.5 mm thick or more, stored horizontally, is insignificant even at high temperatures. At strip thicknesses of less than 2.5 mm, the sagging

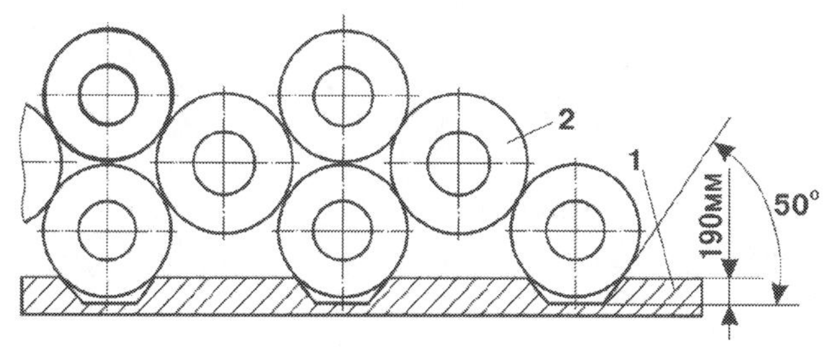

Fig. 9.59. Storage scheme of strip rolls horizontally in several tiers: 1 – lining; 2 – strip rolls.

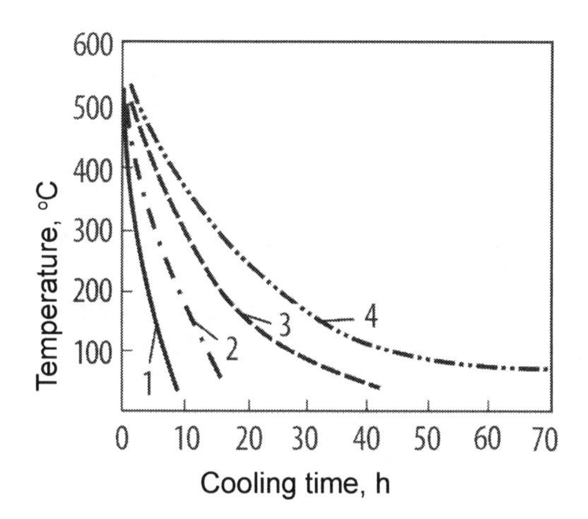

Fig. 9.60. Change of the temperature of the inner surface of the strip rolls at different cooling methods: 1 – immersion in water (water temperature 26°C); 2 – cooling in water mist (water flow of 0.33 m^3/h); 3 – forced air cooling (the distance between the nozzles 500 mm, air flow 8.5 m^3/h); 4 – natural cooling.

of the rolls is greater than permissible. To exclude this negative phenomenon when winding rolls of strips with thicknesses less than 2.5 mm, it is necessary to increase the temperature from the inner turns to the outer ones by 50–160°C.

In the opinion of German specialists, when storing hot coils in a horizontal position in three tiers, the laying is optimal when the angle of inclination of the lodgements (pallets) is 50° (Fig. 9.59). The arrangement of the rolls in three rows, if the angle between the line connecting the centres of the rolls and the horizontal is 30° is undesirable.

Deformation of horizontally located rolls due to high temperature or large mass during their transportation can be prevented by constant rotation of the rolls.

As already noted, a progressive trend in production technology of the hot-rolled strips in the application of accelerated cooling coils. factories of the company Sin Nippon Seitetsu (Japan) tested cooling methods for strip rolls by water mist, immersion in water with additives of inhibitors, natural and forced cooling by air. The cooling curves of the strip rolls in different ways are shown in Fig. 9.60. Experiments have shown that cooling the strip rolls with water is 20 times more efficient than natural air cooling, but it can lead to the appearance of stains and rust on the surface of the strips. Therefore, this method is considered difficult and unsuitable for practical use.

Forced air cooling of the strip rolls is almost 2 times more efficient than natural cooling . With this method, strip rolls are laid in a horizontal position and air is fed to their ends through nozzles located on the floor. It is established that about 90% of the heat is transferred from the strip roll in the axial direction, the rest in the radial direction.

In recent years, an intensive search for effective solutions aimed at the productive use of heat by hot coils of sheet steel has been conducted. It is proposed to cool strip rolls in baths with running water, which is then fed to heat exchangers for heat removal, to use the heat of hot strip rolls to heat pickling solutions in continuously pickling aggregates, etc. [130].

10

Skin pass rolling of sheet steel

10.1. Theoretical basis of skin pass rolling

The purpose of skin pass rolling is the final formation of the mechanical properties (elimination of the yield plateau on the tensile curve), flatness and the surface relief of the steel sheets through the relatively small (typically about 1%) reduction.

Cold-rolled steel sheets are mainly skin pass rolled on single-stand rolling mills. However, the skin pass rolling of especially thin cold rolled strips and tin plate is carried out in two-stand mills. The single-stand mills are used for skin pass rolling strips with a thickness of 0.3–0.38 mm and above, and the two-stand mills – 0.22–0.25 mm thickness or less. The advantage of the two-stand mills in comparison with the single-stand mills is that in skin pass rolling they can generate significant interstand tension – (0.3–0.4)σ_T of the skin pass rolled metal. This makes it possible to apply rolls with large diameters allowing to achieve a high surface quality and flatness of the strips. The two-stand skin pass rolling mills are divided into two types: with the work rolls of the same diameter in the first and second roll stands, and the rolling mills which use in the second stand working rolls of a large diameter than in the first stand. In the latter case, the rolls of the second stand are used for ironing. The skin pass rolling of cold-rolled metal is conducted in quarto mills, hot-rolled metal – in quarto or duo mills. Special features of skin pass rolling consist in the fact that this process is characterized by the non-uniform stress–strain state of the treated metal, large values of the length of the contact area relative to the thickness of the skin pass rolled strips and zones of elastic and elastoplastic deformation at the inlet and outlet of the rolls. Furthermore, the skin pass rolled metal is typically in the annealed condition when its deformation resistance is significantly affected by the rate of deformation.

Depending on the degree of deformation during skin pass rolling the yield strength of steel sheets is initially lowered with respect to its value for the metal not processed by skin pass rolling. Increasing reduction decreases the length of the yield plateau on the stress–strain curve in tensile loading of sheet samples. Skin pass rolling with a reduction of about 1% completely eliminates the yield plateau. This moment corresponds to the minimum value of the yield strength of the steel sheet. With further increase in reduction in skin pass rolling the yield strength of the metal increases again, however, the tensile curve remains smooth without a yield plateau. With increasing degree of deformation during skin pass rolling the tensile strength and hardness of steel sheets increase, while the elongation at fracture decreases.

The described behaviour of the yield strength can be explained using the now classical theory of dislocations. Initially, the yield strength decreases with increasing reduction because due to the deformation in skin pass rolling the dislocations are released from the clouds of impurity atoms of nitrogen and carbon surrounding them. The role of this factor is then overlapped by the metal hardening at higher reductions.

The explanation of the mechanism of elimination of the yield plateau as a result of skin pass rolling sheet steels, based on the provisions of the theory of dislocations, is consistent the interpretation of this effect from the standpoint of the appearance of residual stresses in the metal. Theoretical[1] and experimental studies have shown that compressive residual stresses appear in the surface layers of the skin pass rolled sheets and tensile stresses in the interior parts (Fig. 10.1). In tensile loading of the specimens of the skin pass rolled sheets, characterised by the above diagrams of residual stresses, an inhomogeneous stress–strain state forms in the metal in which plastic flow begins in the stretched central layers of and gradually moves to the surface layers.

This causes a decrease in the observed yield strength. In addition, the plateau and the pronounced yield limit do not appear on the stress–strain curve.

---

[1]E.M. Tret'yakov, Theoretical fundamentals of the process of skin pass rolling of steel sheets and calibration, PhD thesis, Moscow, 1971.

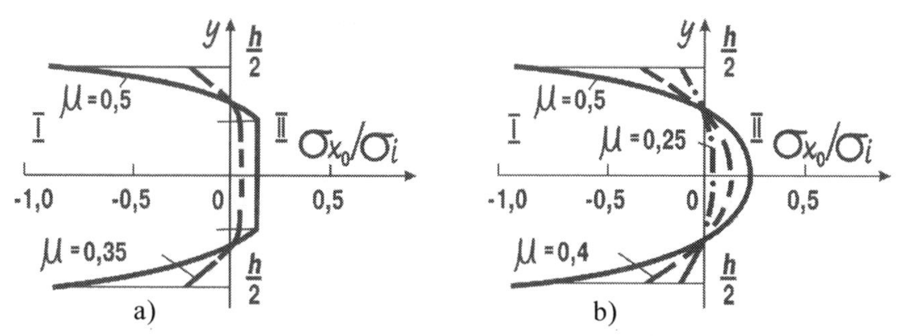

Fig. 10.1 Distribution of residual stresses through the thickness of the skin pass rolled sheet metal (according to E.M. Tret'yakov): *a* – the plastic region did not penetrate through the entire thickness of the strip; *b* – the plastic region penetrated through the entire thickness of the strip (μ is the plastic friction coefficient; *h* is the thickness of the strip; σ_{x_0} is the residual stress acting on the sheet in the rolling direction; $\sigma_i = \sqrt{3}k$ is the stress intensity of the ideal plastic material); I – compression, II – tension.

The skin pass rolling effect is dependent on the factors of the starting material (ferrite grain size in the steel, the content of carbon and nitrogen in the solid solution, the content and the distribution of carbides, nitrides and other non-metallic inclusions) and the technological process conditions (relative reduction, the ratio of the diameter of the rolls to the strip thickness, the surface roughness of the rolls and the strips, applications of lubrication, front and rear tension, temperature and speed of skin pass rolling, etc.).

To achieve the high quality of the sheet steel it is crucial to select the optimum strain in skin pass rolling. Annealed steel must be skin pass rolled at the lowest possible reduction. In this case, the metal retains greater plasticity. However, the degree of deformation must be sufficient to eliminate the yield plateau. In connection with the above, a number of important questions arises: what is the value of the optimal reduction in skin pass rolling? Does the value of the optimal reduction change as a function of the parameters of the skin pass rolling process? Which means can be used to achieve stabilisation of reduction in the optimal range?

By analyzing the deformation intensity distribution over the cross section of the strips, E.M. Tret'yakov put forward the following principle of elimination of the yield plateau of the skin pass rolled metal. The yield plateau in skin pass rolling will be eliminated if the average integral value of the strain ε_{i_c} in the strip thickness becomes equal to or exceeds the value of the strain intensity of deformation ε_u corresponding to the end of the yield plateau on the chart σ_i =

$\sigma_i(\varepsilon_i)$ of tensile loading of the metal in the initial state (before skin pass rolling). This principle can be written as

$$\varepsilon_{i_c} = \frac{1}{h}\int_{h/2}^{h/2} \varepsilon dy \geq \varepsilon_u .$$

Theoretical studies of E.M. Tret'yakov show that the reduction values in skin pass rolling guaranteeing the elimination of a yield plateau, must lie in the following ranges

$$0.522\varepsilon_u \leq \Delta h / h \leq 0.866\varepsilon_u .$$

Thus, the reduction in skin pass rolling sufficient to eliminate the yield plateau of the sheet steel is less in magnitude than the deformation ε_i defining the length of the yield plateau. The same applies to the experimental data.

In the production of cold-rolled sheet the optimal range of reduction in skin pass rolling is usually the interval corresponding to the minimum values of the yield strength on the V-shaped curves of the dependence of the yield strength on deformation (Fig. 10.2). In particular, for cold-rolled steel 08Yu the optimal range is 0.8–1.2% for steel 08kp – somewhat broader: 0.8–1.5 %. When skin pass

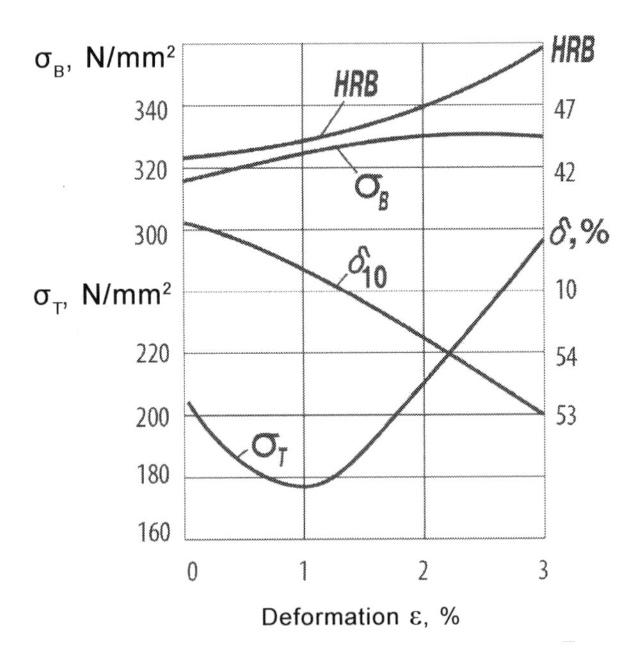

Fig. 10.2. Dependences of the mechanical properties of low-carbon steel sheet 08U on the degree of deformation in skin pass rolling.

rolling rimmed steel sheets the upper limit of permitted reduction is shifted toward larger values because an increase in the degree of deformation reduces the tendency of sheet steel to ageing – the yield plateau on the stress–strain curve forms later.

Maintaining the compression in the said narrow range is difficult because the skin pass rolling process parameters vary for different coils of the steel sheet and also within one coil. The degree of reduction of various sections of the strips in skin pass rolling is unevenly distributed due to the influence of the initial thickness difference of the steel, the heterogeneity of the mechanical properties and other factors. Thus, when skin pass rolling strips with a nominal thickness of 1 mm with an average reduction of 1.0–1.2 % the range of the reduction values within a sheet can reach several percent. The non-uniformity of reduction in skin pass rolling 1.5–2 mm thick strips is 3–4 times higher than for the 0.5–0.65 mm thick strips. According to studies conducted at the 1700 skin pass rolling mill of the Mariupol' Metallurgical Concern, for strips 1 mm in thickness fluctuations of the reduction due to longitudinal thickness difference were on average 0.9% because of the spread of values of the mechanical properties – 0.8 %. The greater sensitivity of the relatively thick (1.5–2 mm) strips to perturbations of the skin pass rolling process is due to the fact that their stiffness coefficients differ significantly ($M_n = \delta P/\delta h$): for $h = 2$ mm $M_n = 40$ MN/mm, and for $h = 0.5$ mm $M_n = 1000$ MN/mm.

According to the experimental data obtained in the skin pass rolling of the sheets varying thicknesses at constant pressure of the screws, decreasing the nominal thickness of the strip at the same amount of reduction causes smaller variations in deformation in comparison with the original thickness difference. The inhomogeneity of the mechanical properties along the length of the strip leads to further unevenness of the degree of deformation.

10.2. Kinematic and power parameters of the skin pass rolling process

Skin pass rolling of the strips in industrial mills is carried out in rough rolls. Under these conditions, the efficiency of the process, the performance of the mill and the quality of the skin pass rolled metal are largely determined by the force parameters of skin pass rolling. The dependences of the force P and the rolling moment M on relative reduction ε in skin pass rolling in the 200 mill of strips 0.8 mm thick

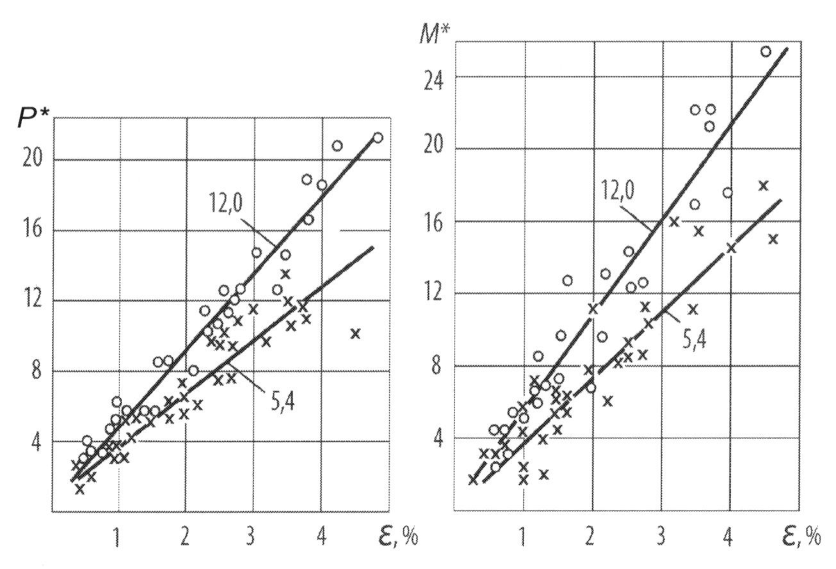

Fig. 10.3. Dependence of the relative pressure of the metal on the rolls $P^* = P/(\sigma_m hV)$ and torque $M^* = M/(\sigma_m h^2 V)$ on relative reduction ε in skin pass rolling strips with initial roughness of $Ra = 0.7 \div 0.9$ μm without tension and lubrication (mill 200; $V = 2$ m/s; $R/h = 140$) in rolls with a roughness Ra of 5.4 μm and 12.0 μm (the numbers on the curves)

with different mechanical properties without tension and lubrication in shot blasted rolls (Ra = 12.0 and 5.4 μm) are shown in Fig. 10.3. According to the graph shown in relative coordinates P^* and M^* using similarity criteria of the process of skin pass rolling, increasing the relative reduction ε and the roughness of the rolls increases force P and rolling moment M. Furthermore, increasing the roughness of the rolls increases the gradient of changes of both dimensionless parameters depending on reduction. Consequently, increase in reduction increases the difference in the force parameters of skin pass rolling of strips with rolls with different surface roughness.

When skin pass rolling the strips without tension with the same speed in the case of constant values of R/h and E/σ_T – the similarity criteria of the process (R – radius of the rolls, h – thickness of the strip, E and σ_T – modulus and yield strength of the rolled metal) the relative force and rolling moment are closely correlated with the magnitude of the relative reduction. Under the conditions of the experiments carried out on the 200 mill (Fig. 10.3), the correlation coefficient between $P^* = P/(\sigma_m hV)$, where B is the strip width and ε is respectively 0.93 and 0.98 at the roll roughness Ra equal to 5.4 and 12 μm. The coefficients of the same order characterize the correlation between the dimensionless moment $M^* = M/(\sigma_m h^2 V)$ and reduction ε. In the investigated range of the parameters of skin pass rolling all these dependences are close to linear and described by the following regression equations:

$$P^* = 0.95 + 2.95\varepsilon, \qquad M^* = 0.38 + 3.61\varepsilon \; (\text{at } Ra = 5.4 \mu\text{m});$$

$$P^* = 0.49 + 4.31\varepsilon, \qquad M^* = -0.07 + 5.4\varepsilon \; (\text{at } Ra = 12 \mu\text{m}).$$

The dependences of the average pressure in skin pass rolling, calculated taking into account the flattening of the rolls and elastic recovery of the strips, on the degree of deformation and roughness of the rolls are similar. The increase of force in skin pass rolling in rougher rollers increases the contact arc length L and its relation to the strip thickness h.

When studying the energy-power and kinematic parameters of the process of skin pass rolling in the 1700 mill at the Zaporozhstal' Metallurgical Concern skin pass rolling was carried out on 0.7–1.5 mm thick strip of 08kp, 08ps and 08Yu steels at speeds of 1.0–17 m/s and different friction conditions: no lubrication, lubrication with Industrial'noe-20 oil and lubrication with 3.5% emulsion E-2 (B) [149]. The rear σ_b and front σ_f tensions were 0.03–0.23 σ_T. Surface roughness *Ra* rolls was varied in the range 2.0–2.8 mm.

It has been found that with increasing reduction increases the deformation, and the greater the increase the smaller the thickness h of the skin pass rolled strips, i.e. the greater the ratio R/h (see Fig. 10.4). When skin pass rolling with E-2(B) lubricant (see Fig. 10.4 *b*)

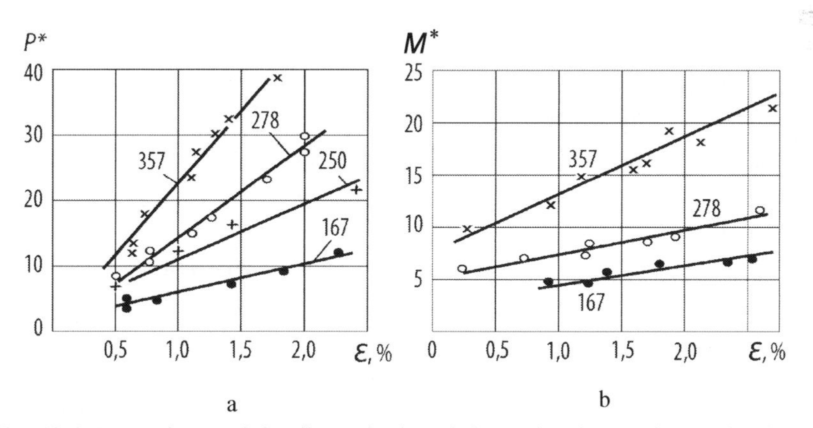

Fig. 10.4. Dependence of the dimensionless deformation force P^* on reduction ε in skin pass rolling without lubrication (a) and lubricated with oil I-20 (b) of the steel 08Yu strips (1700 mill; initial metal roughness Ra = 0.7÷0.9 μm; roll roughness Ra = 2.0 μm; V = 1.2÷1.6 m/s; σ_z/σ_T = 0.08÷0.15; σ_p/σ_T = 0.12÷0.20; numbers on the curves – the ratio R/h).

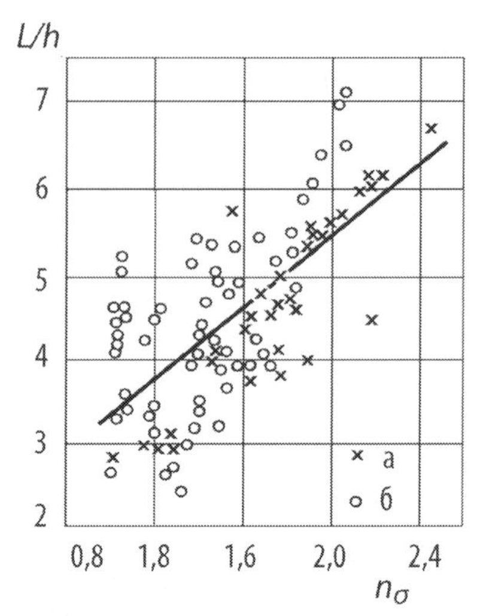

Fig. 10.5. Generalized dependence of the ratio of the length of the deformation zone L to the thickness *of* skin pass rolled strips h on the coefficient of the stress state n_σ (1700 mill) for skin pass rolling without lubrication (*a*) and with I-20 lubricant (*b*).

the P^* values are smaller than in the case of skin pass rolling without lubrication (see Fig. 10.4 *a*) and change less markedly depending on the reduction. The largest deformation force reduction is achieved when using I-20 oil as a lubricant. When skin pass rolling with the application of strips of the emulsion E-2 (B) on the pressure roller is reduced to a lesser extent. The dependences of the skin pass rolling moment on reduction are similar.

Skin pass rolling of low-carbon steel strips takes place at relatively high forces and large length of the deformation zone. So, in skin pass rolling without lubrication average pressure p_{av} in individual cases is 2.5–3 times higher than the yield strength of the metal being rolled, and the ratio L/h reached 7.0–7.5. In constant conditions of skin pass rolling increasing reduction ε increases values of P^*, L/h and the coefficient of the stress state $n_\sigma = p_{av}/\sigma_{T.av}$ (where $\sigma_{T.av}$ is the average value of the yield strength of the metal in the deformation zone). However, due to the fact that the dependences of P^*, n_σ and L/h on reduction ε are very sensitive to changes in other process variables (R/h; E/σ_T; σ_b/σ_T; σ_f/σ_T) and the friction coefficient μ for the general set of values of n_σ, L/H and ε in the appropriate studied range no significant correlation between P^*, n_σ and L/h in dependence on ε was observed. In contrast, a close correlation was found between L/h and n_σ (Fig. 10.5). The regression equation has the form $L/h = 1.28 + 2.1n_\sigma$ with a correlation coefficient of 0.68.

Evaluation of the acceptability of this equation using the criterion F_1, represented by the ratio of the total and residual variances, showed that the value L/h, calculated according to the equation, is close enough to actually observed values ($F_1 = 1.85$).

Changing the roughness of the rolls Ra from 2.0 to 2.8 μmicrons somewhat increases the power parameters of the process of skin pass rolling. However, in this range of variation of Ra for the conditions considered for skin pass rolling the observed effect of the surface roughness is generally insignificant and often overlaps the effect of other influences.

The correlation between P^* and L/h was determined for the set of the experimental data obtained in the 1700 mill. Regression equations were obtained for skin pass rolling without lubrication

$$P^* = -8.96 + 0.094R/h, \left(r = 0.70, F_1 = 1.87\right)$$

and with lubrication using I-20 oil

$$P^* = -5.53 + 0.06R/h, \left(r = 0.81,\ F_1 = 2.83\right),$$

where r is the correlation coefficient.

The results obtained for the given skin pass rolling conditions can be represented most closely by an adequate model expressing the dependence of deformation force P^* on two criteria of similarity of the process: ε and R/h:

$$P^* = -24.1 + 10.1\varepsilon + 0.1R/h\left(r_{mc} = 0.90,\ F_1 = 5.47\right),$$

$$P^* = -10.27 + 2.84\varepsilon + 0.06R/h\left(r_{mc} = 0.93,\ F_1 = 7.0\right),$$

where r_{mc} is the multiple correlation coefficient.

As mentioned above, increase of the speed of skin pass rolling at constant reduction increases P^* (Fig. 10.6) and the rolling torque. When skin pass rolling with lubrication the increase of forces becomes smaller with increase of the efficiency of lubrication.

Increase of the amount of force and the moment of skin pass rolling, observed with increasing reduction and roughness of the rolls, is associated with the influence of these parameters on the friction coefficient. Finding the friction coefficient for the conditions of skin pass rolling is of great practical and scientific interest.

Sufficiently reliable mathematical models of the process of skin pass rolling are based on the assumptions about the constancy of the contact shear stresses in the deformation zone [150]:

$$\tau = \mu 2k,$$

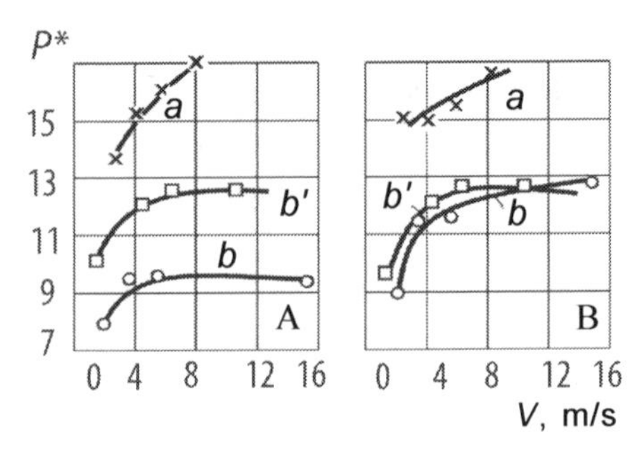

Fig. 10.6. The dependence of the dimensionless pressure P^* rolls on the speed V of skin pass rolling without lubrication (a) and with technological lubricants – emulsion E-2B and oil I-20 (*b, b'*) at constant reduction $\varepsilon = 1.0\%$ (A) and at a constant setting of the rolls (B); 08Yu steel; mill 1700; $R/h = 250$.

where μ is the plastic friction coefficient; k is the resistance of metal to plastic shear.

The relationship of the plastic friction coefficient μ with the parameters of the process of skin pass rolling of strips will be analysed.

The plasctic friction coefficient in skin pass rolling can be most conveniently defined by reverse conversion using experimental values of the pressure of the metal on the rolls. As a mathematical model of the process of skin pass rolling in the calculations the authors of this book used the model developed in detail by E.M.Tret'yakov[150]. The algorithm for calculating the plastic friction coefficient, the stress state coefficient, as well as other parameters of skin pass rolling, takes into account the change in the yield strength of the skin pass rolled metal, depending on the strain rate.

Results show [149, 151] that with increasing reduction ratio the stress state coefficient n_σ increases drastically in skin pass rolling of the strips in rough rolls without lubrication and little in skin pass rolling in the same rolls lubricated with I-20 oil (Fig. 10.7). For other mechanical properties of the skin pass rolled strips the nature of the dependence of n_σ on ε does not change qualitatively. In skin pass rolling of the strips of steel 08Yu with lubrication with I-20 oil the curves n_σ (ε) are located higher than for the steel 08pc strips. n_σ decreases with increasing thickness of the skin pass rolled strips. Compared with skin pass rolling without lubrication and with I-20 oil in skin pass rolling with emulsion E-2 (B) the stress state coefficient takes intermediate values.

Lubrication affects the power and kinematic parameters of skin pass rolling by changing the value of the friction coefficient.

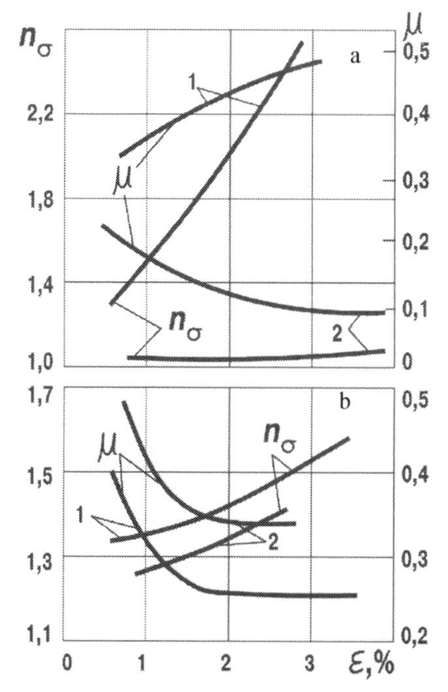

Fig. 10.7. Dependence of the coefficient of the stress state n_σ and the plastic friction coefficient μ on reduction in skin pass rolling (V = 1.1 ÷ 1.3 m/s, and the front and rear tension 20–30 N/mm²): a – without lubrication (1) and with the lubricant I-20 oil (2) of the strips of steel 08pc (σ_T = 265 N/mm²) with a thickness of 1.0 mm with an initial roughness of the metal 0.65 μm Ra; b – strips of steel 08Yu (σ_T = 205 N/mm²) lubricated with I-20 oil, thickness 0.9 (1) and 1.5 mm (2) with the initial surface roughness of the metal Ra = 0.7 μm (1) and Ra = 0.93 μm (2).

When skin pass rolling without lubrication the plastic friction coefficient μ increases with increasing relative reduction (see Fig. 10.7).

Its rate of change decreases with increasing reduction, i.e. friction coefficient tends to its limiting value equal to 0.5. The dependence of the friction coefficient μ in reduction ε has the same character as the curves of changes of the 'imprinting' of the microrelief of the rolls on the strip [151]. Consequently, increasing friction coefficient with increasing reduction in skin pass rolling of the strips in the rough rolls without lubrication due to the increase of actual contact area of the deformable metal and the tool with the introduction of its microprojections into the surface of the strip and flow of the metal into micropits on the roll surfaces. Contact pressure at which the maximum grip of the surfaces of the rolls and the strip is reached corresponds to the limiting value of the friction coefficient.

Since the degree of 'imprinting' of the roughness of the rolls on the surface of the skin pass rolled strips generally defined not so much by the degree of deformation of the strips as by the values n_σ and $n_{\sigma max}$, for the aggregate of all the results of the experiments in skin pass rolling without lubrication no correlation is found between μ and ε.

The effect of various process parameters of skin pass rolling on the friction coefficient can be characterized as follows.

Skin pass rolling of the strips using lubrication with other conditions being the same takes place at lower values of μ than skin pass rolling without lubrication. With increasing compression the friction coefficient is lowered to a certain level, and then in the analyzed degrees of strain the μ value remains virtually constant (see Fig. 10.7). Lubricating with the I–20 oil reduced more significantly the friction coefficient than the emulsion E–2 (B), due to the fact that the friction coefficient with lubricantion during skin pass rolling depends on the thickness of the oil film in the deformation zone which is a function of the rolling speed, viscosity and the piezocoefficient of viscosity of the lubricant, contact pressure, the radius of the rolls, the thickness of the rolled strip, compression [63]. Investigations of the thickness ξ of the lubricating film in the deformation zone during skin pass rolling were carried out by the 'droplet method' and showed that for the considered conditions (roughness of the rolls 2.0 μm, the strips 0.4–1.3 μm, reduction 0.3–2.0%) using oil I-20 $\xi = 2.0 \div 2.5$ μm, emulsion $\xi = 1.0 \div 1.5$ μm. Moreover, with increase of the speed of skin pass rolling from 2 to 15 m/s the lubricant film thickness ξ increases by 15–20%. With increase of the degree of deformation within the given range ξ decreased by 10–15%.

For the same values of n_σ the friction coefficient in skin pass rolling with lubrication is higher for the thicker strips. This is probably due to the fact that as the thickness of the rolled strip increases the absolute reduction and the contact angle increase and, therefore, less lubricant goes to the deformation zone. With increasing speed of skin pass rolling the friction coefficient decreases, which is explained by the increasing amount of lubricant entering the contact zone [63]. The same change of μ is observed at higher tension.

The influence of the technological lubricant on the friction coefficient in the deformation zone appears to limit the actual area of the surfaces of the rolls and the rolled metal and changes the shear stress in the areas of true contact. The shape of surface roughness of the rolls and the strips affects the leakage of the lubricant from the micropits, the actual contact area and the frictional force.

Reduction of the friction coefficient with increase of reduction during skin pass rolling with lubrication can be explained in terms of modern ideas about the mechanism of interaction of the contact surfaces of the rolls and the rolled metal in the presence of the lubricating layer in the deformation zone. Since the height of the microroughness of the rolls and skin pass rolled metal is commensurate with the thickness of the lubricant layer in the deformation zone. at small reductions the surface microprojections 'pierce' the lubrication layer and come into direct contact. As a result, the friction coefficient is higher (see Fig. 10.7). With increasing deformation the surface microasperities of the skin pass rolled metal are removed, so the deformation shows the formation of a dividing lubrication layer and the friction coefficient decreases.

It should emphasized that even small changes in the value of the initial roughness of the strips significantly affect the values of the friction coefficient in skin pass rolling. Thus, in these experiments the increase in initial surface roughness of 0.7 to 1.0 μm *Ra* for the 0.9–0.95 mm thick strip of 08Yu steel lubricated with I-20, the speed of skin pass rolling 1.1–1.3 m/s, the tension of 20–30 N/mm^2 and roll roughness Ra = 2.0 μm caused a 20–25% increase in the friction coefficient.

The generalized dependence of the friction coefficient μ of n_σ in skin pass rolling in the 1700 mill of the strips with different properties and in different conditions (without and with lubrication, with and without tensioning, etc.) is shown in Fig. 10.8. When skin pass rolling with lubrication and without it the friction coefficient is greater the higher the stress state coefficient. Between μ and n_σ and between μ and $n_{\sigma max}$ there is a linear correlation with a correlation coefficient r = 0.74 at F_1 = 2.2 and r = 0.76 at F_1 = 2.4. The graphical interpretation of the linear regression equation

$$\mu = -0.065 + 0.255 n_\sigma$$

and two non-linear approximating relationships between μ and n_σ

$$\mu = -0.066 + 1.04 n_\sigma - 0.246 n_\sigma^2.$$

$$\mu = 0.92 - 0.29 \exp(1/n_\sigma)$$

shown in Fig. 10.8.

For the regression equation, as well as exponential and logarithmic functions $\mu = 0.92–0.29 \exp(1/n_\sigma)$; $\mu = 061–1.24/\exp n_\sigma$;

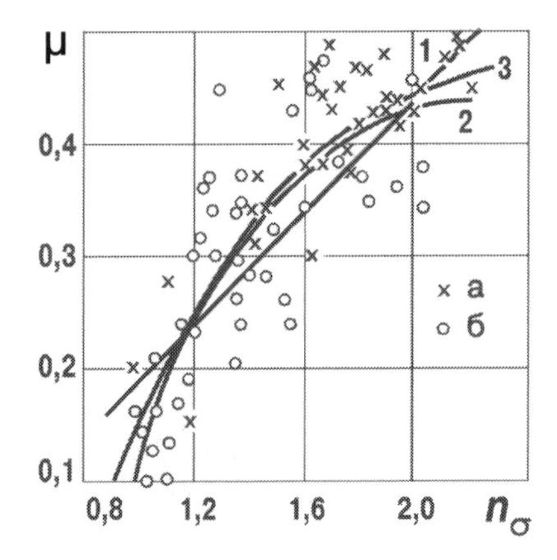

Fig. 10.8. Generalized dependence of the plastic friction coefficient μ on the stress state n_σ (mill 1700) for skin pass rolling without lubrication (a) and with lubricant (b). Approximating functions:

$$1-\mu=-0.065+0.255n_\sigma;2-\mu=-0.66+$$
$$+1.04n_\sigma-0.246n_\sigma^2;\ 3-\mu=0.92-0.29\exp(1/n_\sigma)$$

$\mu = 0.16+0.404 \ln n_\sigma$ the residual variances are smaller than calculated by the linear regression equation. Comparison of the residual variances and the overall variances yields the following criterion values F_1: 2.58; 2.7; 2.5 and 2.42.

The ratio of the length of the deformation zone to the thickness of the skin pass rolled strips L/h increases with increase the degree of deformation, regardless of the nature of changes of μ.

The speed characteristic of the process of skin pass rolling, as well as rolling in general, is the forward slip. In skin pass rolling because of the elastic flattening of the rolls and rolled metal the length of the deformation zone greatly increases, making it difficult to determine the analytical dependences of the geometrical, speed and deformation parameters of the process. The relations between the magnitude of the neutral angle, the friction coefficient and the angle of engagement, known in the theory of rolling, do not apply here. In experimental studies the forward slip in skin pass rolling is generally determined by the reduction measurement system. The speeds of entry and exit of the strip from the rolls are compared.

The dependences of the foward slip on reduction under different conditions of skin pass rolling are shown in Fig. 10.9. According to these data, the forward slip increases with increase of the reduction. Skin pass rolling with lurication takes place at a decrease of the

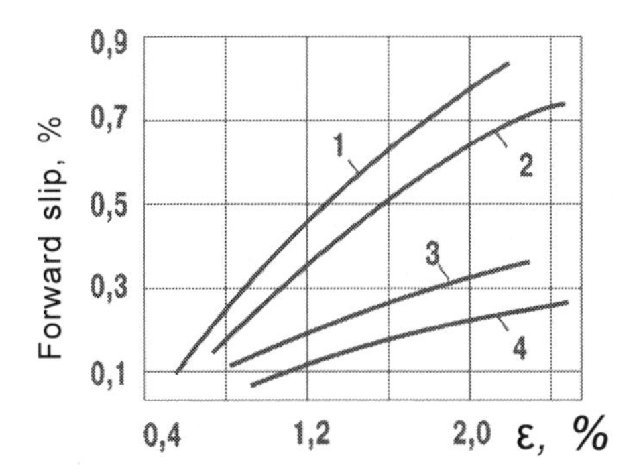

Fig. 10.9. Influence of the degree of deformation in skin pass rolling on forward slip in skin pass rolling without lubrication (1, 3) and in lubrication with oil I-20 (2, 4) of strips of 08Yu steel (σ_T = 205 N/mm²), size 0.9 × 1250 mm (1.2) and 1.5 × 1020 mm (3,4).

forward slip. At the same reduction the forward slip is greater in skin pass rolling of thin strips.

To achieve uniform compression along the length of the strip in the strip roll and homogeneous properties of the metal it is very interesting to study of the impact of the speed skin pass rolling on the forward slip, force and the degree of deformation. At a fixed position of the pressure screws increasing speed of skin pass rolling, along with increasing pressure of the metal on the rolls (see Fig. 10.6) change the amount of deformation of the strips and forward slip (Fig. 10.10). In these conditions, increasing the speed resulted in a slight decrease in compression in skin pass rolling without lubrication and a significant increase in the degree of deformation in skin pass rolling oil lubricated-20. When using an emulsion increasing speed of skin pass rolling results only in a small change of the reduction.

The dependence of the forward slip on the speed reflects the character of the change of the reduction as a function of this parameter: in skin pass rolling without lubrication the forward slip decreased, in lubrication with the I-20 oil increased, and when the emulsion E-2(B) was used it was about the same level (see Fig. 10.10).

Numerous studies have shown that in the course of skin pass rolling at a fixed installation rolls increasing strain rate changes the degree of deformation as the deformation resistance of the metal changes together with the contact friction conditions in the deformation zone. In the mills where the supporting rolls are installed on the fluid friction bearings, an additional influence is exerted by the change in the thickness of the lubricant layer in them. The degree

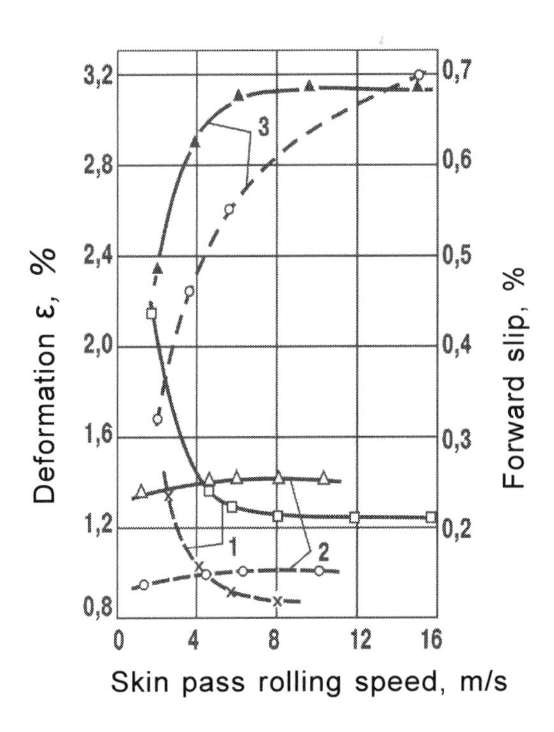

Fig. 10.10. Effect of the skin pass rolling speed without lubrication (1) and with lubrication with emulsion E-2 (B) (2) and oil I-20 (3) while leaving the position of the adjusting screws for the degree of deformation (dashed line) and forward slip (solid line) unchanged. Steel 08Yu (σ_T = 205 N/mm²). 1.0 mm strip thickness. Front and rear tension 25–40 N/mm². The roughness of the rolls Ra = 2.8 μm, the initial surface roughness of strips Ra = 0.8 μm.

of deformation of the strip decreases with increasing speed of skin pass rolling if the pressure of the metal on the rolls is increased rapidly and the increase of the reduction as a result of the increase of the absolute skin pass rolling force $\Delta\varepsilon_r$ is greater than the change in the thickness of the oil wedge $\Delta\varepsilon_{o.w}$ in the fluid friction support bearings of the rolls, i.e. when $\Delta\varepsilon_f > \Delta\varepsilon_{o.w}$. Otherwise, when $\Delta\varepsilon_{o.w.} >$ $\Delta\varepsilon_r$, the degree of deformation of the strips increases with increasing speed.

Experiments conducted on the 1700 mill of the Cherepovets Steel Plant showed that as the speed of skin pass rolling of the steel sheets with a thickness of 0.7–0.8 mm increases there is a monotonic decrease the reduction of the strips despite the constant position of the adjusting screws. During skin pass rolling 0.5–1.0 mm thick strips relatively high skin pass rolling forces are generated. The elastic deformation of the stand reaches a large value that is not offset by 'floating up' of the journals of the supporting rolls. As a result, the reduction decreases with increasing speed of skin pass rolling. When skin pass rolling the strips with a thickness of 1.2–2 mm where the metal pressure on the rolls is relatively small, a reverse phenomenon observed: the reduction increases with the increase of the speed of skin pass rolling. Therefore, in these conditions, the amount of the solution between the work rolls is decisively influenced by the 'floating up' of the journals of the supporting rolls.

The pressure of metal on the rolls increases with increasing speed of skin pass rolling in processing both the 'thin' and 'thick' strips. Qualitative and quantitative changes of the rolling force as a function of the speed are virtually identical with changes in pressure between the mutually pressed rolls (without the metal) when increasing the frequency of their rotation.

The most intense variation of the pressure on the rolls and the degree of deformation of the strips occurs at skin pass rolling speeds up to 10–12 m/s, i.e. in the period, when there is a relatively rapid change in the thickness of the oil layer and a corresponding 'floating up' of the journals of the supporting rolls.

When skin pass rolling sheet steel in the mills equipped with roller bearings of the working and supporting rolls, the speed increase reduces the degree of deformation of the strips of all thickness. Increase of tension leads to a decrease of the skin pass rolling force, and as a consequence, to increase of the degree of deformation (at a constant position of the pressure screws). When skin pass rolling 08Yu steel the front tension variation from 20 to 100 N/mm^2 changes the pressure of the metal on the rolls within 5–20%. The reduction of the initial skin pass rolling force becomes when its magnitude decreases.

Changing the specific tension from 25 to 100 N/mm^2 during skin pass rolling strips with a thickness less than 1 mm more than doubles the reduction. When skin pass rolling the strip thicker than 1 mm the change of the front tension within the above range has no signification effect on reduction.

These patterns are explained by the increase in the deformation resistance of metal with increasing speed, by changing of the thickness of the oil wedge in the fluid friction bearings of the supporting rolls and the friction conditions in the deformation zone.

Given the importance of the issue in question, we emphasize once again that the change of the degree of deformation as a function of speed in skin pass rolling with a lubricant is determined by a

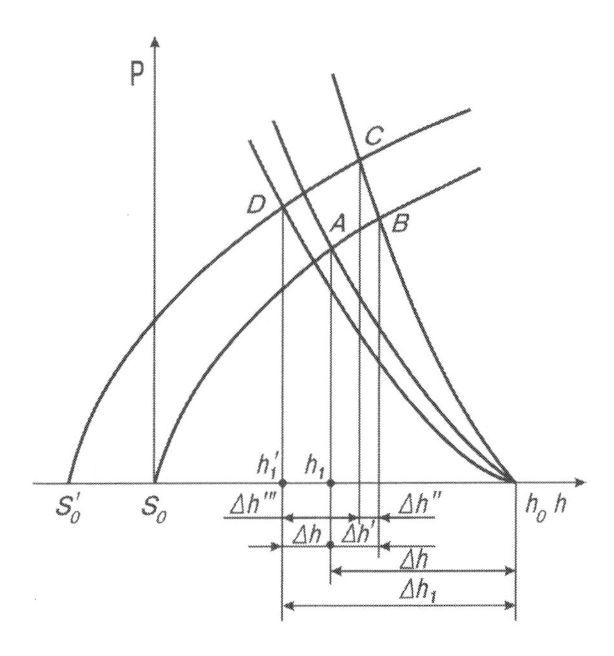

Fig. 10.11. Variation of reduction of strips with increasing speed of skin pass rolling with lubrication. The designations are explained in the text.

joint solution of the equations describing the elastic line of the working stand and the plastic curve of the rolled strip (Fig. 10.11). The abscissa of point A shows the strip thickness h at the outlet of the mill (h_0 is the initial thickness). With increasing skin pass rolling speed, due to changes in the deformation resistance of the steel the pressure of metal on the rolls increases. At the same the roll gap S_0 the process will be characterized by a new point B of intersection of the new plastic curve and the elastic line of the stand S_0A. The reduction in this case changes by the amount $\Delta h'$. Since the increase of the oil film thickness in the fluid friction bearings of the supporting rolls reduces the roll gap of the rolls, as shown in Fig. 10.11 by the displacement of the elastic deformation line of the stand to the position of S_0', point C of intersection of this line with the line h_0B will correspond to the increase of reduction $\Delta h''$, caused by increase in the thickness of the oil wedge. Increasing speed of skin pass rolling increases the amount of the lubricant hydrodynamically drawn into the deformation zone and decreases the friction coefficient. This new value corresponds to the plastic curve h_0D. The abscissa h_1' of point D (characterizing the final state of the process) will show the final strip thickness at a higher speed of skin pass rolling, and the value $\Delta h'''$ – the change of compression due to the decrease of the friction coefficient in the deformation zone.

Obviously, the total change in reduction $\Delta\bar{h}$, due to increasing speed of skin pass rolling with a lubricant is determined by three components:

$$\Delta\bar{h} = \Delta h'' + \Delta h''' - \Delta h'.$$

When skin pass rolling strips with lubrication which significantly lowers the friction coefficient (e.g. using I-20 oil) usually $\Delta h'' + \Delta h''' > \Delta h'$, which means that the degree of deformation increases with the increase of the skin pass rolling speed. When skin pass rolling without lubrication when $\Delta h''' \approx 0$, the reduction may also increase with increasing speed in certain conditions (if $\Delta h'' > \Delta h'$). The effect of the speed is most dramatically evident in the range of small skin pass rolling speeds. The influence of speed factor is enhanced when the thickness of the strips decreases.

The selection of the optimal lubricant can, along with a decrease of the deformation forces, weaken the effect of the speed and stabilize the reduction in skin pass rolling of sheet steel. Using in skin pass rolling of thin strips a processing lubricating emulsion characterized also by good washing properties in addition to good lubricating reduces the extent of contamination of the metal surface. A positive feature of skin pass rolling with lubrication is that by reducing the magnitude of the contact friction forces by the rolls of smaller diameter it is possible to plastically deform thinner strips.

Application of the technological lubricant improves the surface quality of the sheet steel, reduces the energy consumption during skin pass rolling and increases the efficiency of the rolls. When skin pass rolling without lubrication the surface of the work rolls contains soot particles and metal oxides contaminating it and causing prints and other defects on the surface of the skin pass rolled strips. When using the lubricant in skin pass rolling dirt, sand and other particles are washed from the surface of the rolls and the strip. This significantly reduces the number of rejects due to of metal surface defects. Since skin pass rolling with the lubricant takes place at a lower pressure on the metal rolls, the working conditions of the systems of forced bending of the rolls of the mill are improved thus improving the flatness of skin rolled strips.

Skin pass rolling of very thin strips and, in particular, the tin plate is carried out usually on two-stand mills. The technology of skin pass rolling of strips in such systems is different from that in the single-stand quarto or duo mills. Features of the skin pass rolling process of tin plate will be considered on the example of the 1400 two-stand mill of the Karaganda Metallurgical Combine (KarMC) [152].

The 1400 two-stand mill at the KarMC is designed for skin pass rolling of strips 0.18–0.6 mm thick and 700 to 1250 mm wide, in rolls weighing up to 30 t at a speed of up to 40 m/s. The mill has the following automation systems: forced bending of the work rolls (FBWR), adjustment of the relative reduction, slowing down the mill, the positional control of the electric drive of the pressure screws. The design features of the mill are the drive through the supporting rolls, various work roll diameter of the stands 1 and 2 (420–400 and 600–570 mm, respectively), combined (electric and hydraulic) pressure devices. Due to differences in the diameter of the working rolls of the mill stands the main reduction of the strip is performed at the first stand $\varepsilon_1 = (0.7–0.85)\ \varepsilon_\Sigma$. In the second stand the degree of deformation in skin pass rolling of tin plate is usually relatively small.

Values of the plastic friction coefficient μ, obtained by the reverse conversion from the experimental data in skin pass rolling tin plate were 0.20–0.21. Such small values are due to the specifics of skin pass rolling of the strips with small absolute reductions: at thicknesses less than 0.7–0.8 mm and reductions of 1–2% and less the plastic friction coefficient depends on the ratio of the absolute value of reduction Δh and surface roughness Rz. At comparable values of these parameters ($\Delta h/Rz \leq 10$), the friction coefficient is significantly reduced compared to its limiting values [152]. When skin pass rolling tin plate the ratio $\Delta h/Rz$ is 0.4–2.0 for shot blasted rolls and 1.8–3.5 for ground rolls. So the manifestation of this effect is more pronounced in skin pass rolling using the shot blasted rolls. As a result, even for the roll stand 1 with a sufficiently high surface roughness the friction coefficient values are not higher than 0.25.

The skin pass rolling force in almost the entire range of reductions (Fig. 10.12) does not exceed the permissible value of 20 MN. At operating speeds (25–30 m/s), the skin pass rolling force is limited by the temperature conditions of work of the rolls. Applied to the 1400 mill at the KarMC the maximum force in the first stand should not exceed 3.0–3.5 MN. Thus, in accordance with the calculated reduction data the maximum value of reduction in the first stand in the case of the working rolls with a rough (shot blasted) surface (μ = 0.23–0.25) is less than 2.0%, and for the ground or polished rolls (μ = 0.15–0.19) it is 2.5–3.0%.

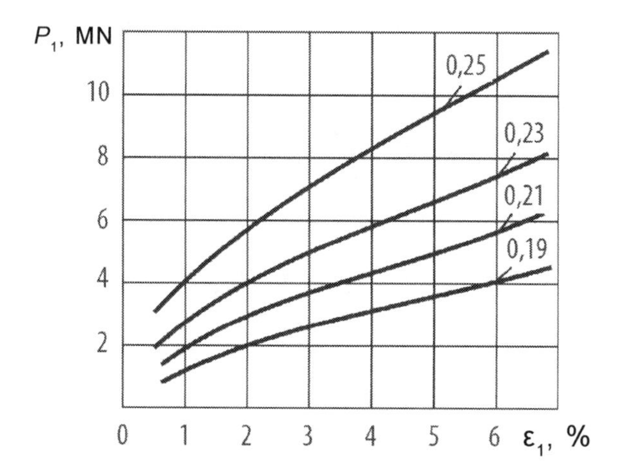

Fig. 10.12. Dependence of the skin pass rolling force P_1 on the degree of deformation and friction coefficient μ_1 (figures at the curves) in stand 1 [152].

The skin pass rolling of tin plate in the 1400 mill at working speeds is implemented at a reduction of more than 0.8–1.0%. At smaller reductions the automatic system for controlling extension does not operate in a stable manner and the required flatness of tin plate is not ensured. To produce the tin plate with hardness A and thickness of 0.20–0.22 mm the reduction should not exceed 1.2%.

Preliminary work hardening of the metal in the stand 1 leads to an increase in the power parameters of the stand 2 (Fig. 10.13), and with greater intensity than in the first cages; this is due to the large diameter of the rolls. When skin pass rolling tin plate strips in the second stand with a reduction of 0.5% and a friction coefficient μ = 0.15–0.19 (ground and polished rolls) the maximum operating force (3.5–4.0 MN) is achieved in the case of preliminary deformation in the first stand equal to 2.0–2.5%.

In general, the theoretical analysis showed that when using the 1400 two-stand skin pass rolling mill the maximum reduction in skin pass rolling tip plate 0.18–0.25 mm thick without lubrication in the first stand is 2.0–2.5% and 0.4–0.6% in the second stand, depending on the state of the surface of the work rolls.

10.3. Features of technology of skin pass rolling thin strips

Consider the features of the technology of skin pass rolling quality sheet steel. The maximum value of tension in skin pass rolling

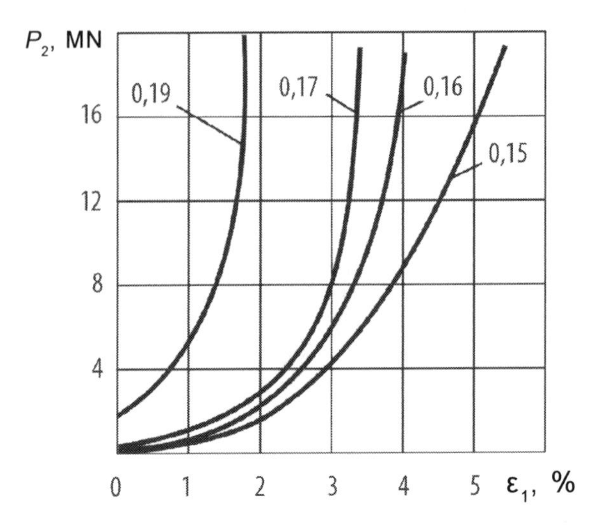

Fig. 10.13. Dependence of the skin pass rolling force in the second stand (ε_2 = 0.5%) on the degree of deformation in the first stand ε_1 and the friction coefficient μ_2 (numbers at the curves) [152]

annealed strips is limited by the risk of fracture on the surface of the metal. Research [72, 153] shows that fractures are the result of plastic deformation of the surface layers of the metal caused by the bending and tension of the strip during the unwinding of the roll. The allowable tension of the strip at its bending or straightening increase with decrease of the curvature of the bent (unbent) strip and the thicker the strip, the higher its yield strength. In practice, during rolling strips of thickness less than 1 mm the tension between the uncoiler and the stand is usually assumed to be between 30–40 N/mm² and between the stand and the coiler – approximately 1.5 times higher (50–60 N/mm²). When skin pass rolling the strips thicker than 1 mm the rear tension is reduced to 20 and the front – to 25 N/mm². It was experimentally proved that if the rear tension in skin pass rolling strips 0.5–2.0 mm thick does not exceed 20 N/mm², no fracture takes place on the metal surface metal. The conditions for skin pass rolling are not degraded.

In the process of skin pass rolling of cold rolled strips in some mills 'ribbing' defects form on the surface of the metal [131]. During further processing of the metal the appearance of finished products is improved. In addition, the appearance of 'ribbing' on the surface of the work and backup rolls leads to their additional handling and regrinding, which results in increased downtime and increased wear of the mill rolls. The 'ribbing' also forms on the surface of the pinch rolls of the pickling lines and cutting units. 'Ribbing' occurs, as a rule, in operation of the system in which the gears in the drive lines are loaded with low forces at very high speeds.

The investigations carried out at the Institute of Ferrous Metallurgy into the sheet surface with the 'ribbing' defects showed that the light bands appear due to slip of the work rolls relative to the surface of the skin pass rolled metal, or due the fact that bands formed in the work roll due to its slippage relative to the support roll and are then again are printed on the surface of the steel sheet.

The number of 'ribbing' bands on the surface of the working rolls is equal to the number of teeth of the drive gear in the drive line of the mill. This led to the conclusion that the engagement in the gearbox is a source of vibration excitation causing the emergence of 'ribbing' on the surface of the rolls and the skin pass rolled metal. When working in the skin pass rolling mills with gearless drives such a defect does not form.

The cause of the vibration is the cyclic error of the circumferential pitch of the gears which is determined by the degree of accuracy of their manufacture. When the teeth are engaged with the inaccurate pitch, many produced gears are relatively displaced on the pitch circle, which leads their uneven rotation. The resultant acceleration in the drive line causes variable inertial forces. The reaction of the drive line to these forces depends on the speed and power modes of operation and the system parameters. There is a combination in which the force in the meshing is zero, and a gap occurs in the contact between the teeth. Calculations show that these discontinuities occur at skin pass rolling speeds greater 7 m/s, and the energy transfer in the engagement in the working mode is pulsed (impact).

The kinetic energy obtained by the driven gear at the moment of break of contact in the engagement changes to the potential energy of spinning the shaft as it moves in the field of the side gap, and then under the effect of elastic forces the contact is restored and repeated collisions occur between the gears and the wheels.

In the collisions of the meshing teeth the energy is scattered at moments of restoration of the contact. Obviously, the steady state of shock vibrations will only be supported if the loss of energy in collisions are compensated by external sources. The energy loss formed when restoring the contact in the engagement are compensated by the drive. Study the shock interaction of the gears showed that the occurrence of such a mode is possible in principle.

This mode has the ability for self-regulation and is sufficiently stable in a range of speeds.

The slip of the work rolls relative to the strip (or backup rolls) comes when the torque applied to the work roll from the spindle reaches the moment of coupling forces between the strip and the roll (or between the work and backup rolls). When the deformation of the shafting reaches a maximum value in movement of the gear in the field of the sid gap, the rate of this deformation becomes zero. For the stability of the vibroimpact process it necessary to ensure that the engagement frequency was equal to or is a multiple of the frequency of impact impulses arising in the system [154].

These conditions allow one to determine the critical speed of rotation of the roll at which ribbing may occur on the strip. The critical speed of skin pass rolling is defined as follows:

$$n_{\text{cr}} = \frac{30\sqrt{C_{\text{sh}} / I_{\text{g}}}\, m}{z_{\text{g}} \arccos \dfrac{M_{\text{sk}}}{M_{\text{eng}}}},$$

where C_{sh} is the shafting rigidity between the gear and the work roll; I_{g} is the moment inertia of the driven gear; m = 1, 2, 3... is the number of critical zones of speed skin pass rolling; z_{g} is the number of teeth of the driven gear; M_{sk} is the skin pass rolling moment; M_{eng} is moment of coupling forces between the work roll and the strip or the backup roll.

For the 2500 mill of the Magnitogorsk Metallurgical Combine, for example, in the steady state $M_{\text{sk}} \ll M_{\text{eng}}$. Therefore, for an approximate calculation of the critical speeds skin pass rolling it can be assumed:

$$M_{\text{sk}} / M_{\text{eng}} \approx 0; \qquad \arccos M_{\text{sk}} / M_{\text{eng}} \approx \pi / 2.$$

The following system parameters are:

$$I_{\text{g}} = 53.5 \text{kg} \cdot \text{m}^2; \; C_{\text{sh}} = 1.5 \cdot 10^7 \,\text{N} \cdot \text{m}; \; Z_{\text{g}} = 54.$$

Substituting these data into the above equation, we get $n_{1\text{cr}}$ = 19.6 rad/s; $n_{2\text{cr}}$ = 39.3 rad/s; $n_{3\text{cr}}$ = 58.7 rad/s; $n_{4\text{cr}}$ = 78.5 rad/s. Among these values $n_{1\text{cr}}$ and $n_{2\text{cr}}$ values can be disregarded, as the mill passes these speeds during acceleration. The drive lines are loaded at these speeds by the moments of inertia forces of the flywheel masses and breaks in the engagement do not occur. The values $n_{3\text{cr}}$ and $n_{4\text{cr}}$ are 'dangerous' so the stationary process of skin pass rolling should not be carried out near these zones of the speed of the work roll.

The above equation indicates that the values of the critical speed of skin pass rolling can be controlled by changing the values of skin pass rolling and coupling moments.

The 'ribbing' defect may form during acceleration or deceleration of the mill. The critical speed at which the ribbing may occur during acceleration or deceleration is defined as follows:

$$n_{cr}^{p} = 30\sqrt{C_{sh} / I_{g}} m / z_{g} \arccos \frac{\frac{GD^2}{375}\frac{dn}{dt} + M_{sk}}{M_{eng}},$$

where dn/dt is the angular acceleration (deceleration) of the stand; GD^2 is the flywheel moment of elements of the drive line reduced to the shaft of the drive..

For example, when skin pass rolling strips of 08Yu steel in the 2500 skin pass rolling mill of the Magnitogorsk Metallurgical Combine we have the following data: dn/dt = 23.9 rad/s^2; GD^2 = 165 kN·m^2; M_{sk} ≡ 20 kN·m. Substituting them into the last equation, we obtain for m = 1 n_{cr}^{p} = 58 rad/s. The other critical areas can be ignored since they are outside the working range of speeds.

According to the last expression the value of the critical speed in the acceleration (deceleration) mode depends on the rate of acceleration (dn/dt) and the grade of the skin pass rolled metal (M_{sk} and M_{eng}). For each grade we can choose the acceleration mode using this relationship in which the value n_{cr}^{p} will lie outside the operating range and hence the 'ribbing' will not form.

If in some cases the process parameters and characteristic skin pass rolling management systems do not make it possible to exclude the mill critical speed when reaching the desired operating speed, this dangerous range should be passed at the maximum possible acceleration during acceleration and deceleration during braking. Regions of the strip with 'ribbing' would be minimal in this case. One can also recommend conducting skin pass rolling at the operating speed lower than critical. Elimination of the critical speed modes from the process of skin pass rolling helps to reduce the rejection of metal due to 'ribbing', to reduce the number of changes of rolls and regrinding, and to reduce roll wear.

Causes and methods to prevent overloads and vibration in the spindles of the skin pass rolling mills with a group drive are partly addressed in the chapter dealing with rolling strips in asymmetric conditions. Here we will focus on the provision of ensuring high quality of the surface of thin sheets skin pass rolled in the mills of this construction. For example, the 1700 skin pass rolling mill of the Karaganda Metallurgical Combine (KarMC) can be considered [71].

The main drive of the work rolls of the 1700 skin pass rolling mill of the KarMC consists of a DC double-armature motor (type 2MP200-330, $N = 2\times1000$ kW, $n = 330/800$ rpm), and the pinion stand spindles, which by means of gear clutches are connected to the work and gear rolls.

In the drive line of this skin pass rolling mill the vibrations with a frequency of 5–20 Hz occur with the maximum amplitude at a skin pass rolling speed of 17–18 m/s. So in this speed range it is impractical to carry out skin pass rolling. The distribution of the moments between the spindles is uneven. In the lower spindle the moment is always positive and in the upper one it depends on the conditions of skin pass rolling and can be positive or negative. The moment at the shaft of the motor is equal to the algebraic sum of the moments on the spindles. In the spindles there are sometimes moments of the opposite sign which are 8–10 times greater than the moment on the motor shaft.

In this mill, the force flow of the drive motor is separated to two work rolls using a gearwheel of the pinion stand with a chevron notched gearing (ratio equal to one). As previously mentioned, usually the diameters of the upper and lower work rolls slightly differ (1 mm). Consequently, the ratio between the working rolls is not equal to unity. Force interaction forms during skin pass rolling between the work rolls and a closed kinematic chain consisting of gear rolls, spindles, work rolls and rolled metal forms.

Because of the inequality of the transfer ratios of the gear and a pair of work rolls the branches of the drive are twisted and high forces form in a closed kinematic chain.

It was established [71] that because of the relative twisting of branches of the closed loop the upper and lower spindles are loaded with additional moments M_a^l *and* M_a^u of the opposite sign, the values of which are determined by the equations

$$M_a^l = R_l K, \quad M_a^u = R_u K, \qquad K = \frac{\gamma_{ol}^2 - \dfrac{R_u^2}{R_l^2}\gamma_{ou}^2 + \dfrac{R_l - R_u}{R_l^2} h_1}{2a_0\left(\gamma_{ol} + \dfrac{R_u}{R_l}\gamma_{ou}\right)},$$

where R_l and R_u are the radii of the lower and upper work rolls ($R_l > R_u$); h_1 is the the thickness of the strip at the exit from the deformation zone:

$$\gamma_{ol} = \frac{\alpha_l}{2}\left(1 - \frac{\alpha_l}{2\mu}\right);\ \gamma_{ou} = \frac{\alpha_u}{2}\left(1 - \frac{\alpha_u}{2\mu}\right);\ a_0 = \frac{1}{4\mu p_{av} R_l b};\ a_l = \sqrt{\frac{\Delta_l h}{R_l}};$$

$$a_u = \sqrt{\frac{\Delta_u h}{R_u}};\ \Delta_l h = \frac{1}{1 + R_l / R_u}\Delta h;\ \Delta_u h = \frac{R_l / R_u}{1 + R_l / R_u}\Delta h;$$

where Δh – absolute compression; μ – friction coefficient in skin pass rolling; p_{av} – average specific pressure; b – the width of the strip.

The values calculated by the above equations coincide with the experimental data (deviation does not exceed 8%).

When twisting a closed circuit of the mechanical system at some point complete unloading of the branches of the smaller diameter roll takes place. The force contact in the gears becomes negligible and the contact in the meshing is broken, leading to the appearance in the driven branch of the vibroimpact processes whose frequency is equal to or a multiple of the frequency of engagement. Therefore, the step of 'ribbing' on the strip and the rolls is equal to or a multiple step of the gear engagement. To prevent the 'ribbing' on the strip and rolls it is advisable to replace the spindle drive with the gear couplings by the spindle with rubber-metal bushings [71].

One of the possible ways to address overloading in the drive line of the 1700 skin pass rolling mill is the replacement of the group drives by individual ones. The emergence of a closed loop in the drive system of the mill will be prevented. It is also rational to change to skin pass rolling with one drive spindle – lower. The upper work roll will rotate due to friction forces in the deformation zone. Calculations show that there is a sufficient reserve of friction forces in the deformation to ensure a normal (without slipping to the surface of the deformed metal of the upper roll) process of skin pass rolling at the required rate of acceleration and deceleration of the mill. In this case, there is almost perfect agreement between the circumferential speeds of the upper and lower rolls, which contributes to a better surface quality of the skin pass rolled metal. The advantage of the drive circuit with one spindle is also a simplification of the drive line – reducing the number of devices (no upper roller gear, spindle, etc are required).

In skin pass rolling thin strips and tin plate in two-stand mills, for example, the 1400 mill of the Karaganda Metallurgical Combine, strip oscillations occur in the interstand gap which lead to a breach of the engagement of the surfaces of the work and backup rolls.

By increasing the speed of skin pass rolling the amplitude of the oscillations increases. The cause of the vibrations of the strip tension between the first and second stand are the vibration of the roll tension meter, installed in the space between the stands of the mill. Fluctuations in the strip tension in the interstand gap cause periodic slipping of the work rolls with respect to the support rolls in the second stand and the occurrence of 'ribbing' on the surface of the supporting rolls. To avoid this phenomenon it is necessary to lower the level of vibroactivity of the tension meter.

10.4. Effect of skin pass rolling conditions on the properties of steel

In the production of thin sheets the reductions in skin pass rolling must be coordinated with deformation modes in the steel during cold rolling and subsequent annealing. The nature of this relationship is concluded that, as was mentioned above, the yield point of the sheet steel is determined by the size of ferrite grains in its structure. For its part, the grain structure is formed in the steel during rolling and annealing.

The dependence of the low yield stress of the on the grain size is determined by the Hall–Petch equation:

$$\sigma_T = \sigma_0 + kd^{-\frac{1}{2}},$$

where σ_T – lower yield stress; σ_0 and k – constants; d – the average grain diameter.

In accordance with the Hall–Petch dependence, the type of stress-strain diagrams in tensile loading coarse and fine-grained steels will be different (Fig. 10.14). Due to the fact that the deformation intensity corresponding to the end of the yield plateau on the tensiles is greater for the fine-grained metal ($\varepsilon_{u_2} > \varepsilon_{u_1}$), to eliminate the yield plateau the skin pass rolling of the fine-grained metal should be carried out so that the average integral value of the deformation intensity ε_{i_c} along the strip thickness is more than is required for the coarse-grained steel. Thus, the reduction in skin pass rolling steel sheets must be greater for the finer ferrite grains in its structure. The yield strength of the steel decreases as a result of skin pass rolling with the reduction of the grain size.

The above described relationships of the influence of the steel microstructure on the mechanism of eliminating the yield plateau through skin pass rolling of the strips reveal the character of the

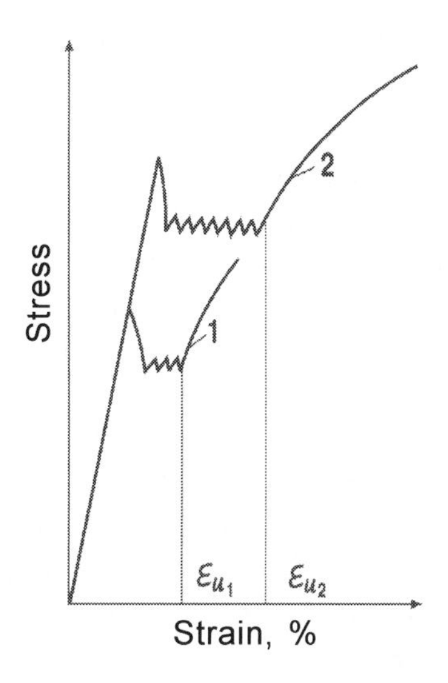

Fig. 10.14. Diagram of tensile loading sheet steel samples of the coarse- (1) and fine-grain (2) structures.

relationship of the skin pass rolling regimes with the regimes of prior cold rolling. It is known that the grain size of the structure of strips of mild steel decreases with an increase in the total cold reduction prior to final annealing. Consequently, the greater the degree of deformation in cold rolling the steel, the greater the reduction which must be applied in skin pass rolling.

Given the above laws, to improve the quality of the skin pass rolled metal, specialists from the Institute of Ferrous Metallurgy and the Karaganda Metallurgical Concern proposed[2] when selecting the skin pass rolling modes to consider the heterogeneity of the structure and mechanical properties along the length of the strips in rolls. This proposal was as follows.

In ingot production technology after hot and cold rolling and recrystallization annealing the structure in the thickness strips of the rimming mild steel is characterized by considerable unevenness of the ferrite grains. At the surface there is a layer of larger grains corresponding to the boiling metal, solidified before introduction of Al into a ladle. Beneath there is a pronounced layer of fine-grained metal (8–10 grain size points). In the central layers the ferrite grains are larger. Along the length of the strip the thickness of the fine-grained layer decreases from a maximum value corresponding to the head of the ingot to zero at the distance of 20–25% of the length

[2]V.L. Mazur, B.A. Fel'dman, P.P. Chernov, et al.

of the strip. In the rest of the length of the strip the steel structure is homogeneous. The fine-grain layer metal has a higher content of aluminum (0.01%).

When skin pass rolling with a reduction of 1.0–1.5% of the cold rolled strips having the above-described multilayer structure, they are characterised by higher microstresses compared with skin pass rolling of the homogeneous metal. As a result, the formability (the hole depth decreases when tested by the Ericksen method) of the sheet steel decreases.

Thus, to achieve the same maximum effect in skin pass the strip portion with a fine-grained interlayer in the structure corresponding the head portion of the ingot, should be skin pass rolled with a smaller reduction than the part of the strip without a fine-grained layer corresponding to the bottom of the ingot. With decreasing thickness of the fine-grained layer the reduction should increase, i.e. varies as a function of the strip length inversely proportional to the thickness of the layer of fine grains in the structure of steel sheets and reaches a maximum after rolling 25% of the length of the strip. The rest of the strip is skin pass rolled with the maximum reduction.

Specifically, first, in the strip portion corresponding to the head portion of the ingot, the deformation of the annealed rimming steel with 0.5–2.0 mm thickness should be 0.6–0.8% and increase as a function of the strip length in proportion to the thickness of the fine-grained layer to 1.0–1.2% for the portion corresponding to the middle and bottom of the ingot. Increasing the degree of deformation from the minimum value equal to 0.6–0.8% up to a maximum equal to 1.0–1.2%, should be performed on the portion constituting 20–25% of the length of the strip. Upper limits in these reduction ranges relate to relatively thin strips of 0.5–1.0 mm thickness, lower – to the strips of greater thickness, 1.2–2.0 mm.

Thus, during skin pass rolling strips, for example, of rimming 08kp steel chemically capped with Al, to increase the quality of the skin pass rolled metal, the reduction must be changed as a function of the strip length inversely proportional to the thickness of the fine-grained layer at the surface of the strip. The technical and economic efficiency of this solution is that the mechanical properties of sheet steel are improved by providing the lowest possible yield strength and its uniformity over the entire length of the strip. At the Karaganda Metallurgical Combine, for example, this method of skin pass rolling has been implemented with the help of a specially developed[3] device.

[3]Yu.M. Krtisky, et al.

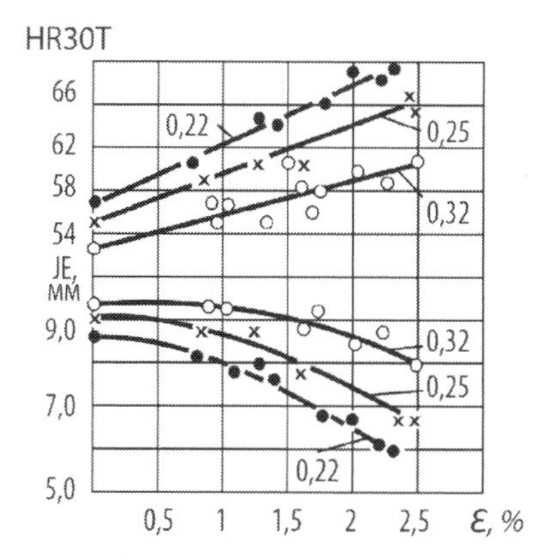

Fig. 10.15. Dependence of certified properties of black tin plate (hardness HR30T and of the depth of the spherical crater according to Erichsen JE) on the total reduction in skin pass rolling in the 1400 mill (figures at the curves – sheet thickness, mm) [152].

The ways to improve the quality of cold-rolled and hot-rolled strip by regulating their skin pass rolling depending on the chemical composition and structure of the metal also apply to steels of higher strength. As an example, one can mention the method (developed by the experts[4] of the Institute of Ferrous Metallurgy of the NAS of Ukraine and the Novolipetsk Metallurgical Concern) of finishing cold-rolled annealed strips of low-alloy higher strength steel of the 08GSYuT type (F), intended for cold forming, according to which the degree of reduction in skin pass rolling is set depending on the thickness of the strip and the carbon equivalent value C_e of the chemical composition of the steel, according to the formula defines

$$C_e = \mathrm{C} + \frac{\mathrm{Mn}}{6} + \frac{\mathrm{Si}}{24} + \frac{\mathrm{Ti(V)}}{5},$$

where C, Mn, Si, Ti(V) are the contents of elements in the steel, wt. %.

The reduction of the strips in skin pass rolling in accordance with this proposal is set in direct proportion to the magnitude of C_e.

The effect of the skin pass rolling conditions on the mechanical properties of tin plate was investiated in industrial environments of the 1400 mill. The strips with a thickness of 0.20–0.32 mm were skin pass rolled with a reduction from 0.7 to 2.8%. The metal temperature before skin pass rolling was 30–50°C.

The results of the experiments show (Fig. 10.15) that the intensity of the change of hardness and of the depth of the spherical dimple

[4]E.S. Kakushkin, et al.

depth in the Erichsen black tin plate in dependence on the total extent of deformation increase with decreasing thickness of the strips.

10.5. Skin pass rolling of hot-rolled steel

In many metallurgical plants the technology of production of cold-rolled sheet steels and tin plate provides that in preparing for the cold rolling of hot-rolled strips (strip plate) the scale is removed from their surface. After unwinding the rolls in the top part of the continuous etching unit (CEU) the strips are compressed in the skin pass rolling mill, if such is installed in the CEU line, and proceed to the pickling baths.

In skin pass rolling the metal is slight stretched (1–5%) which, however, is enough for cracking, loosening and partial destruction of the scale film on the surface of the strips. During subsequent etching the poorly soluble ferrous oxides penetrate through the cracks in the scale layer into the sublayer, which consists mainly of readily soluble oxide FeO, thereby accelerating the etching process with acceleration becoming greater with the increase of the reduction in skin pass rolling. The most significant effect in the rate of pickling of the scale achieved after skin pass rolling with higher reductions (2.5–5.0%). However, in the skin pass rolling of commercial hot-rolled sheets such reductions are undesirable because the plastic properties of steel also decrease.

The relatively rapid removal of the scale from the surface of hot-rolled strip in the pickling baths without deterioration of the steel properties is possible after skin pass rolling with minimal reduction using shot blasted rolls. During skin pass rolling the microprojections of the roll surface become embedded in the scale which cracks due to the drawing of the metal and is refined by the shot blasted surface of the roll. Part of the scale is removed from the metal surface during skin pass rolling, and the rest dissolves rapidly during subsequent pickling, thereby accelerating the passage of the strips through the CEU. However, it should be noted that, as will be shown below, this result is not seen in all the cases and it depends on the composition and properties of the steel, the composition of the scale and the hardness of the surface of skin pass rolled and pickled strips.

In the industrial conditions in the pickling of low-carbon hot-rolled strip steel the factor of the activity of the solution of sulphuric acid in the bath varies continuously within 1.8–0.7. If the degree of deformation bands in skin pass rolling remains unchanged within

1.0–1.5%, the activity of the solution of 0.8–0.85 results in the risk of incomplete pickling, while the activity coefficient of 1.4, for example – results in overpickling of the surface. To improve the quality of the surface of the pickled strips by stabilizing the rate of pickling of the scale at a constant velocity of the strips in the CEU line at the Cherepovets Metallurgical Plant it was proposed[5] to increase (decrease) the degree of deformation in skin pass rolling with a decrease (increase) of the activity coefficient of the pickling solution.

The activity coefficient of the pickling solution K represents the ratio of the total weight percent of the acid in the pickling baths to the ferrous salt content in it: $K = [H_2SO_4]/[FeSO_4]$.

One of the trends in rolling production is to increase the use of skin pass rolling mills for hot-rolled steel finishing. Hot-rolled thin strips rolled in continuous wide-strip mills are skin pass rolled in the mills installed in the pickling lines or transverse cutting mills. The skin pass rolling of hot metal, conducted with nominal reductions of 1–5%, reduces the unevenness, waviness and warpage of the strips, to improve the quality of their surface.

The skin pass rolling condition of hot-rolled strips significantly affect the duration of subsequent pickling of the metal. The influence of reduction and finishing of the surface of rolls in skin pass rolling hot-rolled strips of various steels on the pickling rate of the scale and the quality of the metal surface was investigated[6] in the conditions of the Zaporozhstal' Plant. The effect of skin pass rolling on the duration of pickling of steel 08pc and St3sp strips is shown in Table 10.1. According to the results the skin pass rolling strips of carbon steels in shot blasted (Ra = 15 μm) rolls compared to the polished (Ra = 0.5 μm) rolls is less effective, especially for the steel 08ps strips. This appears to be an unexpected result. However, the cause of the observed relationships is explained by the penetration scale into a relatively soft metal.

With the reduction increasing up to 4% the pickling duration decreases. At larger reductions the duration of pickling does not depend on this parameter.

When pickling strips of Cr19Ni10Ti steel after skin pass rolling, the following results were obtained (Fig. 10.16):

[5]V.P. Sobolenko, et al.

[6]In cooperation with V.T. Tilik, et al.,

Table 10.1 The effect of reduction ε and surface condition (Ra, μm) of rolls in skin pass rolling on duration τ of pickling of hot-rolled strips of 08ps and St3sp steels

Ra = 0.5 μm		Ra = 15 μm		Ra = 0.5 μm		Ra = 15 μm	
ε, %	τ, s	ε, %	τ, s	ε, %	τ, s	ε, %	τ, s
steel 08ps				steel St3sp			
0	124	0	124	0	130	0	130
0.5	75	0.6	77	1.0	85	0.6	70
1.0	75	1.2	90	2.0	60	1.1	65
2.6	59	2.0	89	2.8	54	1.5	63
3.3	55	2.6	84	4.0	50	2.0	60
4.6	53	3.5	68	4.5	50	3.0	55
6.0	51	-	-			4.0	50

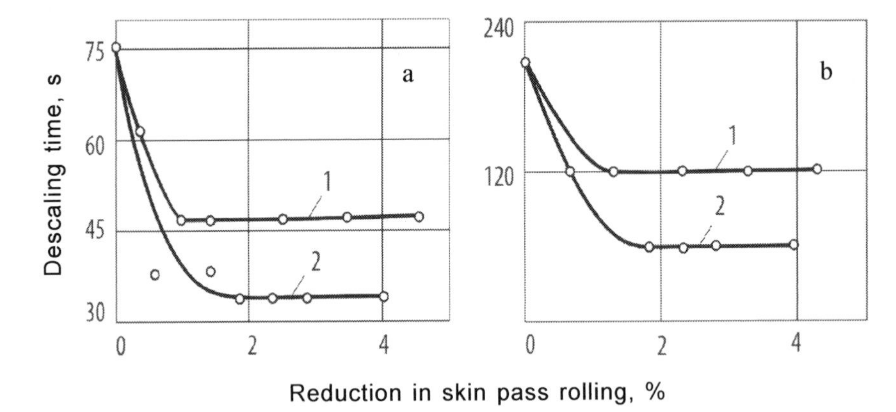

Fig. 10.16. Duration of descaling hot-rolled 18Cr10NiTi steel samples in alkaline (a) and acid (b) baths after skin pass rolling: 1 – in polished rollers; 2 – in the shot blasted rollers.

a) The amount of reduction in skin pass rolling in the polished and shot blasted rolls starting at 1.0–1.5% has almost no effect on the duration of descaling;

b) the time required for processing the metal in alkaline and acid baths decreases after skin pass rolling in the polished rolls by the maximum of 1.7 times;

c) after skin pass rolling in the shot blasted rolls the time required for the treatment of steel in the alkaline bath decreases by no more than 2.5 times, in the acid baths – 3.5 times; the efficiency of skin pass rolling in the shot blasted rolls is 1.5–2.0 times higher than in the polished rolls.

The results of microgeometrical studies on the metal surface after pickling showed that the roughness of the pickled surface of the hot-rolled metal is uniform and homogeneous. The orientation of the microrelief – significant differences in the roughness in the longitudinal and transverse directions of the sheet – was not observed in cases of skin pass rolling both in the smooth and rough rolls. In skin pass rolling of carbon steel in the rough rolls with relatively high (greater than 2%) reductions slurry remnants remained on the surface of the metal in microdepressions after pickling. The pickling time had no appreciable effect on the surface microrelief.

Features of the skin pass rolling strips with the scale are not well understood. In the literature, there are practically no data on the power parameters of skin pass rolling of metal with the scale in both the polished and shot blasted rolls. In studying these issues strips with the scale were skin pass rolled in the rolls 260 mm in diameter with a smooth (polished) or rough (shot blasted) surface in the 200 duo-quarto mill. The surface roughness of the polished rolls was $Ra = 0.8$ μm, the shot blasted rollers $Ra = 5.8$ μm. The reduction in skin pass rolling was varied from 1 to 5%. To ensure the wide range of variation of the mechanical properties, the thickness and structure of the scale on the surface sheets of hot- and cold-rolled annealed stainless steel of various grades were tested.

The dependence of the rolling force as a function of relative reduction in skin pass rolling in the relatively smooth ($Ra = 0.8$ μm) and rough ($Ra = 5.8$ μm) rolls is shown in Fig. 10.17. According to the graph in the considered reduction range the dependence of the skin pass rolling force for the strips with the scale on the degree of deformation is close to linear. Skin pass rolling of the thick hot-rolled strips with loose scale layers takes place with less effort than skin pass rolling of the cold-rolled annealed steel with a thin layer of dense scale. This conclusion is clearly seen in the graphs representing the dependence of the force on reduction in dimensionless coordinates using the similarity criteria of the process skin pass rolling (Fig. 10.18). The difference in the skin pass rolling conditions of the hot- and cold-rolled strips increases with increase of the degree of deformation.

The force in skin pass rolling the strips with the scale increases with increase of the roughness of the roll surface. Comparison of the skin pass rolling conditions in the polished and shot blasted rolls suggests that in the latter case the force is 6–30% higher. A greater

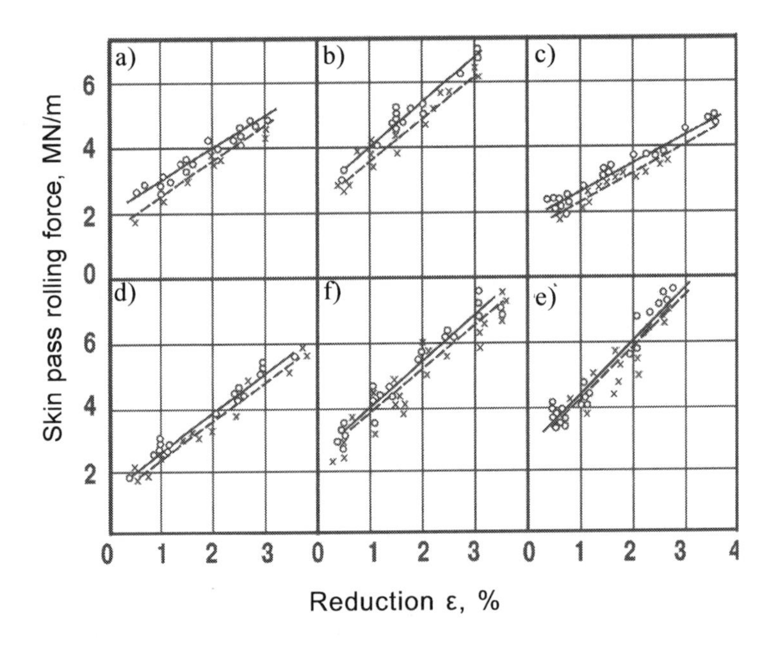

Fig. 10.17. Dependence of the total skin pass rolling force per unit of strip width on relative reduction and the surface condition of the rolls. Hot-rolled steel: a) X6CrTi18; h = 3.2 mm; σ_T = 488 N/mm²; b) X7CrNi17-7; h = 3.2 mm; σ_T = 512 N/mm²; c) X7CrNiTi18-10; h = 3.7 mm; σ_T = 322 N/mm². Cold-rolled annealed steel: g) X6CrTi18; h = 1.8 mm; σ_T = 351 N/mm²; d) X12CrNiTi21-5; h = 1.8 mm; σ_T = 476 N/mm²; e) X7CrNiTi18-10; h = 1.8 mm; σ_T = 303 N/mm². Solid lines – shot blasted rolls, Ra = 5.8 μm; dashed – polished rolls, Ra = 0.8 μm.

difference is observed in skin pass rolling efforts hot metal with thick layers of loose scale on the surface.

Studies have shown that skin pass rolling the annealed cold-rolled sheets with a thin layer of dense oxide scale (annealed in an oxidizing atmosphere) takes place with the same force as the cold-rolled metal without scale. Thick layers of scale in the deformation, zone have effects similar to lubrication, reduce the power parameters of skin pass rolling (Fig. 10.18).

The skin pass rolling of hot-rolled metal is usually carried out in the duo stands with the roll diameter greater than 800 mm. In these circumstances, by choosing the optimum skin pass rolling modes (reduction, speed) it is possible to significantly reduce the longitudinal thickness difference of the strips, the initial value of which is usually in the range of 5–15%.

The effect of the reduction on the equalisation coefficient, which is the ratio of the relative thickness difference of the strip at entry

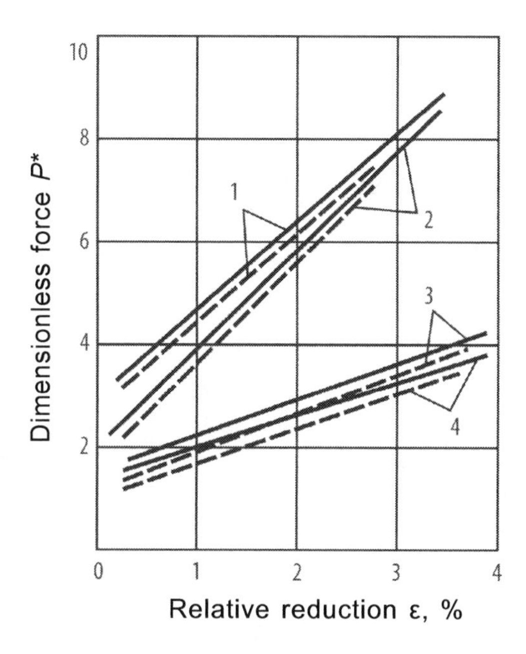

Fig. 10.18. Dependence of dimensionless skin pass rolling foce P^* on the reduction ε. Cold-rolled steel: 1 – X12CrNiTi21-5; 2 – X6CrTi18; hot-rolled: 3 – X7CrNiTi18-10; 4 – X6CrTi18. The other symbols are the same as in Fig. 10.17.

into and exit from the rolls depends on the thickness of the strip plate. The equalisation coefficient increases with increasing reduction in the range 0.5–1.7%, and then decreases with increasing reduction up to 3.5%. With further increase in the reduction the coefficient increases again. In the case of skin pass rolling thicker strips (4 mm) the increase in the equalisation coefficient in the range 0.5–1.7% is more intensive, and the subsequent reduction of the coefficient with increasing reduction occurs less rapidly than for the relatively thin (2 mm) metal. The equalisation coefficient in skin pass rolling increases with increase of the front and rear tension. The same effect is observed with increasing stand rigidity. However, the efficiency of increasing the equalisation capacity of the stand by increasing its rigidity decreases with increasing roll wobbling due to their eccentricity. Starting from a certain value of wobbling the increase of the rigidity of the stand increases the thickness difference of the skin pass rolled strips.

The hot- and cold-rolled annealed steel sheets intended for deep drawing are usually skin pass rolled at a temperature below 80°C. As previously mentioned, during the storage of sheet metal treated in this manner, strain ageing takes place in it which leads to intermitent deformation and the occurrence of slip lines on the thin metal stamped parts. To avoid this negative phenomena, in some cases it is necessary to carry out warm skin pass rolling of the cold-rolled steel intended for deep drawing. In this method, to prevent ageing the steel sheets are skin pass rolled at temperatures of 100–200°C. Skin pass rolling in this temperature range is conducted during cooling of the metal after annealing.

The properties of steel treated by warm skin pass rolling warm remain virtually unchanged if the temperature of the metal does not exceed the temperature of dynamic strain ageing. The tensile diagram of the samples of sheet steel skin pass rolled at 100–200°C, has a monotonic form without the 'tooth' and a yield plateau. Preventing the ageing steel by warm skin pass rolling can sometimes be used to replace killed steel by rimming or semi-killed.

The advantage of the process of warm skin pass rolling and straightening of hot-rolled low-carbon steel strips is that it substantially reduces the cooling time of coils in storage after hot rolling. Furthermore, the deformation resistance of low-carbon steels at temperatures of warm skin pass rolling is significantly (by 30%) lower than at 20–30°C. Due to this the power parameters of the processes of skin pass rolling and subsequent straightening are reduced. When the temperature is increased from 20 to 190°C the force in skin pass rolling strips 1.0–2.0 mm thick is reduced on average by 70% and the torque is straightening by 55%. The stiffness of the strips varies depending on temperature. The minimum stiffness of the 2.0–3.5 mm thick strips is obtained in the range 150–300°C, and for thicker strips (3.5 mm) it is achieved at lower temperatures and lower stiffness values. The length of the deformation zone in skin pass rolling the strips in this temperature range is minimal.

10.6. Relationships governing the formation of metal surface microrelief

Surface quality requirements of cold-rolled sheet of skin pass rolled steel, including its microgeometry (roughness), are continuously tightened. Thus, the range of the surface roughness of the high-quality cold-rolled sheets used for the manufacture of front parts of car bodies has been reduced to $Ra = 0.8 \div 1.2$ μm. As a consequence, all metallurgical companies have intensified research efforts aimed at ensuring the desired surface roughness of the finished sheet products

[155–158]. In connection with this it is necessary to consider at the present level the technology of producing the desired surface roughness of the cold-rolled sheets and strips, to make appropriate generalizations and make recommendations.

The complex of views, concepts, and opinions directed at the interpretation and explanation of the influence of the microrelief (roughness) of the surface of the rolls and strips in rolling and skin pass rolling of the efficiency of plate rolling production and the quality of the finished products is presented in [72, 159]. The possibilities of improving the efficiency of rolling production and the quality of sheet products for various applications through the deliberate action on the surface microrelief of the rolls and rolled metal have been demonstrated. At the current stage of development of plate rolling production these possibilities have been significantly expanded. Tightening and a variety of requirements for the surface roughness of sheet rolled products for various purposes necessitate accounting and implementation of these new requirements, using modern technological solutions and capabilities [160].

When cold rolling and skin pass rolling steel sheets the surface of the deformed metal forms on the already existing surface and therefore the roughness and physical state are largely dependent on this already existing surface. The size and shape of the rolled metal change. The macrogeometrical changes of the surface determined its microgeometrical transformation. The surface roughness resulting from the plastic deformation of the metal is determined by its properties, the nature and extent of deformation, the processing conditions of the rolling process, the microstructure of steel, the surface roughness of the rolls, temperature, lubricant, and many other factors. But one should always distinguish two extreme cases, for which the conditions of formation of the surface microrelief are fundamentally different. In the first case, rolling or skin pass rolling is carried out without lubrication or in the presence of a lubricating film in the deformation zone; the thickness of this film thickness is very small compared with the magnitude of the roughness of the boundary surfaces of the rolls and the strips. In the second case – the surfaces of the rolls and the strips in the deformation zone are separated by the lubricating film the thickness of which is commensurate with the roughness of the metal and the rolls. Thus, under the conditions relevant to both the first and second cases, various combinations of the values of surface roughness of the rolls and rolled strips are possible. Thus, the surfaces of the rolls and the strips may have approximately the same surface roughness, e.g. the surfaces of the rolls and strips are smooth or both surfaces are very rough. The limiting states here are the rolling (skin pass rolling) of relatively smooth strips in very rough rolls and the rolling of rough strips in smooth rolls.

In the production of cold-rolled thin steel sheets the first case is usually dominant – the lubricating film thickness in the deformation zone during rolling is an order of magnitude less than the height of asperities on the surfaces of the work rolls and the rolled strip. At smooth strip surfaces and rough roll surfaces the microrelief is shaped by filling microdepressions on the surface of the roll by the deformed metal.

The depth of flow of metal into microdepressions in the roll surface during the passage of the strip through the deformation zone determines the size and shape of its roughness after rolling. When rolling the rough strip by the smooth rolls the microrelief formation mechanism is different – collapse of the microprojections of the original metal surface takes place in the deformation zone. In general, when rolling without lubrication (or with the negligible thickness of the lubricating film in the deformation zone) rough strips with rough rolls irregularities the strip surface are crushed at first and then the microasperities of the rolls indent the the surface of the metal of the rolls. The controlling role in the formation of the microrelief of the rolled strips and sheets is played by the dominant (largest) microasperities. The direction and nature of changes of the surface of the rolled metal (rise or smoothing of roughness) are defined by the absolute values of the initial roughness of the strips and the rolls.

Experimental studies performed in industrial continuous and 1700 skin pass rolling mills showed that with increasing strain the microgeometry of the steel sheets as regards the size and shape of microasperities approach the microprofile of the roll surface, and the intensity of the changes of the roughness of the strips decreases with increasing reduction. The effect of reduction in skin pass rolling on the sheet steel surface microrelief is manifested by changes of the contact pressure in the deformation zone. In skin pass rolling of metal with high strength properties, the ratio of the roughness of the rolled strip to that of the rolls is higher than for a more ductile metal. The reason for this phenomenon consists in the fact that an increase in the deformation resistance of the skin pass rolled metal increases the flattening of the rolls, the length of the deformation zone and the contact pressure.

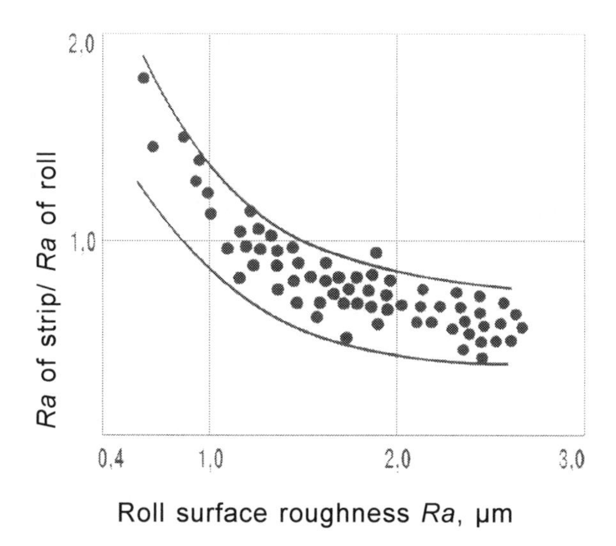

Fig. 10.19. Dependence of the impression coefficient Ra_s/Ra_r on the roughness Ra_g of te rolls (mill 1700, rolls diam. 600 mm, skin pass rolling without lubrication with a reduction of 1.0–1.5% strip thickness 0.5–2.0 mm, the initial roughness of the rolls 0.5–2.5 μm).

Increasing roll diameter increases the extent of impression of their roughness on the surface of skin pass rolled strips. However, in practice its impact is generally not considered, as in the real conditions the sheet rolling and skin pass rolling mills use rolls of the same diameter.

The effect of the degree of roughness of the rolls on the extent of impression of the roughness on the surface of the strips was the main issue in all studies because the results of these studies form the basis of plant regulations governing the shot blasting modes (blasting, shot blasting, electrospark, electroerosion) and operation of the rolls [72,155–159]. It was found with increasing roughness of the work rolls of the cold rolling and skin pass rolling mills decreases the Ra_{strip}/Ra_{roll} ratio, the so-called coefficient of impression of the roughness of the roll on the strip (Fig. 10.19). This ration is greater than unity if the roughness of the skin pass rolling mills is smaller than the initial surface roughness of the strip. If skin pass rolling is carried out on strips with a smooth surface, then at the same thickness of the strips the coefficient of impression of the roughness of the rolls on the surface of the rolled steel is practically independent of the absolute roughness of the rolls.

High surface roughness of the roll typically has a lower density of microprojections and therefore less sharp-angled irregularities. This is reflected in the impression of the roughness of the rolls on the surface of the skin pass rolled steel sheets. One must also consider the indirect effect of the size of the roughness of the rolls on the transfer of roughness onto the surface of the rolled metal. Namely, the magnitude and character of the roughness of the rolls affect the friction coefficient during rolling and skin pass rolling the strips. The friction coefficient significantly affects the magnitude of the normal contact stress (specific pressure) in the deformation zone which is a function of the depth of flowing of the metal into microdepressions in the roll surface and the degree of crushing of the microasperities of the strips. If the rolling conditions or specific skin pass rolling are characterised by an increase of the friction coefficient and contact pressure with increase of the roughness of the rolls, then the extent of impression of the rolls on the strip at the same reduction will be greater as the roughness of the surface of the rolls increases.

Application of the lubricant (usually an emulsion) in cold rolling and skin pass rolling of strips reduces the friction coefficient in the deformation zone. Consequently the level of contact stresses at constant plastic properties of the deformed metal decreases and the impression of the roughness of the rolls on the strip surface becomes less pronounced. This conclusion is also supported by the data of [155] which states that the use of lubricants, for example, in skin pass rolling dynamo steel, reduces the extent of impression of the roughness of the rolls on the surface of the strip by 10%. However, the above explanation of this effect by the formation on the surface of closed chambers from which the lubricant can not leak is ambiguous. A crucial factor in skin pass rolling with a lubricant is the decrease of the friction coefficient and the contact pressure in the deformation zone.

With increasing thickness of the strip, with other conditions being equal; the extent of impression of the roughness of the roll is reduced as the contact pressure when rolling thicker strips is lower. The limiting value of the roughness of the rolled or skin pass rolled strips is achieved faster when their thickness is reduced.

The surface roughness of steel sheets skin pass rolled in the shot blasted rolls depends on the initial surface microrelief of the strips. The effect of the initial roughness (after cold rolling) on the roughness of the skin pass rolled strips in the technology accepted in the industrial conditions is not so great, but it shows up quite clearly

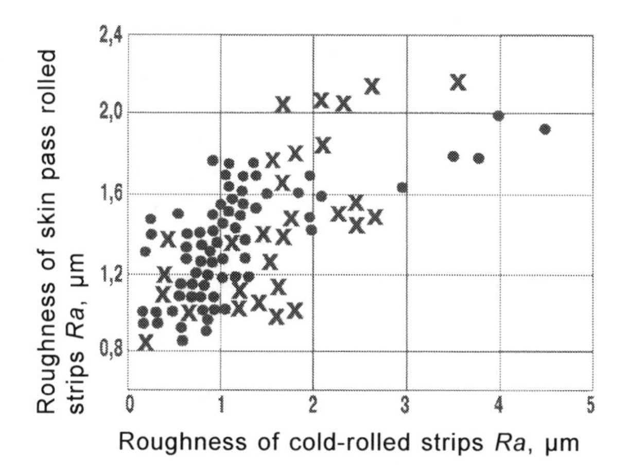

Fig. 10.20. Dependence of the roughness of the skin pass rolled strips Ra_{sk} on the initial roughness of mild steel Ra_{in}, crosses denote the values calculated according to [126] at Ra_{roll} of 2.4 and 3.5 µm.

(Fig. 10.20). So, in skin pass rolling with reductions of 0.8–1.5% of mild steel strips with a roughness Ra = 0.3–5.0 µm the rolls with the surface roughness of Ra = 1.5–3.5 µm relationship between the initial and final roughness can be expressed by the following linear regression equation: $\overline{Ra}_{sk} = 1.1 + 0.22\overline{Ra}_{in}$, where $\overline{Ra}_{in}$ and $\overline{Ra}_{sk}$ are the average values of surface roughness before and after skin pass rolling, respectively.

In [156] the empirical dependence of the surface roughness of skin pass rolled strips not only on the roughness of strip plate but also on the magnitude and roughness of rolls is presented. In our notations it has the form $\overline{Ra} = 0.33 \cdot \overline{Ra}_{roll} + 0.29\overline{Ra}_{in}$. For values $\overline{Ra}_{roll} = 3.5\mu m$ this dependence becomes $\overline{Ra}_{sk} = 1.155 + 0.29\overline{Ra}_{roll}$. The first and last dependences were obtained for different mills and at different skin pass rolling conditions. But they differ only slightly, which once again confirms their reliability.

For the manufacture of skin pass rolled bands with surface roughness Ra of no more than 1.6 µm by the generally accepted process technology the surface roughness of cold-rolled annealed steel must be within Ra = 0.5–1.5 µm. To obtain a skin pass rolled steel roughness Ra = 0.8–1.2 µm the surface of the cold-rolled strips should be as smooth as possible.

The initial state of the metal surface has a strong effect on the density of microprojections on the strip surface after skin pass rolling. The 'dense' surface roughness of the rolls with more than 50 microprojections per 1 cm of the profilogram improves the uniformity of the microrelief in the plane of the sheet. The initial microrelief with the high density of peaks at the same *Ra* value has a greater influence on the final surface condition of the skin pass rolled steel as compared with the flat profile. It is the different form of microasperities and their density that explain a difference in the degree of impressions of the roughness produced by shot blasting and electroerosion machining of the rolls on the surface of skin pass rolled strips.

In accordance with the requirements of the processing technology of cold-rolled sheet steels in automotive or engineering plants the non-uniformity of the roughness in the plane of the sheets should be minimal, and the density microprojections more than 50 cm^{-1}. In general, the unevenness of roughness is defined by the non-uniformity of the contact pressure in rolling (skin pass rolling) of the various portions of the strip. The instability of the contact pressure, caused by non-uniform deformation, thickness differences of the strips, the heterogeneity of the mechanical properties and a number of other factors, degrades uniformity, homogeneity, and the constantct of the roughness of rolled metal.

The effect of the rolling speed and skin pass rolling spped on the formation of roughness of the strips is the following: if the speed increase in skin pass rolling reduces the contact pressure and yield strength (resistance to deformation) of the steel, the impression of the roughness of the rolls on the surface of the strips deteriorates. In cases where the increase in the skin pass rolling speed by increasing the contact pressure increases the stress state coefficient, the transfer of the surface roughness of the roll to the strips improves.

The tension during rolling affects the magnitude of the contact pressure and this is reflected in the formation of the surface microrelief of the strips. With increasing tension the impression capacity of the roll roughness is reduced, but in skin pass rolling, when, due to a large length of the deformation zone the tension affects the pressure only slightly, the change in the surface roughness is small. This is confirmed by the results of experiments performed in the industrial conditions using the rolls after shot blasting with shot and after electroerosion machining [155]. Controlling the level of front and back tension in skin pass rolling one can influence the magnitude of the roughness of the skin pass rolled bands. However, in practice such roughness control capabilities are limited.

According to the results of studies carried out on several laboratory and industrial mills, the impression coefficient Ra_{strip}/Ra_{roll} at the roughness values of the shot blasted rolls over Ra = 2.5 μm is in the range 0.4–0.8 (Fig. 10.19). According to [156] the impression coefficient is 0.23–0.37 when using the rolls treated by electroerosion. In the production of automotive sheets at the Magnitogorsk Metallurgical Concern the impression coefficient was 0.30–0.65 at an average value of about 0.5 [158]. When skin pass rolling annealed cold-rolled steels of various grades (C10, 66Mn4G, 50CrV4, 7CrNiMo4, etc.) on the 630 mill of the same company the impression coefficient of the roughness of the work rolls on the surface of the strip is in the range 0.5–0.9 [158]. Differences in these values of the impression coefficient are small and they are caused by different technological conditions in the production of cold-rolled steel sheets in various factories.

Summarizing the results of theoretical and experimental studies carried out in rolling various metals in different mills, we can make the following conclusion: the formation of the microrelief of rolled metal is a function of the maximum value of the the stress state factor which is the ratio of the contact pressure to the yield strength of deformed metal. When rolling or skin pass rolling sheets with different mechanical properties, different thicknesses and rolls of different diameter and with different reductions, but maintaining the constant stress state factor, the impression of the roughness of the rolls on the strip surface is almost the same.

The generalized dependence of the ratio of the roughness of the strip and the roll on the coefficient of the maximum stress state $n_{\sigma_{max}} = p_{max} / \sigma_{T/p=p_{max}}$, where p_{max} is the maximum specific pressure (normal contact stress) in the deformation zone; $\sigma_{T/p = pmax}$ is the yield strength of the metal in the cross section of maximum pressure, is represented by the field of points shown in Fig. 10.21. The experimental points in this figure correspond to rolling with a reduction to 5% of strips of low-carbon and stainless steel, copper and brass with a thickness of 0.9–2.0 mm. The correlation coefficient between Ra_{strip}/Ra_{roll} and $n_{\sigma_{max}}$ is 0.84, indicating a close relationship between these indicators. The dependence shown in Fig. 10.21 is valid for the range $1.0 \leq n_{\sigma_{max}} \leq 3.2$.

Calculations of the ceofficient of the stress state during rolling metal strips or skin pass rolling do not present any fundamental difficulties as industrial mills are equipped with the systems for

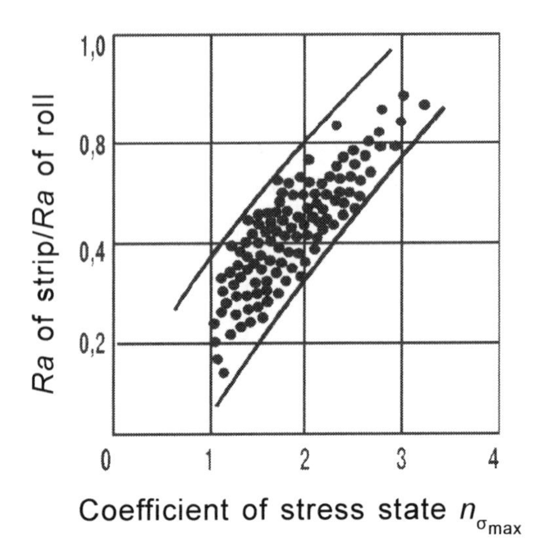

Fig. 10.21. Dependence of the impression coefficient Ra_{strip}/Ra_{roll} on the coefficient of maximum stress state $n_{\sigma_{max}}$.

registration of the process parameters, computer programs for their calculation and control devices. The simplest procedure for calculating the stress state coefficient uses a mathematical model of rolling or skin pass rolling strips and the results of rolling force measurements. Note that the pattern shown in Fig. 10.21 does not change if $n_{\sigma max}$ on the abscissa is replaced with $n_{\sigma_{av}} = p_{av}/\sigma_{T_{av}}$ where p_{av} and σ_{Tav} are the average values of the specific pressure and yield strength of the metal in the deformation zone.

The uniformity of the surface roughness of the rolled and skin pass rolled strips clearly depends on the uniformity of the surface microrelief of the shot blasted rolls. Therefore, the surface roughness of the rolls after grinding should not have the direction of the grooves from the grinding wheel and the magnitude should not exceed Ra = 1.0 μm when the rolls are shot blasted to Ra = 3.0÷4.0 μm. Before electroerosion machining of the rolls the polishing of the surface of the roll should be more accurate ($Ra \leq 0.7$ μm). Obviously, the smaller the roughness of the shot blasted roll, the cleaner should be their surface after polishing (prior to shot blasting) regardless of the processing method (shot blasting or electroerosion).

During operation the roughness of the rolls changes, which leads to variations of the roughness of the strips even within the same batch. Information about the nature of the changes of the microprofile of rolled metal during formation of the roughness of the rolls is needed to select rational modes of roll changing and regulation of the initial state of the surface of the rolls. Modes of roll changing, in turn, determine the park of the rolls and the necessary performance of facilities for roll shor blasting. The required processing performance determines the number and type of installations for roll shot blasting. Therefore, the duration of operation of the roll as one of the key indicators of the technology of production of steel sheets with a predetermined surface roughness is strictly regulated.

The surface roughness of work rolls with the surface produced by shot blasting changes most extensively in the initial period of operation after filling in the rolling or skin pass rolling mills. In this period, the sharply raised and fragile roughness peaks on the surface are crushed and the embedded particle fractions are removed. After electroerosion (electrospark, electropulse) processing the rolls their roughness during operation varies less extensively. The reason for this is that, firstly, such a treatment provides a more uniform surface roughness. The surface microrelief of the rolls after electroerosion machining has no sharp-edged peaks [158]. Second, and perhaps most importantly, when the grooves are produced by creating electrical discharges there is a substantial strengthening of the surface layers of the metal, thereby increasing their durability.

The change of the surface roughness of the rolled and skin pass rolled strips is shown by the field of points in Fig. 10.22. As can be seen, during rolling or skin pass rolling of strips with the rolls with the initial surface roughness (after shot blasting) Ra = 2.5–4.5 μm, which is most often used in the production of low-carbon sheet steel, the magnitude of roughness of the metal decreases as the

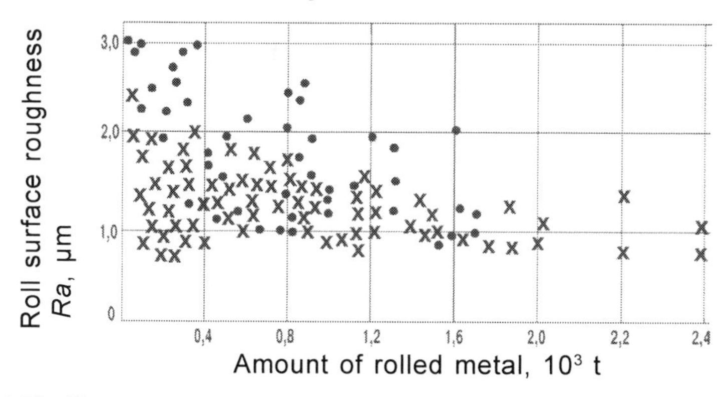

Fig. 10.22. Change of the surface roughness of cold-rolled (crosses) and skin pass rolled (dots) strips 0.5–2.0 mm thick of mild steel in manufacture of the rolls of the last stand of 1680 and 1700 cold rolling mills [127] and also the 1700 skin pass rolling mill.

roll surface wears out. With the growing amount of rolled strips a decrease in the surface roughness changes significantly the nature of the microprofile of the surface of the rolls and rolled (skin pass rolled) metal: the density of peaks decreases, the curvature radius of the tips of the microprojections increase. After rolling 600–800 t of steel sheets the size and density of roughness peaks is stabilized and then retained approximately at the same level [155, 158, 161]. Note that in the case of electrospark grooving the rolls of the last stand of the cold rolling mill the roughness the rolled strip is located at a level not less than $Ra = 1.0$ μm after rolling 1500 t of metal and not less than $Ra = 0.5$ μm after rolling 2500 t of metal. That is, the duration of the campaign work rolls treated by the electroerosion method, increases by 1.5–1.7 times in comparison with the duration of operation of the rolls after grooving by blasting (shot blasting) [155]. According to research results [156] the height of the surface roughness of the strips during their skin pass rolling with the rolls treated by the electroerosion method decreases by only 5–10% and the frequency characteristics do not change substantially. Increase of the wear resistance of the roughness of the work rolls is also observed after application to the surface of a thin layer of 'hard' abrasion resistant chromium.

The wear of the surface roughness of the rolls is a function of the work of friction forces in the deformation zone. The amount of rolled metal in tonnes only approximately characterizes the work of the friction forces. It is why the experimental points in Fig. 10.22 show a considerable scatter. When skin pass rolling the steel sheet in dry shot blasted rolls the surface roughness varies typically faster than in cold rolling. This is due to higher friction in the deformation zone in skin pass rolling. Since the coefficient of the stress state of the metal in the deformation zone during skin pass rolling often takes values higher than in cold rolling, and therefore, the surface roughness of the roll surface is imprinted on the skin pass rolled strip to a considerable larger extent, the points in Fig. 10.22, corresponding to the process of skin pass rolling, are mostly at higher levels.

All types of mills (continuous, reversible, duo, quarto, multiroll, etc.) have a general pattern: during operation of the rolls the microrelief of their surface regardless of the initial value tends to a steady state. The parameters of this steady, equilibrium roughness depend on the mechanical properties of the surface layers of the rolls and the deformed metal, the physical and mechanical properties of the luricant used in rolling, the temperature in the deformation zone, and in general on the friction conditions in the deformation zone. The intensity of the transformation of the surface microrelief of the rolls during running and the heat generated in this process are interconnected. The roughness of the rolls will be transformed so that the bulk temperature in the deformation zone is minimized.

Generalization of may years of experience of the metallurgcial plants and the results of research has shown that the duration of operation of the rolls of the last stand of the continuous mill for one installation should not exceed 800–1000 tons of rolled strips if rolls are notched by the shot (shot blasting) method and the cold- rolled steel is annealed in tightly wound coils at the holding temperature 680°C or more. After electroerosion machining of the rolls and in cases where the cold-rolled metal is annealed at temperatures below 680°C or under the conditions precluding interturn welding the interval between scheduled roll changes can be increased by 1.5–2.0 times.

It should be remembered that the work rolls of the last stands of cold rolling mills are notch to ensure that the roughness formed on the metal surface prevents welding the turns of the strip in coils during their subsequent annealing. A solution of this problem is usually guaranteed for cold-rolled strips of the surface roughness $Ra \geq 1.5$ μm. However, recent trends in the development of cold rolling steel sheets indicate that welding of the strip in coils is prevented provide by selecting the special modes of tensioning the coiled strips and the surface roughness of cold-rolled steel is reduced, which respectively reduce the roughness value of the work rolls. It is recommended in the last stand of the continuous cold rolling mills to groove the rolls to obtain a roughness $Ra = 1.5$–2.5 μm.

When skin pass rolling thin strips (tin plate) on two-stand mills the roughness value of the skin pass rolled metal depends strongly on the distribution of total strain between the reductions in the first and second stands. This effect is particularly strong in the mills where the work rolls of the first and second stands have different diameters, such as the 1400 mill of the Karaganda Metallurgical Concern (KarMC) [152].

For small values of absolute and relative reductions characteristic of the process of skin pass rolling of tin plate, the main influence on the formation of the microrelief of the strips is exerted by the surface roughness of work rolls and the skin pass rolling force. The curves in Fig. 10.23 show the same effect of the microgeometry of

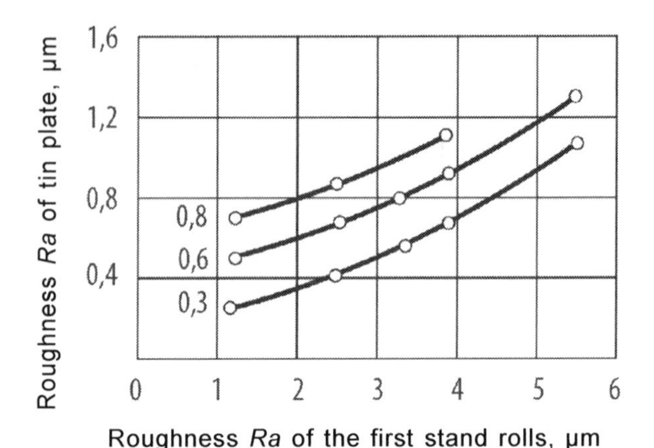

Fig. 10.23. Dependence of roughness *Ra* of skin pass rolled tin plate on the roughness of the rolls of the first and second stands (numbers on the curves – roughness of rolls of stand 2) [152].

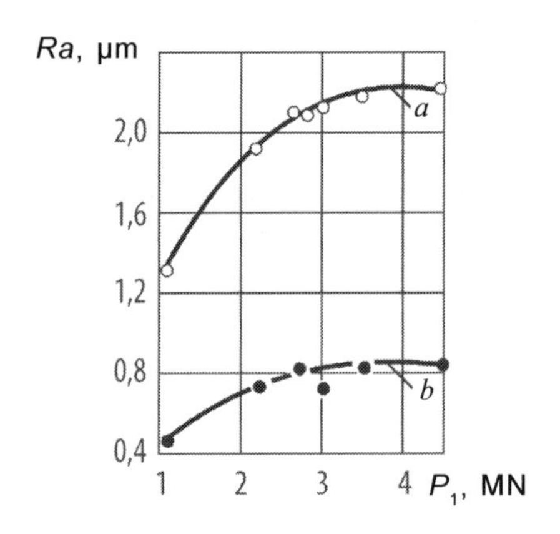

Fig. 10.24. Dependence of surface roughness *Ra* of 0.2 mm thick strips after the first (a) and second (b) stands on skin pass rolling force in the first stand (skin pass rolling force in the second stand 2.5 MN) [152].

the work rolls of the stands 1 and 2 on the final roughness of tin place at uniform loading of the stands.

The dependence of the roughness of the strip after the first and second stands on the dressing is shown in Fig. 10.24. The growth of the skin pass rolling force in the stand increases the roughness of the strip which asymptotically approaches the value corresponding to the maximum impression of the microgeometry of the rolls. Thus, redistributing reductions and, consequently, the skin pass rolling force between the first and second stands the microgeometry of skin pass rolled tin plate can be efficiently regulated. Of course, the

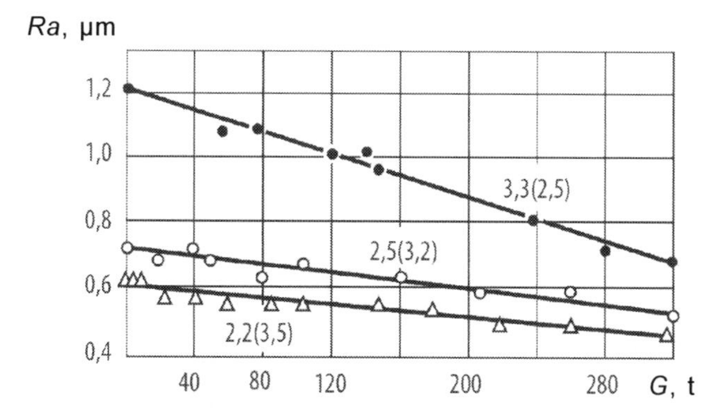

Fig. 10.25. Dependence of roughness *Ra* of skin pass rolled 0.2 mm thick tin plate on the amount *G* rolled after complete changes work rolls (roughness *Ra* of the rolls of the first and second stands 3.0 and 0.45 μm, respectively; figures on the curves – forces in the first (second) stand, MN) [152].

controlling role is also played here by the roughness values of the work rolls of each stand of the skin pass rolling mill.

The accepted technology of production of cold-rolled sheet steel, including tin plate, should maintain continuity of the roughness of the skin pass rolled strips throughout the campaign of the work rolls. The dependence of the roughness of the strip on the amount of metal rolled by the work rolls after changing the rolls at different levels of the skin pass rolling force in the first and second stands of the 1400 mill is shown in Fig. 10.25. The minimum average roughness and the highest stability of this roughness during the campaign of the work rolls are achieved with maximum ratio P_2/P_1 (Fig. 10.26).

Based on the studies the authors of [152] have chosen the following technological modes of skin pass rolling of tin plate in the 1400 mill:

Surface roughness is one of the most important indicators of quality of the skin pass rolled tin plate designed for electrolytic tinning. The magnitude and nature of the microgeometry of the steel

Stands	1	2
Roughness of rolls shot blasted with shot *Ra*, μm	2.5–3.5	≤0.5
Skin pass rolling force, MN	2.0–2.8	3.0–4.0

substrates significantly affect the corrosion resistance of the coating of tin plate and determine its marketability. According to the results of experimental studies carried out in the industrial conditions of the KarMC conditions [152], reduction of the surface roughness (*Ra*) of tin plate from 1.0–1.1 to 0.63–0.72 μm greatly increased

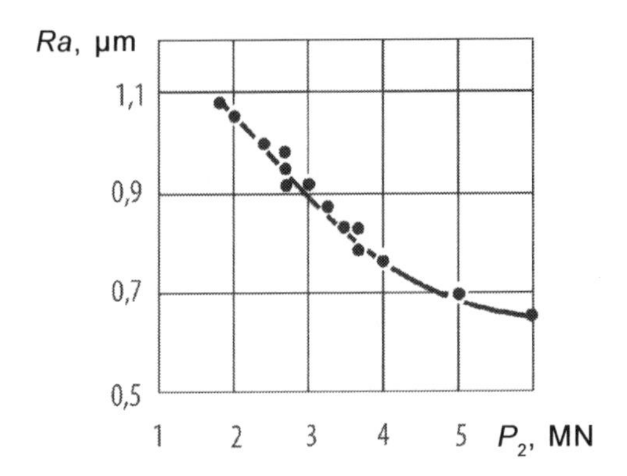

Fig. 10.26. Dependence of strip roughness *Ra* after skin pass rolling on force in the stand 2 (strip thickness 0.2 mm, the roughness of the rolls of the second stand 0.65 µm, the force in the first stand 3.0 MN) [152].

the corrosion resistance of tinplate. Therefore, in the production of tin plate the surface roughness (*Ra*) of the steel basis should be restricted to no more than 0.6–0.7 µm.

It is known [72] that the rolling and skin pass rolling the strips using lubrication even the initially smooth surface of the rolled (skin pass rolled) metal becomes rough. Below we consider a possible mechanism of the formation of microirregulaties of the surface of the strips due to changes in the thickness of the lubricant layer in the deformation zone [162].

Fluctuations in the yield strength of rolled metal σ_T cause the uneven thickness ξ of the lubricating film in the inlet section of the deformation zone, accompanied by changes of the surface microrelief of the strips. The dependence $\xi = \varphi(\sigma_T)$ is represented in the form $\xi = \frac{C}{(\sigma_T)^n}$, where C – coefficient; n – the exponent. Consider σ_T as a random variable with the known expectation $M(\sigma_T)$ and dispersion $D(\sigma_T)$ of its distribution. To find the expression of the expectations $M(\xi)$ and dispersion $D(\xi)$ of the values we apply the linearization method, under which $M(\xi) \approx \varphi(M(\sigma_T))$ and $D(\xi) \approx [\varphi'(M(\sigma_T))]^2 D(\sigma_T)$, where $\varphi'(M(\sigma_T))]$ – derivative of $\frac{\partial\varphi}{\partial\sigma_T}$ calculated for σ_T, equal to $M(\sigma_T)$.

Assuming that the increase in the standard deviation of the roughness profile ΔRq of the surface of the rolled strip corresponds to the standard deviation of the lubricating film thickness from its average value, i.e. $\Delta Rq = \sqrt{D(\xi)}$, and considering that $Ra \approx 0.8Rq$, we

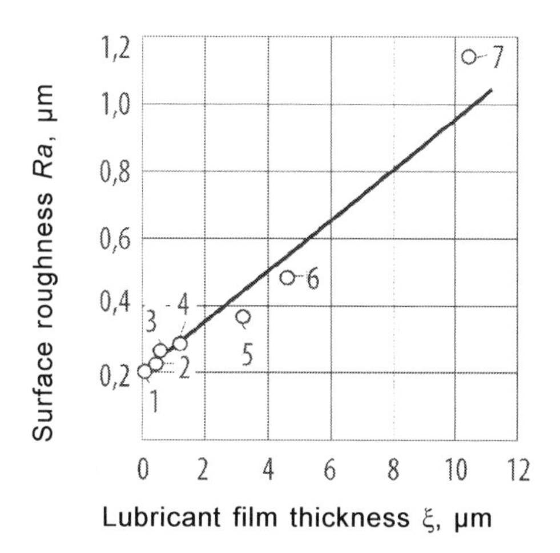

Fig. 10.27. The experimental points (according to Yu.B. Sigalov) and calculated dependences of the surface roughness values after rolling strips on lubricant film thickness in the deformation zone: 1 – spindle oil; 2 – cotton seed oil; 3 – PKS oil; 4 – castor oil; 5 – braystock oil; 6 – viscosine; 7 – vapor.

get $\Delta R\alpha = 0.8\ nM(\xi)\ \upsilon_{\sigma_T}$, where $\Delta R\alpha$ – increment of the arithmetic mean deviation of the roughness profile; $\upsilon_{\sigma_T} = \dfrac{\sqrt{D(\sigma_T)}}{M(\sigma_T)}$ – coefficient of variation of σ_T; $R\alpha$ and Rq – surface roughness parameters.

The magnitude of the surface roughness of the metal after rolling is $R\alpha = Ra_{in} + \Delta Ra$, where Ra_{in} is the surface roughness of the metal before rolling. Comparison of calculated and experimental dependences for C10 steel (thickness 2.2 mm, $Ra_{in} = 0.2$ μm, $\upsilon_{\sigma_T} = 0.1$) showed satisfactory agreement (Fig. 10.27). Roll roughness $Ra = 0.11$ μm, the degree of deformation 10% during rolling. The indicator $n = 1$.

The linearization method was used above to show the influence of the variability of the yield strength of rolled metal on the thickness of the lubricant layer ξ in the deformation zone and the resultant roughness of the surface strips. In the actual conditions all the rolling process variables affecting ξ are random variables whose distributions are characterized by their mean values and variances. Therefore, it is desirable to assess the simultaneous effect the non-constancy of all variables on ξ and the roughness of the metal surface. As shown in our work [48], such analysis can be performed most efficiently using the Monte Carlo method.

11

Energy saving when rolling strips

11.1. Saving energy in wide-strip hot-rolling mills

The energy consumption in the rolling mill and. in particular, existing wide-strip hot rolling mills (WSHRM) is usually reduced by reducing the cost of heating of slabs and ingots before hot rolling; except where possible, by intermediate heating of the metal due to 'direct' rolling; choice of a rational distribution of degrees of deformation between roughing and finishing trains; reducing heat losses by strip plate in the mill line [163–170]. A special place here is occupied by the theme of saving energy in the production of rolled steel in sheet-casting modules (complexes) using 'thin' slabs. It should be emphasized that, as stated in [165], the problem of reducing energy costs in rolling has been ignored for a long time. Today, the priority in the industry is the significant reduction of the energy consumption of steel production. Implementation of the energy-saving direction of metallurgy is considered as a relevant way to solve the energy dependence number of many countries. Trends in the global energy market indicate that the price of gas, oil, coal will increase continuously. Most likely the volume of production and extraction of these energy sources in the world will not increase in the future. The cost of electricity will rise. Therefore, the problem of heat and energy saving in metallurgy and rolling mills will become more urgent.

We consider the potential for energy savings in the process line of the WSHRM using the guidelines given in [165–167].

The energy in rolling the steel in WSHRM is consumed first, in heating the slabs in reheat furnaces and, secondly, in metal deformation. The heating costs account for 55–60% of all energy consumption in WSHRM. Note that for small-section mills for heating billets before rolling the energy consumption is 70–90%. The total energy consumption depends on the layout of the roughing and finishing trains of a specific mill and can be minimized in certain ranges of temperature of heating of slabs and billets and accepted rolling modes. In rolling production a large amount of energy is also spent on heating and heat treating the billets before the rolling of pipes directly.

Energy in the production of sheets and strips can be saved significantly in the direct supply of continuously cast slabs to the WSHRM bypassing the heating furnace. However, the implementation of this solution is not always possible due to inconsistencies in the arrangement of the steel melting and rolling shops. In organizing the hot charging of continuously cast slabs into heating furnace the fuel consumption in heating them is reduced by 12% at a charging temperature of 300°C and by 60% at 900°C [166]. Optimization of the slab heating modes can reduced by 14% the fuel consumption in heating furnaces. It is believed that lowering the temperature of slab heating in furnaces by 10°C allows to reduce the fuel consumption by almost 2%.

At metallurgical plants where steel is cast into ingots and then rolled in the slabbing mills it is easier to arrange direct ('transit') feeding of the slab in the WSHRM. However, the problem here lies in the possibility of equipment WSHRM to roll the slab at a relatively low temperature. In cases of any delay in transporting slabs and reducing their temperature below the permissible level there is a risk of rolling equipment of the WSHRM breaking down.

The possibility of reducing the temperature in slab heating in the furnaces or carrying our 'transit' rolling increases with the implementation of measures to reduce the heat loss by the strip plates as they move in the WSHRM line. Such measures include, firstly, installing a rewinder in the WSHRM (coilbox) on the intermediate roller table and use heat-retaining devices, shielding the surface of strip plates from the interaction with the environment.

Coilboxes have been installed in dozens of the WSHRM. When coiling strip plate on the intermediate roller table of the WSHRM the losses are significantly reduced due to heat accumulating properties of the roll. Due to the fact that the rear end of the strip plate becomes the front end in unwinding the roll and fed into the finishing train, the unevenness of the temperature decreases along the length of the rolled strip.

Heat-retaining devices (screens), located at the intermediate roller table of the WSHRM, reduce heat loss by radiation of the strip plate. The heat-retaining devices are continually being improved. The classification and the stages of construction of such systems is discussed in detail in [166]. It can be expected the improvement of the heat-retaining installations and systems will continue on the path of increasing their efficiency in terms of heat saving and reliability under heavy use.

In the reconstruction and modernization of the existing WSHRM special attention is paid to finding such solutions, primarily the composition of the stands of the roughing group of the mill which provide on the one hand the minimum energy consumption for heating and rolling bands, and on the other – the temperatures of the end of rolling and coiling of the strips needed to obtain the desired properties of hot-rolled steel, as discussed in previous sections of the book. An example analysis of different variants of reconstruction of the WSHRM in the transition to continuously billets is shown in [171]. More detailed recommendations on the choice of equipment to reconstruct the existing and designing new WSHRM in terms of ensuring the minimum energy consumption of the process of production of sheet steel are given in [165].

11.2. Reduction of energy consumption in the production of cold rolled steel sheet and tin plate

Among the different types of rolled products the most energy is consumed by cold-rolled steel and tin plate. Experience shows that the total energy consumption for hot- and cold-rolling mills can be reduced by decreasing the thickness of hot-rolled strip plate [165]. As will be shown below, this is not necessarily true. Furthermore, reducing the thickness of strips to 2 mm in the majority of the hot rolling mills is limited by the requirements on the deformation temperature which defines the structure and mechanical properties [98]. Using the strip plate of increased thickness is in some cases limited by the level of admissible values of power parameters in the cold rolling mills. The thickness variation of the strip plate affects the performance on the mills. Thus, the solution of reducing the energy consumption for different sheet mills is ambiguous because of the unequal level of their technical and economic indicators. Nevertheless, general patterns can be identified for all the mills together with special features having a fundamental nature in both the technical and economic aspects.

The numerical solution to reduce the energy consumption for hot and cold rolling mills in the production of tin plate will be illustrated by the example of the Karaganda Metallurgical Plant (KarMK) [168, 172 pp. 186–189].

The results of studies on the 1700 WSHRM and the six-stand hot rolling mill in 1700 and the 1400 six-stand cold rolling mill at the KarMK showed [98] that the strip plate with a thickness of 0.18–0.28 mm can be produced using strip plate with thicknesses from 1.8 to 2.8–3.0 mm. And in terms of the power and temperature parameters of the process of cold rolling tin plate 0.20–0.25 mm thick it is most rational to roll strip plate with a thickness of 2.4–2.5 mm. Taking this into account, we estimate the rate of energy expended on the production of tin plate, depending on the thickness of hot-rolled strip plate.

The deformation and power parameters of rolling tin plate on the 1400 mill were calculated using a mathematical model described in the previous sections of this book. The reduction schedule of the mill was chosen based on the condition of approximately equal rolling forces in the stands, because this condition is most appropriate to achieve high accuracy of the flatness and thickness of the bands and also for the stability of the process. Strip tension between all stands was taken equal to 150 N/mm^2, the rolling speed was 20 m/s. The energy consumption of each stand was calculated as the product of rolling power by its computer time, the specific consumption – as the quotient of the division of the total flow of all mill stands by the weight of the rolled metal.

The energy consumption during hot rolling strips at the 1700 WSHRM mill was calculated using a mathematical model of the process and algorithms developed at the Institute of Ferrous Metallurgy of the National Academy of Sciences of Ukraine[1]. Calculations were performed for the conditions of rolling slab sizes 190–215×1100–1200×8200–9500 mm weighing 15.7–17.8 t. The thickness of the strip plate between the roughing and finishing mill groups changed from 34 to 40 mm, and of the hot-rolled strip from 2.0 to 2.8 mm. The rolling speed of the slabs in the last roughing stand was 3.14 m/s. The charging rate was 9.5 m/s; working – 14.0 m/s; the acceleration during acceleration of the finishing group stands after gripping the front end of the strip by the coilers was 0.25 m/s^2.

[1]The model was developed by S.A. Vorobey.

The length of the front end of the strip, rolled at the charge speed, was 170 m. It was assumed that the hot-rolled steel sheet is produced by the ingot–slab–roll procedure. The hot-rolled strip plates from ingots weighed 15.7 tons. The length of the 190 mm thick slabs was 9.5 m; 200 mm – 9.1 m; 210 mm – 8.6 mm, and 215 m – 8.2 m (width of the slabs 1110 mm). Since the mass of the slabs does not change, regardless of the thickness of the slab the rolling time in the last roughing stand and the length of the same thickness were constant peals; in the finishing mill – the same.

The slab heating temperature before rolling was 1250°C. The temperature of the strip plate after exiting from the last roughing stand, depending on the thickness slabs (190–215 mm) and strip plate (34–40 mm) for the front end was changed from 1082 to 1105°C for the rear end – from 1072 to 1097°C. Lowering the temperature of the strip plate in the intervening roller table of the 1700 mill depends on the thickness of the strip plate and the transport speed on the roller table, which is a function of the thickness of the finished strip. Moreover, if the temperature of the front and rear ends at exit from the finishing group differs by 7–11°C, on entering the first finishing stand this difference increased by almost 10 times.

At the same deformation–speed modes of hot rolling steel in the WSHRM the rolling end temperature increases with increasing thickness of the strips. In the 1700 mill of the KarMK example, the rolling end temperature of the 2.8 mm thick strips was 40–50°C higher (839–857°C) than of the strips with a thickness of 2.0 mm (798–815°C). Increasing the thickness of strip plate between the roughing and finishing mill groups from 34 to 40 mm enables the temperature of the end of rolling the strips of this range to be increased on average by 15°C.

When calculating the energy consumption in the roughing group of the 1700 mill the rolling power was determined for each stand and averaged over three points (beginning, middle, end) of the strip plate with regard to their temperature, and the rolling time. The general and specific energy consumption were also determined.

In the finishing group stands the average value was represented by the half sum of the rolling powers of the front and rear ends of the strip, taking into consideration the degree of cooling in the intermediate roller conveyor prior to entering the first finishing stand.

The results of calculations of the energy consumption in the rolling of the strip plate in the roughing stands of the 1700 mill are presented in Table 11.1. According to information received,

Table 11.1. Energy consumption* in rolling in roughing train of the 1700 WSHRM mill of strip plate 34–40 mm thick from slabs 1110 mm wide weighing 15.7 t

Strip plate thickness, mm	Slab thickness, mm			
	190	200	210	215
34	270.09 / 17.20	282.26 / 17.98	286.98 / 18.27	307.63 / 19.59
38	237.63 / 15.13	248.81 / 15.84	252.53 / 16.08	281.02 / 17.89
40	223.41 / 14.22	234.41 / 14.93	237.94 / 15.15	263.10 / 16.75

* The numerator and denominator give the total (MJ) and specific (MJ/t) energy consumptions respectively.

increasing the thickness of the slab provides an increase in the temperature of strip plate, but the power consumption in rolling also increases. Consequently, the decrease of the deformation resistance of steel due to the increase of its temperature in the case of application of thicker slabs, which contributes to a reduction of the rolling power, does not compensate the increase of power due to the increase of the total reduction in the stands. When the thickness of the slabs changes from 190 mm to 215 mm the energy consumption increases by 14–18%. With the growth of the total deformation (reduction) of the steel in the roughing stands the effect of the thickness of the slabs on energy weakens (Table 11.1). When the thickness of strip plate is increased from 34 to 40 mm the energy consumption in the finishing train is reduced by 15–17%.

When hot rolling thin strips the specific energy consumption in the finishing stands of the 1700 KarMK mill is 7–10 times higher than in the roughing stands. Increasing the thickness of the strips from 2.0 to 2.8 mm reduces the power consumption in the finishing train by 25%. In these conditions, the thickness of the strip plate has virtually no effect on the energy consumption in the finishing stands, as the increase of the total reduction of the strip plate in the case of increasing their thickness is compensated by the effect of increasing the temperature of the rolled metal.

The total specific energy consumption for rolling strips in the roughing and finishing stands of the 1700 mill from the slabs with the dimensions 210×1110×8600 mm is characterized by the following values (left of the slash – at a strip plate thickness of 34 mm, right – 40 mm):

Hot-rolled strip thickness, mm	2.0	2.2	2.4	2.8
Specific energy consumption, MJ/m	171/168	158/155	148/145	132/129

The energy consumption in the stands of the finishing group was determined by taking the mean value for the front and rear ends of the strips. As can be seen, increasing the thickness of the strips from 2.0 to 2.8 mm reduces the total energy consumption for hot rolling at 23%.

Because the rolling end temperature of thin strips at the 1700 mill of the KarMK decreases with decreasing thickness, the yield strength of the hot-rolled steel with a thickness of 2.4–2.5 mm is on average 2.5–5.0 N/mm^2 lower than for the 2.0–2.2 mm thick steel. Studies have shown [98] that, depending on the temperature rolling conditions, the hot-rolled strip plate has a different microstructure and shows different hardening during subsequent cold deformation. Therefore, the influence of the thickness of strip plate on the energy consumption in cold rolling the tin plate in the six-stand 1400 mill were analyzed taking into account different starting properties and hardenability of the steel in the cold rolling process (Table 11.2).

Table 11.2. The dependence of the energy consumption during the cold rolling of tin plate in the 1400 KarMK mill on the thickness and mechanical properties of hot-rolled tin plate*

Strip plate thickness, mm	Tin plate thickness, mm	σ_{T_0}, N/mm^2	n	Specific energy consumption, MJ/m
2.0	0.20	240/325	0.627/0.644	160.3/189.6
	0.25			138.9/164.6
2.2	0.20	240/325	0.627/0.644	169.3/200.2
	0.25			147.6/175.4
2.4	0.20	250/300	0.629/0.644	181.8/203.6
	0.25			159.5/178.6
2.5	0.20	250/300	0.629/0.644	185.7/208.0
	0.25			163.4/183.1
2.8	0.20	250/300	0.629/0.644	199.4/220.1
	0.25			178.4/195.9

* The width of the strips 1000 mm, weight of rolls 30 t. Hardening of steel in cold rolling described by equation $\sigma_t(\varepsilon) = \sigma_{T_0} + 33.5\ \varepsilon^n$, where σ_{T_0} – initial yield strength of steel; ε – the degree of deformation; n – hardening index. Left of the slash – when rolling relatively 'soft' strip plate; right – 'hard'.

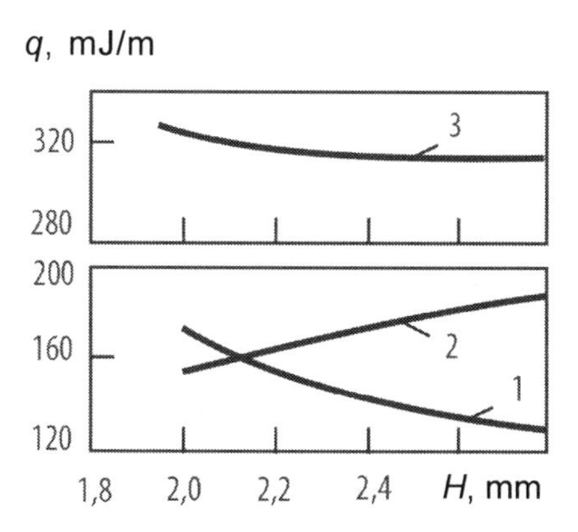

Fig. 11.1. The dependence of the specific energy consumption (q) in hot rolling in the 1700 mill and the 1400 six-stand cold rolling mill of 0.25 mm thick tin plate rolled from strip plate with a thickness (H) of 2.0–2.8 mm: 1 – specific energy consumption for hot rolling, 2 – cold rolling, 3 – total; hardening of steel in the cold rolling process is characterized by the equation $\sigma_T(\varepsilon) = 281 + 33.5\ \varepsilon^{0.639}$ for a strip plate thickness of 2.0–2.2 mm, $\sigma_T(\varepsilon) = 277 + 33.5\ \varepsilon^{0.638}$ at 2.4–2.8 mm, where $\sigma_T(\varepsilon)$ – yield strength in N/mm^2; ε – the degree of deformation, %.

With increasing thickness of the hot-rolled strip plate the energy consumption in rolling tin plate increases (Fig. 11.1, Table 11.2). However, the power consumption depends on the initial properties of strip plate. Thus, in the rolling of tin plate 0.2 mm in thickness the increase of the thickness of strip plate from 2.0 to 2.5 mm (25%) leads to an increase in the specific consumption energy by 15.5% for the 'soft' strip plate and 9.7% for the 'hard' strip plate. When producing 0.25 mm thick thin plate, this difference is respectively 17.6 and 11.2%. Increasing the thickness of strip plate by 40% (from 2.0 to 2.8 mm) in the manufacture of tin plate 0.25 mm thick at an average initial plasticity of the strip plate increases the energy consumption by 22.2% (Fig. 11.1).

The energy consumption when rolling tin plate is strongly influenced by differences in the properties of the strip plate of the same thickness. In particular, for the 0.20–0.25 mm tin plate the change in the properties of the 2.0–2.5 mm strip plate from 'soft' to 'hard' increases the energy consumption by 18–20% (Table 11.2). Fluctuations of the properties from 'soft' to 'hard' for the strip plate with a thickness of 2.4–2.8 mm changes the energy consumption by 10–12%. Therefore, by increasing the thickness of the strip plate the differences in its properties have a weaker effect on the consumption, since in this case the instability of the properties is smaller.

Table 11.3. Specific energy consumption (MJ/t) when rolling strip plate in the roughing and finishing stands of the 1700 mill millstands and cold rolling of tin plate in the 1400 KarMK mill*

Thickness of tin plate. mm	Strip plate thickness, mm			
	2.0	2.2	2.4	2.8
0.20	328/357	324/355	327/349	328/349
0.25	307/333	303/330	305/324	307/325

*Numerator – for 'soft' strip plate. the denominator – for 'hard' strip plate.

The distribution of reductions in the stands of the 1400 mill at constant total strain of the metal has little effect on the total energy consumption. For example, an increase in the degree of deformation of the first stand from 28.2 to 32.2% during rolling 0.25 mm tin plate from 'soft' strip plate 2.2 mm thick increases the total specific energy consumption by only 0.2%. The force and power in the rolling mill stands change as follows:

Stands	1	2	3	4	5	6
Strain,%	28.2/32.2	30.6/29.9	29.3/29.1	30.2/29.3	31.7/30.6	32.4/31.0
Rolling force, MN	7.77/8.47	7.64/7.76	8.34/8.31	8.23/8.15	7.92/7.80	7.73/7.64
Rolling power, kW	68.7/193.2	992.0/994.0	932.5/935.3	1053.7/1037.2	1192.9/1142.0	1533.0/1459.6

In modern mills for cold rolling sheets and tin plate the reduction schedules is selected on the basis of the process conditions: the minimum probability of rupture of strips through welds; favourable conditions for the operation of automatic systems for controlling thickness, tension, strip flatness; providing high process stability. In this regard, the possibility of a significant (more than 10%) redistribution of the relative reduction in the mill stands is limited.

The overall energy losses in two mills were calculated for the conditions in the 1700 mill for the strip plate 40 mm thick. In this case, the temperature control of rolling thin strips is more efficient and less power consumption. The results showed (Table 11.3) that by using a 'soft' strip plate the total energy consumption for the two mills at 0.20–0.25 mm tin plate production is almost independent of the thickness of hot-rolled strip plate. In the case of the 'hard' tin plate the total energy consumption decreases with increasing thickness. When the initial plastic properties and hardenability of the strip plate during cold deformation are characterized by average values, the total energy consumption in the production of tin plate is roughly the same for the strip plate thicknesses of 2.2–2.8 mm thickness and slightly higher for the 2.0 mm thickness (Fig. 11.1). If we compare the energy consumption for hot rolling and cold rolling in the production of tin plate 0.20–0.25 mm thick, using thin (2.0–2.2 mm) and a 'hard' or thicker (2.4 mm or more) and the 'soft' strip plate, which is usually observed in practice [98], it can be concluded that the use of relatively thick (2.4–2.8 mm) strip plate provides a reduction in the total energy consumption by 8–10% (Table 11.3).

To reduce the energy consumption in the production of tin plate is necessary to determine the upper limit of the thickness of hot-rolled strip plate, necessary to ensure the stability of the process of cold rolling strip in the 1400 mill. The results of [98] led to the conclusion about the principal possibility of rolling in the 1400 six-stand KarMK mill tin plate with a thickness of 0.18–0.25 mm from the strip plate up to 3.0 mm thick. Subsequent experiments conducted on this mill, confirmed the validity of this conclusion. Studies have shown that the temperature at the end of hot rolling in the 1700 KarMK mill of the strip plate 2.8–2.9 mm thickness was in the range 850–890°C and the coiling temperature of the strip rolls was 640–710°C (thickness of strip plate between the roughing and finishing group mills was 38–40 mm; charging speed 9.0–9.5 m/s^2). The profile of the cross section of the hot-rolled strips was lenticular with a convexity of 0.025–0.06 mm. The microstructure of chemically capped 08kp steel after rolling at these temperatures consisted of ferrite grains, size points 7–8, yield strength was 255–270 N/mm^2 tensile strength 340–370 N/mm^2, hardness 48–55 HRB.

Cold rolling in the 1400 mill of tin plate 850 mm wide and 0.18–0.20 mm thick was performed at 15–17% relative reduction in the first stand and 36–38% in the second to sixth stands. The total interstand strip tension was adjusted to be equal (kN) interval I – 320; II – 220; III – 170; IV – 110, V – 70; between the last stand and the coiler – 20. When rolling the tin plate with a thickness of 0.18 mm at a speed of 21 m/s (numerator) and 0.20 mm with a speed of 24 m/s (denominator), the rolling forces were as follows:

Stands	1	2	3	4	5	6
Rolling force, MN	$\frac{6.4}{6.2}$	$\frac{10.8}{10.0}$	$\frac{7.6}{7.6}$	$\frac{10.0}{10.5}$	$\frac{7.5}{7.5}$	$\frac{9.2}{9.8}$

The rolling of tin plate from 2.8 mm strip plate increases the reduction of metal in each stand by 1–3% compared to the strip plate 2.4 mm thick, but the rolling force increases only slightly. The stability of the rolling process of 0.18–0.20 mm tin plate from the 2.8 mm strip plate is satisfactory at speeds of 18–25 m/s. The durability of welds was at the same level as when rolling tin plate from the strip plate with a thickness of 2.4 mm.

In the analysis of the energy consumption for hot- and cold-rolling the fuel consumption for heating the slabs was disregarded, as this consumption item, although very significant (in wide-strip hot-rolling mills it is 55–60% of the energy), but is constant in the manufacture of strips of the same size. Nevertheless, it is necessary to bear in mind that as a result of the growth of the productivity of the hot rolling mill with increasing thickness of the strip plate the specific fuel consumption for its production decreases.

Thus, in the manufacture of strip plate for the cold rolling of sheets and tin plate the increase the thickness of the slab increases the temperature of the strip plate between the roughing and finishing hot rolling mills, but the energy consumption in rolling also increases. In the finishing stands of the rolling mill in rolling at the operating speed of the rear ends of the thin strips the energy consumption is about two times greater than in rolling at the charging speed at the front ends. The thickness of the strip plate has almost no effect on the energy consumption in the finishing stands of the WSHRM. Increasing the thickness of the strips from 2.0 to 2.8 mm reduces the total energy consumption for the hot rolling mill by 23%.

The energy consumption during the cold rolling of tin plate substantially depends not only on thickness but also on the plastic properties of hot-rolled strip plate. Application of thicker (2.4–2.8 mm) and thus 'softer' strip plate in the production of tin plate 0.20–0.25 mm thick leads to a reduction in the total energy consumption in hot- and cold-rolling by 8–10%.

It is important to emphasize that the analysis which takes into account the differences in the properties of strip plate of different thicknesses, led to the opposite conclusion regarding the effect of the thickness strip plate on the total energy consumption in the rolling thin strips compared with the data in Fig. 1 of [165]. Perhaps the significant difference in the conditions of the examined examples (thicknesses of strip plate, hot-rolled and cold-rolled strip steel) exerted a strong effect in this case.

A mathematical model of the energy consumption in the process of cold rolling of the strips was proposed in [167], which takes into account heating of the slabs before hot deformation considering thermal changes in slabs before placing them in a furnace. This model takes into account the proportion of 'transit' rolling slabs in the total production volume. The advantages of this approach selected by the authors relate to the possibility of selecting the optimum between the reduction of energy consumption during heating of slabs by lowering their temperature and, as a consequence, an increase in the energy consumption for metal deformation in the stands of the WSHRM. Restrictions include the possibilities of electromechanical equipment and power transmission lines stands of the mill, as well as the requirements for the temperature and speed conditions of rolling and coiling the strips in terms of ensuring the specified mechanical properties of hot-rolled steel.

In general, it can be concluded that the implementation of the above calculations and the use of mathematical models and algorithms for energy optimization in the production of sheet steel are undoubtedly promising for all metallurgical complexes.

11.3. Thermal insulation and heat saving in rolling mills

Unfortunately the rolling plants of the metallurgical plants in Ukraine do no pay not enough to the reduction of energy consumption for heating and heat treatment of metal before and after rolling. This also applies in the same measure to other metallurgical processing (blast furnace, steel melting, etc.). Moreover, the effect resulting from the use of advanced thermal insulation materials in industrial facilities of metallurgical enterprises exceeds opportunities to reduce the energy costs by choosing rational modes of the production processes, including the reduction schedules in sheet and section mills [169, 170].

The majority of metallurgical and machine-building factories in Russia and Ukraine use traditionally refractory brick masonry, including in the production of rolled steel, in various types of heating devices, thermal, heating furnaces and shafts, boilers, chimneys and other equipment as a working unprotected lining layer and intermediate (protected) lining layer to solve the problem of isolation of the effect of high temperatures. Disadvantages of using brickwork to solve problems of heat insulation and saving are well known (increase in the size and cost of units, high heat capacity of the refractory brickwork at a relatively inefficient insulation, thermal and temperature inertia of masonry, the energy consumption for heating up to its operating temperature etc.).

In modern steelworks the refractory brick masonry is used only in conditions where there is direct contact of the lining with the liquid metal or the masonry may be subjected to shock loads. In heating devices of the rolling mills of the enterprises the brick masonry is more and more displaced by porous or fibrous refractory materials of a new generation.

The advantages of modern fibrous materials, which primarily include materials and refractory insulating mullite–silica (aluminosilicate) glass fibre include: relatively low density (density of the roll material is not greater than 150 kg/m^3, mullite–silica products of complex configuration with an apparent density of not more than 500 kg/m^3), low thermal conductivity (5–6 times lower than that of the refractory bricks), heat resistance is practically unlimited, including the ability to withstand cyclic variations of temperature with periodic heating and cooling of the lining, there is no danger of destruction of the lining), low specific heat, which allows to quickly implement heating and cooling heating devices, systems, e.g. heat treatment furnaces. The advantages of refractory materials and products based on mullite–silica fibres also relate to the diversity of their species (cotton, rolled material, felts, plate, cardboard), which determines the ease of application in industry. Furthermore, by varying the chemical composition (the content of aluminum, chromium, zirconium) it is possible to change the maximum possible temperature of operation of the mullite–silica materials and products from them.

In accordance with current standards, the heat-insulating refractory mullite–silica glass fibre materials and their products can be used at temperatures up to 1350°C as an insulating compensating material for thermal insulation, heating, vertical-section, cylindrical and other types of furnaces, pit furnaces and other thermal units of the rolling mills.

The most widely used fibrous refractory mullite–silica materials are used for lining the walls and arches of the bell furnaces in rolling production. This reduces the heat flow through the furnace roof and walls and also the heat capacity of the lining. This provides a reduction in fuel consumption and increases the furnace productivity by shortening the heating and cooling cycle of the system.

One example of the effective use of such material in the 'Ural Steel' (OKhMK), where the reconstruction of the flue hog of the thermal roller furnaces in the sheet rolling plant was carried out using refractory fiber materials for the lining [173]. The Termostal' and OKhMK companies developed solutions for introducing the fibrous refractories at all furnaces of the concern. In this case, the fibrous lining is primarily recommended for heat treatment furnaces of the sheet and section rolling mill shops. In continuous heating furnaces it is recommended [173] to use ceramics–fibrous blocks (plates, complex shapes) for the 'ends' of the upper zones, and use of fibrous roll material and mats (felt) for the insulation of dry skids and support tubes, wall seals, crowns and burner units. In cogging shops the greatest effect is provided by the replacement of the brick lining of the covers soaking pits by fibrous refractory materials.

The replacement of the refractory brickwork by the fibrous lining for one heat oven with a bogie hearth with an area of 27 m^2 reduces fuel consumption 20 times. Therefore, the crown and walls in the heat-treatment furnaces of different types (with stationary and portable hearth, vertical, conveyor, with walking beams or hearth, etc.) are recommended to be constructed completely from fibrous refractory materials. Generally, in ferrous metallurgy 1 t of refractory fibrous materials replaces from 10 to 23 t of traditional refractories (refractory bricks).

Particularly effective is the application of fibrous refractory material in batch electric furnaces. When using a fibrous refractory material the heat loss is reduced by one third and, accordingly, the costs of electricity is also reduced. In the systems of this type the fibrous refractories are used for all elements of the lining except the loading windows and furnace hearth. The developed new fibrous insulation materials, given their high thermal resistance, allow to perform the energy-saving reconstruction of the existing furnaces and construct more efficient thermal units. The walling made of mullite–silica materials have several advantages over the refractory brickwork: firstly, a small mass of the fibrous lining of the furnace units due to the small apparent density of high-porosity insulation, and secondly, a small heat accumulation. In periodically operating furnaces the heat loss due to heat accumulation of the fibrous materials of brickwork is 20–40% less than in the brick lining. In the furnaces operating in the continuous mode, the heat energy savings by reducing the heat flow through the furnace wall can reach 8–10 %. In continuously operating heating pusher or walking beam furnaces of the rolling mills the heat loss through the brickwork can be smaller, in the range of 2–5 %. However, the absolute values of the energy savings in high-performance furnaces for heating billets are quite impressive.

It should also be noted that the use of the considered refractory materials simplifies the metal constructions (the mass of metal frames is reduced by 15–20%) for the newly designed furnaces, accelerates the process of heating the metal and makes it easier to control it. The refractories consumption decreases up to 10–12 times and the labour installation costs of the masonry 2–3 times. In the rolling mills of the ferrous metallurgy plants the application of fibrous mullite–silica materials and products in often limited by their use in insulating layers of the lining of the furnaces only in combination with the brick and concrete masonry. The products from fibrous materials, which permit higher maximum application temperatures (up to 1600°C), allow to achieve a greater effect due to the possibility of using them in the working layer of the thermal barriers of the furnaces.

The advanced complex producing mullite–silica refractory heat insulating materials and products based on them is operated by the Sintiz Company (Sinel'nikovo, Dnieperpetrovsk region) [164, 170]. Significantly, the Sintiz Company used 300×300×70 mm blocks of mullite–silica fibres for lining part of the working space of the continuous tunnel furnace 120 m long, designed for firing products at 1000°C. As a result, the gas consumption for heating the furnace was reduced by 45%. For example, this tunnel furnace is similar to furnaces for the normalization of hot-rolled thick sheets exploited in many rolling shops of metallurgical and engineering plants.

12

Preventing defects in thin sheet steel

Several generations of researchers have developed the basic production technology for cold rolled steel [41,42, 50, 51, 67, 74, 89, 96, 99 and others]. Key considerations here are the maintenance of sheet quality and the prevention of surface defects on sheets, strips, and tinplate [42, 153, 174 and others].

On all technological redistribution of thin-sheet steel production, there are various defects. Consideration of the reasons, the mechanism of formation and the possibilities of preventing absolutely all observed (observed) defects [153, etc.] of steel sheets and strips is not provided for in the plan of this book, since this topic is unlimited and requires a special presentation in a separate independent monograph. However, given the urgency of the problem of improving the quality of thin-sheet products, this chapter of this book will analyze the causes and ways to prevent one of the most frequent and hardly preventable defects – kink (breaks) of sheet steel.

The continuous evolution of steel production constantly poses new problems in terms of maintaining defect-free sheet surfaces with specified microrelief. Kinking defects are common (State Standard GOST 21014–88): the sharp bends that appear when coiling or uncoiling thin cold-rolled steel strip are associated with light transverse bands of increased surface roughness. Such bands may also be observed with change in the flexure of sheet assemblies on lifting or transportation [72, 113, 114, 122, 126, 153, 176–178 and others].

In the literature and plant standards, these are sometimes referred to as kinks ('cross breaks', 'coil breaks') or kinking defects. According to the dictionary definitions, a kink is a bend, while a defect is a flaw; this term is readily applicable to the quality assessment of thin cold-rolled steel sheet.

Considerable efforts have been made to find means of preventing kinks. In the present work, we summarize the results of our research, with analysis of the concepts found in the literature. While various terms are used in the literature, we will describe this defect as a kink.

The prevention of kinks is a priority at all metallurgical enterprises producing thin cold-rolled steel sheet. Aside from foreign plants, these are the Zaporozhstal' [72, 153], Karagandinsk [153, 174], Cherepovetsk [42], and Magnitogorsk [43, 50, 126, 153] steel works. Despite the successes that have been achieved, further efforts are necessary [179, 180]. Kinks may pose problems in the processing not only of thin strip but of thick sheet. In particular, the risk of kinking at the edges of relatively thick sheet (~6 mm) used in the production of welded oil pipe (diameter 530 mm) was noted in [180]. Accordingly, available experience must constantly be reassessed in the light of recent developments, the requirements on sheet quality, and promising measures to minimize kinking. In that context, it is expedient to further develop the concepts in [174, 175]. In the present work, we consider the prospects for such measures at the cold-rolling, annealing, and temper-rolling stages in the production of rolled steel sheet.

12.1. Defect morphology. External features of kinking

In kinking, we observe light bands (width ~2.0 mm) on the darker sheet surface; they are usually perpendicular to the direction of rolling [72, 153, 176]. As a rule, the metal is thinner in the defect region. The defect is the result of local plastic deformation of the sheet steel on account of extension or combined extension and flexure of the cold-rolled strip.

The surface roughness in the kinking zone is somewhat different from that in the adjacent sections. Some changes in the microirregularities are seen in the defect zone, according to microphotographs of the surface microrelief on 0.76 × 1250 and 0.76 × 1325 mm samples of 08ps steel strip, using 3*D* criteria [179]. The bulk characteristics of the roughness are practically the same in the defect and the adjacent regions. The microirregularities within the defect are sharper than usual.

The role of thinning in the production and processing of steel sheet was not addressed in [179]. It is possible that the data in [179] are partial and do not apply to other steel-sheet samples.

We accept that the surface microrelief in the defect region must differ from that in adjacent sections of the sheet. In plastic deformation, the surface relief of the metal is formed either without direct surface contact of the tool and the sheet (free deformation) or with such contact (coupled deformation). If the defects on the sheet surface arise on account of the extension of local sections of strip in kinking – that is, in free deformation – the microstructure (the ferrite grain size) will be modified in the defect zone. If the size of the microirregularities exceeds the grain size in the initial steel structure, the surface roughness in the defect will be somewhat reduced, as shown in [127]; otherwise, it will be increased. However, if the defects are formed as a result of flexure with extension – for example, when the strip bends around the tension rollers when the steel coil is unwound in the temper-rolling mill, the surface roughness in the defect zone will be formed in coupled deformation or semicoupled deformation (coupled in one direction). In such conditions, the deformation of the steel's microstructure will have less influence on the surface roughness of the defects. In all cases, kinking is clearly seen in the form of light bands on the dark surface.

The defects may appear at the edges or over the whole width of the strip. At the edges, the defects may extend over 250–300 mm. The appearance of the defects, their length and frequency, and their periodicity over the length of the strip depend on its thickness, the thickness of the tension rollers, the degree of welding of the turns, the tensile forces in the strip on uncoiling, the temperature, and the metal's yield point. These correlations have not been fully explored in the known studies [42, 122, 126, 153, etc.].

The kinking defects observed on thin cold-rolled steel sheet resemble, in their appearance and origins, another defect: slip bands and lines, which take the form of dark bands and branched lines that appear on the surface of cold-rolled sheet and strip on account of local stress exceeding the yield point of the metal. Slip lines are usually inclined at 45° to the rolling direction; in other words, they are in the direction of the primary tangential stress. These defects are known as Lüders bands. As a rule, they appear in parallel groups. This is their main difference from kinks, which are always transverse to the longitudinal rolling axis. Kinks and Lüders bands are often observed at the same time and are prevented by the same means [153].

12.2. Defect formation. Causes and mechanism of kinking

Despite extensive research on kinking defects, their formation is still poorly understood. We now consider their formation in detail. In unwinding, according to Fig. 12.1, the upper turn of the strip separates from the coil at point A in the ideal case – that is, without adhesion or welding of the surfaces of adjacent turns in the coil. In this case, the strip trajectory to the tension rollers (shown by the dashed line) does not depend on the tension T and the force required to straighten the unwound strip and is tangential to the surfaces of the coil at point A and the lower tension roller at point C. Since there is no welding of the adjacent turns, the separation of the upper turn from the coil at point A does not require any force perpendicular to the tangent at that point. Nevertheless, some bending torque M_S must be applied to straighten the upper turn of the strip and create the opposite curvature. Since heat-treated steel strip has some elasticity, the actual trajectory will be somewhat different.

In unwinding the coil, when the strip is supplied to the rollers of temper rolling, its tension is created by the motor turning the unwinding drum, operating in the generator mode. The torque due to the strip tension T is $M_{str} = TR$, where R is the radius of the upper turn on the coil. Steel strip (thickness h) is weakened after recrystallizing annealing. Straightening of the upper turn by plastic

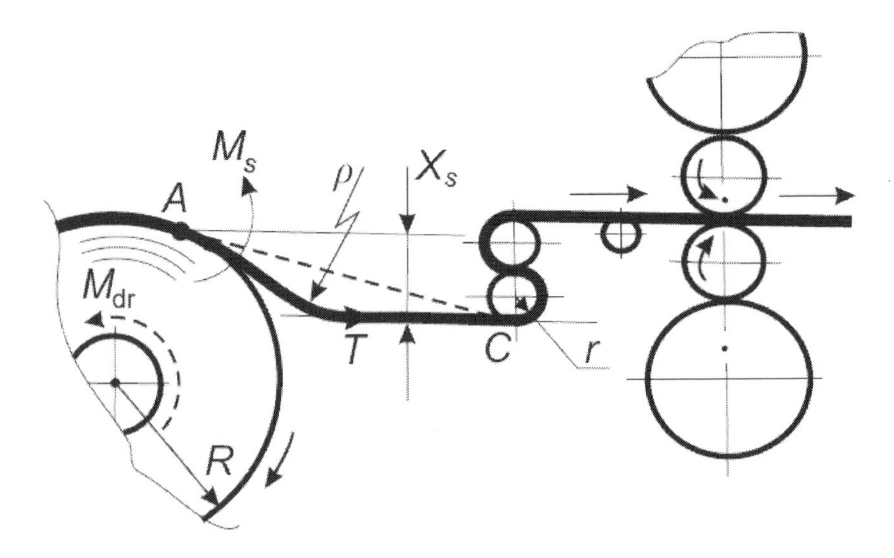

Fig. 12.1. Plastic flexure of strip in unwinding a coil with tension T created by the braking torque M_{dr} of the uncoiling drum.

deformation calls for the creation of stress equal to or exceeding the metal's yield point on account of the tension T in the strip. We require that $T \geq \sigma_y bh$, where b is the strip width.

In industrial practice, the tension T on the uncoiling device is significantly less than $\sigma_y bh$. Production experience is usually employed in selecting the factor k by which the tension is reduced: $T \geq k\sigma_y bh$. According to publications of the Research Institute of Heavy Machinery, Ural Heavy-Machinery Plant, the range of k at different temper-rolling mills and finishing systems is fairly broad: 0.005–0.040. When the tension T is insufficient for equalization of the strip at point A (Fig. 12.1), the upper turn, as it separates from the coil, will be in the elastic state for some time and will retain curvature $1/R$. The bending torque M_s created by force T for flexure of the upper turn acts at a distance X_s (Fig. 12.1), which increases gradually as long as M_s is no less than the plastic-flexure torque of the strip $\sigma_y bh^2/4$. In other words, the upper turn of the coil is straightened when $M_s \geq \sigma_y bh^2/4$.

Under the action of the tensile force T and the bending torque M_s, the strip is first straightened out and then acquires the opposite curvature as it wraps around the lower tension roller. The total strip strain is $\varepsilon_\Sigma = \varepsilon_1 + \varepsilon_2 + \varepsilon_3$, where $\varepsilon_1 = \sigma/E$ is the elastic strain according to Hooke's law; $\varepsilon_2 = h/(2R)$ is the strain of surface layers of the strip in the external turn of the coil (radius R) on account of plastic flexure on straightening; $\varepsilon_2 = h/(2r)$ is the strain of the strip (thickness h) as it wraps around the lower tension roller (radius r): E is the elastic modulus of the strip; σ is the tension.

On separating from the coil, the upper turn of heat-treated strip is first in an elastic state with curvature $1/R$. Its straightening begins when the flexure torque M_s reaches or exceeds the plastic-flexure torque of the strip – that is when $M_s = \sigma_y bh^2/4$. Here b is the width of the strip; σ_y is its yield point. On the other hand, the flexure torque created by the tensile force T is $M_s = TX_s$, where X_s is the distance at which torque M_s acts (Fig. 1). Then $TX_s = \sigma_y bh^2/4$ and $X_s = \sigma_y bh^2/(4T)$. Accordingly, X_s declines with increase in T, and hence the distance from the onset of strip straightening to point A decreases. The flexure spacing is reduced. With increase in σ_y, the point at which the strip curvature is reversed becomes more distant.

Thus, in order to the straighten the strip, the stress in its surface layer must be at least equal to the yield point of the metal. Note that kinking defects do not always appear in straightening and reversal of

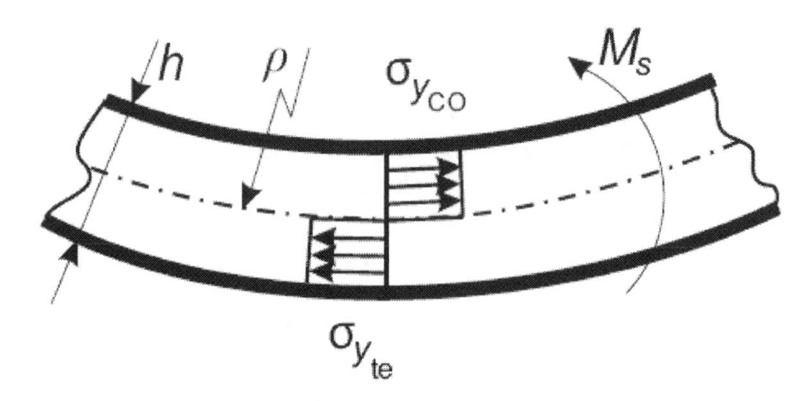

Fig. 12.2. Stress distribution in the upper turn of the strip in plastic flexure during the unwinding of a coil: σ_{Yco} – compressive yield point; σ_{Yte} – tensile yield point.

strip curvature. They only appear if the bending torque M_s converts the strip completely to the plastic state.

Plastic flexure of a strip offers a useful model here (Fig. 12.2). In the flexure of a smooth strip, we may assume that its neutral layers retain the same length; the outer layers are extended; and the inner layers are compressed. The strain of the outer layers is $\varepsilon = h/2\rho$, where ρ is the radius of curvature of the neutral cross section over the strip thickness. According to Hooke's law, $\varepsilon = \sigma/E$. In the limiting case, $h/2\rho = \sigma/E$ and $\rho = hE/2\sigma_y$.

This is the condition for transition of the metal to the plastic state over the whole strip thickness. However, this is insufficient for the formation of kinking defects. As a rule, local thinning of the strip occurs in the vicinity of the defect: a neck is formed. (This is the limit of sample failure in fracture tests.) Therefore, we may assume that kinking defects arise when the plastic strain of the metal on flexure exceeds the critical strain that corresponds to (is equal or close to) the uniform elongation in extension. Many researchers [181, 182, etc.] believe that the critical strain corresponding to kinking is twice the strain at the onset of plastic flow: $\varepsilon_{cr} = 2\sigma_y/E$.

According to Fig. 12.1, the tensile strain ε of the lower layers of the strip at the upper turn of the coil is the sum of the elastic strain σ/E of the strip, the plastic strain $h/2R$ in straightening of the upper turn, and the strain $h/2r$ created in the reversal of strip curvature as it passes around the lower tension roller (radius r). These components are of approximately the same order. Obviously, ε increases with decrease in R as the coil unwinds. Hence, the risk of kinking defects increases. Likewise, the probability of such defects declines with increase in radius r of the lower tension roller.

Analysis of the equation $\rho = hE/2\ \sigma_y$ indicates that kinking defects are more likely with increase in strip thickness h and decrease in σ_y, since the elastic strain in the metal becomes plastic strain at large flexure radius ρ in uncoiling. This is confirmed by production experience. Kinking defects are less common in relatively thin strip such as tinplate and strip of special hard-to-deform steel, for which σ_y is greater. With decrease in ρ, tensile outer layers of the strip will be obtained at higher plastic strain. Consequently, the risk of kinking defects is higher.

This description of the unwinding of steel coils does not include adhesion or welding of the surfaces of adjacent turns in the coil.

Welding of contacting turns in the coil on heat treatment (annealing) has a decisive influence on the appearance of coil breaks. The fundamental mechanism by which coil breaks are formed in the unwinding of coils is the same in the presence of adhesion or welding of adjacent turns. However, this factor is often decisive.

Taking such behaviour into account does not fundamentally change Fig. 12.1, but the analysis is more complex. In particular, we must take account of the force required to separate the upper turn from the one below. In Fig. 12.3, we show the separation of the upper turn of the strip welded to the lower layer; the two adjacent layers are separated at point B. The x axis runs along the tangent to the

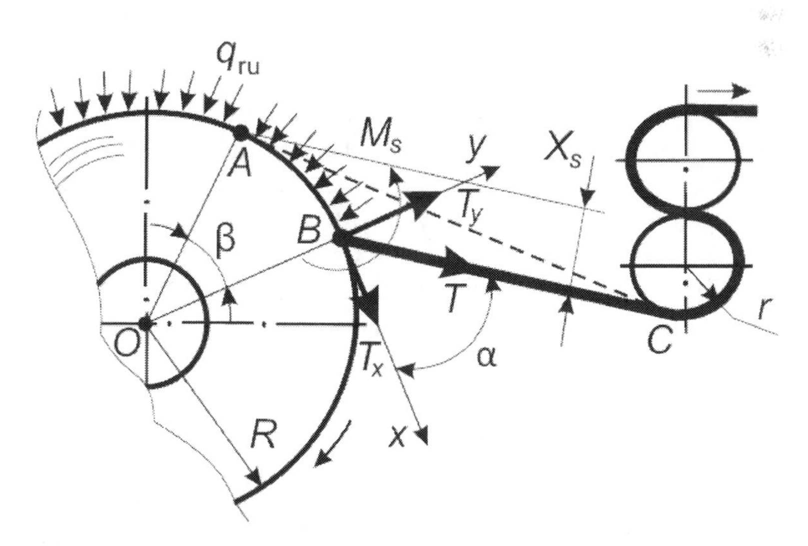

Fig. 12.3. Separation of a welded upper layer from the coil in unwinding: q_{ru} is the force required for separation of the welded surfaces of adjacent turns in the coil.

upper turn of the coil, while the y axis is perpendicular to the x axis at the origin (point B). As point B moves, the coordinate system xy changes its orientation relative to the coil and the tension rollers in the temper-rolling mill. The unit force required to separate the welded surfaces of the adjacent turns is q_{ru}. The total rupture force corresponding to the arc AB of the upper turn of the strip (width b) is $Q = q_{ru}bl_{AB}$, where $l_{AB} = R\beta$ is the length of arc AB; the angle β is measured in radians (Fig. 12.3).

In the first approximation, we may assume that the breakaway of the upper turn from the coil occurs over arc AB under the action of force T_y applied at point B along the y axis (perpendicular to the tangent to the coil surface). The component T_y of the tensile force T in the strip must be no less than Q. We see that T_y and Q depend on the interrelated angles α and β, which are determined by the position of point B (Fig. 12.3). Knowing the radius R of the coil's outer turn, the radius of the tension rollers, and the distance between the vertical and horizontal axes of the coil and the lower tension roller—that is, the configuration of the mechanisms in the temper-rolling mill—we may determine the angle α (or $90° - \alpha$) determining the strip flexure in the region of the breakaway point B of the upper turn welded to the adjacent layer, for fixed q_{ru} and T.

In uncoiling sheet steel both Figs. 12.1 and 12.3 apply. In the unwinding of a coil in which adjacent turns are welded together, we note partial or complete straightening of the strip on path BC under the action of tension T. Then the upper turn breaks away from the adjacent strip within an area corresponding to arc AB. In some cases, separation of the upper turn may only occur over part of the arc AB. The distance between defects will be less in those circumstances.

Analysis confirms that the appearance of kinking defects depends primarily on the adhesion or welding of the adjacent turns in the coil. With stronger welding of the adjacent turns on annealing, separation of the upper turn from the coil will be delayed. This is accompanied by increase in strip flexure and by higher risk of kinking defects.

With increase in the tensile force T of the strip as the coil is unwound, the adhesion or welding of the turns in the coil will have less influence on the formation of coil breaks. As the coil is unwound, the tensile force T in the strip will vary in direct proportion to the force q_{sep} required to separate the welded surfaces of the adjacent turns. Prevention of coil breaks requires the elimination of adhesion or welding of the turns in the coil during the annealing of cold-rolled steel.

In turn, the degree of adhesion or welding of adjacent turns in the coil depends on the radial contact stress and the annealing conditions. This is discussed in Chapter 9 of this book.

12.3. Welding of adjacent turns during the annealing of cold-rolled steel coils

The adhesion between adjacent turns in coils of cold-rolled steel depends on many factors, as noted in [113–115, 122, 126, 176, 177, 181– 185, etc.]: the annealing conditions (the heating rate, temperature, holding time, and cooling time); the yield point of the metal at the annealing temperatures; the surface roughness of the strip; and the stress–strain state of the coil (with redistribution of the radial stress as a function of the temperature).

In determining the stress in a coil of cold-rolled strip as it is wound on a drum, after removal from the winding machine, on heating in annealing, and during subsequent cooling, we must establish the dependence of the interturn gap on the compressive forces and the temperature. For the compression of stacks of steel sheet of different thickness and roughness (Fig. 12.4), such results may be found in [127, 183]. Their analysis permits the following conclusions.

For ideally smooth surfaces of the sheet in the coil, the distance between adjacent turns depends only on the pliability of the microrelief. In practice however, the strip wound into the coil is characterized by non-planarity, on account of warping, undulation, and local distortion. In the compression of stacks of steel sheet, the contacting surfaces are pushed together on account of equalization of

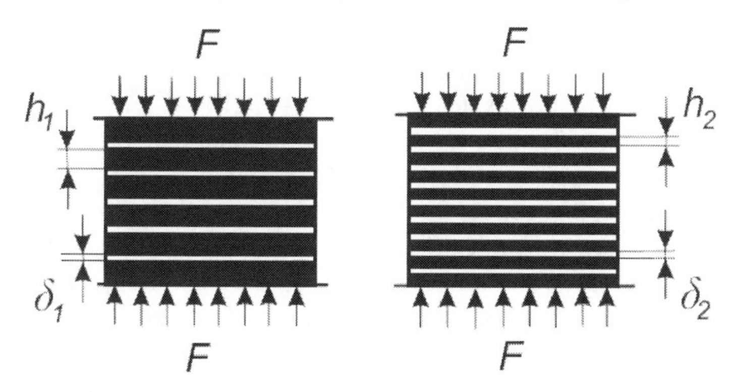

Fig. 12.4. Compressive tests of stacks of steel sheet of different thickness h ($h_1 \approx 2h_2$); δ_1 and δ_2 are the gaps between the plates ($\delta_1 \approx d_2$).

the plates and flattening of the microirregularities. In stacks of thin sheets (thickness h_2), the number of gaps δ and their total volume will be less than for stacks of relatively thin sheets (thickness h_1), other conditions being equal (Fig. 12.4). However, on compression, the decrease $\Delta\delta$ of the gaps (absolute strain) is greater for the stacks of thicker sheets. In experiments at the Institute of Ferrous Metals, Ukrainian Academy of Sciences and previously at the State Research Institute of Metallurgical Machinery, it was found that $\Delta\delta/h$ declines with increase in the sheet thickness h in the stack [127]. The decrease in distance between the contacting surfaces is directly proportional to the compressive force on the stack, the sheet thickness, and the surface roughness and is inversely proportional to the yield point of the metal. That must be taken into account in assessing the change in stress–strain state of the coils on annealing.

As already mentioned in section 9.2 of this book, the dependence of $\Delta\delta$ in the stack on the compressive force is significantly non-linear, especially in the section where compression begins. The dependence on the surface roughness is close to linear. For the production of thin steel sheet, we may write $\Delta\delta_{Ra} = (0.91 + 0.09Ra)\,\Delta\delta$.

Empirical formulas for $\Delta\delta/h$ were presented in section 9.2.

No significant difference in the deformation was noted in the compression of stacks with different numbers of sheets (disks) in [184, Fig. 14 *a*]. The lack of information regarding the experimental conditions hinders the analysis of these results. The behavior of $\Delta\delta$ in stacks of steel sheet under load is basically the same in [127, 183].

As already mentioned in section 9.2 of this book, extensive experimental and theoretical research on the interaction of rough surfaces (including elastoplastic deformation on compression) and the influence of the contact microgeometry and duration on the thermal conductivity – by Demkin, Kragel'skii, and others– has produced formulas in which the roughness of the contacting surfaces is characterized by parameters associated with considerable indeterminacy and complexity, with poor accuracy. Nevertheless, analysis of the theoretical literature permitted refined approaches to the thermal flux through the contact zone of adjacent turns in strip coils on heating and cooling [187, 188]. The mathematical model of the stress–strain state of coils is enhanced by the findings in [188].

The actual heat-conducting contact area of the rough surfaces of adjacent turns in strip coils is formed predominantly by plastic deformation in one-time loading. With repeated analogous loads, the deformation of the microirregularities is elastic. The thermal contact resistance falls with increase in the rated pressure in elastic and plastic contact of the microirregularities of adjacent turns in the coil. The actual contact area increases with increase in the contact time under load, on account of creep of the metal, especially at relatively high temperatures. The corresponding decrease in distance between the surfaces increases the heat conduction.

Research shows that non-ideal thermal contact of rough surfaces in the coil may be taken into account through the thermal resistivity, which depends on the microrelief, the surface contamination, and the compressive forces [113, 114]. The mathematical models of the stress–strain state of the coils on heating and cooling in the course of annealing employ an empirical formula presented in [113].

The interturn welding in the coil determines the type of kinking defects, their position, and their frequency in the production of cold-rolled steel. The actual contact area of adjacent turns and their welding on annealing depend on the surface roughness of the strip, its temperature, and the interturn contact pres-sure. An inverse relation is observed. The temperature and compressive force on the turns and also the surface microrelief of the metal influence the contact area of adjacent turns and change the heating and cooling rate of the coils. Theoretical description of this behavior is very complex, and therefore experiments are conducted.

The results of research at various steel works were summarized in [72, 127, 153, 186]. The conclusions obtained are generally applicable. It is found that reduction in roughness and in the density of surface microirregularities facilitates the welding of adjacent turns in the coil on annealing and the formation of kinking defects in subsequent temper rolling. Direct laboratory research regarding the influence of the compression, temperature, and residence time at high temperatures on the welding of contacting samples and the rupture force are of most value. It is found that the force required to separate welded contact surfaces increases sharply at annealing temperatures above 700°C [184]. The role of the strip's surface roughness increases with increase in contact pressure. The same conclusions may be drawn from similar experiments conducted by Pargamonov at the Zaporozhstal' Metallurgical Works [178].

These findings are largely confirmed by later research [122]. The welding of adjacent turns in the coil mainly occurs on cooling. With increase in the residence time of the coil under load, more welding occurs, and the force required for their separation increases [122,

184]. (The increase is sharper at higher temperatures.) That confirms the significant role of creep of the surface microirregularities in elastoplastic contact of the rough surfaces under load [188].

Summarizing all the research, we may say that the welding of adjacent turns in the coils depends greatly on the surface roughness of the metal. This dependence is greater at higher temperatures. With increase in the welding when the contact surfaces are less, greater force is required for the separation of the turns. Accordingly, the risk of kinking defects increases.

Thus, the experimental data presented in [122, 184] regarding the influence of the annealing temperature and the compressive force due to the cold-rolled steel plates on the welding of contacting strip surfaces were presented in. Analogous findings were reported in [183]. The welding of adjacent turns in steel coils on annealing mainly accompanies high-temperature holding and cooling of the steel.

Significant results were obtained in the experiments of [122, 183, 184]. However, they had the following shortcomings.

(1) The compressive forces employed were significantly less than in industrial conditions.

(2) The assessment of the adhesion of contacting surfaces by shifting one sample relative to another is somewhat arbitrary, since it does not adequately simulate the separation of the turns in the unwinding of coils. In the unwinding of coils on a temper mill, the upper turn is separated from the body of the coil.

So that the experiments[1] more closely resemble production conditions, samples of cold-rolled 08ps, 1ps, 2ps, 20, 30KhGSA, and 12Cr18Ni10Ti steel strip (thickness 0.3–2.0 mm, width 30 mm, length 60–70 mm) are held in sets of three or four pairs under a force of up to 100 N/mm^2, in special C clamps whose thermal linear expansion matches that of the tested steel [178]. A circular contact spot of these samples (diameter 30 mm) is ensured by means of a circular gasket of the same diameter (Fig. 12.5).

The clamped samples are roasted in a single pile in industrial cupola furnaces used for the heat treatment of coils of cold-rolled steel strip in a protective atmosphere (nitrogen). The annealing conditions are as follows: the furnace temperature is 700–710°C; and the duration of the process is 68–72 h (heating, holding for 8–15 h at the annealing temperature, and cooling). After annealing, the pairs

[1]Conducted by E.A. Pargamonov.

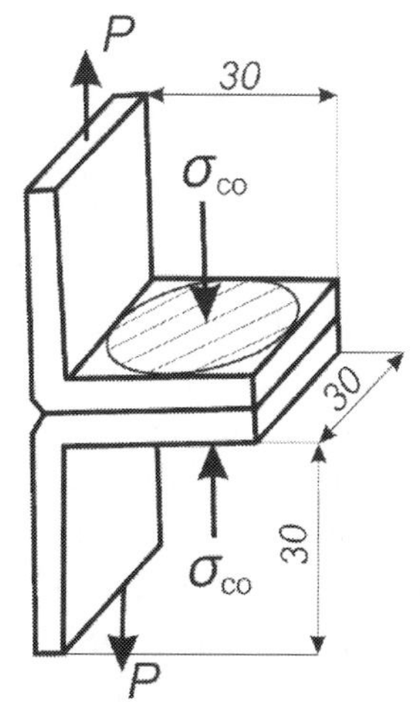

Fig. 12.5. Loading of samples prior to annealing and their subsequent separation: σ_{co} – compressive stress before annealing; P – force separating the welded adjacent surfaces.

of samples are separated on a special machine with an extension rate of 0.33 mm/s.

The adhesion of the samples is estimated in terms of the ratio P/d, where P is the rupture force of the samples and d is the width of the welded section of the contacting surfaces (the diameter of the contact spot). Assessment on this basis is selected because the separation of samples under the action of force P occurs along the rupture line, which is parallel to the samples' flexure line (Fig. 12.5). This line gradually moves over the samples' adhesion plane. The force required to separate the samples depends on the adhesion area of the contacting surface and hence on the width d of the rupture line. Thus, the ratio P/d characterizes the separation force q_{sep} for the welded adjacent layers. In the tests, each measurement is repeated many times, and the results are averaged.

Influence of surface roughness. In the production of thin steel sheet, the strip surface is roughened by cold rolling in incised rollers so as to prevent adhesion of the adjacent turns in the coil on annealing [189]. As a rule, the roughness is different on opposite surfaces of the cold-rolled steel strip, since the upper and lower working rollers in the continuous mill will have different wear. To take account of this factor, the contact of rough surfaces is assessed by the sum of the roughness for the upper and lower sides of the strip: $\Sigma Ra = Ra_L + Ra_U$. This characteristic use-fully characterizes the contact with surface roughness in the range $Ra = 1.0–6.0$ μm.

In the interaction of two rough surfaces, their contact may be characterized by

$$Ra_{\Sigma} = (Ra_{L}{}^{2}+Ra_{U}{}^{2})^{1/2}.$$

However, in qualitative analysis, the conclusions are basically the same whether ΣRa or Ra_{Σ} is employed. It is simpler to use ΣRa.

The dependence of the separation force P/d on ΣRa is shown in Fig. 12.6.

As expected, P/d decreases with increase in ΣRa and rises with increase in compression of the samples. With relatively small compressive force, the surface roughness has practically no influence on the adhesion in annealing. The roughness plays a greater role with increase in the compressive contact stress, especially at relatively small Ra. With increase in ΣRa, the compressive force has less influence on the adhesion (Fig. 12.7). These findings are very important in selecting the tensile stress for coiled cold-rolled strip so as to minimize the radial contact stress.

The variation of P/d as a function of ΣRa (Fig. 12.6) is the opposite of the variation of $\Delta\delta$, which denotes the decrease in the distance between the contact surfaces under compression in the course of annealing shown in section 9.2 of this book. Since the contacting surfaces move closer together under the action of the normal load, their actual contact increases and hence the adhesion of the surfaces on annealing will increase.

The experimental results show that the direction of the surface microrelief on the strip affects the welding of adjacent turns in the coil. For example, if the compressive stress σ_{co} = 50–100 N/mm^2,

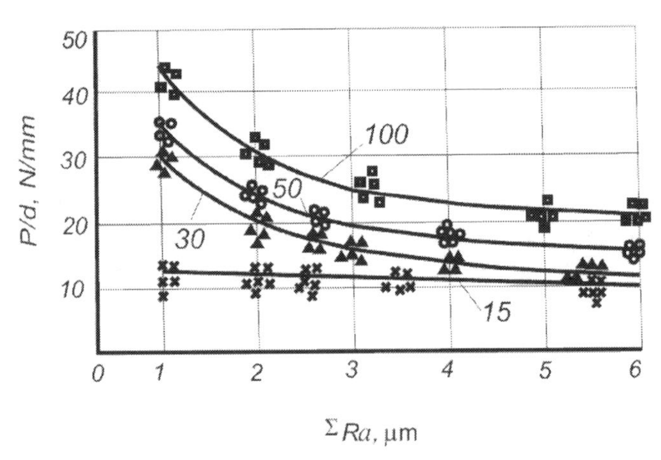

Fig. **12.6.** Dependence of the unit separation force P/d on the total roughness ΣRa of the contacting surfaces. The numbers on the curves specify the compressive contact force applied (N/mm^2).

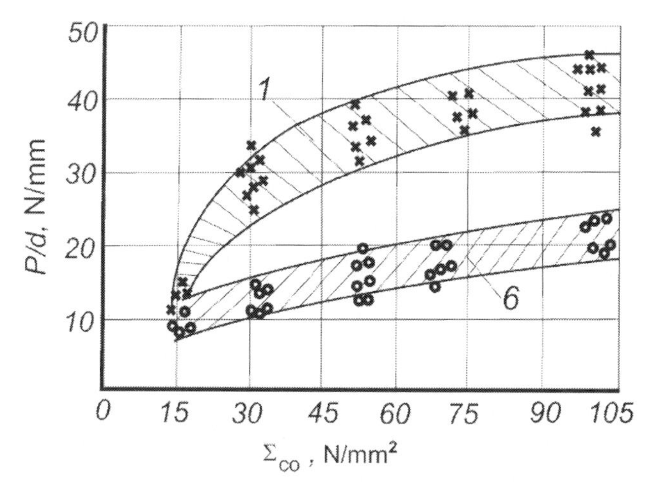

Fig. 12.7. Dependence of the unit separation force P/d on the compressive stress σ_{co}. The numbers on the curves specify the total roughness ΣRa of the contacting surfaces (μm.).

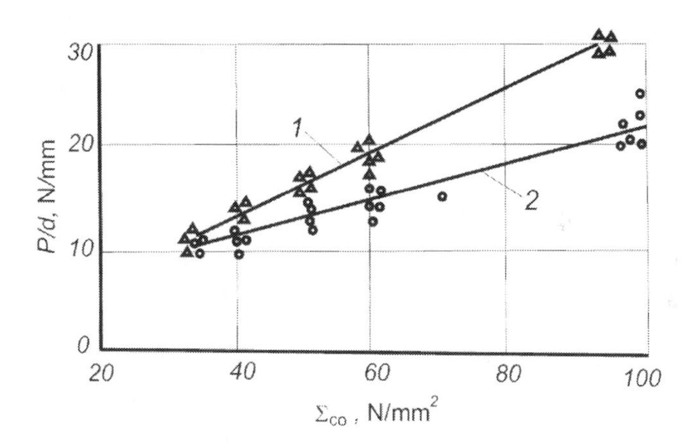

Fig. 12.8. Dependence of the unit separation force P/d on the compressive stress σ_{co} for 08ps steel strip (thickness 1.0 mm), with different surface preparation: (*1*) when one strip surface is formed by a ground roller, and the other by an incised roller (total roughness ΣRa = (0.8 + 2.0) μm); (*2*) when one both strip surfaces are formed by ground rollers (total roughness ΣRa = (1.6 + 1.2) μm).

when one surface has unidirectional microrolling (after rolling in ground rollers) while the other has arbitrary relief (after rolling in incised rollers), the unit separation force between adjacent samples after annealing is ~30% greater than in the contact of two samples with arbitrary microrelief (Fig. 12.8). With increase in compressive force, the direction of the surface microrelief will have a greater influence.

Influence of the lubricant in cold rolling. The lubricant (emulsion) affects the distance between the contacting turns under load and hence the stress–strain state of the coil of cold-rolled steel strip. In addition, the decomposition products of the lubricant formed on annealing affect the adhesion of the contacting turns. Attempts have been made to reduce adhesion in steel coils by introducing special additives in the lubricant [183, pp. 373–374]. However, secondary effects prevent the widespread adoption of such measures.

In our research, lubricant is applied to the sample surface in rolling [178]. The experimental results show that the lubricant at the contact surface may either reduce or intensify the adhesion on annealing. The screening effect depends on the strength of the film that arises at the contact surfaces on account of decomposition of the lubricant. The welding of the turns in the coils on annealing is reduced if lubricants that form oxide films at the steel surface are added in cold rolling. As a rule, additives that improve the washing properties of the lubricant and thus help remove contaminants from the adjacent surfaces increase the welding of turns in the coil.

Influence of the reduction in cold rolling. In modern cold-rolling mills for the production of thin steel sheet and tinplate, the total reduction of the strip is 30–93%. With increase in the strain, the scale of the microstructure is reduced, and the lattice distortion is increased. Correspondingly, recrystallization of the cold-rolled steel is intensified in subsequent heat treatment. Structural changes in cold-rolled steel in the course of annealing significantly affect the adhesion of the turns in the coil. According to our results, the separating force P/d increases in direct proportion to the total reduction ε_Σ in the cold rolling of low-carbon steel strip (Fig. 12.9).

Metallographic study of the microstructure of annealed steel shows that ferrite grains are present on both surfaces in the contact zone. The quantity of such grains is increased if the total strain of the strip in cold rolling increases. After separation of the welded samples, points from which metal has been excavated are seen on their surfaces at contact points. This is also observed in metallographic surface analysis at welded sections for strip rolled and annealed in industrial conditions. The influence of the strain of the steel in cold rolling on the likelihood of diffusional welding at the contacting strip surfaces is intensified at high temperatures and with prolonged holding of the coils in the course of annealing.

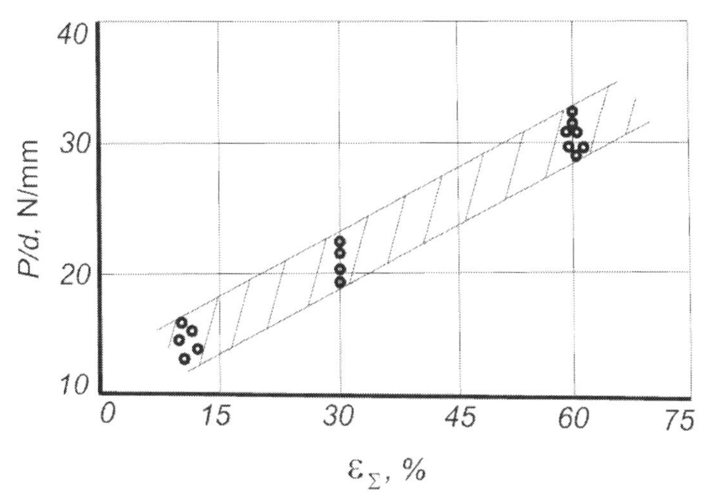

Fig. 12.9. Dependence of the unit separation force P/d on the total strain ε_Σ in the cold rolling of the strip: 08ps steel, strip thickness 0.8 mm, ΣRa = 0.5 μm, σ_{co} = 50 N/mm^2.

Influence of the mechanical properties of cold-rolled steel. It has been suggested that the chemical composition of the steel affects the welding of turns in coils of cold-rolled steel strip on annealing. However, the available information is inadequate. We do not understand how this influence might be exerted.

The chemical composition of the steel has a decisive influence on its mechanical properties and, in particular, on its yield point σ_y. As has already been shown, coil breaks are more likely in the unwinding of coils at a temper mill for smaller values of σ_y, other conditions being equal. However, the yield point of steel also affects the likelihood of welding within steel coils on annealing.

The classical work by Demkin, Kragel'skii, Rudzit, and others indicates that the interaction of rough surfaces is a complex elastoplastic process and depends on the yield point of the contacting materials [187]. As already mentioned, the actual contact area of the turns within coils of cold-rolled steel strip depends on the elastic and plastic deformation of the contacting surface microirregularities. Plastic deformation of the microirregularities declines with increase in yield point of the steel. With increase in the interturn pressure, the actual contact area of rough surfaces increases in coiled strip. The actual contact area also increases with increase in contact time of the surfaces on account of creep, especially at relatively high temperatures during recrystallizing annealing. The corresponding decrease in distance between the surfaces increases the heat conduction at contact, which will affect the temperature field of the coils on heating or cooling.

With increase in yield point of the steel, the actual contact area of rough surfaces within the coil will decrease at both low and high annealing temperatures, since larger forces are required for plastic deformation of the contacting surface microprojections. As a result, for stronger steel, with higher yield point, especially at high temperatures, welding of adjacent turns in the coil is less likely. According to experimental data, the welding seen in samples annealed under a load of 50 N/mm^2 declines significantly with increase in the steel's content of carbon and alloying elements. For example, *P*/*d* falls from 19.0–22.3 N/mm for 08ps steel to 15.2–19.5 N/mm for 2ps steel and to 10.9–11.6 N/mm for 20ps steel. The smallest *P*/*d* values correspond to annealing for 59 h with 14 h holding at 700°C; the largest P/d values correspond to annealing for 73 h with 14 h holding at 710°C. In those conditions, no welding of adjacent turns is seen in 12Cr18Ni10Ti steel samples.

With increase in the steel's content of carbon and alloying elements, σ_Y rises. According to our experimental results, σ_Y = 30–35 N/mm^2 at 700–710°C for 08ps steel; 40–43 N/mm^2 for 2ps steel; 50–52 N/mm^2 for 20ps steel; and 180–190 N/mm^2 for 12Cr18Ni10Ti steel. The results indicate an inverse dependence of *P*/*d* on the yield point of steel at the annealing temperatures. This finding is consistent with the data in [153, 181, 183, 185, etc].

12.4. Effect of coiling technology of cold rolled steel and of unwinding of strip rolls on the formation of kinking defects

Influence of the structure of the temper mill. In production conditions, the losses of the enterprise on account of the reduced quality of thin sheet associated with flexure defects over the operating life of the sheet-rolling system, which is usually a few decades, may be as much as the initial cost of the temper mill. Therefore, in selecting the configuration and parameters of the temper mill during the design process or reconstruction, careful attention must be paid to the prevention of flexure defects in the production of cold-rolled steel.

In Fig. 12.10, we show two systems for uncoiling the strip and supplying it to the temper mill. It is clear that the direction of the tensile force in the strip and the radius of curvature in bending at

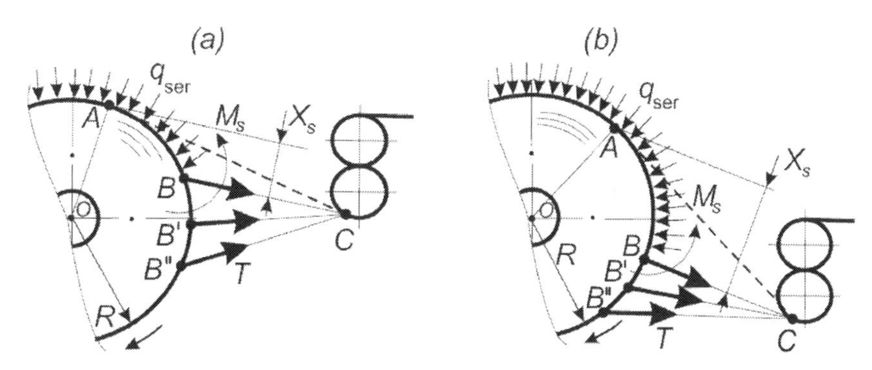

Fig. 12.10. Influence of the relative position of the unwinding drum and the tension rollers on the separation of the upper turn from the roller.

point B depend on the distance between the vertical and horizontal axes of the coil and the rollers of the mill's tension system.

For the sake of clarity, the horizontal axis of the lower tension roller is above that of the unwinding drum in Fig. 12.10, *a* but is significantly lower in Fig. 12.10, *b*. The different configuration of this system in the mill changes the forces on the upper turn as it separates from the coil.

The mutual position of the unwinding drum and the tension rollers primarily affects the position of the points A and B: in Fig. 12.10 *a*, these points are higher than in Fig. 12.10 *b*. The length of arc AB is less in Fig. 12.10 *a* than in Fig. 12.10, *b*. Accordingly, with identical values of q_{sep} and T, the separation of the upper turn from the coil calls for smaller torque M_s and correspondingly smaller distance X_s for Fig. 12.10 *a* than for Fig. 12.10 *b*. Since the point B is higher with respect to the drum's horizontal axis in Fig. 12.10 *a* than in Fig. 12.10 *b*, the flexure of the strip at point B will be smaller for Fig. 12.10 *a*. That helps to prevent coil breaks.

The two configurations in Fig. 12.10 are both schematic. In considering real systems, they must be refined, and the analysis must be expanded; in particular the distance of the tension rollers from the unwinding drum must be fully taken into account. Such analysis is essential, since the recommendations for specific conditions may change after all of the relevant factors have been taken into account.

As the unwinding of the coils progresses and the radius of the external turn is reduced, point A in Fig. 12.11 will be displaced toward the coil's vertical axis. With constant q_{sep} and T, the creation of the torque M_s required to separate the upper turn from the coil calls for displacement of point B downward, so as to ensure the

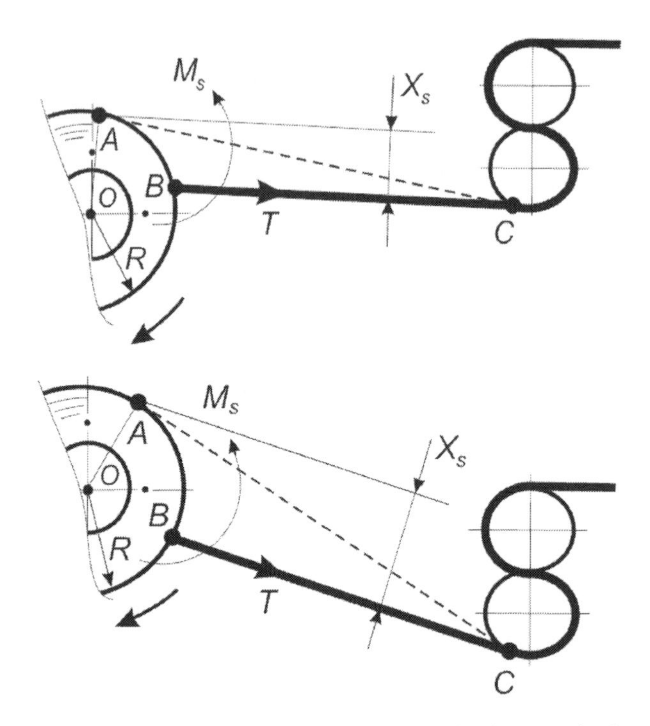

Fig. 12.11. Variation in the force on the upper turn with decrease in the coil diameter.

required distance X_s at which torque M_s acts. In that case, there will be greater flexure of the strip at point *B*, and the risk of coil breaks increases, for both configurations of the temper mill.

The likelihood of coil breaks depends not only on the external radius of the coil, which is determined by its mass and the dimensions of the unwinding drum, but also on the diameter of the tension rollers (the ratio of the strip thickness to the roller diameter), as well as the direction in which the coil is unwound.

When the coil is unwound from above (as shown in Figs. 12.1, 12.3, 12.10), the strain due to straightening of the strip is summed with the strain of the same sign from its flexure at the lower tension roller. As a result, the risk of coil breaks increases; this effect is greatest for thick strip. When the coil is unwound from below, the coil turns in the direction opposite to that shown in Figs. 12.1, 12.3, 12.10; the internal layers, which are extended in equalization of the strip, are compressed as they wrap around the lower tension roller. As a result, the total tension is reduced, and the risk of coil breaks decreases. The risk of coil breaks is also reduced if the strip is supplied to the temper mill without tension rollers. However, this requires relative proximity of the unwinding drum to the temper mill. These factors must be taken into account in the design and modernization of temper mills.

Preventive measures. On the basis of Fig. 12.3, we may make certain conclusions even without mathematical analysis. Obviously, the position of point B, at which the upper turn separates from the coil, depends on many factors: the force q_{ru} required to separate the welded surfaces of the adjacent turns; the strip tension T; the radius of the upper turn, determined by the initial coil dimensions and the number of turns already unwound; and the relative position of the unwinding unit and the tension rollers in the temper-rolling mill. With constant tensile force T, its component T_y will increase as point B is moved away from point A and the angle β increases. If the upper turn separates from the coil at point A, we find that $T_y = 0$ and $\alpha = 0$ on account of the change in direction of coordinate axes xy and hence the forces (Fig. 12.3). With fixed q_{ru}, larger T results in earlier separation of the upper turn from the coil. In that case, point B moves close to point A. With stronger adhesion of the adjacent turns and higher q_{ru}, the force T_y required to separate the surfaces of the adjacent turns will be larger.

With increase in q_{ru} or decrease in T, point B descends, and the upper turn separates later from the coil. Correspondingly, with later separation of the upper turn from the coil, the strip's flexure angle will increase, the radius of curvature of the strip at point B will be smaller, and the probability of kinking defects will be greater. The degree of flexure and the radius of strip curvature at point B are determined by the angle α (Fig. 12.3). With increase in α, the risk of kinking defects increases. Hence, with increase in q_{ru} and decrease in T, the deviation α of the strip trajectory from the tangent to the coil surface at point B will increase. This means that, at smaller T and larger q_{ru}, the strip curvature at point B increases and the risk of kinking defects is higher.

In the unwinding of the coil, the radius of the upper turn declines; point B is shifted downward; and point A is shifted upward. On account of the decrease in the external radius of the coil on unwinding, β will increase. Change in the position of point B affects the force T_y, which is a function of the radius of the upper turn in unwinding and q_{ru}. Note that our conclusions are not fundamentally changed if we consider the bending torque M_s rather than T_y (Fig.

12.3). The distance X_s at which M_s acts is equal to the distance between the line BC and the parallel line through point A.

The adhesion (welding) of the contacting surfaces varies over the thickness of the coil. That corresponds to the dependence of q_{ru} on the external coil radius on unwinding in the temper-rolling mill. The variation in the surface welding over the coil thickness depends on the strip tension selected on coiling in the cold-rolling mill; the cooling time before annealing; and the annealing conditions for the coil (the heating rate, the holding temperature, the holding time, and the cooling rate). At each enterprise, the production of thin cold-rolled steel sheet is different. Therefore, it is difficult to offer general recommendations. However, we may state a general conclusion: to prevent kinking defects, the strip tension $T(R)$ in unwinding the coils in the temper-rolling mill must be consistent with the winding conditions of the cold-rolled strip and the annealing conditions. In particular, since q_{ru} varies over the coil thickness, the tensile force $T(R)$ in unwinding must vary in direct proportion to $q_{ru}(R)$. Examples of this approach in production conditions may be found in [113].

In the course of uncoiling, the point C at which strip contact with the tension rollers begins will also be shifted. However, this has no significant influence on the formation of kinking defects.

In industrial practice, kinking defects are often formed over only part of the strip width: for example, at the edges or at the center, on account of undulation or camber. In that case, their formation follows the same principles. Note, however, that the tensile force will be required to separate the upper turn from the coil not over the whole width of the strip but only over the area characterized by welding of the contacting surfaces. In that case, with constant total strip tension between the uncoiling unit and the tension rollers, the unit tension at point B will increase in proportion to the decrease in welding area of the contacting surfaces.

Correspondingly, point B will move toward point A, and the spacing between the kinking defects – the distance between successive lines of strip flexure – will decrease. Camber, undulation, crescent distortion, thickness fluctuation, and other shape errors of the cold-rolled strip facilitate the formation of kinking defects.

Note that strip flexure at point B, where the upper turn separates from the coil, does not necessarily lead to kinking defects. Research by Kudin at the Institute of Ferrous Metallurgy, Ukrainian Academy of Sciences, shows that the critical radius of curvature ρ_{cr} at which kinking defects appear (Figs. 12.1 and 12.2) depends on σ_y, h, and T for the strip. In particular, to prevent kinking defects in the unwinding of steel-strip coils with σ_y =240 N/mm^2, when the tension σ_p = 70 N/mm^2, the recommended value is $\rho_{cr} \geq 235h$ [72, 174]. The radius ρ of strip flexure must be no less than ρ_{cr}.

In the case of strip flexure (or straightening) with extension, the permissible radius of curvature declines with increase in yield point of the steel and decrease in unit tension in the strip. With increase in temperature, σ_y declines. Therefore, to prevent kinking defects, the coils of cold-rolled strip must be cooled to the lowest possible temperatures prior to unwinding.

Influence of the tension in the winding and unwinding of coils. In chapter 9 of this book it is shown that the stress–strain state of coils of thin cold-rolled steel strip is determined by the tension in the winding process. At current levels of automation, a control algorithm tension in the strip as it is coiled may be regulated so as to minimize the compressive radial stress in coils at all stages of subsequent manipulation, including the annealing process. At the cold-rolling mill, the strip is wound into a coil with constant tensile force. Winding conditions for cold-rolled steel strip such that the risk of stability loss (sagging) of the coils after removal from the winding drum is minimized, on the other hand, and the welding of adjacent coils is minimized, on the other, were recommended in [119, 190]. Design principles for a mathematical model that may be used to calculate the stress–strain state of coils of thin cold-rolled steel strip, in the presence of temperature variation, were outlined in detail in chapter 9 of this book[2]. Implementation of those recommendations on industrial rolling mills at various steel plants indicates that this is a promising approach to improving rolling technology and reducing coil breaks in thin steel sheet [113–115].

The next step is to combine the algorithms and subsystems for regulation of the strip tension in the unwinding of coils at the temper mill (with allowance for all the relevant parameters in cold rolling and annealing of the metal) into a single integrated production-control system.

[2]The model was developed by V.L. Mazur, V.I. Timoshenko, I.Yu. Prikhod'ko, and M.V. Timoshenko at the Institute of Ferrous Metallurgy and Engineering Mechanics, Ukrainian Academy of Sciences.

References

1. Polukhin P.I., Zinoviev A.V., Some patterns of distribution of contact stresses in cold rolled sheet, Plastic deformation metals and alloys: Proceedings of MISiS, Moscow: Metallurgiya, 1975, 80, P. 66–76.
2. Vassilev Ya.D., Shuvyakov V.G., Izv. AN SSSR. Metally, 1980, No. 6, P. 110–115.
3. Belosevich V.K., Friction, lubrication, heat transfer during cold rolling sheet steel, Moscow: Metallurgiya, 1989.
4. Chekmarev A.P., Klimenko P.L., Izv. VUZ. Chern. Metallurgiya, 1961, No. 2, P. 68–76.
5. Godunov S.K., Ryaben'kii V.S., Difference schemes, Moscow: Nauka, 1977.
6. Nogovitsyn A.V., Izv. AN SSSR. Metally, 1989, No. 6, P. 59–65.
7. Nogovitsyn A.V., Izv. VUZ. Chern. Metallurgiya, 1992, No. 1, P. 38–43.
8. Rokotyan E.S., Rokotyan S.E., Energy and force parameters of cogging and sheet mills, Moscow: Metallurgiya, 1968.
9. Matsuda S., Okumura N., Trans. Iron and Steel Inst. Japan, 1978, V. 18, P. 198–205.
10. Matrosov Yu.I., et al., Steel for transmission pipelines, Moscow: Metallurgiya, 1985.
11. Sellars C.M., Whiteman J.A., Metal Science, March, April, 1979, P. 187–194.
12. Liska S., Wozniak J., Kovove materialy, 1982, V. 20, No. 5, P. 562–571.
13. Liska S., Zela L., Pionek E. et al. Rolling load calculation in hot strip with respect to restoration processes, Int. Conf. Steel Rolling Tokyo, 1980, Proc. Vol. 2, 1980, P. 840–857.
14. Kobajashi X., Trans Iron and Steel Inst. Japan, 1977, V. 62, No. 1, P. 73–79.
15. Shtremel' M.A., et al., MiTOM, 1984, No. 6, P. 2–5.
16. Zheleznov Yu.D., et al., Izv. VUZ. Chern. Metallurgiya. 1979, No. 1, P. 64–67.
17. Cuddy L.J., Bauwin J.J., Raley J.C., Metallurgical Transactions, 1980, V. 11, No. 3, P. 381–386.
18. Liska S., Wozniak J., Hutnicke aktuality, 1981, V. 22, No. 9, P. 1–49.
19. Popova A.A., Popov A.E., Isothermal and thermokinetic diagrams of breakdown of supercooled austenite. Moscow–Sverdlovsk: Mashgiz, 1961.
20. Metals science and heat treatment of steel: 3rd ed., In 3 Vols. V. II. Fundamentals of thermal treatment (Eds. Bernstein M.L., Rahshtadta A.G.), Moscow: Metallurgiya, 1983.
21. Suechiro M., Trans. Iron and Steel Inst. Japan, 1987, V. 27, No. 6, P. 439–445.
22. Shtremel' M.A., et al., Stal', 1983, No. 3, P. 69–71.
23. Mazanec K., Energy in the physico-metallurgical processes, Papers held at the UPM CSAV, Brno, 1971, P. 449–490.
24. Starodubov K.F., et al., Thermal hardening of rolled material. M.: Metallurgiya, 1970.
25. Polukhin V.P., et al., Stal', 1983, No. 9, P. 68–71.

26. Polukhin P.I., et al., Resistance to plastic deformation of metals and alloys. Handbook., Moscow: Metallurgiya, 1976..
27. Levchenko G.V., Nogovitsyn A.V., Stal', 1977, No. 4, P. 336–338.
28. Zhuchin V.N., et al., Force calculations in continuous hot rolling, Moscow: Metallurgiya, 1986..
29. Solod V.S., et al., Metall i lit'e Ukrainy, 2006, No. 7–8, S. 52–56.
30. Song R., Ponge D., Kaspar R., Steel Research, 2004, No. 1, P. 33–37.
31. Neimark B.E. (ed), Physical properties of steels and alloys used in the energy sector. Moscow and Leningrad: Energiya, 1967.
32. Shulkovsky R.A., Microstructure Evolution Model Used For Hot Strip Rolling, Materials Science & Technology Conference, November 2003, Copyright, 2003, ISS and TMS., P. 1–8.
33. Andorfer J., Auzinger D., Hubmer G., AISE Steel Technology, 2000, No. 7–8, P. 43–46.
34. Tret'yakov V.A., Implementation of automated forecasting system structure and properties of hot-rolled products, Chern. metallurgiya. Bulletin of Chermetinformatsiya, 2004, No. 4, P. 34–41.
35. Frantsenyuk L.I., Bogomolov I.V., Metallurg, 1999, No. 10, P. 40–45.
36. Efron L.I., et al., Metallurg, 2001, No. 10, P. 47–49.
37. Kohlmann R., et al., MPT Int., 2000, No. 2, P. 56–62.
38. Peu C., Varo R., MPT Int., 1999, No. 4, P. 88–90.
39. Matveev B., Zagotovit. Proiz. Mashinostr., 2006, No. 2, P. 38–47.
40. Levchenko G,V., et al., Metall i lit'e Ukrainy. 1996, No. 1–2, P. 41–44.
41. Polukhin P.I., et al., The quality of the sheet and continuous rolling modes, Alma Ata: Nauka, 1974.
42. Benyakovsky M.A., Mazur V.L., Meleshko V.I., Production of automotive sheet, Moscow: Metallurgiya, 1979.
43. Mazur V.L., et al., Improving the quality of sheet metal, Kiev: Tekhnika, 1979.
44. Shmitkhams P.Ch., Grass E., Chernye metally,1966, No. 12, P. 3–8.
45. Grashof G.V., *ibid,* 1969, No. 15, P. 18–29.
46. Tret'yakov A.V., Zyuzin V.I., The mechanical properties of metals and alloys in pressure treatment, Moscow: Metallurgiya, 1973.
47. Prikhod'ko E.V., Metal chemistry of complex alloying, Moscow: Metallurgyia, 1983.
48. Mazur V.L., et al., Reliability of the process of production of rolled sheets. Kiev: Tekhnika, 1992.
49. Mazur V.L., et al., Stal', 1988, No. 3, P. 50–54.
50. Pimenov A.F., et al., Cold rolling and finishing of tin plate, Moscow: Metallurgiya, 1980.
51. Konovalov Yu.V., et al., Increasing the accuracy of rolling sheet and strip. Kiev: Tekhnika, 1987, 144.
52. Grigor'yan G.D., Elements of process reliability, Kiev, Vishcha shkola, 1984.
53. Romanovsky D.L. Mazur V.L., Vorobey S.A., in: Scientific and technical progress in rolling production, Moscow: Metallurgiya, 1988, P. 5–11.
54. Brovman M.Ya., Application of the theory of plasticity in rolling, Moscow: Metallurgiya, 1991.
55. Nikolaev V.A., Theory and practice of rolling processes, Zaporozh'e: ZGIA, 2002.
56. Sinitsyn V.G., Asymmetrical rolling of sheets and strips, Moscow: Metallurgiya, 1984.

57. Golubchenko A.K., Mazur V.L., Binkevich E.V., Metallurg. Gornorud. Promst., 1994, No. 3, P. 20–24.
58. Mazur V.L., et al., Stal', 1994, No. 1, P. 39–41.
59. Svichinsky A.G., et al., Stal', 1992, No. 11, P. 41–44.
60. Svichinsky A.G., et al., Izv. RAN, Metally, 1993, P. 70–79.
61. Nikolaev V.A., Mazur V.L., Holubchenko A.K., Binkevich E.V., Theory and technology of asymmetric rolling. Moscow: Informat Agency, 1996.
62. Nikolaev V.A., et al., Stal', 1992, No. 11, P. 45–47.
63. Mazur V.L., Tymoshenko V.I., Rolling theory (hydrodynamic effects of lubrication), Moscow: Metallurgiya, 1989.
64. Grudev A.P., et al., Friction and lubrication in metal forming: a handbook, Moscow: Metallurgiya, 1982.
65. Mazur V.L., Mazur S.V., Stal', 2009, No. 1, P. 58–60.
66. Grishkov A.I., Izv. AN SSSR, Metally, 1976, No. 5, P. 117–123.
67. Tselikov A.I., Grishkov A.I., Theory of rolling, Moscow: Metallurgiya, 1970.
68. Mazur V.L., Izv. VUZ. Chern. Met., 1981, No. 10, P. 66–72.
69. Mazur V.L., et al., *ibid*, 1977, No. 2, P. 54–59.
70. Mazur V.L., Metallurg. Gornorud. Promst., 1979, No. 3, P. 13–15.
71. Leepa I.I.,, et al., *ibid*, 1978, No. 4, P. 58–61.
72. Meleshko V.I., et al., Processing the sheet surface. Moscow: Metallurgiya, 1975.
73. Borisov L.P., et al., Izv. VUZ, Chern. Met., 1979, No. 2, P. 92–94.
74. Tretyakov A.V., Theory, research and studies of cold rolling mills, Moscow: Metallurgiya, 1986
75. Belosevich V.K. Netesov N.P., Improving the process of cold rolling, Moscow: Metallurgiya, 1971.
76. Chekmarev A.P., et al., in: Pressure treatment of metals, Moscow: Metallurgiya, 1970, No. 54, P. 35–39.
77. Roberts V.J., Iron and Steel Eng., 1968, V. 45, No. 5, P. 123–134.
78. Edwards V.J., Fuller N.A. Influence of strip velocity on tandem cold rolling mill performance, Automation of tandem mills, 1973, P. 213–244.
79. Tret'yakov A.V., et al., Stal', 1973, No. 3, P. 248–251.
80. Hennig S., Weber K.-H., Neue Hutte, 1977, No 22, P. 551–553.
81. Gokyu J., et al., J. Japan Soc. Technol. Plast., 1973, V. 14, No. 145, P. 160–167.
82. Grudev A.P., Sigalov Yu.B., in: Pressure treatment of metals, Moscow: Metallurgiya, 1971, No. 56, P. 47–56.
83. Garber E.A., et al., Technical progress of cooling system for the rolling mills, Moscow: Metallurgiya, 1991.
84. Tret'yakov A.V., et al., Calculation and study of rolling rolls, Moscow: Metallurgiya, 1976.
85. Funke N., Kottman K., Chernye Metally, 1973, No. 15, P. 15–24.
86. Grudev A.P., External friction during rolling, Moscow: Metallurgiya, 1973.
87. Grudev A.P., Zil'berg Yu.V., in: Pressure treatment of metals, Moscow: Metallurgiya, 1971, No. 56, P. 184–191.
88. Prikhod'ko I.Yu., et al., Stal', 2006, P. 87–93.
89. Polukhin V.P., Mathematical modeling and computer calculation of sheet rolling mills, Moscow: Metallurgiya, 1972.
90. Skichko P.Ya., et al., in: Sheet rolling production, Moscow: Metallurgiya, 1975, No. 4, P. 69–73.
91. Druzhinin N.N., Continuous mills as the automation object, Moscow: Metallurgiya, 1967.

92. Mazur V.L., et al., Rolling metal with welded joints. Moscow: Metallurgiya, 1985.
93. Nogovitsyn A.V., et al., Chern. Metallurg.. Bull. n-t. informatsii. 1983, No. 14, P. 42–44.
94. Gorbunkov S.G. et al., Metall i lit'e Ukrainy, 1995, No. 6, P. 23–28.
95. Billigmann J., Pomp A., Stahl und Eisen, 1954, No. 8, V. 74, P. 441–461.
96. Meerovich I.N., et al., Increased accuracy of rolled sheet metal, Moscow, Metallurgiya, 1969.
97. Moskvin V.M., et al., Izv. VUZ, Chern. Metall., 1978, No. 2, P. 79–82.
98. Zenchenko F.I., et al., Stal', 1984, No. 7, P. 40–44.
99. Mazur V.L., et al., Controlling the quality of rolled thin sheets, Kiev, Tehnika, 1997.
100. Golubchenko A.K., Stal', 1996, No. 10, P. 32–36.
101. Golubchenko A.K., Metallurg. Gornorud. Promst., 1994, No. 4, P. 19–24.
102. Garber E.A., et al., Proizvodstvo prokata, 2004, No. 6, P. 34–41.
103. Kolpakov S.S., et al., Stal', 1993, No. 1, P. 47–52.
104. Prikhod'ko I.Yu., et al., In.: Foundations and applied problems of the steel industry. Sb. Nauchn. Tr. IChM, No. 12, 2006, P. 232–244.
105. Leepa I.I., et al., Stal', 1979, No. 8, P. 614–616.
106. Leepa I.I., et al., Stal', 1987, No. 4, P. 67–71.
107. Mazur V.L., Tymoshenko V.I., Izv. VUZ. Chern. Metall., 1979, No. 4, P. 55–59.
108. Mazur V.L., Tymoshenko V.I., Izv. VUZ. Chern. Metall., 1979, No. 6, P. 52–55.
109. Ochan M., Mekhanika polimerov, 1975, No. 6, P. 1011–1020.
110. Meleshko V.I., et al., Listoprokat. Proiz., 1972, No. 1, P. 46–52.
111. Timoshenko M.V., Inzh. Fiz. Zh., 1996, No. 5, P. 773–778.
112. Timoshenko M.V., Tekh. mekh.,1999, No. 2, P. 53–61.
113. Prikhod'ko I.Yu., et al., Metallurg. Gornorud. Promst., 2002, No. 8–9, P. 92–101.
114. Chernov P.P., et al., *ibid,* 2002, No. 8–9, P. 102—108.
115. Prikhod'ko I.Yu., et al., In: Proceedings of the Fifth Congress of Rolling Experts. Cherepovets, 21–24 October 2003, Moscow: Chermetinformatsiya, 2004, P. 124–127.
116. Koshkin V.K., et al., Aviats. tekh., 1971, No. 5, P. 75–83.
117. Popov V.M., Heat transfer in the contact zone of temporary and permanent connections, Moscow: Energiya, 1971.
118. Kovalenko A.D., Fundamentals of thermoelasticity, Kiev, Naukova Dumka, 1970.
119. Mazur V.L., Stal', 1980, No. 7, P. 591–596.
120. Solov'ev P.I., et al., Tr.. VNIImetmash, 1962, No. 6, P. 54–87.
121. Mazur V.L., Dobronravov A.I., Metallurg, 1979, No. 2, P. 34–36.
122. Pavelski O., et al., Chernye metally, 1989, No. 4, P. 12–20.
123. Chernov P.P., et al., Stal', 1983, No. 2, P. 34–38.
124. Chernov P.P., et al., Stal', 1982, No. 7, P. 46–47.
125. Water M., Bander Bleche Rohre, 1996, No. 3, P. 135–141.
126. Zlov V.E., Stal', 1991, No. 3, P. 45–47.
127. Mazur V.L., Sheet production with high quality surface, Kiev, Tehnika, 1982.
128. Mazur V.L., et al., Stal', 1987, No. 9, P. 60–64.
129. Kostyakov V.V., et al., Metallurg. Gornorud. Promst., 1986, No. 3, P. 25–26.
130. Mazur V.L., Kostyakov V.V., Chern. Metall., Bull. n-t. informatsii, 1985, No. 4, P. 16–25.
131. Leepa I.I., et al., Stal', 1978, No. 7, P. 634–635.
132. Ashikhmin G.V., Iroshnikov S.A., Prokat. proizvod., 2002, No. 9, P. 14–17.
133. Ashikhmin G.V. Iroshnikov S.A., *ibid,* 2002, No. 11, P. 16–22.
134. Zheleznov Yu.D., et al., Auth. cert. 1219201 USSR. A method of winding cold-

rolled strip at coiling drum. Otkrytiya. Izobr., 1986, No. 11.
135. Bozhkov A.I., et al., Proizvod. prokata, 2003., No. 3, P. 9–15.
136. Bozhkov A.I., et al., *ibid,* 2003, No. 4, P. 14–18.
137. Mazur V.L., et al., Stal', 1989, No. 4, P. 44–48.
138. Mazur V.L., et al., Izv. VUZ, Chern. Met., 1983, No. 3, P. 60–63.
139. Mazur V.L., et al., *ibid,* 1983, No. 5, P. 73–77.
140. Berkovskii B.M., Nogotov E.F., Difference methods for the study of heat transfer problems, Moscow, Nauka, 1976.
141. Zenkiewich O., The finite element method in technology, New York, Wiley, 1975.
142. Meleshko V.I., et al., in: Rolling production, Moscow, Metallurgiya, 1971, V. XXXV, P. 14–26.
143. Saf'yan M.M., et al., Technology of rolling and drawing. Plate rolling production, Textbook for high schools, Kiev, Vishcha shkola, 1988.
144. Mazur V.L., Timoshenko V.I., Mekh. Kompozit. Mater., 1982, No. 5, P. 880–886.
145. Gorshkov Yu.F., et al., Stal', 1977, No. 4, P. 373–374.
146. Timoshenko V.I., Goodier J., Theory of elasticity, Moscow, Nauka, 1975.
147. Mazur V.L., in: Current problems in metallurgy (in Ukrainian). Volume 5. The plastic deformatino of metals. Dnipropetrovsk, Sistemni tehnologiï, 2002, P. 33–35.
148. Mazur V.L., et al., Auth. cert. 1199322 USSR. A method for determining the temperature of the cross section of a roll of strip material, Otkrytiya. Izobr., 1985, No. 47, P. 25.
149. Mazur V.L., Kolesnichenko B.P., Stal', 1975, No. 9, P. 821–824.
150. Tret'yakov A.V., et al., Dressing and quality of thin sheet, Moscow, Metallurgiya, 1977.
151. Meleshko V.I., et la., Progressive methods of rolling and finishing rolled steel, Moscow, Metallurgiya, 1980.
152. Chernov P.P. et al., Stal', 1986,, No. 8, P. 56–60..
153. Mazur V.L., et al., Preventing sheet metal defects, Kiev, Tekhnika, 1986.
154. Prazdnikov A.V., et al., Metallurg. Gornorud. Promst., 1976, No. 5, P. 68–70.
155. Raimbekov A.M., et al., Stal', 2006, No. 2, P. 38–41.
156. Bodyaev Yu.A., Stal', 2006, No. 5, P. 90–94.
157. Garber E.A., et al., Stal', 2007, No. 1, P. 48–50.
158. Smirnov P.N., et al., Stal', 2007, No. 2, P. 79–80.
159. Nikolaev V.O., Mazur V.L., Production of flat strips, Zaporozh'ye, Vidavnitstvo ZDIA, 2010.
160. Mazur V.L., Stal', 2007, No. 12, P. 35–39.
161. Tishchenko O.I., et al., Stal', 1979, No. 5, P. 355–358.
162. Mazur V.L., Izv. VUZ, Chern. Metall., 1981, No. 3, P. 186.
163. Mazur V.L., Stal', 2008, No. 7, P. 113–117.
164. Mazur V.L., et al., Proizv. prokata, 2009, No. 1, P. 34–37.
165. Ostapenko A.A., Reducing the energy losse in strip rolling. Kiev, Tekhnika, 1983.
166. Khloponin V.N., In: Advanced technologies OMD. Teaching aid., Moscow, IRIAS, 2009.
167. Orobtsev V.V., Konovalov Yu.V., Proizvod. prokata, 2003, No. 8, P. 14–18.
168. Mazur V.L., et al., Stal', 1989, No. 1, P. 50–54.
169. Mazur V.L., et al., Metallurg. Gornorud. Promst., 2007, No. 1, P. 68–71.
170. Mazur V.L., *ibid*, 2008, No. 2, P. 82–86.
171. Mazur V.L., *ibid,* 1994, No. 4, P. 24–26.
172. Mazur V.L., in: Current problems in metallurgy (in Ukrainian). Volume 8. The plastic deformation of metals. Dnipropetrovsk, Sistemni tehnologiï, 2005.

173. Nikolaev O.N., et al., Metallurg, 2004, No. 7, P. 55–56..
174. Soskovets O.N., Thin steel sheet production. Fundamentals of technology. Moscow: Informart, 1995.
175. Mazur V.L., in: Fiziko-tekhnicheskie problemy sovremennogo materialovedeniya (Physical and Technical Problems of Modern Material Science), Kyiv: Akademperiodika, 2013, vol. 1, P. 289–301.
176. Mazur V.L. Steel in translation, 2015, V. 45, No. 12, P. 959–966.
177. Mazur V.L., Pargamonov E.A., Metall. Gornorudn. Promst, 2015, No. 6, P. 39–47.
178. Mazur V.L., Pargamonov E.A., Steel in translation, 2016, vol. 46, No. 8, P. 595–601.
179. Belov, V.K., et al., Steel in translation, 2014, V. 44, No. 12, P. 906–909.
180. Khlybov et al., Stal', 2015, No. 5, P. 67–70.
181. Trost A., Holman F., Chern. Met., 1964, No. 7, P. 51–54.
182. Meleshko V.I., et al., Stal', 1969, No. 6, P. 537–540.
183. Yuen W.Y.D., Cozijnsen V., Optimum tension profiles to prevent coil collapses, SEAISI 2000 Australia Conf. on Improving the Cost Competitiveness of the Iron and Steel Industry, 2000, V. 2, P. 1/1–1/10.
184. Gol'dfrab E.M., et al., Stal', 1971, No. 6, P. 532–533.
185. Burns R., Lataur, H., Blast Furnace Steel Plant, 1969, No. 12, P. 49.
186. Mazur V.L., Timoshenko V.I., Theory and technology of rolling (hydrodynamic effects of lubrication and surface microrelief). Kiev. I=D ADEF Ukraina, 2018.
187. Demkin N.B., Contacting of rough surfaces), Moscow, Nauka, 1970.
188. Demkin N.B., et al., Influence of microgeometry and contact time on thermal conductivity of a contact, in: Metrologicheskie i tekhnologicheskie issledovaniya kachestva poverkhnosti (Metrological and Technological Studies of Surface Quality), Riga: Zinatne, 1976, P. 64–72.
189. Mazur V.L., Steel in translation, 2015, V. 45, No. 5, P. 371–377.
190. Mazur V.L., Steel in translation, 2011, V. 41, No. 9, P. 756–760.
191. Mazur V.L., Nogovitsyn A.V., Theory and Technology of Thin Sheet Rolling: Numerical Analysis and Engineering Applications. Dnepropetrovsk: Dnipro-Val, 2010.

Index

F

H

I

K

L

M

V

Z